Hazard Analysis
Techniques for
System Safety

Hazard Analysis Techniques for System Safety

Second Edition

Clifton A. Ericson, II
Fredericksburg, Virginia

Library of Congress Cataloging-in-Publication Data
Ericson, Clifton A., II.
   Hazard analysis techniques for system safety / Clifton A. Ericson, II. –
Second edition.
      pages cm
   Includes index.
   ISBN 978-1-118-94038-9 (hardback)
 1.   Industrial safety–Data processing. 2.   System safety.   I. Title.
   T55.E72 2015
   363.11–dc23

                                            2015016350

Printed in the United States of America

10 9 8 7 6 5 4 3 2 1

Contents

Preface

During my 50 year career in system safety, there have been two things about hazard analysis that have always bothered me. First, there has never been a formal description of hazard theory that defines the components of a hazard and the hazard–mishap actuation process. This is significant because risk cannot be determined unless the hazard is fully understood and described. Second, there is a lack of good reference material describing in detail how to perform the most relevant hazard analysis techniques or methodologies. This too is significant because hazard analysis is more complex than most people think, thus good descriptions and reference material are needed. I wrote this book to resolve these issues for system safety engineers and practitioners. The material in this book is applicable to both experienced professionals and those analysts just starting out in the field.

One of the main features of this book is that it describes hazard theory in detail. The hazard–risk–mishap connection is explained, with illustrations and examples provided. In addition, the three required components of a hazard are presented, along with the hazard triangle model.

Another primary feature of this book is that it describes 28 of the most commonly used hazard analysis methodologies in the system safety discipline. Each of the 28 hazard analysis methodologies covered in this book is given an entire chapter devoted to just that technique. In addition, each methodology chapter is organized in a similar pattern that is intended to provide consistency in answering the most common questions that an analyst might have. Detailed examples are provided to help analysts learn and understand these methodologies.

System safety is a proven engineering discipline that is applied during system development to identify and mitigate hazards, and in so doing eliminate or reduce the risk of potential mishaps and accidents. System Safety is ultimately about savings lives. It is my greatest hope that the readers of this book can use the material contained herein to better understand hazard identification and analysis. This in turn will help in designing and constructing systems that are safe, thereby saving many lives.

This revised version of the book has added eight new chapters, six of which are additional hazard analysis techniques. Also, this updated version has added new and revised material to reflect changes made as a result of the new MIL-STD-882, version E, which was released in 2012.

Acknowledgments

In a book of this undertaking, there are naturally many people to acknowledge. This book reflects my life's journey through 50 years of engineering in the system safety discipline. My life has been touched and influenced by many people, far too many people to list and credit. For those whom I have left out I apologize. But it seems that there are a few people that always remain in the forefront of one's memory.

First and foremost, I would like to dedicate this book to my parents, Clifton Ericson I and Margaret Ericson. They instilled in me many good qualities that I might not have found without them, particularly the values of reading, education, science, religion, morality, and a work ethic.

I would like to acknowledge and dedicate this book to the Boeing System Safety organization on the Minuteman Weapon System development program. This was the crucible where the experiment of system safety really started, and this is where I started my career in system safety engineering. This group has provided my most profound work-related memories and probably had the greatest influence on my life. It was led by Niel Classon, who was an early visionary and leader in the system safety field. Other people in this organization who helped in my development included Dave Haasl, Gordon Willard, Dwight Leffingwell, Kaz Kanda, Brad Wolfe, Joe Muldoon, Harvey Moon, and Bob Schroder. Another Boeing manager who provided system safety guidance early in my career was Hal Trettin.

Later in my career, Perry D'Antonio of Sandia National Laboratories pushed me to excel in the System Safety Society and to eventually become president of this international organization. Paige Ripani of Applied Ordnance Technology, Inc. helped turn my career in a new direction, consulting for the Navy. And, last but not least, Ed Kratovil of the Naval Ordnance Safety and Security Activity (NOSSA) provided me with the opportunity to work on special Navy system and software safety projects.

In addition, I would like to acknowledge and thank the following individuals for reviewing early drafts of this manuscript: Jim Gerber, Sidney Andrews, Dave Shampine, Mary Ellen Caro, Tony Dunay, Chuck Dorney, John Leipper, Kurt Erthner, Ed Nicholson, William Hammer, and Jerry Barnette. Many of their comments and suggestions proved invaluable.

Chapter *1*

System Safety and Hazard Analysis

1.1 INTRODUCTION

We live in a world comprised of systems. When viewed from an engineering perspective, most aspects of life involve systems. For example, houses are a type of system, automobiles are a type of system, and electrical power grids are another type of system. Commercial aircraft are systems that operate within a larger transportation system that in turn operate in a larger worldwide airspace system. Systems have become a necessity for modern living.

As a result of living in a system-centric world, we also live in a world comprised of hazards and risk. With systems and technology also comes exposure to hazards and potential mishaps. A hazard is a potential condition existing within a system, which when actuated becomes an actual mishap event resulting in damage, loss, injury, and/or deaths. Risk is the probability that a hazard occurs accompanied by the severity of the resulting outcome.

Hazard risk is a metric that predicts the likelihood and severity of a possible mishap. We live with risk, and make risk decisions, on a daily basis. For example, there is the hazard that a traffic light will fail, resulting in the mishap of another auto colliding with your auto. Automobiles, traffic, and traffic lights form a unique system that we use daily and accept the hazard risk potential because the risk is small. There is the danger that the gas furnace in our house will fail and explode, thereby resulting in the mishap of a burned house, or worse. This is another unique system, with known adverse side effects that we choose to live with because the mishap risk is small and the benefits are great. We live in a world comprised of many different systems with many different risks.

Our lives are intertwined within a web of different systems, each of which can affect our safety. Each of these systems has a unique design and a unique set of components. In addition, each of these systems contains inherent hazards that present unique mishap risks. We are

Hazard Analysis Techniques for System Safety, Second Edition. Clifton A Ericson, II.
© 2016 John Wiley & Sons, Inc. Published 2016 by John Wiley & Sons, Inc.

always making a tradeoff between accepting the benefits of a system and the mishap risk they present. As we develop and build systems, we should be concerned about eliminating and reducing mishap risk. Some risks are so small that they can easily be accepted, while other risks are so large that they must be dealt with immediately. Risks are akin to the invisible radio signals that fill the air around us, in that some are loud and clear, some very faint, and some are distorted and unclear. Life, as well as safety, is a matter of knowing, understanding, and choosing the risk to accept.

System safety is the formal engineering discipline and process for identifying and controlling hazards, and the risk associated with these hazards. As systems become more complex and more hazardous, more effort is required to understand and manage system mishap risk. Hazard (and mishap) risk can be intentionally reduced and controlled to a small and acceptable level through the system safety process.

The key to system safety and effective risk management is the identification and mitigation of hazards. To successfully control hazards, it is necessary to understand hazards and know how to identify them. The purpose of this book is to better understand hazards and the tools and techniques for identifying them, in order that they can be effectively controlled during the development of a system. The system safety process is sometimes referred to as *design for safety* (DFS).

1.2 THE NEED FOR HAZARD ANALYSIS

Forensic engineering is the detailed investigation of a mishap after it has occurred, performed to determine the specific causes for the mishap in order that corrective action can be applied to prevent reoccurrences. System safety, on the other hand, is a form of preemptive forensic engineering, whereby potential mishaps are identified, evaluated, and controlled before they occur. Potential mishaps and their causal factors are anticipated and identified during the design stage, and then design safety features are incorporated into the design to control the occurrence of the potential mishaps – safety is intentionally designed in and mishaps are effectively designed out. This proactive approach to safety involves hazard analysis, risk assessment, risk mitigation through design, and testing to verify the design results. Potential mishaps are recognized and identified by the hazards that ultimately cause them. System safety is a proactive approach to affecting the future (i.e., preventing mishaps before they occur) by identifying hazards and then eliminating or controlling the risk they present.

Systems are intended to improve our way of life, yet they also contain the inherent capability to spawn many different hazards that present us with mishap risk. It is not that systems are intrinsically bad; it is that systems can go awry, and when they go awry they typically result in mishaps. System safety is about determining how systems can go bad and implementing design safety mitigations to eliminate, correct, or work around safety imperfections in the system.

Murphy's law states that "if anything can go wrong, it will." This truism illustrates that the unexpected and undesired must be anticipated and controlled in order to prevent mishaps, and this can be achieved only through the system safety process. Hazards and risk often cannot be eliminated; however, hazards and risk can be anticipated and mitigated via safety design features, thereby preventing or reducing the likelihood of mishaps. If system safety is not applied, accidents and loss of life will not be prevented. System users are typically not aware of the actual risk they are exposed to, and without system safety this risk may be much higher than the users realize.

Hazard analysis is the basic key component of the system safety process. Therefore, it is necessary to fully understand the hazard theory and the hazard analysis process in order to develop safe systems.

1.3 SYSTEM SAFETY BACKGROUND

The primary guidance document for system safety is MIL-STD-882, System Safety Standard Practice. Version E was released on May 11, 2012. This standard has been in existence since 1969; its predecessor MIL-S-38130 was released in 1963.

MIL-STD-882 and its predecessor MIL-S-38130 are the genesis of system safety. The US military, along with US aerospace companies, saw the need for a holistic and proactive "systems" approach for the design, development, test, and manufacture of "safe" systems. Working together, these two groups developed the system safety methodology and discipline. MIL-S-38130 was originally released on September 30, 1963 and replaced by MIL-STD-882 on July 15, 1969. System safety was actually documented as a process prior to any formal documentation of the systems engineering discipline. System safety as a formal discipline was originally developed and promulgated by the military-industrial complex to prevent aircraft and missile mishaps that were costing lives, dollars, and equipment loss. As the effectiveness of the discipline was observed by other industries, it was adopted and applied to these industries and technology fields, such as commercial aircraft, nuclear power, chemical processing, rail transportation, the FAA, and NASA, to name a few.

The ideal objective of system safety is to develop a system free of hazards. However, absolute safety is not possible because complete freedom from all hazardous conditions is not always possible, particularly when dealing with complex inherently hazardous systems, such as weapon systems, nuclear power plants, commercial aircraft, etc.

Since it is generally not possible to eliminate all hazards, the realistic objective becomes that of developing a system with acceptable mishap risk. This is accomplished by identifying potential hazards, assessing their risks, and implementing corrective actions to eliminate or mitigate the identified hazards. This involves a systematic approach to the management of mishap risk. Safety is a basic part of the risk management process.

Hazards will always exist, but their risk can and must be made acceptable. Therefore, safety is a relative term that implies a level of risk that is measurable and acceptable. System safety is not an absolute quantity, but rather an optimized level of mishap risk management that is constrained by cost, time, and operational effectiveness (performance). System safety requires that risk be evaluated and the level of risk accepted or rejected by an appropriate decision authority. Mishap risk management is the basic process of system safety engineering and management functions. System safety is a process of disciplines and controls employed from the initial system design concepts, through detailed design and testing to system disposal at the completion of its useful life (i.e. "cradle to grave" or "womb to tomb").

The fundamental objective of system safety is to identify, eliminate or control, and document system hazards. System safety encompasses all the ideals of mishap risk management and design for safety; it is a discipline for hazard identification and control to an acceptable level of risk. Safety is a system attribute that must be intentionally designed into a product.

From a historical perspective, it has been learned that a proactive preventive approach to safety during system design and development is much more cost-effective than trying to add safety to a system after the occurrence of an accident or mishap. System safety is an initial investment that saves future losses that could result from potential mishaps.

1.4   SYSTEM SAFETY OVERVIEW

System safety is effectively a design-for-safety process, discipline, and culture. DFS means that the design process utilizes the system safety process to intentionally design-in safety. This process anticipates potential safety problems (i.e., hazards) and eliminates them or reduces the risk they present. Safety risk is calculated from the identified hazards, and risk is eliminated or reduced by eliminating or mitigating the appropriate hazard causal factors. System safety, by necessity, considers function, criticality, risk, performance, and cost parameters of the system. Risk mitigation is achieved through a combination of design mechanisms, design features, warning devices, safety procedures, and safety training to counter the effect of hazard causal factors.

System safety involves a systems approach, which accounts for the distinctive name. System safety is the art and science of looking at all aspects and characteristics of a system as an integrated whole, rather than looking at individual components in isolation from the system. System safety is a holistic approach that considers the subject as an integrated sum-of-the-parts combination, rather than a piecemeal approach of looking at separate individual and solitary pieces of the system.

Often, system safety is not fully appreciated for the contribution it can make to creating safe systems that present a minimal chance of deaths and serious injuries. System safety applies a planned and disciplined methodology for purposely designing safety into a system. A system can be made safe only when the system safety methodology is consistently and properly applied. Safety is more than eliminating hardware failure modes; it involves designing the safe system interaction of hardware, software, humans, procedures, and the environment, under all normal and adverse failure conditions. Safety must consider the entirety of the problem, not just a portion of the problem, that is, a systems perspective is required for full safety coverage. System safety anticipates potential problems and either eliminates them or reduces their risk potential, through the use of design safety mechanisms applied according to a safety order of precedence.

System safety is the process of managing the system, personnel, environmental, and health mishap risks encountered in the design development, test, production, use, and disposal of systems, subsystems, equipment, materials, and facilities.

A system safety program (SSP) is a formal approach to eliminate hazards through engineering, design, education, management policy, and supervisory control of conditions and practices. It ensures the accomplishment of the appropriate system safety management and engineering tasks. The formal system safety process has been primarily established by the US Department of Defense (DoD) and its military branches and promulgated by MIL-STD-882. However, the same process is also followed in private industry for the development of commercial products, such as commercial aircraft, rail transportation, nuclear power, and automobiles, to mention a few.

The goal of system safety is the protection of life, systems, equipment, and the environment. The basic objective is the elimination of hazards that can result in death, injury, system loss, and damage to the environment. When hazard elimination is not possible, the next objective is to reduce the risk of a mishap through design control measures. Reducing mishap risk is achieved by reducing the probability of the mishap and/or the severity of the mishap.

This objective can be attained at minimum cost when the SSP is implemented early in the conceptual phase and is continued throughout the system development and acquisition cycle. The overall complexity of today's systems, particularly weapon systems, is such that system safety is required in order to consciously prevent mishaps and accidents. Added to complexity is the inherent danger of energetic materials, the effects of environments, and the complexities of operational requirements. In addition, consideration must be given to hardware failures, human error, software interfaces, including programming errors, and vagaries of the environment.

The following definitions from MIL-STD-882E help define the system safety concept:

System Safety   The application of engineering and management principles, criteria, and techniques to achieve acceptable risk within the constraints of operational effectiveness and suitability, time, and cost throughout all phases of the system life cycle.

System Safety Engineering   An engineering discipline that employs specialized knowledge and skills in applying scientific and engineering principles, criteria, and techniques to identify hazards and then to eliminate the hazards or reduce the associated risks when the hazards cannot be eliminated.

System Safety Management   All plans and actions taken to identify hazards; assess and mitigate associated risks; and track, control, accept, and document risks encountered in the design, development, test, acquisition, use, and disposal of systems, subsystems, equipment, and infrastructure.

The intent of system safety is mishap risk management through hazard identification and mitigation techniques. System safety engineering is an element of systems engineering involving the application of scientific and engineering principles for the timely identification of hazards and initiation of those actions necessary to prevent or control hazards within the system. It draws upon professional knowledge and specialized skills in the mathematical and scientific disciplines, together with the principles and methods of engineering design and analysis to specify, predict, evaluate, and document the safety of the system.

System safety management is an element of program management that ensures accomplishment of the correct mix of system safety tasks. This includes identification of system safety requirements; planning, organizing, and controlling those efforts that are directed toward achieving the safety goals; coordinating with other program elements; and analyzing, reviewing, and evaluating the program to ensure effective and timely realization of the system safety objectives.

The basic concept of system safety is that it is a formal process of intentionally designing-in safety by designing-out hazards or reducing the mishap risk of hazards. It is a proactive process performed throughout the system life cycle to save lives and resources by intentionally reducing the likelihood of mishaps to an insignificant level. The system life cycle is typically defined as the stages of concept, preliminary design, detailed design, test, manufacture, operation, and disposal (demilitarization). In order to be proactive, safety must begin when system development first begins at the conceptual stage.

The goal of system safety is to ensure the detection of hazards to the fullest extent possible, and provide for the introduction of protective measures early enough in system development to avoid design changes late in the program. A safe design is a prerequisite for safe operations. Things that can go wrong with systems are predictable, and something that is predictable is also preventable. As Murphy's law states "whatever can go wrong, will go wrong," the goal of system safety is to find out what can go wrong (before it does) and establish controls to prevent it or reduce the probability of occurrence. This is accomplished through hazard identification and mitigation.

Figure 1.1   *Core system safety process.*

1.5   SYSTEM SAFETY PROCESS

MIL-STD-882 establishes the core system safety process in eight principal steps, which are shown in Figure 1.1.

The core system safety process involves establishing an SSP to implement the mishap risk management process. The SSP is formally documented in the system safety program plan (SSPP), which specifies all of the safety tasks that will be performed, including the specific hazard analyses, reports, etc. As hazards are identified, their risk will be assessed, and hazard mitigation methods will be established to mitigate the risk as determined necessary. Hazard mitigation methods are implemented into system design via system safety requirements (SSRs). All identified hazards are converted into hazard action records (HARs) and placed into a hazard tracking system (HTS). Hazards are continually tracked in the HTS until they can be closed.

It can be seen from the core system safety process that safety revolves around hazards. Hazard identification and elimination/mitigation is the key to this process. Therefore, it is critical that the system safety analyst understands hazards, hazard identification, and hazard mitigation.

The core system safety process can be reduced to the process shown in Figure 1.2. This is a mishap risk management process whereby safety is achieved through the identification of hazards, the assessment of hazard mishap risk, and the control of hazards presenting unacceptable risk. This is a closed loop process, whereby hazards are identified and tracked until acceptable closure action is implemented and verified. It should be performed in conjunction with actual system development, in order that the design can be influenced during the design process, rather than trying to enforce design changes after the system is developed.

System safety involves a life cycle approach, based on the idea that mishap and accident prevention measures must be initiated as early as possible in the life of a system and carried through to the end of its useful life. It is usually much cheaper and more effective to design

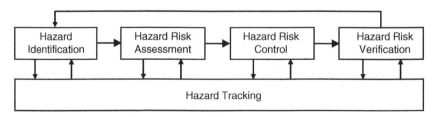

Figure 1.2   *Closed loop hazard control process.*

safety features into an item of equipment than it is to add the safety features when the item is in production or in the field. Also, experience indicates that some of the hazards in a newly designed system will escape detection, no matter how aggressive the safety program. Therefore, the safety program for a system must remain active throughout the life of the system to ensure that safety problems are recognized whenever they arise and that appropriate corrective action is taken.

The key to system safety is the management of hazards. To effectively manage hazards, one must understand hazard theory and the identification of hazards. The purpose of this book is to better understand hazards and the tools and techniques for identifying them. When hazards are identified and understood, they can then be properly eliminated or mitigated.

1.6 SYSTEM SAFETY STANDARDS

System safety is a process that is formally recognized internationally and used to develop safe systems in many countries throughout the world. MIL-STD-882 has long been the bedrock of system safety procedures and processes, and the discipline has grown and improved with each improvement in MIL-STD-882. Since the advent of MIL-STD-882, many other agencies, organizations, and standards groups have developed their own variation on the system safety process. These agencies have recognized that the system safety process is successful in making systems safe and they have tailored the methodology to reflect their particular industry.

Key reference standards and guideline documents that help define the system safety process and methodology include the following:

1. MIL-STD-882E, System Safety Standard Practice, May 11, 2012.
2. ANSI/GEIA-STD-0010-2009, Standard Best Practices for System Safety Program Development and Execution, February 12, 2009.
3. NAVSEA SWO20-AH-SAF-010, Weapon System Safety Guidelines Handbook, February 1, 2006.
4. *System Safety Handbook: Practices and Guidelines for Conducting System Safety Engineering and Management*, Federal Aviation Administration (FAA), December 30, 2000.
5. NPR 8715.3, NASA General Safety Program Requirements, March 12, 2008.
6. *Air Force System Safety Handbook*, July 2000.
7. MIL-HDBK-764, System Safety Engineering Design Guide for Army Materiel, January 12, 1990.
8. SAE/ARP-4754, Certification Considerations for Highly-Integrated or Complex Aircraft Systems, Aerospace Recommended Practice, 1996.
9. SAE/ARP-4761, Guidelines and Methods for Conducting the Safety Assessment Process on Civil Airborne Systems and Equipment, 1996.

1.7 SYSTEM SAFETY PRINCIPLES

The following are a set of system safety principles; note that they all relate to hazards, hazard identification, and hazard mitigation, demonstrating the importance of hazard analysis:

Principle 1   Hazards, mishaps, and risk are a reality of life; they are typically the by-product of man-made systems.

Principle 2   Hazards are the *thing* that create risk and lead to mishaps. This thing, or entity, is a unique built-in design element that can be identified and controlled through design.

Principle 3   Systems can, and will, fail in various different key modes that often initiate or contribute to system hazards.

Principle 4   Safety is a resultant systems property, which changes with system design changes. This property is intentionally improved with design-for-safety activities.

Principle 5   Hazard risk is the metric that provides a measure of the safety property.

Principle 6   Risk should always be known; unnecessary or unacceptable safety risk should never be accepted.

Principle 7   System safety is the process for identifying and controlling hazards, mishaps, and risk. Safety is achieved through the hazard identification and hazard mitigation process.

Principle 8   Hazards are identified, and risk determined, from hazard analysis.

Principle 9   Safety is best achieved through a dedicated system safety program based on hazard analysis and risk control.

Principle 10   Safety is not free, but it costs less than the direct, indirect, and hidden costs associated with mishaps.

Principle 11   An absence of mishaps does not necessarily equate to a safe system. There could be hidden hazards that have not yet occurred; this is why hazard analysis should always be performed.

Principle 12   Safety must be earned; it does not happen by accident. This requires a system safety program based on hazard analysis and hazard mitigation.

1.8   KEY TERMS

For purposes of clarity and understanding, some key safety terms used throughout this book are defined as follows:

Safety-Significant	A term applied to a condition, event, operation, process, or item that is identified as either safety-critical or safety-related.
Safety-Related (SR)	A term applied to a condition, function, event, operation, process, or item whose mishap severity consequence is either Marginal or Negligible (see HRI criteria).
Safety-Critical (SC)	A term applied to a condition, function, event, operation, process, or item whose mishap severity consequence is either Catastrophic or Critical (see HRI criteria). SC is a term indicating a high consequence or high risk safety concern.
Safety-Critical Item (SCI)	SCI is a designation given to any item that has been identified as being safety-critical or part of a safety-critical function. Safety-critical items can be any system level of hardware, software or human tasks involving an event, operation, process, or procedure

that is safety-critical. SCIs can be (but are not limited to) functions, requirements, paths, tasks, procedures, components or component tolerances.

Safety-Critical Function (SCF) · A term applied to any function involved in a hazard whose mishap severity consequence is either Catastrophic or Critical (see HRI criteria).

1.9 SUMMARY

This chapter discussed the basic concept of system safety. The following are the basic principles that help summarize the discussion in this chapter:

1. The goal of system safety is to save lives and preserve resources by preventing mishaps.
2. Mishaps can be predicted and controlled through the system safety process.
3. The focus of system safety is on hazards, mishaps, and mishap risk.
4. Hazard analysis is essential because hazards are the key to preventing or mitigating hazards.
5. System safety should be consistent with mission requirements, cost, and schedule.
6. System safety covers all system life cycle phases, from "cradle to grave."
7. System safety must be planned, proactive, integrated, comprehensive, and system oriented.
8. The system safety process contains eight core steps, all based on hazard analysis:
   a. Plan
   b. Hazard identification
   c. Risk assessment
   d. Identify hazard mitigation methods
   e. Reduce risk with mitigation methods
   f. Verify risk reduction
   g. Accept residual risk
   h. Track hazards through system lifecycle

Chapter *2*

Systems

2.1 SYSTEM CONCEPT

As the name itself implies, *system safety* is involved with "systems." System safety is an engineering process applied specifically to a system in order to ensure the entire system is safe. Therefore, in order to effectively apply the system safety process, it is necessary to completely understand the term "system" and all of its ramifications, characteristics, and attributes, especially as they apply to the system safety process. This includes understanding what comprises a system, how a system operates, system analysis tools, the life cycle of a system, and the system development process. A proactive and preventive safety process can be effectively implemented only if the proper system-oriented safety tasks are performed during the appropriate system life cycle phases, in conjunction with utilizing the appropriate system engineering tools. The timing and content of safety tasks must coincide with certain system development domains to ensure safety success.

The standard definition of a system provided by MIL-STD-882E is "The organization of hardware, software, material, facilities, personnel, data, and services needed to perform a designated function within a stated environment with specified results." In other words, a system is a composite, at any level of complexity, of personnel, procedures, materials, tools, equipment, facilities, and software. The elements of this composite entity are used together in the intended operational or support environment to perform a given task or achieve a specific purpose, support, or mission requirement. Essentially, a system is a combination of subsystems and functions interconnected to accomplish the system objective.

A subsystem is a subset of the system that could include equipment, components, personnel, facilities, processes, documentation, procedures, and software interconnected in the system to perform a specific function that contributes to accomplishing the system objective. To some degree, a subsystem is a minisystem. The system objective is a desired result to be

Hazard Analysis Techniques for System Safety, Second Edition. Clifton A Ericson, II.
© 2016 John Wiley & Sons, Inc. Published 2016 by John Wiley & Sons, Inc.

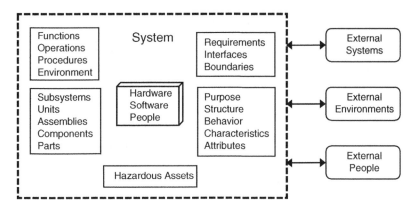

Figure 2.1   *System characterization.*

accomplished by the system; it defines the purpose for the system. System function are the operations the system must perform in order to accomplish its objective. Generally performed by subsystems, system functions define how the system operates. Functions form the corner-stone of a system and are very important in developing safe system designs.

In short, a system is any group of interrelated, interacting, and interdependent parts that form a complex and unified whole that has a specific function or purpose that is greater than that of the individual parts. If the parts are not interacting and interdependent, it is not truly a system; it is merely a collection of parts.

Figure 2.1 characterizes the general aspects and components of a system. The key components comprising a system are hardware, software, and people. A system has purpose and structure, and the components and architecture establish certain qualities and character-istics unique to the system design. System objectives are achieved via the operations, procedures, and functions designed into the system. The system has its own internal environ-ment, and is impinged upon by external environments, systems, and people. A very important safety aspect of a system involves the particular hazardous assets within the system.

Figure 2.2 depicts the generic concept of a system. This diagram shows that a system that is comprised of many subsystems, with an interface between each subsystem. The system has an

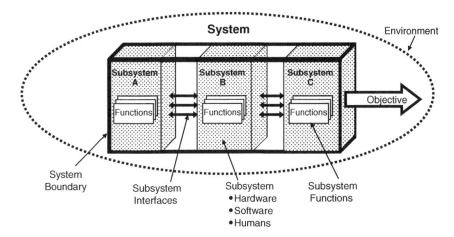

Figure 2.2   *System model.*

objective and is surrounded by a boundary and an environment. The system has functions that must be performed, and it is comprised of many different component types, such as hardware, software, humans, environments, procedures, and so on. System safety analysis involves evaluation of all these system aspects.

2.2 SYSTEM ATTRIBUTES

Systems have many different attributes of interest to system safety. Defining and understanding a system's key attributes is necessary, because they provide the framework for designing, building, operating, and analyzing systems. Key system attributes are shown in Table 2.1, where the major attribute categories are listed on the top row, with subelements identified below. Each of these attributes is usually addressed in system safety hazard analyses at some point in the system development program.

Each one of the system attributes must be understood in order to perform a complete and thorough hazard analysis. Examples of safety considerations for the elements attribute are as follows:

- *Hardware*   Failure modes and hazardous energy sources
- *Software*   Design errors and design incompatibilities
- *Personnel*   Human error, human injury, and human control interface
- *Environment*   Weather and external equipment
- *Procedures*   Instructions, tasks, and warning notes
- *Interfaces*   Erroneous input/output and unexpected complexities
- *Functions*   Fail to perform or performs erroneously
- *Facilities*   Building faults, storage compatibility, and transportation faults

An SSP must adequately address each of these system attributes in order to ensure that an optimum degree of safety is achieved in the system. All of the system elements and their interrelationships must be considered in the safety analyses, assessments, and evaluations to ensure it truly is a *system* safety analysis. For example, it is possible for each operational phase or mode to have a different and/or significant impact upon system safety. Different functions are performed during each phase that could have a direct impact on subsequent phases. During certain phases, safety barriers or interlocks are often removed, making the system more susceptible to the occurrence of a hazard; for example, at one point in the operational mission

TABLE 2.1   Key System Attributes

Hierarchy	Elements	Domains	Operations	Types
Systems Subsystems	Hardware	Boundaries	Functions	Static
Units	Software	Complexity Criticality	Task modes	Dynamic
Assemblies	Humans		Phases	Robotic
Components	Procedures			Process
Piece part	Interfaces			Weapon
	Environments			Aircraft
	Facilities			Spacecraft
	Documentation			—

5

of a missile, the missile is powered and armed. This means that fewer potential failures are now necessary in order for a mishap to occur and there are fewer safeguards activated in place to prevent hazards from occurring.

2.3 SYSTEM TYPES

The types of systems dealt with in system safety are typically physical, man-made objects comprised of hardware, software, user interfaces, and procedures. These types of systems include ships, weapons, electrical power, railroads, aircraft, and so on, which are used for some specific purpose or objective. Table 2.2 provides some example systems, showing their intended purpose and some of the subsystems comprising these systems. It is interesting to note that many of the systems are comprised of similar types of subsystems, which means that they may have similar types of hazards.

Understanding system type and scope is very important in system safety and hazard analysis. The system type can be an indication of the safety-criticality involved. The scope of the system boundaries establishes the size and depth of the system. The system limitations describe basically what the system can and cannot safely do. Certain limitations may require the system to include special design safety features. Every system operates within one or more different environments. The specific environment establishes what the potential hazardous impact will be on the system. System criticality establishes the overall safety rating for the system. A nuclear power plant system has a high-consequence safety rating, whereas a TV set as a system has a much lower safety-criticality rating.

2.4 SYSTEM LIFE CYCLE

The system life cycle involves the actual phases a system goes through from concept to its disposal. This system life cycle is analogous to the human life cycle of conception, birth, childhood, adulthood, death, and burial. The life cycle of a system is very generic and generally a universal standard. The system life cycle stages are generally condensed and summarized into the five major phases shown in Figure 2.3. All aspects of the system life cycle will fit into one of these major categories.

TABLE 2.2 Example System Types

System	Objective	Subsystems
Ship	Transport people/ deliver weapons	Engines, hull, radar, communications, navigation, software, fuel, and humans
Aircraft	Transport people/ deliver weapons	Engines, airframe, radar, fuel, communications, navigation, software, and humans
Missile	Deliver ordnance	Engines, structure, radar, communications, navigation, and software
Automobile	Transportation	Engine, frame, computers, software, fuel, and humans
Nuclear power plant	Deliver electrical power	Structure, reactor, computers, software, humans, transmission lines, and radioactive material
Television	View video media	Structure, receiver, display, and electrical power
Toaster	Browning of bread	Structure, timer, electrical elements, and electrical power
Telephone	Communication	Structure, receiver, transmitter, electrical power, and analog converter

Figure 2.3   *Major system life cycle phases.*

The life cycle stages of a system are important divisions in the evolution of a product and are, therefore, very relevant to the system safety process. Safety tasks are planned and referenced around these five phases. In order to proactively design safety into a product, it is essential that the safety process start at the concept definition phase and continue throughout the life cycle.

> **Phase 1: Concept Definition**   This phase involves defining and evaluating a potential system concept in terms of feasibility, cost, and risk. The overall project goals and objectives are identified during this basic concept evaluation phase. Design require- ments, functions, and end results are formulated. The basic system is roughly designed, along with a thumbnail sketch of the subsystems required and how they will interact. During this phase, safety is concerned with hazardous components and functions that must be used in the system. The system safety program plan (SSPP) is generally started during this phase to outline the overall system risk and safety tasks, including hazard analyses that must be performed.
>
> **Phase 2: Development and Test**   This phase involves designing, developing, and testing the actual system. Development proceeds from preliminary to detailed tasks. The development phase is generally broken into the following stages:
>
> - *Preliminary Design* Initial basic design
> - *Detailed Design* Final detailed design
> - *Test* System testing to ensure all requirements are met

The preliminary design translates the initial concept into a workable design. During this phase, subsystems, components, and functions are identified and established. Design requirements are then written to define the systems, subsystems, and software. Some testing of design alternatives may be performed. During this phase, safety is concerned with hazardous system designs, hazardous components/materials, and hazardous functions that can ultimately lead to mishaps and actions to eliminate/mitigate the hazards.

The preliminary design evolves into the final detailed design. The final design of the system involves completing development of the design specifications, sketches, drawings and system processes, and all subsystem designs. During the final design phase, safety is concerned with hazardous designs, failure modes, and human errors that can ultimately lead to mishaps during the life cycle of the system.

The system test phase involves verification and validation testing of the design to ensure that all design requirements are met and are effective. In addition, safety is concerned with potential hazards associated with the conduct of the test and additional system hazards identified during testing.

> **Phase 3: Production**   The final approved design is transformed into the operational end product during the production phase. During this phase, safety is concerned with safe production procedures, human error, tools, methods, and hazardous materials.

Phase 4: Operation   The end product is put into actual operation by the user(s) during the operation phase. This phase includes use and support functions such as transportation/handling, storage/stowage, modification, and maintenance. The operational phase can last for many years and during this phase performance and technology upgrades are likely. Safe system operation and support are the prime safety concerns during this phase. Safety concerns during this phase include operator actions, hardware failures, hazardous system designs, and safe design changes and system upgrades.

Phase 5: Disposal   This phase completes the useful life of the product. It involves disposing of the system in its entirety, or individual elements, following completion of its useful life. This stage involves phaseout, deconfiguration, or decommissioning where the product is torn down, dismantled, or disassembled. Safe disassembly procedures and safe disposal of hazardous materials are safety concerns during this phase.

Normally, each of these life cycle phases occurs sequentially, but occasionally development tasks are performed concurrently, spirally, or incrementally to shorten the development process. Regardless of the development process used, sequential, concurrent, spiral, or incremental, the system life cycle phases basically remain the same.

2.5   SYSTEM DEVELOPMENT

System development is the process of designing, developing, and testing a system design until the final product meets all requirements and fulfils all objectives. System development consists primarily of phases 1 and 2 of the system life cycle. It is during the development stages that system safety is "designed into" the product for safe operational usage. Figure 2.4 shows the five system life cycle phases, with phase 2 expanded into preliminary design, final design, and test. These are the most significant phases for applying system safety.

There are several different models by which a system can be developed. Each of these models has advantages and disadvantages, but they all achieve the same end – development of a system. These development models include the following:

Engineering Development Model   This is the standard traditional approach that has been in use for many years. This method performs the system life cycle phases sequentially. The development and test phase is subdivided into preliminary design, final design, and test for more refinement. Under this model, each phase must be complete and successful before the next phase is entered. This method normally takes the longest length of time because the system is developed in sequential stages. Four major reviews are conducted for exit from one phase and entry into the next. These are the system design review (SDR), preliminary design review (PDR), critical design review (CDR), and test readiness review (TRR). These reviews are an important aspect of the hazard analysis process and timing. Figure 2.4 depicts the traditional engineering development model.

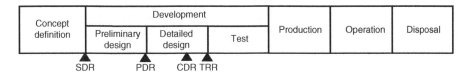

Figure 2.4   *Engineering development model.*

Concurrent Engineering   This method performs several development tasks concurrently, in an attempt to save development time. This method has a higher probability for technical risk problems since some items are in preproduction before full development and testing.

Spiral Development   In the spiral development process, a desired capability is identified, but the end-state requirements are not known at program initiation. Requirements are refined through demonstration, risk management, and continuous user feedback. Each increment provides the best possible capability, but the requirements for future increments depend on user feedback and technology maturation.

Incremental Development   In the incremental development process, a desired capability is identified, an end-state requirement is known, and that requirement is met over time by developing several increments, each depending on available mature technology. This method breaks the development process into incremental stages, in order to reduce development risk. Basic designs, technologies, and methods are developed and proven before more detailed designs are developed.

2.6   SYSTEM DEVELOPMENT PROCESS

Depending on the type of system, the development process can take months or years. Design requirements are developed to show how the system is intended to work, and not to work in the case of negative requirements. Design requirements are translated into design drawings, which show how the system actually works. Design functional diagrams show how the system is supposed to work, along with interrelationships. Software design requirements lay out the way the system software is intended to work as it interfaces with the system hardware. Software code defines how the system software actually works. Procedures show how system users interface with the system and operate the system.

The system development process is typically a dual development involving hardware and software development paths. System design and development is an involved process. System safety engineering is a component of systems engineering and part of the system design and development process. System development goes through the phases of *specify, design, build, and test* as shown in Figure 2.5. The information resulting from each of these processes is used by the hazard analyst.

2.7   SYSTEM HIERARCHY

Systems vary in size, shape, function, criticality, and complexity. A system can be small, such as a toaster that consists of fewer than 50 parts. A system can be very large, comprising hundreds of subsystems, thousands of assemblies, and millions of components, such as a commercial aircraft or a ship. Large complex systems can easily become overwhelming for human comprehension. In order to more easily visualize and understand a system, the system is typically broken down or subdivided in a hierarchical manner into manageable pieces that can be more easily grasped by the human mind. Referred to as a *system equipment table (SET)*,[1] this

---

[1]The SET hierarchy table also goes by other names, such as indentured equipment list (IEL), master equipment list (MEL), and system hierarchy table.

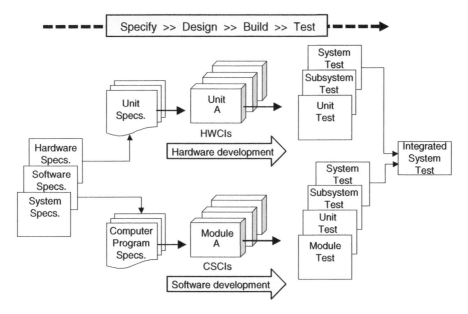

Figure 2.5 *System development.*

view of the system defines the system structure in an orderly, linked, and comprehensible manner.

The system equipment table establishes the organizational structure defining dominant and subordinate relationships between subsystems, down to the lowest component/piece part level. This table comprises a list of all the systems, subsystems, units, assemblies, and components in the major system, with each item in the list indented to reflect its hierarchy and ownership level. The indenture level also identifies or describes the relative complexity of assembly or function. The levels progress from the more complex (system) to the simpler (part) divisions. System design data and drawings will usually describe the system's internal and interface functions beginning at the system level and progressing to the lowest indenture level of the system.

The following is a typical breakdown of successive indenture level groupings in the SET hierarchy:

- *System*   The system of interest
- *Subsystem*   The major subsystems comprising the system
- *Unit*   The major units comprising each subsystem
- *Assembly*   The assemblies comprising a unit
- *Component*   The components comprising an assembly
- *Part*   The lowest level of separately identifiable items, for example, a bolt

Figure 2.6 demonstrates the hierarchy of elements in a system along with an example SET hierarchy for an aircraft system.

When performing an HA, all of the system components and functions must be considered to ensure a complete *systems* analysis. It is recommended that a SET be established early in the program and used by the system safety practitioner in the performance of an HA. The following are some benefits gained from utilizing a SET in safety analyses:

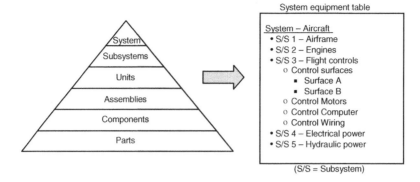

Figure 2.6   *System hierarchy representation.*

- It provides a complete list of equipment and can be expanded to include functions.
- It helps ensure that all of the system hardware and functions have been adequately covered by the hazard analyses.
- It helps establish and set the level of details for a particular safety analysis.
- It helps establish at what level in the hierarchy hazards should be identified.
- It helps in determining at what system level risk should be assessed.

2.8   SYSTEM VIEWS

A system can be viewed from several different perspectives. It is important that hazard analysis consider and evaluate a system from each of these perspectives in order to ensure complete safety coverage. This also explains why hazard analysis requires more than one type of methodology in order to identify all hazards; one hazard analysis type alone does not typically provide sufficient hazard identification coverage for all the many facets of a system.

The primary system viewpoints include the following:

1. *Physical*   This is the actual physical system (i.e., hardware) and the drawings and schematics that depict the physical architectural view (i.e., how the hardware elements are interconnected together).
2. *Functional*   The system functions describe what the system must do in order to produce the required system behavior. Functions are decomposed into subfunctions with input, output, and transformation rules.
3. *Operational*   This aspect defines how the user will interface with and operate the system, including instructions, conditions, parameters, and limitations. This also includes maintenance and repair aspects of the overall system operation.
4. *Software*   This view involves the system software that operates and controls computer-controlled systems. Software is complex, large in size, and more abstract, therefore, making it more difficult to fully evaluate.
5. *Environment*   This view looks at the various environments that the system will encounter, including both internal and external sources. It includes generation and vulnerability.

6. *Human*   This view looks more closely at human performance in the system and the effect of potential human errors. Often, safety-critical human errors can be avoided or deflected using special design measures.

7. *Organizational*   This view considers the organizational and management aspects influencing the decisions made affecting safety and hazards.

Systems can be perceived from these different perspectives, each of which provides a different viewpoint and understanding of the system. Chapanis [1] recognized the first three perspectives in his 1996 book.

Two distinguishing characteristics of systems are that (a) the whole is more than the sum of the parts and (b) what is best for the parts is not necessarily best for the overall system. As might be expected, these two system characteristics are important in safety analysis and in the identification of hazards. System safety works well because it looks at both the system and the interrelated aspects of the parts. Although system safety must focus on the system, it must also evaluate both the parts and the system as a whole; nothing is evaluated in isolation. This is why hazard analysis must consider and focus on system functions, system operations, and system components, including hazardous energy sources.

2.9 SYSTEM DEVELOPMENT ARTIFACTS

System development is the process of designing, developing, and testing a system design until the final product meets all requirements and fulfils all objectives. The system engineering development process produces many artifacts that are useful in the system safety process.

In the process of developing large and highly complex systems, many tools are used by systems engineering and other related disciplines. Some systems engineering tools that greatly aid the system safety analyst in performing an HA include the following:

- System specification and design requirements
- Drawings/schematics
- Simplified system diagram
- Functional diagrams
  - Function flow diagrams
  - Functional flow block diagrams (FFBD)
  - Functional dependency diagrams
- System hierarchy table or SET
- Reliability block diagram (RBD)
- Failure mode and effects analysis (FMEA)
- Timeline analysis

These artifacts assist the safety analyst in understanding the system architecture and how a system operates. They provide useful system information, such as system interrelationships, system interfaces, redundancies, and dependencies. By using this information, the safety analyst can perform a hazard analysis to identify system hazards, interface hazards, and functional hazards. These artifacts are not always meant to be detailed but rather to show basic system content, interfaces, and factors to consider and subsystem relationships.

2.10 SYSTEMS COMPLEXITY AND SAFETY

Systems tend to have various levels of complexity. System complexity is an individual system variable based on many different factors. The complexity of a system can have a significant effect on the safety of a system. System complexity has a very direct relationship with system safety. Generally, the more complex the system, the more likely it is to have safety concerns or problems. Also, more complex systems are more difficult to understand, analyze, and verify for safety. System complexity is a very difficult concept to completely define because complexity is comprised of many factors, such as size, abstraction, coupling, interconnectivity, criticality, and so on.

Understanding system type and scope is very important in system safety as it can be an indication of the system's complexity and safety-criticality. The scope of the system boundaries establishes the size and depth of the system. The system limitations describe basically what the system can and cannot safely do. Certain limitations may involve the ability of the system to include design safety features. Every system operates within one or more different environments. The specific environment establishes what the potential hazardous impact will be on the system. System criticality establishes the overall safety rating for the system. A nuclear power plant system has a high-consequence safety rating (i.e., safety-critical), whereas a TV set as a system has a much lower safety-criticality rating.

There are four basic system models that describe almost all types of systems and their relative complexity. Each model varies depending upon factors such as composition, relationships, intent, and environment. These system types are (a) static, (b) dynamic, (c) homeostatic, and (d) cybernetic [2].

Static Systems Static systems are simple, essentially "dumb" systems that operate only as built or designed. They receive no input information to modify their process and they are unable to modify their process on their own. They have fixed goals, but no means of internal control to ensure that the system goals are met. An example of a static system would be a watch or a clock. The functional purpose is to keep accurate time, but the system has no internal control devices for ensuring this goal is achieved. An ordinary, analog watch has no way to readjust itself if the time is incorrect. Also, the watch has no awareness of changes, such as changing the time to daylight savings time. The environmental effect on the watch may produce different outcomes. Figure 2.7a is a model representing this type of system.

Dynamic Systems Dynamic systems are a little more complex, but are still essentially dumb systems, in that they can only respond as they are programmed to function according to given input information. The system merely provides output according to the input received. An example of a dynamic system would be a computer display. Its goal is to provide a specific output according to the input. It operates according to a fixed set of unchangeable rules. Environmental obstacles, such as water on the electronics, might produce different outcomes. Figure 2.7b is a model representing this type of system.

Homeostatic Systems Homeostatic systems are a step up in complexity with some level of "intelligence." These systems respond to environmental changes and have effective internal control devices that enable the system to meet its fixed system goals. Figure 2.7c is a model representing this type of system. An example of a homeostatic system would be the internal temperature control system in a house. The house temperature depends upon two main inputs (1) the temperature outside the house and (2) how much heating or

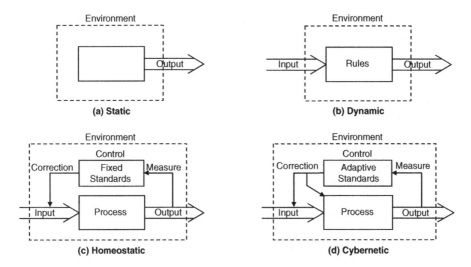

Figure 2.7 *System types.*

cooling the central system is providing. The output depends on the combined inputs compared to an established standard. If the internal house temperature does not meet the established standard, then the system makes the necessary adjustments until the standard is achieved. This type of system functions to meet the fixed system goals. Another example of this type of system would be a guided missile system.

Cybernetic Systems Cybernetic systems are the most complex and intelligent systems. These systems are affected by environmental shifts and have effective internal control devices that enable the system to meet its fixed system goals. In this type of system, the goals are not rigidly fixed but are adaptable to changing environments and conditions. Figure 2.7*d* is a model representing this type of system. An example of a cybernetic system would be a manufacturing process system that has adaptable standards and can adjust or correct both the input and the process itself. These systems tend to typically be safety-critical and require significant system safety and software safety effort.

2.11 SYSTEM REQUIREMENTS

In engineering, a design requirement is a statement that specifies or describes what the design must do, or what the design must include, in order for an item to meet its objectives. It is an essential condition, state, or form that the design must satisfy. A single requirement is an incremental design definition. A set of design requirements establishes the framework for designing and building a system. For example, one requirement in a series of requirements for a weapon system might be "The Safe and Arm device shall have a window whereby the actual safe/arm state can visually be viewed by the operator." Note that this requirement implies that there must be another design requirement specifying the use of a Safe and Arm device.

A requirement is a singular documented need of what a particular product or service should be or perform. It is most commonly used in a formal sense in systems engineering or software engineering. It is a statement that identifies a necessary attribute, capability, characteristic, or quality of a system in order for it to have value and utility to a user. An interface requirement is a requirement that specifies the necessary attributes between two interfacing system elements, such as systems, subsystems, software modules, and so on. In the systems engineering approach to system development, sets of requirements are used as inputs into the design stages of product development. Requirements are also an important input into the verification process, since tests should trace back to specific requirements. Requirements show what elements and functions are necessary for the particular project.

There are two basic types of requirements, function and nonfunctional. A functional requirement is a description of what a system must do. This type of requirement specifies something that the delivered system must be able to perform or accomplish. Nonfunctional requirements specify something about the system itself, and how well it performs its functions. Such requirements are often called performance requirements. Examples of such requirements include safety, usability, availability, reliability, supportability, testability, maintainability, and so on. These types of requirements should be defined in a way that is verifiably measurable and unambiguous.

A collection of requirements defines the characteristics or features of the desired system. A good list of requirements generally avoids saying how the system should implement the requirements, leaving such decisions to the system designer. Describing how the system should be implemented may be known as implementation bias or "solution engineering," However, implementation constraints on the solution may be validly expressed by the future owner if desired.

Requirements typically fall into the following categories:

- Functional requirements describe the functionality that the system is to execute; for example, formatting some text or modulating a signal. They are also known as capabilities.
- Nonfunctional requirements are the ones that act to constrain the solution. Nonfunctional requirements are known as quality requirements or "ility" requirements (e.g., reliability).
- Constraint requirements are the ones that act to constrain the solution. No matter how the problem is solved, the constraint requirements must be adhered to.

Nonfunctional requirements can be further classified according to whether they are usability requirements, look and feel requirements, humanity requirements, performance requirements, maintainability requirements, operational requirements, safety requirements, reliability requirements, or one of many other types of requirements.

In software engineering, only functional requirements can be directly implemented in software. The nonfunctional requirements are controlled by other aspects of the system. For example, in a computer system, reliability is related to hardware failure rates, and performance is controlled by CPU and memory. Nonfunctional requirements can in some cases be decomposed into functional requirements for software. For example, a system-level non-functional safety requirement can be decomposed into one or more functional requirements. In addition, a nonfunctional requirement may be converted into a process requirement when the requirement is not easily measurable. For example, a system-level maintainability requirement may be decomposed into restrictions on software constructs or limits on lines or code. Design requirements have many different aspects, which are summarized in Table 2.3.

Requirements are usually written as a means for communication between the different stakeholders. This means that the requirements should be easy to understand both for normal users and for developers. One common way to document a requirement is stating what the

TABLE 2.3   Various Aspects of Requirements

Aspect	Description
Defines the design	• Functional – things the product must do, an action that the product must take • Nonfunctional – properties, or qualities, the product must have • Constraint – a limitation that constrains the design options
Defines the major system elements	• Hardware – specifies hardware design • Software – specifies software design • HMI – specifies human–machine interface design • Environment – specifies environmental constraints and design • Interfaces – specifies design of system and subsystem interfaces
Specify user and life cycle needs	• Performance – specifies system performance needs and objectives • Operations – specifies system operational needs • Support – specifies system support needs • Test – specifies system test needs and objectives
Defines system abstraction levels	• Physical – defines physical attributes of system • Functional – defines functional attributes of system
Defines system qualities	• Safety – to mitigate hazards and establish system mishap risk level • Reliability – to establish system reliability level • Quality – to establish system quality level
Defines source	• Prescribed – defined by customer, standards, guidelines, and so on • Derived – developed to expand higher requirements or solve problems
Defines basic type	• Qualitative – qualitative attribute and format • Quantitative – quantitative objective and format

system shall do; for example, "The contractor shall deliver the product no later than xyz date." Use Cases are another way to document requirements.

Writing good design requirements is an art and a skill. Table 2.4 lists some typical characteristics of good requirements.

Requirements generally change with time. Once requirements are defined and approved, they should fall under formal change control. For many projects, requirements are altered before the system is complete. This is partly due to the complexity of the system and the fact that users do not know what they want before they see it. Requirement changes should always be evaluated by system safety for possible safety impact on the system design.

Safety design requirements are an intrinsic element of the system safety program. The purpose of a system safety requirement (SSR) is to provide design guidance for intentionally designing safety into a system or product. An SSR is a design requirement that is primarily intended to enhance the safety quality of the system under development. SSRs generally focus on eliminating or mitigating a hazard. Requirements are the lifeblood of a system; however, requirements do not guarantee the system is hazard and risk free; they only assure protection from known hazards. Requirements analysis without hazard analysis achieves only half the safety job.

The components and functions comprising a system are highly interrelated and complex, which means they must be well understood and well defined in order to properly build the system. The requirement and specification process must correctly define the system as a whole, including architectures, functions, interrelationships, constraints, and so on.

Figure 2.8 is a diagram illustrating the role of the requirements management in the system development life cycle. It should be noted that requirements management does not stop after the initial requirements are established, as it is an ongoing process.

TABLE 2.4   Characteristics of Good Requirements

Characteristic	Explanation
Unitary (cohesive)	The requirement addresses one and only one thing
Complete	The requirement is fully stated in one place with no missing information
Consistent	The requirement does not contradict any other requirement and is fully consistent with all authoritative external documentation
Nonconjugated (atomic)	The requirement is *atomic*, that is, it does not contain conjunctions. For example, "The postal code field must validate American *and* Canadian postal codes" should be written as two separate requirements
Traceable	The requirement meets all or part of a business need as stated by stakeholders and authoritatively documented
Current	The requirement has not been made obsolete by the passage of time
Feasible	The requirement can be implemented within the constraints of the project
Unambiguous	The requirement is concisely stated without recourse to technical jargon, acronyms (unless defined elsewhere in the requirements document), or other esoteric verbiage. It expresses objective facts, not subjective opinions. It is subject to one and only one interpretation. Vague subjects, adjectives, prepositions, verbs, and subjective phrases are avoided. Negative statements and compound statements are prohibited
Mandatory	The requirement represents a stakeholder-defined characteristic the absence of which will result in a deficiency that cannot be ameliorated. An optional requirement is a contradiction in terms
Verifiable	The implementation of the requirement can be determined through one of the four possible methods: inspection, demonstration, test, or analysis
Concrete	A requirement should be a concrete prerequisite rather than a goal. A goal-oriented requirement is a vague requirement that usually cannot be verified or tested, and usually provides no firm design guidance
Designer flexible	A requirement should avoid specifying a design implementation, unless specifically intended or necessary. It is better to leave implementation options open to the designer whenever possible
Realizable/viable	A requirement must be realistically achievable within the constraints of time, cost, and technology
Nonconflicting/ Nonoverlapping	Requirements should neither conflict with one another nor overlap one another. Every requirement should stand on its own merit
Nonduplicated	Requirements should not be duplicated, either partially or fully. This leads to possible conflicts, confusion, and traceability problems
Must/should clarity	Requirements have to be clear on "must" and "should." Design requirements are generally mandatory; therefore, the term "must" is necessary. Terms such as "may," "should," or "could" are not mandatory properties and thus do not necessarily require implementation. These nonmandatory terms imply that something would be nice to have, but that it is not positively demanded
Consistent terminology	Where domain or specialist terms, names, or documents are mentioned in requirements, ensure they are used consistently and correctly. This is often achieved by having a separate section of the requirements specification that contains a glossary, abbreviations, and definitions

The process of establishing and implementing design requirement is probably one of the most important tasks in the system development process. Requirements management is the process of developing and implementing design requirements for a system or product. This process seeks to minimize design and development problems by using systematic and structured methods to establish and control design requirements. Since it is only logical and feasible to have design requirements prior to the design process, requirements development must begin at an early stage in the development life cycle.

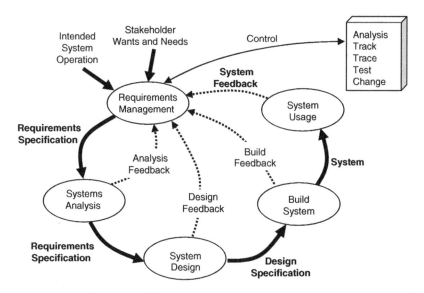

Figure 2.8 *The requirements management process.*

The requirements management process is a gathering and developing process. This process starts by gathering requirements from user needs, wants, and intent. Requirements are developed to provide the best system solution in the form of design requirements. The output from this process is the requirements specification, which is a complete description of the system, including its functions, data, and behavior. It goes without saying that requirements should not be missing. It may be difficult to know when a requirement is missing, but system analysis and traceability analysis help identify missing links. Attempting to build a systems analysis model from existing requirements may reveal where requirements are missing. A review of other requirements for implied requirements that cannot be found helps in identifying missing requirements.

Requirements management involves communication between the program team members and the stakeholders, and adjustment to requirements changes throughout the course of the project. To prevent one class of requirements from overriding another, constant communication among members of the development team is critical. Requirements management involves the following aspects:

- *Requirements Analysis*   Acquire user needs and translate into design.
- *Track*   An electronic database to record and maintain history of requirements.
- *Trace*   Linkage of requirements from inception through test verification.
- *Test*   Test all requirements for verification and validation; include evidence.
- *Change*   Change management system to control modification of requirements.

Requirements traceability is part of the requirements management process. Requirements traceability is concerned with documenting the life of a requirement and to provide bidirectional traceability between various associated requirements. It enables users to find the origin of each requirement and track every change that was made to this requirement.

Not only the requirements themselves should be traced but also the requirements relationship with all the artifacts associated with it, such as models, analysis results, test cases, test procedures, test results, and documentation of all kinds. Even people and user groups associated with requirements should be traceable. Traceability should encompass a trace of where requirements are derived from, how they are satisfied, how they are tested, and what impact will result if they are changed. Electronic database traceability tools are available to assist in this task, which can become very large and complex for large systems.

In the software safety domain, requirements traceability is an important element in ensuring safety-critical functions are safely implemented. In addition, software requirements should be traced to hazards in order to ensure the hazards are properly and completely mitigated.

2.12 SYSTEM LAWS

Systems have certain inherent laws, which include the following:

- *Law 1* Systems are entities comprised of hardware, software, humans, rules/laws, procedures, and environments.
- *Law 2* Systems are designed to work in an intended manner for an intended purpose.
- *Law 3* Systems react predictably to input; they also react predictably to failures (determining these responses is the task of system safety and reliability).
- *Law 4* System size and complexity directly impact system understanding, reliability, and safety.
- *Law 5* Systems are designed to perform specific functions when intended during normal operation; however, failure modes can cause these functions to occur inadvertently when not intended.
- *Law 6* Because of component interdependencies, faults can be propagated through the system, so failure in one component can affect the operation of other components.
- *Law 7* System failures often occur because of unforeseen interrelationships between components.
- *Law 8* Systems often have unintended functions accidentally designed into the system.
- *Law 9* If anything can go wrong . . . it will (Murphy's law).
- *Law 10* Everything is a system, and every system is part of a larger system.
- *Law 11* Systems have emergent properties (e.g., safety, reliability, and quality) that can vary and change depending upon the system design.

Emergent properties relate to the behavior of the system in its operational environment. These are properties of the system as a whole rather than properties that can be derived from the properties of system components. Emergent properties are a consequence (result) of the relationships between system components. They can, therefore, be assessed and measured once the components have been integrated into a system. They can change when system configuration changes (note that failures often indirectly change the system configuration). Examples include safety, reliability, and quality.

2.13 SUMMARY

This chapter discussed the basic concept of what comprises an actual system and what occurs during the system development process. The following are basic principles that help summarize the discussion in this chapter:

1. Systems are a collection of components that work together to achieve a greater goal than any one component.
2. The major components of a system include hardware, software, humans, procedures, and environments.
3. All systems have a life cycle with the following phases: concept definition, development and test, production, operation, and disposal.
4. The system engineering development process produces many artifacts that are useful in the system safety process.
5. Systems are developed consistent with requirements, cost, and schedule.
6. System safety covers all system life cycle phases, from "cradle to grave."
7. The system safety process must be planned and executed around the system engineering development process.

REFERENCES

1. Alphonse Chapanis, *Human Factors in Systems Engineering*, John Wiley & Sons, Inc., New York, 1996.
2. T.H. Athey, *Systematic Systems Approach*, Prentice-Hall Inc., 1974.

Chapter **3**

Hazards, Mishap, and Risk

3.1 INTRODUCTION

Hazard analysis requires a thorough understanding and comprehension of hazards and hazard theory. If system hazards cannot be properly and correctly identified, then the safety process is compromised. And it is difficult to identify hazards if one does not understand what actually comprises a hazard. Hazard theory seems simple, but in reality it is a very complex concept that requires a good technical engineering breakdown of the specific components of a hazard. The focus of this chapter is on hazard theory, mishap theory, and risk theory, which are all closely interrelated. A thorough understanding of this material is important for the identification and control of hazards.

In order to design-in safety, hazards must be designed-out (eliminated) or mitigated (designed to reduce risk). Hazard identification is a critical system safety function, and thus, the correct understanding and appreciation of hazard theory is critical. This chapter focuses on what constitutes a hazard, in order that hazards can be recognized and understood during the hazard identification, evaluation, and mitigation processes.

Hazard analysis provides the basic foundation for system safety. Hazard analysis is performed to identify hazards, hazard effects, and hazard causal factors. Hazard analysis is used to determine system risk, to determine the significance of hazards, and to establish design measures that will eliminate or mitigate the identified hazards. Hazard analysis is used to systematically examine systems, subsystems, facilities, components, software, personnel, and their interrelationships, with consideration given to logistics, training, maintenance, test, modification, and operational environments. In order to effectively perform hazard analyses, it is necessary to understand what comprises a hazard, how to recognize a hazard, and how to define a hazard. To develop the skills needed to identify hazards and hazard causal factors, it is necessary to understand the nature of hazards, their relationship with mishaps, and their effect upon system design.

Hazard Analysis Techniques for System Safety, Second Edition. Clifton A Ericson, II.
© 2016 John Wiley & Sons, Inc. Published 2016 by John Wiley & Sons, Inc.

Many important hazard-related concepts will be presented in this chapter, which will serve as building blocks for hazard analysis and risk evaluation. In sum and substance, humans inherently create the potential for mishaps. Potential mishaps exist as hazards, and hazards exist in system designs. Hazards are actually designed into the systems we design, build, and operate. In order to perform hazard analysis, the analyst must first understand the nature of hazards. Hazards are predictable, and what can be predicted can also be eliminated or controlled.

3.2 HAZARD, MISHAP, AND RISK DEFINITIONS

The overall system safety process is one of mishap risk management, whereby safety is achieved through the identification of hazards, the assessment of hazard mishap risk, and the control of hazards presenting unacceptable risk. This is a closed loop process, where hazards are identified, mitigated, and tracked until an acceptable closure action is implemented and verified. System safety should be performed in conjunction with actual system development, in order that the design can be influenced by safety during the design development process, rather than trying to enforce more costly design changes after the system is developed.

In theory, this sounds simple, but in actual practice a common stumbling block is the basic concept of what comprises a hazard, a hazard causal factor (HCF) and a mishap. It is important to clearly understand the relationship between a hazard and an HCF when identifying, describing, and evaluating hazards. To better understand hazard theory, let us start by looking at some common safety-related definitions.

Accident

a. An undesirable and unexpected event, a mishap, and an unfortunate chance or event (dictionary).

b. Any unplanned act or event that results in damage to property, material, equipment, or cargo, or personnel injury or death when not the result of enemy action (Navy OP4 & OP5).

Mishap

a. An unfortunate accident (dictionary).

b. An unplanned event or series of events resulting in unintentional death, injury, occupational illness, damage to or loss of equipment or property, or damage to the environment (MIL-STD-882E).

c. An unplanned event or series of events resulting in death, injury, occupational illness, or damage to or loss of equipment or property, or damage to the environment. Accident. (MIL-STD-882C). (Note the last word "accident" is contained in this definition.)

Hazard

a. To risk; to put in danger of loss or injury (dictionary).

b. Any real or potential condition that can cause injury, illness, or death to personnel; damage to or loss of a system, equipment, or property; or damage to the environment (MIL-STD-882D).

c. A real or potential condition that could lead to an unplanned event or series of events (i.e., mishap) resulting in death, injury, occupational illness, damage to or loss of equipment or property, or damage to the environment (MIL-STD-882E).

d. A condition that is a prerequisite for an accident (Army AR 385-16).

Risk

a. Hazard, peril, jeopardy (dictionary).

b. An expression of the impact and possibility of a mishap in terms of potential mishap severity and probability of occurrence (MIL-STD-882D).

c. A combination of the severity of the mishap and the probability that the mishap will occur (MIL-STD-882E).

Note both the differences and the similarities between the above definitions. It should be apparent from these definitions that there is no significant differentiation between a mishap and an accident, and these terms can be used interchangeably. To be consistent with the standard system safety terminology, the term *mishap* will be preferred to the term *accident*; however, it should be not that they are synonymous.

The dictionary definition states that an accident or mishap is a random chance event, which gives a sense of futility by implying that hazards are unpredictable and unavoidable. System safety, on the other hand, is built upon the premise that mishaps are not random events; instead, they are deterministic and controllable events. Mishaps and accidents do not just happen; they are the result of a unique set of conditions (i.e., hazards), which are predictable when properly analyzed. A hazard is a potential condition that can result in a mishap or accident, given that the hazard occurs. This means that mishaps can be predicted via hazard identification. And mishaps can be prevented or controlled via hazard elimination, control, or mitigation measures. This viewpoint provides a sense of control over the systems we develop and use.

3.3   ACCIDENT (MISHAP) THEORY

Hazards lead to accidents; there is a direct link and relationship between hazards and accidents. Hazards occur, or more correctly are created, due to many different factors. In order to effectively understand hazard theory, it is important to understand how accidents occur. Accidents can best be understood by looking at accident models and timelines, which in turn will provide background information for better understanding the hazard theory and the hazard–mishap relationship.

An accident (mishap) occurs when a hazard is actuated or, more specifically, when all of the hazard causal factors occur together to fulfill the required hazard scenario conditions. This scenario is depicted in Figure 3.1. This illustration depicts three HCFS; however, an accident can be the result of one or more HCFs. HCFs are sometimes referred to as events, but they really are factors, because they can be many different things, such as an event, condition, failure, time, and so on. When all of the HCFs for a particular hazard are satisfied (i.e., occur), the hazard scenario conditions are fulfilled and a fusion of all the HCFs transforms the situation into an actual accident event. The time element is a unique function for each hazard. Also, the critical fusion point depends on the hazard.

Understanding the sequence of events that occur during an accident helps in the understanding of hazards. Figure 3.2 depicts the relative timeline of an accident.

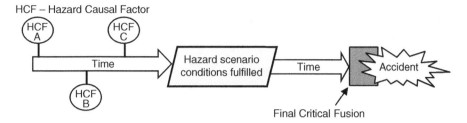

Figure 3.1 *Accident description.*

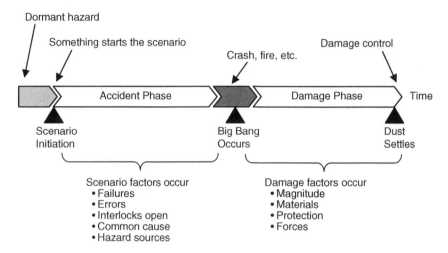

Figure 3.2 *Accident timeline.*

This timeline depicts that, first, a hazard must exist, which is in the dormant state. Then, some initial HCF occurs to start the accident phase. When all the events and conditions necessary to fulfill the hazard statement occur together, the accident occurs. This is the *big bang*, exemplified by a crash, explosion, fire, and so on. This could be a very fast occurring event or it could involve some element of time. Following the accident is the damage phase, where all of the undesired consequences from the accident occur. This typically occurs over some element of time until it is all finally over and the dust settles. Note that for each hazard-accident, both the causal factors and the time elements are different.

3.4 THE HAZARD–MISHAP RELATIONSHIP

Per the system safety definitions, a *mishap* is an *actual event* that has occurred and resulted in death, injury, and/or loss, and a *hazard* is a *potential condition* that can potentially result in death, injury, and/or loss. These definitions lead to the principle that a hazard is the precursor to a mishap; a hazard *defines* a potential event (i.e., mishap), while a mishap is the occurred event. This means that there is a direct relationship between a hazard and a mishap, as depicted in Figure 3.3.

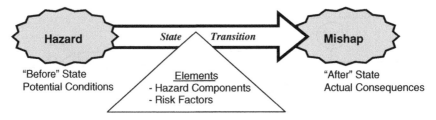

Figure 3.3 *Hazard–mishap relationship.*

The concept conveyed by Figure 3.3 is that a hazard and a mishap are two separate states of the same phenomenon, linked by a state transition that must occur. You can think of these states as the *before* and *after* states. A hazard is a "potential event" at one end of the spectrum, which may be transformed into an "actual event" (the mishap) at the other end of the spectrum, based upon the state transition. An analogy might be water, where water is one entity, but it can be in a liquid state or a frozen state, and temperature is the transitional factor. This concept is in alignment with the timeline of an accident, as depicted in Figure 3.2.

Figure 3.4 illustrates the hazard–mishap relationship from a different perspective. In this viewpoint, a hazard and a mishap are at opposite ends of the same entity. Again, some transitional event causes the change from the conditional hazard state to the actualized mishap state. Note that both states look almost the same, the difference being that the verb tense has changed from referring to a future potential event to referring to the present actualized event, where some loss or injury has been experienced. A hazard and a mishap are the same entity, only the state has changed, from a dormant state to an active state.

Mishaps are the immediate result of actualized hazards. The state transition from a hazard to a mishap is based on two factors: (1) the unique set of *hazard components* involved and (2) the *mishap risk* presented by the hazard components. The hazard components are the items comprising a hazard, and the mishap risk is the probability of the mishap occurring and the severity of the resulting mishap loss.

An accident can be defined as a scenario of events and conditions. An accident scenario is equivalent to a hazard; scenario frequency is equivalent to hazard probability; scenario outcome is equivalent to hazard severity. Identifying and developing accident scenarios are fundamental to the concept of hazard analysis. The process begins with a set of initiating events (IEs) that perturb the system (i.e., cause it to change its operating state or configuration).

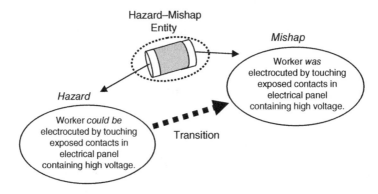

Figure 3.4 *Same entity, different states.*

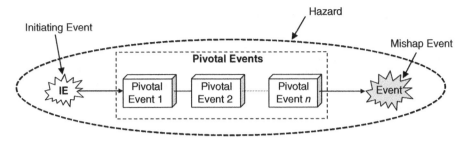

Figure 3.5 *Accident/hazard scenario.*

For each IE, the analysis proceeds by determining the additional failure modes that are necessary to lead to the undesirable consequences. The consequences and frequencies of each scenario are computed for the individual IEs and the collection of probabilities form a risk profile for the system.

Since hazards and accidents are directly linked, an accident can be resolved to a hazard, and in turn, a hazard can be identified as a potential accident. Hazards are used to model accident scenarios. The scenario starts with the IE and progresses through the scenario via a series of pivotal events (PEs) until an end state is reached. The PEs are failures or events that aggravate or contribute to the scenario. The frequency (i.e., probability) of the PE can be obtained from an FTA of the event. An accident scenario typically contains an IE and (usually) one or more pivotal events leading to an end state as shown in Figure 3.5.

Note that a hazard defines and describes the entire accident scenario.

3.5 HAZARD RISK

Hazard risk (and mishap risk) is a fairly straightforward concept, where risk is defined as

$$\text{Risk} = \text{Probability} \times \text{Severity}$$

The mishap probability factor is the probability of the hazard components occurring and transforming into the mishap. The mishap severity factor is the overall consequence of the mishap, usually in terms of loss resulting from the mishap (i.e., the undesired outcome). Both probability and severity can be defined and assessed in either qualitative terms or quantitative terms. Time is factored into the risk concept through the probability calculation of a fault event, for example, $P_{\text{FAILURE}} = 1.0 - e^{-\lambda T}$, where $T = $ exposure time and $\lambda = $ failure rate.

Please note that hazard risk and mishap risk are essentially the same thing. The only difference is context; in one case, the discussion is about hazards and in the second case, the discussion is about mishaps, but they both cover the same undesired outcome.

3.6 THE COMPONENTS OF A HAZARD

The hazard component concept is a little more complex in definition. A hazard is an entity that contains only the elements *necessary and sufficient* to result in a mishap. The components of a hazard define the necessary conditions for a mishap and the end outcome or effect of the mishap.

TABLE 3.1   Examples of Hazard Components

Hazardous Element	Initiating Mechanism	Target/Threat Outcome
• Ordnance	• Inadvertent signal; RF energy	• Personnel/explosion; death/injury
• High-pressure tank	• Tank rupture	• Personnel/explosion; death/injury
• Fuel	• Fuel leak and ignition source	• Personnel/fire; loss of system; death/injury
• High voltage	• Touching an exposed contact	• Personnel/electrocution; death/injury

A hazard is comprised of the following three basic components: source (S), mechanism (M), and outcome (O). This concept was developed by Pat Clemens [1]. I have found it useful to change the wording slightly as follows for increased descriptive comprehension:

1. *Hazard Source (HS)*   This is the basic hazardous resource creating the impetus for the hazard, typically a hazardous energy source such as explosives being used in the system. This can be thought of as the basic hazardous component that produces a source of danger.

2. *Initiating Mechanism (IM)*   This is the trigger or initiator event(s) causing the hazard to occur. The IMs cause actualization or transformation of the hazard from a dormant state to an active mishap state. This can be thought of as the hazard trigger or trigger events that transform the hazard to a mishap.

3. *Target/Threat Outcome (TTO)*   This is the person or thing that is vulnerable (target), the threat to that target (threat), and the resulting severity outcome (outcome) when the mishap event occurs. The outcome is the expected consequential damage and loss. This can be thought of as the basic hazard threat.

Either set of wording is acceptable, S–M–O or HS–IM–TTO; both have equivalent meaning.

Table 3.1 provides some example items and conditions for each of the three hazard components.

To demonstrate the hazard component concept, consider a detailed breakdown of the following example hazard: "Worker could be electrocuted by touching exposed contacts in electrical panel containing high voltage." Figure 3.6 shows how this hazard is divided into the three necessary hazard components.

Note in this example that all three hazard components are present and can be clearly identified. In this particular example, there are actually two IMs involved. The TTO defines the mishap outcome, while the combined HS and TTO define the mishap severity. The HS and IM are the HCFs and define the mishap probability. If the high-voltage component can be

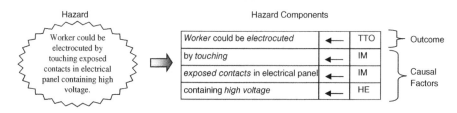

Figure 3.6   *Example of hazard components.*

Figure 3.7 *Hazard triangle.*

removed from the system, the hazard is eliminated. If the voltage can be reduced to a lower, less harmful level, then the mishap severity is reduced.

Key hazard theory concepts to remember are as follows:

- Hazards result in (i.e., cause) mishaps.
- Hazards are (inadvertently) built into a system.
- Hazards are recognizable by their components.
- A design flaw can be a mishap waiting to happen.
- A hazard will occur according to the hazard components involved.
- A hazard is a deterministic entity, and not a random event.
- Hazards (and mishaps) are predictable, and therefore, are preventable or controllable.

3.7   HAZARD TRIANGLE

The three components of a hazard form what is known in system safety as the *hazard triangle*, as illustrated in Figure 3.7. The hazard triangle illustrates that a hazard consists of *three necessary* and *coupled* components, each of which forms the side of a triangle.

All three sides of the hazard triangle are essential and required in order for a hazard to exist. Remove any one of the triangle sides and the hazard is eliminated because it is no longer able to produce a mishap (i.e., the triangle is incomplete). Reduce the probability of the IM triangle side and the mishap probability is reduced. Reduce an element in the HE or the TTO side of the triangle and the mishap severity is reduced. This aspect of a hazard is useful when determining where to mitigate a hazard.

3.8   HAZARD ACTUATION

Mishaps are the immediate result of actualized hazards. The state transition from a hazard to a mishap is based on two factors: (1) the unique set of *hazard components* involved and (2) the *mishap risk* presented by the hazard components. The hazard components are the items comprising a hazard, and the mishap risk is the probability of the mishap occurring and the severity of the resulting mishap loss.

Figure 3.8 depicts a hazard using the analogy of a molecule, which is comprised of one or more atoms. The molecule represents a hazard, while the atoms represent the three types of components that make up the molecule. The concept is that a hazard is a unique entity,

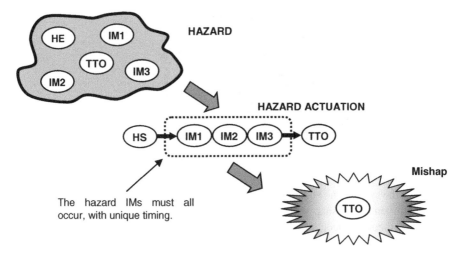

Figure 3.8   *Hazard–mishap actuation (view 1).*

comprised of a unique set of components, similar to a molecule. This set of components consists of three specific required elements; HS, IM, and TTO. All three specific elements are required, and the IM can involve more than one mechanism. The molecule model indicates that there is no particular order between the hazard components; they all merely exist within the hazard.

When the components comprising the hazard are in alignment, the hazard then transitions from a conditional state to a mishap event state. This viewpoint shows that all randomly oriented hazard components must line up (or occur) in the correct sequence before the mishap actually occurs. The HS is always present, and the mishap occurs only when the IMs force the transition; the TTO determines the mishap severity.

Figure 3.9 presents another way of viewing the hazard–mishap relationship. The spinning wheels represent all the components forming a particular hazard. Only when the timing is right, and all the holes in the wheels line up perfectly, does the hazard move from a potential to an actual mishap.

There are two key points to remember about the hazard–mishap transition process. One, there is generally some sort of energy buildup in the transition phase, which ultimately causes

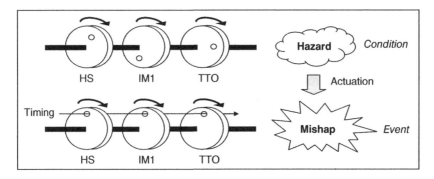

Figure 3.9   *Hazard–mishap actuation (view 2).*

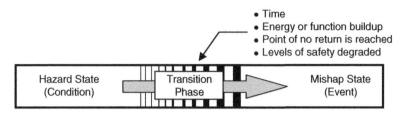

Figure 3.10 *Hazard–mishap actuation transition.*

the mishap damage. Two, there is usually a point of no return for the mishap, where there is no possibility of it being reversed. Each individual hazard is unique and therefore this time period is unique to every hazard.

Figure 3.10 illustrates the hazard–mishap state transition. During the transition phase, the energy buildup occurs as the IMs are occurring. This could also be viewed as the elements of a function being completed, or a functional buildup occurring in a rapid or slow process. It is during this time period that the levels of safety are degrading and the transition process reaches a point of no return, where the hazard becomes irreversible.

A system is designed and built to a specification, for the purpose of performing an intended function or functions. But, a system can also contain an inherent design flaw that is capable of performing an unintended and undesirable function. It is this design flaw that provides the necessary events and conditions that comprise a hazard. Quite often, this design flaw (hazard) remains hidden from the designers because it is not always obvious. These hazards can be discovered and identified only through hazard analysis.

Mishaps do not just happen, they are the result of design flaws inadvertently built into the system design. Thus, in a sense mishaps are predictable events. If a hazard is eliminated or mitigated, the corresponding mishap is also eliminated or controlled. Therefore, hazard identification and control, via hazard analysis, is the key to mishap prevention.

3.9 HAZARD CAUSAL FACTORS

There is a difference between why hazards exist and how they exist. The basic reasons why hazards exist are as follows: (1) they are unavoidable because hazardous elements must be used in the system and/or (2) they are the result of an inadequate design safety consideration. An inadequate design safety consideration results from poor or insufficient design or the incorrect implementation of a good design. This includes inadequate consideration given to the potential effect of hardware failures, sneak paths, software glitches, human error, and the like. HCFs are the specific items responsible for the existence of a unique hazard in a system.

Figure 3.11 depicts the overall HCF model. This model correlates all of the factors involved in hazard–mishap theory. The model illustrates that hazards create the potential for mishaps, and mishaps occur according to the level of risk involved (i.e., hazards and mishaps are linked by risk). The three basic hazard components define both the hazard and the mishap. These components can be further broken into major hazard causal factor categories, which include (1) hardware, (2) software, (3) humans, (4) interfaces, (5) functions, and (6) the environment. Finally, the causal factor categories are refined even further into the actual specific detailed causes, such as a hardware component failure mode.

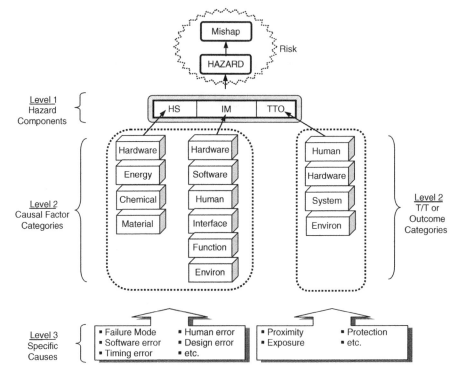

Figure 3.11 *Hazard causal factor model.*

Figure 3.11 illustrates how HCFs can be viewed at three different levels:

Level 1: *Top Layer* The three hazard *components* (HS, IM, and TTO)

Level 2: *Mid Level* The HCF *categories* (hardware, software, HSI, environment, functions, interfaces, etc.)

Level 3: *Bottom Level* The detailed *specific causes* (failure modes, errors, etc.)

The top-level HCF categories define the basic root cause sources of all hazards. The first step in HCF identification is to identify the category and then identify the detailed specifics in each category, such as specific hardware component failures, operator errors, software error, and the like. High-level hazards in a PHA might identify root causes at the HCF category level, while a more detailed analysis, such as the SSHA, would identify the specific detailed causes at the component level, such as specific component failure modes. A hazard can be initially identified from the causal sources without knowing the specific detailed root causes. However, in order to determine the mishap risk and the hazard mitigation measures required, the specific detailed root causes must eventually be known.

In summary, the basic principles of hazard–mishap theory are as follows:

1. Hazards cause mishaps; a hazard is a condition that defines a possible future event (i.e., mishap).

2. A hazard and a mishap are two different states of the same phenomenon (before and after).

3. Each hazard/mishap has its own inherent and unique risk (probability and severity).
4. A hazard is an entity comprised of three components (HS, IMs, and TTO).
5. The HE and IMs are the HCFs and they establish the mishap probability risk factor.
6. The TTO along with parts of the HS and IM establish the mishap severity risk factor.
7. HCFs can be characterized on three different levels.
8. The probability of a hazard existing is either 1 or 0; however, the probability of a mishap is a function of the specific HCFs.

A hazard is like a minisystem; it is a unique and discrete entity comprised of a unique set of HCFs and outcomes. A hazard defines the terms and conditions of a potential mishap; it is like a wrapper that contains the entire potential mishap description. The mishap that results is the consequence of the hazard components.

3.10 HAZARD–MISHAP PROBABILITY EXAMPLE

A hazard has a probability of either 1 or 0 of existing (it either exists or it does not; the three components are present or they are not present). A mishap, on the other hand, has a probability somewhere between 1 and 0 of occurring, based on the HCFs. The HE component has a probability of 1.0 of occurring, since it must be present in order for the hazard to exist. It is, therefore, the IM component that drives the mishap probability, that is, when the IMs occur the mishap occurs. The IMs are factors such as human error, component failures, timing errors, and the like.

The HCFs are the *root cause* of the hazard. The HCFs are, in fact, the hazard components that provide a threat and a mechanism for transitioning from a hazard to a mishap. The HS and IM components of a hazard are the HCFs. These two HCFs establish the probability of the hazard becoming a mishap. The combined effect of the T/T and parts of the HS and IM components determines the severity of the hazard. For mishap severity, the HS amount is usually of concern, and the IM factor that places the target in proximity of the HS.

This concept is illustrated in Figure 3.12.

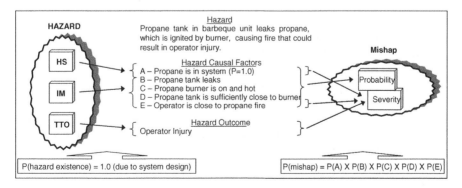

Figure 3.12 *Hazard component and probability example.*

3.11 RECOGNIZING HAZARDS

Hazard identification is one of the major tasks of system safety, and hazard identification involves the task of hazard recognition. Hazard recognition is the cognitive process of visualizing a hazard from an assorted package of design information. In order to recognize or identify hazards, the following four things are necessary:

1. An understanding of the hazard theory
2. A hazard analyses technique to provide a consistent and methodical process
3. An understanding of hazard recognition methods
4. An understanding of the system design and operation

This chapter has discussed hazard theory, in order that the safety analyst better understands the various aspects of a hazard, such as the hazard–mishap relationship, the three components of a hazard, and the types of hazard causal factors. Chapter 4 discusses the system safety hazard analysis types that are necessary to provide a complete coverage for identifying various types of hazards. The rest of this book describes a number of different hazard analysis techniques that provide different levels of hazard analysis structure, rigor, and details. This section concentrates on how to cognitively recognize a hazard.

One of the annoying truisms of system safety is that it is usually easier to mitigate a hazard than it is to recognize or find it in the first place. An analyst can fully understand the hazard theory and have a hazard analysis tool available, yet there is still the vexing reality of actually recognizing a hazard. Hazard recognition is a key aspect of the hazard identification process and therefore a key element of all hazard analyses. Hazards sometimes seem like ubituqous and elusive little creatures that must be hunted and captured, where the hunt is akin to the recognition process.

Another truism of system safety is that hazard recognition is somewhat an art more than a science (this may be one reason why skilled system safety analysts are invaluable). However, there are some key recognition factors that can help the safety analyst visualize hazards, such as follows:

1. Utilizing the hazard triangle components:
   a. HS – use hazardous element/component checklists.
   b. IM – evaluate trigger events and causal factors.
   c. TTO – evaluate possible threats and mishaps.
2. Utilizing past knowledge from experience and lessons learned
3. Analyzing good design practices
4. Review of general design safety criteria, precepts, and principles
5. Review and analysis of generic level 2 hazard causal factors
6. Key failure state questions
7. Evaluation of top-level mishaps (TLMs) and safety-critical functions (SCFs)

The concept of hazard triangle provides the best hazard recognition resource, by evaluating individually each of the three hazard component categories (HS, IM, and TTO) from a system's context. This means, for example, identifying and evaluating all the HE components in the unique system design as the first step. Subsequently, all system IM sources are evaluated for hazards and then all TTO sources.

When considering the HS component of the hazard triangle, focus on hazardous element checklists. The HE component involves items known to be a hazardous source, such as explosives, fuel, batteries, electricity, acceleration, chemicals, and the like. This means that generic hazard source checklists can be utilized to help recognize hazards within the unique system design. If a component in the system being analyzed is on one of the hazard source checklists, then this is a direct pointer to potential hazards that may be in the system. Speculate all the possible ways that the hazardous element can be hazardous within the system design. There are many different types of hazard source checklists, such as those for energy sources, hazardous operations, chemicals, and so on. Refer to Appendix C for some example checklists.

Hazards can be recognized by focusing on known consequential hazard-triggering mechanisms (the IM hazard component). For example, in the design of aircraft it is common knowledge that fuel ignition sources and fuel leakage sources are initiating mechanisms for fire/explosion hazards. Therefore, hazard recognition would benefit from a detailed review of the design for ignition sources and leakage sources when fuel is involved. Component failure modes and human error are common triggering modes for hazards.

Hazards can be recognized by focusing on known or preestablished undesired outcomes or mishaps (the target/threat hazard component). This means considering and evaluating known undesired outcomes within the system. For example, a missile system has certain undesired outcomes that are known right from the conceptual stage of system development. By following these undesired outcomes backward, certain hazards can be more readily recognized. In the design of missile systems, it is well accepted that inadvertent missile launch is an undesired mishap and, therefore, any conditions contributing to this event would formulate a hazard, such as autoignition, switch failures, and human error.

Another method for recognizing hazards is the use of past knowledge from experience and lessons learned. Mishap and hazard information from a previously developed system that is similar in nature or design to the system under analysis will aid in the hazard recognition process. For example, by reviewing the mishaps of an earlier spacecraft system, it might be discovered that the use of a particular seal design on a specific part of the spacecraft resulted in several mishaps. The seal design on this part of the spacecraft should then be recognized as a potential hazardous area requiring special focus in a new spacecraft design.

Another method of recognizing hazards is the review and analysis of good design practices. By reverse engineering good design practices found in various design documents and standards, hazardous design situations can be identified. For example, a good design practice for interrupting DC circuits is to place the switch or circuit breaker on the power branch, rather than the ground branch of the circuit. This prevents an inadvertent operation should a fault to ground branch occur. A hazard analysis should look for this and similar hazards in a new design.

Another method of recognizing hazards is the review of general design safety criteria, precepts, and principles. By considering the reasoning and logic behind specific safety criteria, precepts, and principles, some types of hazards can be more easily recognized. For example, there is a good safety reason for the following safety criterion: "Do not supply primary and redundant system circuits with power from the same bus or circuit." This safety criterion is a clue that aids in recognizing hazards in systems involving redundancy. A hazard analyst should look for such hazards in a new design.

Another method of recognizing hazards is the review and analysis of generic level 2 system hazard causal factors. Considering the various causal factor categories at a broad level can aid in recognizing hazards. The broad causal factor categories include hardware, software, humans, interfaces, and the environment. For example, by evaluating all the potential environments the system will operate in, or to which the system will be exposed, can assist in recognizing system hazards caused by environmental factors.

Another method of recognizing hazards is the use of key state questions. This method involves a set of clue questions that must be answered, each of which can trigger the recognition of a

TABLE 3.2   Failure State Checklist

1.	Fails to operate
2.	Operates incorrectly/erroneously
3.	Operates inadvertently
4.	Operates at wrong time (early or late)
5.	Unable to stop operation
6.	Receives erroneous data
7.	Sends erroneous data
8.	Conflicting data or information
9.	The component is exposed to fluid from external source
10.	The component is exposed to heat from external source

hazard. The key states are potential states or ways the subsystem could fail, or operate incorrectly, and thereby result in creating a hazard. For example, when evaluating each subsystem, answering the question "what happens when the subsystem *fails to operate*?" may lead to the recognition of a hazard. Table 3.2 contains a partial failure state checklist that provides some example key questions that should be asked when identifying hazards. Each system development SSP should develop their specialized and unique list of key failure state questions.

Table 3.3 demonstrates the hazard recognition process by listing the key considerations involved in the recognition of a hazard. The left-hand column of the table states the identified hazard, while the right-hand column lists the key factors used to help recognize the hazard.

It should be noted that it is very likely that additional hazards could probably be identified by using the listed hazard recognition keys, other than the single hazard shown in each of the above examples.

TABLE 3.3   Examples of Hazard Recognition

Hazard	Hazard Recognition Keys
Missile battery fluid leaks into electronics installed below the battery; electronics has hot surface that could cause ignition of fluid and a fire/explosion, resulting in equipment damage and/or personnel injury	1. A battery is a hazard checklist item, so the battery should be looked at from all possible hazard aspects, such as leakage, fire, and toxicity 2. Battery fluid is a hazard checklist item, so the battery fluid should be looked at from all possible hazard aspects, such as chemical damage, and fire 3. High temperature is a hazard checklist item, so components with high surface temperature should be evaluated as potential ignition sources 4. Experience and lessons learned from other missile programs would show that battery leakage is a safety concern and therefore should be considered a hazard source 5. Since a fluid is involved, key state question 9 (from Table 3.2) should be considered
Premature in-bore explosion of artillery round inside gun barrel could occur due to excessive heat, resulting in operator death or injury.	1. Explosives are a hazard checklist item; so, the explosives should be looked at from all possible hazard aspects, such as excessive barrel heat, jamming, and so on 2. The potential mishap consequence of personnel injury from premature explosion of artillery round is a known mishap to be prevented; therefore, all possible causes should be considered 3. Experience and lessons learned from other gun programs would show that in-bore explosion is a safety concern and therefore should be considered a hazard source 4. Since heat is always a concern with explosives, key state question 10 (from Table 3.2) should be considered

TABLE 3.4 Examples of Hazard Descriptions

Poor Examples	Good Examples
Repair technician slips on the oil	Overhead valve V21 leaks oil on the walkway below, the spill is not cleaned, the repair technician walking in the area slips on the oil and falls on the concrete floor, causing serious injury to him
Signal MG71 occurs	Missile launch signal MG71 is inadvertently generated during a standby alert, causing inadvertent launch of missile and death/injury to personnel in the area impacted by the missile launch
Round premature	Artillery round fired from gun explodes or detonates prior to the safe separation distance, resulting in death of or injury to the personnel within the safe distance area
Ship causes oil spill	The ship operator allows the ship to run aground, causing catastrophic hull damage, leading to a massive oil leakage, and resulting in a major environmental damage

Another method of recognizing hazards is the evaluation of TLMs and SCFs. Looking for potential causal factors of already established TLMs is a way of identifying hazards. Also, a technique for identifying hazards is to carefully examine what might cause the inadvertent or premature occurrence of each of the elements in an SCF.

3.12 HAZARD DESCRIPTION

Correctly describing the hazard is a very important aspect of hazard theory and analysis. The hazard description must contain all three components of the HCM (hazard source, initiating mechanism, and target/threat outcome). The hazard description should also be clear, concise, descriptive, and to the point.

If the hazard description is not properly worded, it will be difficult for anyone other than the original analyst to completely understand the hazard. If the hazard is not clearly understood, the concomitant risk of the hazard may not be correctly determined. This can lead to other undesirable consequences, such as spending too much time mitigating a low-risk hazard, when it was incorrectly thought to be a high-risk hazard.

Table 3.4 shows some good and poor examples of hazard descriptions. Note that the good examples contain all three elements of hazard: hazard source, initiating mechanism, and target/threat outcome.

3.13 HAZARD THEORY SUMMARY

This chapter discussed the basic hazard–mishap–risk concept. The following are the basic principles that help summarize the discussion in this chapter:

1. A hazard is potential condition built into the system design that, if it occurs, will result in an undesired event (i.e., mishap or accident).
2. Hazards are essentially the cause, or the precursor, of accidents and mishaps.
3. A hazard describes the special circumstances, conditions, and components that can result in accidents and mishaps. A hazard defines and predicts a potential future mishap.

4. A hazard consists of three required components, referred to as the *hazard triangle*:

   - Hazard source (source)
   - Initiating mechanism (mechanism)
   - Target/threat outcome (outcome)

5. A hazard is a deterministic entity that has its own unique design, components, characteristics, and properties.

6. If any side of the hazard triangle is eliminated through design techniques, the hazard and its associated risk are also eliminated.

7. If a hazard cannot be eliminated, then its risk can be reduced (mitigated or controlled) by reducing the hazard's probability of occurrence and/or the mishap severity through design techniques.

8. When a hazard is created, its outcome is almost always fixed and unchangeable. For this reason, it is very difficult to reduce hazard severity; however, it is much easier to reduce the hazard probability (mishap risk probability). The risk severity of a hazard almost always remains the same after mitigation.

9. It is important that a hazard description is clear, concise, and complete; it must contain all three components of the hazard triangle (HS, IM, and TTO).

10. Hazards occur in a somewhat predictable manner, based on the necessary and sufficient conditions of their design. What is predictable (hazard) is also preventable or controllable (mishap). By focusing on hazards and hazard analysis, mishap risks can be eliminated or reduced.

11. When identifying hazards, the following factors help trigger the recognition of a hazard:

    - Evaluation of the hazard triangle components
    - Utilizing past knowledge from experience and lessons learned
    - Analyzing good design practices
    - Review of general design safety criteria, precepts, and principles
    - Review and analysis of generic level 2 hazard causal factors
    - Key state questions

12. Just as a picture is worth a thousand words, a good hazard description creates a scenario picture that is invaluable in assessing hazard risk and identifying where to implement hazard mitigation measures.

HAZARD THEORY REFERENCES

The following are some references regarding hazard theory and hazard analysis:

1. Pat Clemens, System Safety Scrapbook, sheet 98-1, *Describing Hazards*, Sverdrup Corp., 1998.

FURTHER READINGS

C. Ericson, *Hazard Analysis Primer*, CreateSpace, 2012.

C. Ericson, *Clif's Notes on System Safety*, CreateSpace, 2012.

Chapter *4*

Hazard Analysis Features

4.1 INTRODUCTION

Hazard analyses are performed to identify hazards, hazard effects, and hazard causal factors. These analyses are used to determine system risk, and thereby ascertain the significance of hazards, in order that safety design measures can be established to eliminate or mitigate the hazard. Hazard analyses are performed to systematically examine the system, subsystem, facility, components, software, personnel, and their interrelationships.

In the system safety discipline, there are many different hazard analysis techniques that exist. Each of these techniques has a different purpose, focus, and methodology. The *System Safety Analysis Handbook* published by the International System Safety Society (ISSS) has listed over 100 different techniques. It should be noted that this large number of methodologies creates some confusion, as some of the techniques are not valid and some are merely modifications of other techniques. It is, therefore, important for the safety analyst to understand each technique and the unique features of each technique.

This chapter presents some of the unique features that all hazard analysis methodologies contain. The primary features or attributes that are discussed include the following:

- Type versus technique
- Primary versus secondary
- Inductive versus deductive
- Quantitative versus qualitative

Hazard Analysis Techniques for System Safety, Second Edition. Clifton A Ericson, II.
© 2016 John Wiley & Sons, Inc. Published 2016 by John Wiley & Sons, Inc.

4.2 TYPES VERSUS TECHNIQUE

There are two categories of hazard analyses: *types* and *techniques*. Hazard analysis type defines an analysis category (e.g., detailed design analysis) and technique defines a unique analysis methodology (e.g., fault tree analysis (FTA)). The type establishes analysis timing, depth of detail, and system coverage. The technique refers to a specific and unique analysis methodology that provides specific results. System safety is built upon *seven* basic types, while there are well over 100 different techniques available.[1] In general, there are several different techniques available for achieving each of the various types. The overarching distinctions between type and technique are summarized in Table 4.1.

Hazard analysis type describes the scope, coverage, detail, and life cycle phase timing of the particular hazard analysis. Each type of analysis is intended to provide a time- or phase-dependent analysis that readily identifies hazards for a particular design phase in the system development life cycle. Since more detailed design and operation information is available as the development program progresses, so in turn more detailed information is available for a particular type of hazard analysis. The depth of detail for the analysis type increases as the level of design detail progresses.

Each of these analysis types define a point in time when the analysis should begin, the level of detail of the analysis, the type of information available, and the analysis output. The goals of each analysis type can be achieved by various analysis techniques. The analyst needs to carefully select the appropriate techniques to achieve the goals of each of the analysis types.

There are seven hazard analysis types in the system safety discipline:

1. Conceptual design hazard analysis type (CD-HAT)
2. Preliminary design hazard analysis type (PD-HAT)
3. Detailed design hazard analysis type (DD-HAT)
4. System design hazard analysis type (SD-HAT)
5. Operations design hazard analysis type (OD-HAT)
6. Health design hazard analysis type (HD-HAT)
7. Requirements design hazard analysis type (RD-HAT)

An important principle of hazard analysis is that one particular hazard analysis type does not necessarily identify all the hazards within a system; identification of hazards may take more than one analysis type (hence the seven types). A corollary to this principle is that one particular hazard analysis type does not necessarily identify all of the hazard causal factors; more than one analysis type may be required. After performing all seven of the hazard analysis types, all

TABLE 4.1 Hazard Analysis Type versus Technique

Type	Technique
• Establishes where, when, and what to analyze	• Establishes how to perform the analysis
• Establishes a specific analysis task at a specific time in program life cycle	• Establishes a specific and unique analysis methodology
• Establishes what is desired from the analysis	• Provides the information to satisfy the intent of the analysis type
• Provides a specific design focus	

[1]Refer to the *System Safety Analysis Handbook* published by the ISSS for a detailed list.

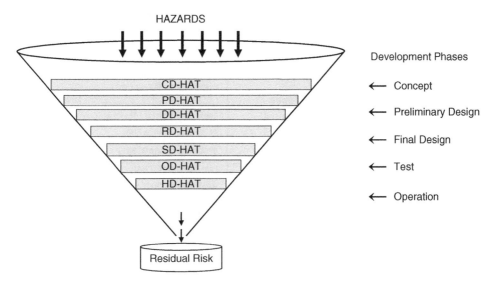

HAZARDS

	Development Phases
CD-HAT	← Concept
PD-HAT	← Preliminary Design
DD-HAT	
RD-HAT	← Final Design
SD-HAT	← Test
OD-HAT	
HD-HAT	← Operation

Residual Risk

Figure 4.1 *Hazard filter.*

hazards and causal factors should have been identified; however, additional hazards may be discovered during the test program.

Figure 4.1 conveys the filter concept behind the seven hazard analysis types.

In this concept, each hazard analysis type acts like a filter that identifies certain types of hazards. Each successive filter serves to identify hazards missed by the previous filter. The thick dark arrows at the top of the filter signify hazards existing in the system design. When all of the hazard analysis types have been applied, the only known hazards remaining have been reduced to an acceptable level of risk, denoted by the smaller thin arrows. The use of all seven hazard analysis types is critical in identifying and mitigating all hazards, and reducing system residual risk.

Each hazard analysis type serves a unique function or purpose. For a best practice, system safety program (SSP), it is recommended that all seven of these hazard analysis types be applied; however, tailoring is permissible. If tailoring is utilized, the specifics should be spelled out in the system safety management plan (SSMP) and/or the system safety program plan (SSPP).

Figure 4.2 depicts the relationship between hazard types and techniques. In this relationship, the seven hazard analysis types form the central focus for SSP hazard analysis. There are many different analysis techniques to select from when performing the analysis types and there are many different factors that must go into the hazard analysis, such the system life cycle stages of concept, design, test, manufacture, operation, and disposal. The system modes, phases, and functions must be considered. The system hardware, software, firmware, human interfaces, and environmental aspects must also be considered.

Some textbooks refer to the seven types as PHL, PHA, SSHA, SHA, O&SHA, HHA, and RHA. These names are, however, the same names as the basic hazard analysis techniques established by MIL-STD-882, versions A, B, and C. The concept of analysis types is a good concept, but having types and techniques with the same name is somewhat confusing. The approach recommended in this book ensures there are no common names between types and techniques, thus avoiding much confusion.

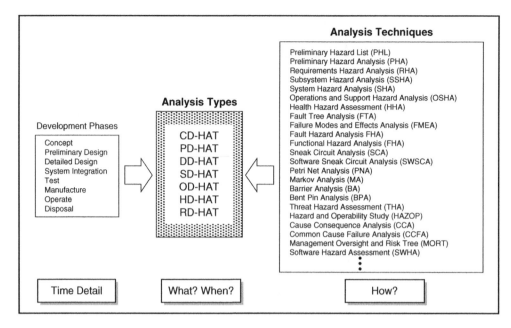

Figure 4.2 *Type–technique relationship.*

4.3 DESCRIPTION OF HAZARD ANALYSIS TYPES

4.3.1 Conceptual Design Hazard Analysis Type

The CD-HAT is a high-level (low level of detail) form of hazard analysis that identifies top-level hazards that can be recognized during the conceptual design phase. The CD-HAT is the first analysis type performed and is the starting point for all subsequent hazard analyses. The CD-HAT provides a basis for initially estimating the overall SSP effort.

The purpose of the CD-HAT is to compile a list of hazards very early in the product or system development life cycle to identify potentially hazardous areas. These hazardous areas identify where management should place design safety emphasis. The CD-HAT searches for hazards that may be inherent in the design or operational concept. It is a brainstorming, "what-if" analysis. A hazard list is generated from the brainstorming session, or sessions, where everything conceivable is considered and documented. The topics include review of safety experience on similar systems, hazard checklists, mishap/incident hazard tracking logs, safety lessons learned, and so on to identify possible hazards.

The key to a successful SSP is involvement early in the development program, beginning during conceptual design. The CD-HAT is started when the concept definition for a product or system begins, and carries into the preliminary design phase. It is performed early in the program life cycle in order to influence design concepts and decisions for safety as early as possible. The CD-HAT is the first analysis type performed and precedes the PD-HAT since it provides input for the PD-HAT.

If the CD-HAT is not performed during concept definition, it should be performed prior to, and as part of, any PD-HAT effort since it is an essential precursor to the PD-HAT. Once the initial CD-HAT is completed and documented, it is rarely updated as additional hazard identification analysis is achieved via the PD-HAT. In general, the CD-HAT supports the system design review (SDR), and CD-HAT effort ends at the start of the PD-HAT.

The following are the basic requirements of a comprehensive CD-HAT:

1. The CD-HAT will be applied during the design concept phase of system development.
2. The CD-HAT will be a high-level analysis (low level of detail) based on conceptual design information.
3. The CD-HAT will identify system hazards and potential mishaps.
4. The CD-HAT will consider hazards during system test, manufacture, operation, maintenance, and disposal.
5. The CD-HAT will consider system hardware, software, firmware, human interfaces, and environmental aspects.

Input information for the CD-HAT includes everything that is available during conceptual design. Experience has shown that generally the CD-HAT can be performed utilizing the following types of information:

1. Design concept
2. SOW, specification, and drawings (if available)
3. Preliminary (conceptual) indentured equipment list
4. Preliminary (conceptual) functions list
5. Energy sources in the system
6. Hazard checklists (generic)
7. Lessons learned (similar systems)

The primary purpose of the CD-HAT is to generate a list of system-level hazards, which can be used as an initial risk assessment and as the starting point for the subsequent hazard analysis types. As such, the following information is typically output from the CD-HAT:

1. System hazards
2. Top-level mishaps (TLMs)
3. Information to support the PD-HAT

4.3.2 Preliminary Design Hazard Analysis Type

The PD-HAT is a preliminary-level form of analysis that does not go into extensive detail; it is preliminary in nature. The PD-HAT is performed to identify system-level hazards and to obtain an initial risk assessment of a system design. It is performed early, during the preliminary design phase, in order to affect design decisions as early as possible to avoid future costly design changes.

The PD-HAT is the basic hazard analysis that establishes the framework for all of the follow-on hazard analyses. It provides a preliminary safety engineering evaluation of the design in terms of potential hazards, causal factors, and mishap risk. The intent of the PD-HAT is to recognize the hazardous system states and to begin the process of controlling hazards identified by the CD-HAT. As the design progresses in detail, more detailed analyses are performed to facilitate the elimination or mitigation of all hazards.

Identification of safety-critical functions (SCFs) and TLMs is a key element of the PD-HAT. The specific definition of what constitutes classification as safety-critical (SC) is generally program specific, as different types of systems may warrant different definitions based on the hazardous nature of the system.

The PD-HAT should be started during the design conceptual stage (after the CD-HAT) and continued through preliminary design. If the PD-HAT is not initiated during conceptual design, it should be initiated with the start of preliminary design. It is important that safety considerations identified in the PD-HAT are included in trade studies and design alternatives as early as possible in the design process.

Work on the PD-HAT usually concludes when the DD-HAT is initiated. In general, the PD-HAT supports all preliminary design reviews. The PD-HAT may also be used on an existing operational system for the initial examination of proposed design changes to the system.

The following are the basic requirements of a comprehensive PD-HAT:

1. The PD-HAT will be applied during the design concept and preliminary design phases of system development.
2. The PD-HAT will focus on all system hazards resulting from the preliminary design concept and component selection.
3. The PD-HAT will be a high-to-medium-level analysis (low-to-medium level of detail) that is based on preliminary design information.
4. The PD-HAT will identify hazards, potential mishaps, causal factors, risk, and SCFs. It will identify applicable safety guidelines, requirements, principles, and precepts to mitigate hazards. It will also provide recommendations to mitigate hazards.
5. The PD-HAT will consider hazards during system test, manufacture, operation, maintenance, and disposal.
6. The PD-HAT will consider system hardware, software, firmware, human interfaces, and environmental aspects.

Input information for the PD-HAT consists of the preliminary design information that is available during the preliminary design development phase. Typically, the following types of information are available and utilized in the PD-HAT:

1. Results of the CD-HAT
2. SOW
3. System specification
4. Design drawings and sketches
5. Preliminary indentured equipment list
6. Functional flow diagrams of activities, functions, and operations
7. Concepts for operation, test, manufacturing, storage, repair, and transportation
8. Energy sources
9. Hazard checklists (generic)
10. Lessons learned from experiences of similar previous programs or activities
11. Failure modes review
12. Safety guidelines and requirements from standards and manuals

The primary purpose of the PD-HAT is to perform a formal analysis for identifying system-level hazards, and evaluating the associated risk levels. As such, the following information is typically output from the PD-HAT:

1. System level hazards
2. Hazard effects and mishaps
3. Hazard causal factors (to subsystem identification)
4. SCFs
5. TLMs
6. Safety design criteria, principles, and precepts for design guidance in hazard mitigation
7. Risk assessment (before and after design safety features for hazard mitigation)
8. Safety recommendations for eliminating or mitigating the hazards
9. Information to support DD-HAT, SD-HAT, and OD-HAT

4.3.3 Detailed Design Hazard Analysis Type

The DD-HAT is a detailed form of analysis, performed to further evaluate hazards from the PHA with new detailed design information. The DD-HAT also evaluates the functional relationships of components and equipment comprising each subsystem. The analysis will help identify all components and equipment whose performance degradation or functional failure could result in hazardous conditions. Of particular concern is the identification of single-point failures (SPFs). The DD-HAT is also used to identify new hazards that can be recognized from the detailed design information that is available and to identify the hazard causal factors of specific subsystems and their associated risk levels.

The DD-HAT is an analysis of the detailed design and can, therefore, run from the start of detailed design through completion of final manufacturing drawings. Once the initial DD-HAT is completed and documented, it is not generally updated and enhanced, except for the evaluation of design changes.

The following are the basic requirements of a comprehensive DD-HAT:

1. The DD-HAT will be a detailed analysis at the subsystem and component level.
2. The DD-HAT will be applied during the detailed design of the system.
3. The DD-HAT will identify hazards, resulting mishaps, causal factors, risk, and SCFs. It will also identify applicable safety recommendations for hazard mitigation.
4. The DD-HAT will consider hazards during system test, manufacture, operation, maintenance, and disposal.
5. The DD-HAT will consider system hardware, software, firmware, human interfaces, and environmental aspects.

Input information for the DD-HAT consists of all detailed design data. Typically, the following types of information are available and utilized in the DD-HAT:

1. PD-HAT results
2. System description (design and functions)
3. Detailed design information (drawings, schematics, etc.)

4. Indentured equipment list
5. Functional block diagrams
6. Hazard checklists

The primary purpose of the DD-HAT is to evaluate the detailed design for hazards and hazard causal factors, and the associated subsystem risk levels. As such, the following information is typical output from the DD-HAT:

1. Subsystem hazards
2. Detailed causal factors
3. Risk assessment
4. Safety-critical subsystem interfaces
5. Safety design recommendations to mitigate hazards
6. Special detailed analyses of specific hazards using special analysis techniques such as fault tree analysis
7. Information to support the RD-HAT and SD-HAT

4.3.4 System Design Hazard Analysis Type

The SD-HAT assesses the total system design safety by evaluating the *integrated* system design. The primary emphasis of the SD-HAT, inclusive of both hardware and software, is to verify that the product is in compliance with the specified safety requirements at the system level. This includes compliance with acceptable mishap risk levels. The SD-HAT examines the entire system as a whole by integrating the essential outputs from the DD-HAT. Emphasis is placed on the interactions and the interfaces of all the subsystems as they operate together.

The SD-HAT provides determination of system risks in terms of hazard severity and hazard probability. System hazards are evaluated to identify *all* causal factors, including hardware, software, firmware, and human interaction. The causal factors may involve many interrelated fault events from many different subsystems. Thus, the SD-HAT evaluates all the subsystem interfaces and interrelationships for each system hazard.

The SD-HAT is system oriented, and therefore it usually begins during preliminary design and is complete by the end of final design, except for closure of all hazards. The SD-HAT is finalized at completion of the test program when all hazards have been tested for closure. SD-HAT documentation generally supports safety decisions for commencement of operational evaluations. The SD-HAT should be updated as a result of any system design changes, including software and firmware design changes to ensure that the design change does not adversely affect system mishap risk.

The following are the basic requirements of a comprehensive SD-HAT:

1. The SD-HAT will be applied primarily during the detailed design phase of system development; it can be initiated during preliminary design.
2. The SD-HAT is a detailed level of analysis that provides focus from an *integrated system* viewpoint.
3. The SD-HAT will be based on detailed and final design information.
4. The SD-HAT will identify new hazards associated with the subsystem interfaces.

5. The SD-HAT will consider hazards during system test, manufacture, operation, maintenance, and disposal.
6. The SD-HAT will consider system hardware, software, firmware, human interfaces, and environmental aspects.

Typically, the following types of information are available and utilized in the SD-HAT:

1. PD-HAT, DD-HAT, RD-HAT, OD-HAT, and other detailed hazard analyses
2. System design requirements
3. System description (design and functions)
4. Indentured equipment and function lists
5. System interface specifications
6. Test data

The primary purpose of the SD-HAT is to perform a formal analysis for identifying system-level hazards and evaluating the associated risk levels. As such, the following information is typically output from the SD-HAT:

1. System interface hazards
2. System hazard causal factors (hardware, software, firmware, human interaction, and environmental)
3. Assessment of system risk
4. Special detailed analyses of specific hazards using special analysis techniques such as FTA
5. Information to support the safety assessment report (SAR)

4.3.5 Operations Design Hazard Analysis Type

The OD-HAT evaluates the operations and support functions involved with the system. These functions include use, test, maintenance, training, storage, handling, transportation, and demilitarization or disposal. The OD-HAT identifies operational hazards that can be eliminated or mitigated through design features and through modified operational procedures when necessary. The OD-HAT considers human limitations and potential human errors (human factors). The human being is considered an element of the total system, receiving inputs and initiating outputs.

The OD-HAT is performed when operations information becomes available and should start early enough to provide inputs to the design. The OD-HAT should be completed prior to the conduct of any operating and support functions.

The following are the basic requirements of a comprehensive OD-HAT:

1. The OD-HAT will be performed during the detailed design phases of system development when the operating and support procedures are being written.
2. The OD-HAT will focus on hazards occurring during operations and support.
3. The OD-HAT will provide an integrated assessment of the system design, related equipment, facilities, operational tasks, and human factors.
4. The OD-HAT will be a detailed analysis based on final design information.

5. The OD-HAT will identify hazards, potential mishaps, causal factors, risk and safety-critical factors, applicable safety requirements, and hazard mitigation recommendations.

6. The OD-HAT will consider hazards during system use, test, maintenance, training, storage, handling, transportation, and demilitarization or disposal.

The following types of information are utilized in the OD-HAT:

1. PD-HAT, DD-HAT, and SD-HAT and any other applicable hazard analyses
2. Engineering descriptions/drawings of the system, support equipment, and facilities
3. Available procedures and operating manuals
4. Operational requirements, constraints, and required personnel capabilities
5. Human factors, engineering data, and reports
6. Lessons learned, including human factors
7. Operational sequence diagrams

The OD-HAT focus is on operating and support tasks and procedures. The following information is typically available from the OD-HAT:

1. Task-oriented hazards (caused by design, software, human, timing, etc.)
2. Hazard mishap effect
3. Hazard causal factors (including human factors)
4. Risk assessment
5. Hazard mitigation recommendations and derived design safety requirements
6. Derived procedural safety requirements
7. Cautions and warnings for procedures and manuals
8. Input information to support the SD-HAT

4.3.6 Human Health Design Hazard Analysis Type (HD-HAT)

The HD-HAT is intended to systematically identify and evaluate human health hazards, evaluate proposed hazardous materials, and propose measures to eliminate or control these hazards through engineering design changes or protective measures to reduce the risk to an acceptable level.

The HD-HAT assesses design safety by evaluating the human health aspects involved with the system. These aspects include manufacture, use, test, maintenance, training, storage, handling, transportation, and demilitarization or disposal.

The HD-HAT concentrates on human health hazards. The HD-HAT is started during preliminary design and continues to be performed, as more information becomes available. The HD-HAT should be completed and system risk known prior to the conduct of any of the manufacturing, test, or operational phases.

The following are the basic requirements of a comprehensive HD-HAT:

1. The HD-HAT will be applied during the preliminary and detailed design phases of system development.

2. The HD-HAT will focus on the human environment within the system.

3. The HD-HAT will be a detailed analysis based on system design and operational tasks affecting the human environment.

4. The HD-HAT will identify hazards, potential mishaps, causal factors, risk and safety-critical factors, and applicable safety requirements.

5. The HD-HAT will consider human health hazards during system test, manufacture, operation, maintenance, and demilitarization or disposal. Consideration should include, but is not limited to, the following:

   a. Materials hazardous to human health (e.g., material safety data sheets)

   b. Chemical hazards

   c. Radiological hazards

   d. Biological hazards

   e. Ergonomic hazards

   f. Physical hazards

Typically, the following types of information are available and utilized in the HD-HAT:

1. CD-HAT, PD-HAT, DD-HAT, SD-HAT, and OD-HAT and any other applicable detailed hazard analyses

2. Materials and compounds used in the system production and operation

3. Material safety data sheets

4. System operational tasks and procedures, including maintenance procedures

5. System design

The following information is typically available from the HD-HAT:

1. Human health hazards

2. Hazard mishap effects

3. Hazard causal factors

4. Risk assessment

5. Derived design safety requirements

6. Derived procedural safety requirements (including cautions, warnings, and personal protective equipment)

7. Input information for Occupational Safety and Health Administration and environmental evaluations

8. Information to support the OD-HAT and SD-HAT

4.3.7 Requirements Design Hazard Analysis Type (RD-HAT)

The RD-HAT is a form of analysis that verifies and validates the design safety requirements and ensures that no safety gaps exist in the requirements. The RD-HAT applies to hardware, software, firmware, and test requirements. Since the RD-HAT is an evaluation of design and test safety requirements, it is performed during the design and test stages of the development program. The RD-HAT can run from mid-preliminary design through the end of testing.

Safety design requirements are generated from three sources: (1) the system specification, (2) generic requirements from similar systems, subsystems, and processes, and (3) requirements derived from recommendations to mitigate identified system unique hazards. The intent of the RD-HAT is to ensure that all of the appropriate safety requirements are included within the design requirements and that they are verified and validated through testing, analysis, or inspection. Applicable generic system safety design requirements are obtained from such sources as federal, military, national, and industry regulations, codes, standards, specifications, guidelines, and other related documents for the system under development.

The RD-HAT supports closure of identified hazards. Safety requirements levied against the design to mitigate identified hazards must be verified and validated before a hazard in the hazard tracking system can be closed. The RD-HAT provides a means of traceability for all safety requirements, verifying their implementation and validating their success.

The RD-HAT is an evolving analysis that is performed over a period of time, where it is continually updated and enhanced as more design and test information becomes available. The RD-HAT is typically performed in conjunction with the PD-HAT, DD-HAT, SD-HAT, and OD-HAT. The RD-HAT should be complete at the end of testing.

The following are the basic requirements of a comprehensive RD-HAT:

1. The RD-HAT will be applied from the preliminary design phases through testing of the system.
2. The RD-HAT will be focus on safety requirements intended to eliminate and/or mitigate identified hazards.
3. The RD-HAT will be a detailed analysis that will be based on detailed design requirements and design information.
4. The RD-HAT will ensure that all identified hazards have suitable safety requirements to eliminate and/or mitigate the hazards.
5. The RD-HAT will ensure that all safety requirements are verified and validated through analysis, testing, or inspection.

Typically, the following types of information are available and utilized in the RD-HAT:

1. Hazards without mitigating safety requirements
2. Design safety requirements (hardware, software, and firmware)
3. Test requirements
4. Test results
5. Unverified safety requirements

The primary purposes of the RD-HAT are to establish traceability of safety requirements and to assist in the closure of mitigated hazards. The following information is typically output from the RD-HAT:

1. Traceability matrix of all safety design requirements to identified hazards
2. Traceability matrix of all safety design requirements to test requirements and test results

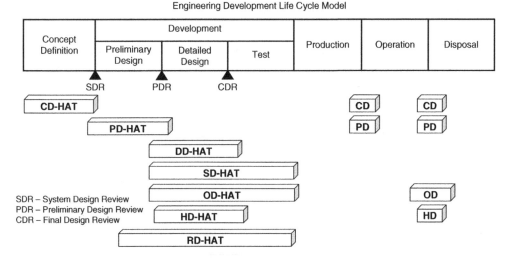

Figure 4.3 *Overall timing of hazard analysis types.*

3. Identification of new safety design requirements and tests necessary to cover gaps discovered by 1 and 2 above
4. Data supporting closure of hazards

4.4 THE TIMING OF HAZARD ANALYSIS TYPES

Figure 4.3 contains a consolidated view of the time period over which the hazard analysis types are typically performed. This schedule shows the most typical timing that has been found practical through many years of trial and error. The system development phases shown are from the standard engineering development life cycle model.

The time period for performing the hazard analysis is not rigidly fixed but depends on many variables, such as size of the system and project, safety criticality of the system, personal experience, common sense, and so on. The time period is shown as a bar because the analysis can be performed at any time during the period shown. Specifying the time period for a hazard analysis is part of the safety program tailoring process, and should be documented in the SSPP. Each of the hazard analysis types has a functional time period when it is most effectively applied to achieve the desired intent and goals.

Note how each of the hazard analysis types correlates very closely to its associated development phase. Also, note that some of the analysis types should be performed in a later development phase if that phase was not specifically covered by the original analysis.

4.5 THE INTERRELATIONSHIP OF HAZARD ANALYSIS TYPES

Figure 4.4 shows the relative relationship of each of the hazard analysis types and their interdependencies. This figure shows how the output of one analysis type can provide input data for another analysis type.

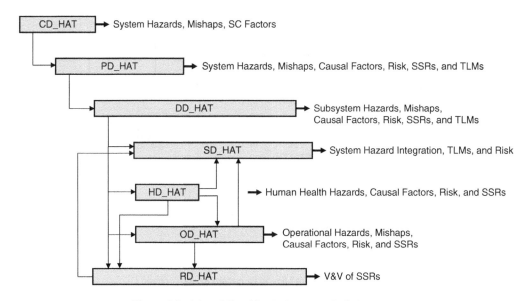

Figure 4.4 *Interrelationships between analysis types.*

TABLE 4.2   Hazard Analysis Techniques Presented in This Book

No.	Title	Chapter
1	Functional hazard analysis	6
2	Preliminary hazard list analysis	7
3	Preliminary hazard analysis	8
4	Subsystem hazard analysis	9
5	System hazard analysis	10
6	Operating and support hazard analysis	11
7	Health hazard analysis	12
8	Requirements hazard analysis	13
9	Environmental hazard analysis	14
10	Fault tree analysis	15
11	Failure mode and effects analysis	16
12	HAZOP analysis	17
13	Event tree analysis (ETA)	18
14	Cause–consequence analysis	19
15	Common cause failure analysis (CCFA)	20
16	Software hazard analysis (SwHA)	21
17	Process hazard analysis	22
18	Test hazard analysis (THA)	23
19	Fault hazard analysis	24
20	Sneak circuit analysis (SCA)	25
21	Markov analysis (MA)	26
22	Petri net analysis (PNA)	27
23	Barrier analysis	28
24	Bent pin analysis (BPA)	29
25	MORT analysis	30
26	Job hazard analysis	31
27	Threat hazard analysis (THA)	32
28	System of systems hazard analysis	33

4.6   HAZARD ANALYSIS TECHNIQUES

Hazard analysis technique defines a unique analysis methodology (e.g., fault tree analysis). The technique refers to a specific and unique analysis methodology that is performed following a specific set of rules and provides specific results.

As previously mentioned, there are over 100 different hazard analysis techniques in existence, and the number continues to slowly grow. Many of the techniques are minor variations of other techniques. And, many of the techniques are not widely practiced. This book presents 28 of the most commonly used techniques by system safety practitioners. Table 4.2 lists the hazard analysis techniques covered in this book, along with the corresponding chapter number describing the technique.

Each of these hazard analysis techniques is important enough to warrant an individual chapter devoted to describing just that technique. The system safety engineer/analyst should be thoroughly familiar with each of the analysis techniques presented in this book. They form the basic building blocks for performing hazard and safety analysis on any type of system.

4.7   HAZARD ANALYSIS TECHNIQUE ATTRIBUTES

Hazard analysis techniques can have many different inherent attributes, which make their utility different. The appropriate technique to use can often be determined from the inherent attributes of the technique. Table 4.3 contains a list of the most significant attributes for a hazard analysis methodology.

Table 4.4 summarizes some of the select attributes for the analysis techniques presented in this book. These attributes will be covered in greater detail in each chapter dealing with a particular technique.

4.8   PRIMARY AND SECONDARY TECHNIQUES

There is some confusion regarding which hazard analysis techniques are truly hazard analyses and which are not. Within the system safety discipline, there are over 100 different HA

TABLE 4.3   Major Attributes of Analysis Techniques

	Attribute	Description
1	Qualitative/quantitative	Analysis assessment is performed qualitatively or quantitatively
2	Level of detail	The level of design detail that can be evaluated by the technique
3	Data required	The type and level of design data required for the technique
4	Program timing	The effective time during system development for the technique
5	Time required	The relative amount of time required for the analysis
6	Inductive/deductive	The technique uses inductive or deductive reasoning
7	Complexity	The relative complexity of the technique
8	Difficulty	The relative difficulty of the technique
9	Technical expertise	The relative technical expertise and experience required
10	Tools required	The technique is stand-alone or additional tools are necessary
11	Cost	The relative cost of the technique
12	Primary safety tool	The technique is a primary or secondary safety tool

TABLE 4.4 Summary of Select Attributes for Analysis Techniques

Technique	Type	Identify Hazards	Identify Root Causes	Life Cycle Phase	Qualitative/ Quantitative	Skill	Level of Detail	I/D
PHL	CD-HAT	Y	N	CD-PD	Qual.	SS	Minimal	I
PHA	PD-HAT	Y	P	CD-PD	Qual.	SS	Moderate to In-depth	I-D
SSHA	DD-HAT	Y	Y	DD	Qual.	SS, Engr., M&S	In-depth	I-D
SHA	SD-HAT	Y	Y	PD-DD-T	Qual.	SS, Engr., M&S	In-depth	I-D
O&SHA	OD-HAT	Y	Y	PD-DD-T	Qual.	SS, Engr., M&S	In-depth	I-D
HHA	HD-HAT	Y	Y	PD-DD-T	Qual.	SS, Engr., M&S	In-depth	I-D
RHA	RD-HAT	P	N	PD-DD	Qual.	SS	In-depth	N/A
FTA	SD-HAT, DD-HAT	P	Y	PD-DD	Qual./quant.	SS, Engr., M&S	Moderate to In-depth	D
ETA	SD-HAT	P	P	PD-DD	Qual./quant.	SS, Engr., M&S	Moderate to In-depth	D
FMECA	DD-HAT	P	P	PD-DD	Qual./quant.	SS, Engr., M&S	In-depth	I
FaHA	DD-HAT	Y	P	PD-DD	Qual.	SS, Engr., M&S	In-depth	I
FuHA	SD-HAT, DD-HAT	Y	P	CD-PD-DD	Qual.	SS, Engr., M&S	Moderate to In-depth	I
SCA	SD-HAT, DD-HAT	P	Y	DD	Qual.	SS, Engr., M&S	Moderate to In-depth	D
PNA	SD-HAT, DD-HAT	P	N	PD-DD	Qual./quant.	SS, Engr., M&S	In-depth	D
MA	SD-HAT, DD-HAT	P	N	PD-DD	Qual./quant.	SS, Engr., M&S	Moderate to In-depth	D

BA	SD-HAT	Y	P	PD-DD	Qual.	SS, Engr.	Moderate to In-depth	I
BPA	DD-HAT	Y	P	PD-DD	Qual.	SS, Engr., M&S	In-depth	D
HAZOP	SD-HAT, DD-HAT	Y	P	PD-DD	Qual.	SS, Engr., M&S	Moderate to In-depth	I
CCA	SD-HAT, DD-HAT	Y	P	PD-DD	Qual./quant.	SS, Engr., M&S	Moderate to In-depth	D
CCFA	SD-HAT, DD-HAT	Y	P	PD-DD	Qual.	SS, Engr., M&S	Moderate to In-depth	D
MORT	SD-HAT, DD-HAT	Y	P	PD-DD	Qual./quant.	SS, M&S	Moderate to In-depth	D
SWSA	SD-HAT, DD-HAT	Y	P	CD-PD	Qual.	SS, Engr., M&S	Moderate to In-depth	N/A

Y = yes, N = no, P = partially
Skill required: SS = system safety; Engr. = engineering – electrical/mechanical/software; M&S = math and statistics
Life cycle phase: CD = conceptual design; PD = preliminary design; DD = detailed design; T = testing
I = inductive; D = deductive
FaHA = functional hazard analysis/FuHA = functional hazard analysis

techniques that have been proposed, some of which are unique, some of which are variants of others, some of which are extremely useful, and some of which are not useful at all. Some of the techniques are not true HAs, and many are merely variations of other HA techniques. There are only about 15–20 HA techniques that are commonly used by system safety experts.

Essentially, there are *primary* and *secondary* hazard analysis techniques. The primary hazard analysis techniques are full-fledged, or complete, formal methodologies that are designed for identifying all, or most of the, system hazards. The secondary hazard analysis techniques are limited in their hazard identification ability; typically, they are not designed to identify all hazards. The secondary techniques essentially provide support for the primary techniques; many help identify the root causal factors of already identified hazards.

It should be noted that many analysis techniques are incorrectly used for hazard analysis, such as failure mode and effects analysis (FMEA). FMEA results can be used as resource information for an HA, but the FMEA does not suffice for an HA because it does not thoroughly cover system hazard–mishap scenarios and it does not cover the combined effect of multiple simultaneous failures that often cause hazards. An FMEA is not a hazard analysis or a safety analysis and should not be used in place of either, but it can be used to supplement them as a secondary supporting technique; it is an excellent source for failure modes and failure rates.

Table 4.5 contains a list of the major HA techniques utilized in the system safety discipline. This list is provided at the front of this chapter in order to identify the techniques and their

TABLE 4.5   Hazard Analysis Type versus Technique

Technique	Acronym	Primary	Secondary	Type	Prob.
Preliminary hazard list	PHL	X		X	QL
Preliminary hazard analysis	PHA	X		X	QL
Subsystem hazard analysis	SSHA	X		X	QL
System hazard analysis	SHA	X		X	QL
Operations and support hazard analysis	O&SHA	X		X	QL
Health hazard assessment	HHA	X		X	QL
Functional hazard analysis	FHA	X			QL
Hazard and operability analysis	HAZOP	X			QL
Threat hazard assessment	THA	X			QL
Fault tree analysis	FTA		X		B
Event tree analysis	ETA		X		B
Failure mode and effects analysis	FMEA		X		B
Bent pin analysis	BPA		X		QL
Sneak circuit analysis	SCA		X		QL
Requirements hazard analysis	RHA		X	X	QL
Barrier analysis	BA		X		QL
Interlock analysis			X		B
Code safety analysis			X		QL
Particular risk analysis			X		QL
Markov analysis	MA		X		QT
Test hazard analysis	THA		X		QL
Common cause failure analysis	CCFA		X		B
Probabilistic risk assessment	PRA		X		QT
Safety assessment report	SAR	—	—		QL

Prob. = probability calculation is QL, QT, or B

acronyms. The various technique parameters in the table will be explained in the following sections. These parameters include the following categories:

- Primary technique or secondary technique
- If the methodology is also recognized as a HA type
- Qualitative, quantitative, or both

The "prob" category indicates if the probability calculation for the technique is qualitative (QL), quantitative (QT), or both (B). Many of the techniques are identified as qualitative, which means they are primarily qualitative in nature. However, they could easily become quantitative if more research and probabilistic math is applied to each hazard. All of the methods are HA techniques; the "type" category means they are considered as both a type and a technique.

4.9 INDUCTIVE AND DEDUCTIVE TECHNIQUES

System safety hazard analysis techniques are quite often labeled as being either an inductive or a deductive methodology. For example, an FMEA is usually referred to as an inductive approach, while an FTA is referred to as a deductive approach. Understanding how to correctly use the terms inductive and deductive is often confusing, and even sometimes incorrectly applied. The question is: what do these terms really mean, how should they be used, and does their use provide any value to the safety analyst?

The terms deductive and inductive refer to forms of logic. Deductive comes from deductive reasoning and inductive comes from inductive reasoning. The great detective Sherlock Holmes was a master of both deductive and inductive reasoning in solving criminal cases from clues, premises, and information.

Deductive reasoning is a logical process in which a conclusion is drawn from a set of premises and contains no more information than the premises taken collectively. For example, all dogs are animals; this is a dog; therefore, this is an animal. The truth of the conclusion depends upon the premises; the conclusion cannot be false if the premises on which it is based are true. For example, all men are apes; this is a man; therefore, this is an ape. The conclusion seems logically true; however, it is false because the premise is false. In deductive reasoning, the conclusion does not exceed or imply more than the premises it is based upon.

Inductive reasoning is a logical process in which a conclusion is proposed that contains more information than the observation or experience on which it is based. For example, every crow ever seen was black; therefore, all crows are black. The truth of the conclusion is verifiable only in terms of future experience and certainty is attainable only if all possible instances have been examined. In the example, there is no certainty that a white crow will not be found tomorrow, although past experience would make such an occurrence seem unlikely. In inductive reasoning, the conclusion is broader and may imply more than the known premises can guarantee with the data available.

Given these definitions, an inductive hazard analysis might conclude more than the given data intends to yield. This is useful for general hazard identification. It means a safety analyst might try to identify (or conclude) a hazard from limited design knowledge or information. For example, when analyzing the preliminary design of a high-speed subway system, a safety analyst might conclude that the structural failure of a train car axle is a potential hazard that could result in car derailment and passenger injury. The analyst does not know this for sure, there is no conclusive evidence available, but the conclusion appears reasonable from past knowledge and experience. In this case, the conclusion seems realistic but is beyond any factual knowledge or proof available at the time of the analysis; however, a credible hazard has been identified.

A deductive hazard analysis would conclude no more than the data provides. In the above example, the analysis must go in the opposite direction. The specific causal factors supporting the conclusion must be identified and established, and then the conclusion will be true. It seems like reverse logic; however, the hazard can be validly deduced only from the specific detailed causal factors.

Deductive and inductive qualities have become intangible attributes of hazard analysis techniques. An inductive analysis would be used to broadly identify hazards without proven assurance of the causal factors, while a deductive analysis would attempt to find the specific causal factors for identified hazards. An inductive analysis can be thought of as a "what-if" analysis. The analyst asks, "what if" this part failed, what are the consequences. A deductive analysis can be thought of as a "how-can" analysis. The analyst asks, if this event were to occur, "how can" it happen or what are the causal factors?

In system safety, inductive analysis tends to be for hazard identification (when the specific root causes are not known or proven), and deductive analysis for root cause identification (when the hazard is known). Obviously, there is a fine line between these definitions, because sometime the root causes are known from the start of an inductive hazard analysis. This is why some analysis techniques can actually move in both directions. The preliminary hazard analysis (PHA) is a good example of this. Using the standard PHA worksheet, hazards are identified inductively both by asking "what if" this component fails and by asking "how can" this undesired event happen.

Two additional terms that are confusing to system safety analysts are top-down analysis and bottom-up analysis. In general, top-down analysis means starting the analysis from a high-level system viewpoint, for example, a missile navigation system, and continually burrowing into deeper levels of detail until the discrete component level is reached, such as a resistor or diode. A bottom-up analysis moves in the opposite direction. It begins at a low system level, such as the resistor or diode component, and moves upward until the system top level is reached.

These definitions are illustrated in Figure 4.5.

Some system safety practitioners advocate that a deductive analysis is always a top-down approach, and that an inductive analysis is always a bottom-up approach. This may be a good generalization but is likely not always the case.

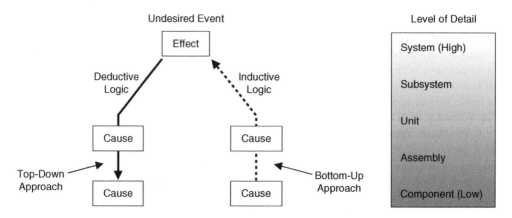

Figure 4.5   *Inductive and deductive analysis relationship.*

TABLE 4.6   Inductive and Deductive Analysis Characteristics

	Inductive	Deductive
Methodology	• What-If • Going from the specific to the general	• How-Can • Going from the general to the specific
General characteristics	• System is broken down into individual components • Potential failures for each component are considered (what can go wrong?) • Effects of each failure are defined (what happens if it goes wrong?)	• General nature of the hazard has already been identified (fire, inadvertent launch, etc.) • System is reviewed to define the cause of each hazard (how can it happen?)
Applicability	• Systems with few components • Systems where SPFs are predominant • Preliminary or overview analysis	• All sizes of systems • Developed for complex systems • Designed to identify hazards caused by multiple failures
Potential pitfalls	• Difficult to apply to complex systems • Large number of components to consider • Consideration of failure combinations becomes difficult	• Detailed system documentation required • Large amount of data involved • Time-consuming
Examples	• Failure mode and effects analysis • HAZOP	• Fault tree analysis • ETA • CCFA

Table 4.6 summarizes some of the characteristics of inductive and deductive analysis techniques.

The bottom line is that in the long run it does not really matter to the safety analyst if a hazard analysis technique is inductive or deductive. An analyst does not select an analysis technique based on whether its methodology is inductive, deductive, top-down, or bottom-up. What is important is that there are various techniques available for identifying hazards and hazard causal factors and that the safety analyst knows how to correctly use and apply the appropriate techniques. An analyst is more concerned with the actual task of identifying and mitigating hazards.

4.10   QUALITATIVE AND QUANTITATIVE TECHNIQUES

System safety analysts are often in a quandary as whether to use a qualitative analysis technique or a quantitative analysis technique. Understanding which analysis type to use, and when, often seems more of an art than a science. The qualitative–quantitative factor is one of the basic attributes of a hazard analysis technique.

Most hazard analysis techniques are performed to identify hazards and then determine the mishap risk of the hazard, where mishap risk is defined as Risk = Probability × Severity. The probability risk factor means the probability of the mishap actually occurring, given the latent hazard conditions, and the severity factor means the damage and/or loss resulting from the mishap after it occurs. To determine the risk of an identified hazard, a risk characterization methodology must be utilized for the probability and severity parameters. Both quantitative and qualitative risk characterization methods have been developed for use in the system safety

TABLE 4.7 Differences between Quantitative and Qualitative Techniques

	Attribute	Qualitative	Quantitative
1	Numerical results	No	Yes
2	Cost	Lower	Higher
3	Subjective/objective	Subjective	Objective
4	Difficulty	Lower	Higher
5	Complexity	Lower	Higher
6	Data	Less detailed	More detailed
7	Technical expertise	Lower	Higher
8	Time required	Lower	Higher
9	Tools required	Seldom	Usually
10	Accuracy	Lower	Higher

discipline. Both approaches are useful, but each approach contains inherent unique advantages and disadvantages.

Qualitative analysis involves the use of a qualitative criterion in the analysis. Typically, this approach uses categories to separate different parameters, with qualitative definitions that establish the ranges for each category. Qualitative judgments are made as to which category something might fit into. This approach has the characteristic of being subjective, but it allows more generalization and is, therefore, less restricting. For example, arbitrary categories have been established in MIL-STD-882 that provides a qualitative measure of the most reasonable likelihood of occurrence of a mishap. For example, it the safety analyst assesses that an event will occur frequently, it is assigned an index level A and if it occurs occasionally it is given an index level C. This qualitative index value is then used in qualitative risk calculations and assessments.

Quantitative analysis involves the use of numerical or quantitative data in the analysis and provides a quantitative result. This approach has the characteristic of being more objective and possibly more accurate. It should be noted, however, that quantitative results can be biased by the validity and accuracy of the input numbers. For this reason, quantitative results should not be viewed as an exact number but as an estimate with a range of variability depending upon the goodness quality of the data.

Table 4.7 identifies some of the attributes that can be used to judge the strengths and weaknesses of qualitative and quantitative approaches.

The system safety discipline primarily uses the qualitative risk characterization approach for a majority of safety work, based on the advantages provided. This approach has been recommended in MIL-STD-882 since the original version in 1969.

One reason why the qualitative risk characterization method is preferred in system safety is that for a large system with many hazards, it can become cost prohibitive to quantitatively analyze and predict the risk of each and every hazard. In addition, low-risk hazards do not require the refinement provided by quantitative analysis. It may be necessary to conduct a quantitative analysis only on a select few high-consequence hazards. Experience over the years has proven that qualitative methods are very effective and in most cases provide decision-making capability comparable to quantitative analysis.

Qualitative risk characterization provides a very practical and effective approach when cost and time are concerns, and/or when the availability of supporting data is low. The key to developing a qualitative risk characterization approach is to carefully define severity and mishap probability categories.

Quantitative risk characterization provides a useful approach when accuracy is required. Occasionally, a numerical design requirement must be met, and the only way to provide evidence that it is met is through quantitative analysis. Probabilistic risk assessment (PRA) is a quantitative analysis that estimates the probability factor of mishap risk. For high-consequence systems, it is often necessary to conduct a PRA to determine all of the causal factors for a given mishap and their total probability of causing the mishap to occur.

Scientific theory teaches that when something can be measured (quantitatively) more is known about it and, therefore, numerical results provide more value. This is generally true; however, the strict use of quantitative methods must be tempered by utility. Sometimes, qualitative judgments provide useful results at less time and expense. In a risk assessment, precise numerical accuracy is not always necessary. Mishap risks are not easily estimated using probability and statistics when the hazard causal factors are not yet well understood. Qualitative measures provide a useful and valid judgment at much less expense than quantitative measures, and they can be obtained much earlier in the system development life cycle. It makes sense to first evaluate all identified hazards qualitatively, and then, for high-risk hazards conduct a quantitative analysis for more precise knowledge.

In any evaluation of mishap risk assessment, the question of measure and acceptability parameters arises. There is always the danger that system safety analysts and managers will become so enamored with probability and statistics that simpler and more meaningful engineering processes will be ignored. Before embarking in a particular direction, be sure that the limitations and principles of both approaches are well understood, so are the actual analysis needs. Quantitative models are useful, but do not ever confuse mathematical model results with reality.

4.11 SUMMARY

This chapter discussed the basic concept of hazard analysis types and techniques. The following are basic principles that help summarize the discussion in this chapter:

1. A hazard analysis type defines the analysis purpose, timing, scope, and level of detail and system coverage; it does not specify how to perform the analysis.
2. A hazard analysis technique defines a specific and unique analysis methodology that provides a specific methodology and results.
3. The seven hazard analysis types, their coverage, and their intended focus are summarized in Table 4.8. In addition, the primary technique for satisfying the type requirements is listed.

TABLE 4.8   Hazard Analysis Type/Technique Summary

Type	Coverage	Hazard Focus	Primary Technique
CD-HAT	Conceptual design	System hazards	PHL
PD-HAT	Preliminary design	System hazards	PHA
DD-HAT	Detailed subsystem design	Subsystem hazards	SSHA
SD-HAT	Integrated system design	System hazards	SHA
OD-HAT	Operational design	Operational hazards	O&SHA
HD-HAT	Human health design	Human health hazards	HHA
RD-HAT	Design requirements	Requirements/testing	RHA

4. There are seven hazard analysis types in the system safety discipline that, together, help ensure identification and resolution of system hazards. There are over 100 different analysis techniques that can be used to satisfy the analysis type requirements.

5. One particular hazard analysis type does not necessarily identify all the hazards within a system; it may take more than one type, and usually all seven types.

6. A best practice SSP includes all seven hazard analysis types to ensure complete hazard coverage and provide optimum safety assurance.

7. The seven primary hazard analysis techniques are generally the best option for satisfying the corresponding analysis type requirement in an SSP.

FURTHER READINGS

Ericson, C., *Hazard Analysis Primer*, CreateSpace, 2012.
Ericson, C., *Clif's Notes on System Safety*, CreateSpace, 2012.

Chapter 5

Hazard Recognition and Management

5.1 INTRODUCTION

Hazard analysis is very similar to root cause analysis (RCA), except that while RCA looks for causal factors to a known problem, hazard analysis looks for both the potential problem and the causal factors involved, making the process slightly more complicated. The idea of hazard analysis is to proceed logically through the system identifying system hazards and the root causes that can cause the hazards. First, the specific undesired mishap event must be postulated and then the analysis addresses the questions of what, how, why, and when. When all these factors are combined together the hazard can be formulated with its specific causal factors.

Although there are many different hazard analysis techniques available, in general each methodology follows a basic pattern. There is a somewhat generic set of steps for hazard analysis that are typically followed for most methodologies. In addition, there are certain tools that aid in the recognition of hazards during the hazard analysis process. All of these considerations are covered in this chapter.

5.2 HAZARD ANALYSIS TASKS

Hazard identification, evaluation, and control constitute the backbone of the system safety process, which is achieved through a formal hazard analysis process. Understanding and thoroughly analyzing the system is a key aspect of effective hazard analysis. The hazard analysis process is performed on a specific system design configuration, since all of the aspects of that unique design must be considered in order to identify and mitigate the system's unique

Hazard Analysis Techniques for System Safety, Second Edition. Clifton A Ericson, II.
© 2016 John Wiley & Sons, Inc. Published 2016 by John Wiley & Sons, Inc.

TABLE 5.1   Hazard Analysis (HA) Tasks and Output

Steps		Tasks	Output
1	Plan HA	• Evaluate HA requirements • Select HA technique • Establish ground rules • Establish risk criteria	- HA plan - HA ground rules - Risk criteria
2	Understand system design	• Obtain design data • Understand design	- Design questions - System boundaries
3	Acquire HA tools	• Identify tools • Acquire tools	- Hierarchy table - SMM - TLMs
4	Identify hazards	• Apply HA technique • Follow ground rules • Recognize hazards • Test credibility	- Hazards - Hazard causal factors - Hazard reports
5	Validate hazards	• Peer reviews • SSWGs	- Credible hazards - Concurrence
6	Assess Risk	• Obtain failure rate data • Obtain severity data • Calculate risk	- Hazard risk
7	Mitigate risk	• Develop hazard defenses • Establish safety design requirements	- Safety requirements
8	Verify mitigation	• Test result reports • Requirements verification	- Hazard closure if pass - Hazard update if failed
9	Accept risk	• Package hazards • Obtain approvals	- Risk acceptance letter - Signatures
10	Track hazards	• Record hazards • Track hazards/risks • Close hazards	- Hazard database - Safety case - Lessons learned

hazards. If the design changes, the hazard analysis must change accordingly in order for the hazard analysis to correctly model the system.

Strictly speaking, hazard analysis could be considered as only identifying hazards. However, the hazard analysis process is tightly coupled with the system safety process, thus added steps are involved in the complete hazard analysis process. Table 5.1 lists the generic steps and specific tasks that are performed during a hazard analysis, along with representative output from each step. In general, these steps are performed regardless of the specific hazard analysis technique that is applied.

5.2.1   Plan the Hazard Analysis

Properly planning the hazard analysis process for a program is one of the most important and critical steps in the hazard analysis process. Performing a good hazard analysis is not a simple or trivial task; it requires foresight, planning, methodology, organization, and a total systems viewpoint in order to achieve uniformity, consistency, and full system coverage. It also requires someone with experience and skill in safety and hazard analysis. Quite often, analysts just jump into a system and immediately start identifying what they think are hazards, without considering the system architecture and the overall risk objectives and relationships for the hazard analysis. This approach often leads to confusion, overlap, gaps, and hazard–risk mismatches.

The hazard analysis plan should define and document the many aspects and parameters of a hazard analysis, such as follows:

- Hazard and hazard analysis definitions
- How to write hazards for the hazard analysis (context)
- Hazard analysis guidelines
- Hazard analysis scope, ground rules, and schedule
- Risk rating criteria
- Risk acceptance criteria
- Hazard analysis techniques to be used
- System level at which hazards–risk should be written
- The program risk management process
- Hazard analysis tools to be used

Since there are so many complexities involved with hazards and hazard analysis, a system-oriented hazard architecture should be developed as part of the planning process. This architecture models the hazard space of the system and is referred to as the system mishap model (SMM). It should define top-level mishaps (TLMs), system hierarchy, and hazard paths into the system. Hazards should not be developed in an arbitrary manner; they should be developed in an organized and consistent process.

5.2.2 Understand the System Design

Each hazard analysis is performed on a specific system design configuration, in order to make that specific design safe. If the system design and method of operation is not thoroughly understood, the hazard analysis will likely not be effective and may not even be entirely correct. This step involves obtaining, reviewing, and understanding system drawings, schematics, and design documents. It also involves attending design meetings and reviews during the system development.

5.2.3 Acquire Hazard Analysis Tools

Certain tools are typically utilized during the performance of a hazard analysis. These tools are typically identified during the planning process and acquired prior to initiating the hazard analysis. Hazard analysis tools aid in understanding system design and operation, and in recognizing and identifying hazards. Engineering tools that can be used in the hazard analysis process include

- Simplified system diagrams
- Functional flow diagrams
- Reliability block diagrams
- Timing diagrams
- System hierarchy diagram
- Hazard checklists
- Historical failure/accident data

- Failure mode and effects analysis (FMEA)
- Fault tree analysis computer program
- Event sequence diagrams
- Fishbone (Ishikawa) diagrams

5.2.4 Identify Hazards

Identifying, or recognizing, hazards is the key to hazard analysis and probably the most difficult and involved task of all. Hazard recognition is not a trivial process, and it is made even more difficult by the many hazard analysis complexities involved. Hazard analysis requires a thorough understanding of what exactly comprises a hazard (see Chapter 3), as well as a system hazard roadmap (Section 5.8). An agreed upon definition for a hazard should be included in the hazard analysis plan. In many respects, hazard analysis is a skill and art that requires a knowledgeable and experienced analyst. The following sections provide further information on hazard recognition.

5.2.5 Validate Hazards

Hazard validation involves reviewing postulated hazards and obtaining concurrence that the hazards are credible and real. Typically, hazards are reviewed by a separate group of individuals after they have been identified and established by the safety group. A peer review of hazards is conducted by technical areas experts, or subject matter experts (SMEs), in the technical areas covered by the hazard analysis. For example, after completing the initial hazard analysis of an aircraft system, SMEs on aircraft engines would review all engine-related hazards for concurrence. The SMEs might recommend rewording hazards, deleting hazards, or adding hazards. System safety working groups (SSWGs) are often held to review postulated hazards in the same manner. An SSWG can be comprised of any stakeholders in the system design, development, and operation. This step is important and should not be taken lightly or skipped over.

5.2.6 Assess Risk

Mishap risk is the safety metric characterizing the amount of danger presented by a hazard. Risk is the likelihood that a hazard will result in an actual mishap, multiplied by the mishap severity or amount of expected loss to be incurred from that mishap. Risk provides a predictive measure that system safety uses to rate the safety significance of a hazard and the amount of improvement provided by hazard mitigation.

A risk assessment process must be established for use during the hazard analysis. The risk management process used by system safety typically utilizes a hazard risk index (HRI) matrix to identify and assess the amount of safety risk presented by a hazard. The HRI matrix is a risk ranking scheme used in system safety to distinguish lower risk hazards from higher risk hazards. The HRI matrix ranks the safety criticality of each hazard through the use of a risk index and a risk level. Once an HRI matrix has been established, hazard risk can be evaluated according to the HRI matrix definitions. The HRI matrix is a tool for evaluating, judging, and accepting or rejecting risk.

5.2.7 Mitigate Risk

The risk management process is a key element in developing a system that presents acceptable mishap risk resulting from residual hazards. An effective hazard–mishap risk management process must be formulated for each program. The program HRI matrix is used to rank the safety criticality of each hazard through the use of an HRI and a risk level. The HRI determines if the hazard risk must be mitigated, and the hazard risk level establishes who accepts the risk.

When the risk presented by a hazard is determined to be unacceptable, it must be mitigated. Mitigation is typically done through the implementation of a design safety feature (DSF), such as a redundant component, fail-safe mechanism, interlock, and so on. System safety requirements (SSRs) are the vehicle that establishes the DSFs for the system design. By implementing the DSFs contained in the SSRs, the mishap risk is reduced to an acceptable level. But, it does not end there; the SSRs must be verified and validated (V&V) to ensure they are implemented and effectively eliminate the hazard or reduce it to an acceptable level of risk. Mitigation success is verified through appropriate analysis, testing, demonstration, or inspection. The use of a traceability matrix provides a trace from the hazard to the mitigating SSRs, to the test requirements, and to the test results.

SSRs are a combination of generic and derived safety requirements. Generic SSRs typically come from existing safety documents and standards, such as special safety requirements for laser safety. These are SSRs established from historical hazards from other systems. Derived SSRs are established for the mitigation of hazards identified for the system under development. SSRs are included in the system specification.

5.2.8 Verify Mitigation

The purpose of this step is to ensure that the planned hazard mitigation is carried out. It is also to validate the success of the implemented risk mitigation measures. System safety must take action if the mitigation plans are not being followed. System safety must also perform follow-up analysis and data reviews to ensure that the implemented mitigations are effective. This step involves ensuring that the safety design requirements are documented, tested, and successfully pass testing. This step is important and should not be taken lightly or skipped over; no hazard should be closed until this step is successfully satisfied.

5.2.9 Accept Risk

A risk assessment system must be established for use during the hazard analysis. The risk assessment system utilized depends upon the industry involved for the particular product involved. DoD and OSD policy mandates that the risk for all identified hazards must be accepted by the appropriate risk acceptance authority per MIL-STD882E. If the project is a commercial product and not a DoD program, risk acceptance should still be performed by the company developing the product to ensure complete management oversight.

5.2.10 Track Hazards

Hazard tracking is the process of systematically recording all identified hazards and the data associated with these hazards as they progress through the life cycle of a hazard: identification, assessment, mitigation, verification, acceptance, and closure. It is typically achieved through the use of a formal hazard tracking system (HTS). Hazard tracking is a basic required element of an effective system safety program (SSP), essential to knowing the status of all hazards at any point in time, in order to ensure that none goes unresolved. It also provides a record of all activity and information associated with each hazard. It is usually referred to as closed loop hazard tracking because it is similar to a control system with a feedback loop that is active and allowing for changes and updates, and the reiterative performance of some risk management tasks. As changes are incorporated into the system, the HTS is updated to reflect changes in hazards and residual risk. The HTS is a living document that needs to be revisited periodically throughout the system life cycle to determine if accepted hazards can be eliminated or further reduced through new technology or process changes. It is important to remember that a closed hazard does not mean no risk exists; it is an acknowledgement and acceptance of a system hazard and its residual mishap risk.

5.3   HAZARD RECOGNITION

5.3.1   Hazard Recognition Introduction

Hazard identification, or hazard recognition, is not a trivial process; it requires a thorough understanding of what exactly comprises a hazard and a grounded understanding of the system under investigation. In many respects, hazard analysis is a skill and art that requires a knowledgeable and experienced safety analyst. Hazard identification must be taken seriously because of the safety-related repercussion involved.

Hazard recognition is the cognitive process of visualizing a hazard from an assorted package of design information and hazard knowledge. In order to recognize or identify hazards, four things are necessary:

1. An understanding of hazard theory
2. A hazard analysis technique to provide a consistent and methodical process
3. An understanding of hazard recognition methods
4. An understanding of the system design and operation
5. An understanding of system hazard organization

Hazard analysis involves the recognition and identification of system hazards and causal factors using a systematic and methodical approach, which is typically achieved through the application of several different system safety hazard analysis techniques. The hazard analysis plan should describe the combination of analyses that will be used to identify hazards for the particular project. Hazard analysis should evaluate hardware, software, firmware, procedures, and Human System Integration (HSI). Hazard analysis should consider all life cycle phases of the system, as well as environmental factors. Consideration should also be given to proposed design changes, software trouble reports, and technology upgrades, to name a few.

5.3.2   Hazard Recognition: System Perspectives

When performing a hazard analysis, the system must be viewed from different perspectives (discussed in Chapter 2), each of which provides a different viewpoint and understanding of the system. These system viewpoints include:

1. *Physical*   This view involves the various architectural views that depict what the system contains and how it is constructed. This view establishes subsystems, assemblies, components, and the overall system equipment hierarchy.
2. *Functional*   This view evaluates what the system must do in order to produce the required system behavior, broken down into functions with input, output, and transformation rules.

3. *Operational* This view defines how the user will view and operate the system, including instructions, tasks, conditions, parameters, and limitations.

4. *Software* This view looks at the system software, which is difficult to fully grasp and is somewhat abstract.

5. *Environment* This view looks at the various environments that the system will encounter (internal and external). This includes both natural environments (e.g., weather, tornadoes, etc.) and system environments (e.g., heat, EMI, etc.).

6. *Human* This view looks more closely at human performance in the system and the effect of potential human errors. It also includes user interfaces with the system and their potential impact on the user.

7. *Organizational* This view looks at the organizational and management causal factors affecting a hazard.

It is important that hazard analysis consider and evaluate a system from each of these perspectives in order to ensure complete safety coverage of the system. This is why hazard analysis must consider and focus on system functions, system operations, and system components, including hazardous energy sources. This also explains why more than one type of hazard analysis must be applied in order to identify all hazards, because one hazard analysis type alone does not typically provide sufficient hazard identification coverage.

5.3.3 Hazard Recognition: Failure Perspectives

All of the basic system components and characteristics must be understood in order to perform a complete and thorough hazard analysis. Examples of typical safety considerations for various system elements include the following:

- *Hardware* Failure modes and hazardous energy sources
- *Software* Design errors and design incompatibilities
- *Personnel* Human error, human injury, and human control interfaces
- *Environment* Weather and external equipment (e.g., radiation, chemicals, etc.)
- *Procedures* Instructions, tasks, and warning notes
- *Interfaces* Erroneous input/output and unexpected complexities
- *Functions* Fails to perform, performs erroneously and performs inadvertently
- *Facilities* Construction factors, storage factors and transportation factors

A hazard analysis must adequately consider and address each of these system attributes and their interrelationships in order to ensure that all possible hazards are identified. For example, it is possible for different operational phases to have different safety impacts; different functions performed during a phase could have a direct impact on subsequent phases. During certain phases, safety barriers or interlocks are often removed, making the system more susceptible to the occurrence of a hazard; for example, at one point in the operational mission of a missile, the missile is powered and armed. This means that fewer potential failures are now necessary in order for a mishap to occur and there are fewer safeguards activated in place to prevent hazards from occurring.

Hazards are typically actuated as the result of any of the following system-initiating mechanism factors or combinations thereof:

- Hardware failures (aging, wear, and random failure)
- Software errors/flaws (functional failure)
- Human errors (performance, decisions, and judgment)
- Design errors/flaws (interface errors, sneak paths, etc.)
- External environmental factors (EMI, lightning, etc.)
- Maintenance flaws/errors (resulting in failures)
- Manufacturing flaws/errors (resulting in failures)
- Particular risk events (events occurring outside of a subsystem, such as a flood, fire, etc.)

In addition to typical hazard sources, failures, flaws, and errors can be also created or perpetuated by the following organizational factors:

- Poor design/development/manufacturing processes
- Organizational errors (performance, decisions, and safety as a core value)
- Lack of safety culture in overall organization/company
- Safety organizational level of competence

It is important to note that hazards can exist and occur without the presence of a hardware failure mode or software error. Subtle design flaws can produce hazards, such as a sneak path in an electrical circuit. When performing an HA, the following standard considerations for identifying hazards are taken into account:

- Safety-related functions (SRFs) and safety-critical functions (SCFs) (hardware/software)
- Hazard sources (e.g., energy sources, SCFs, environments, and particular risk sources)
- Hazardous assets
- Hardware failures (component failure modes)
- Sneak paths
- Safety-related (SR) and safety-critical (SC) human tasks
- Design requirement errors and conflicts

5.3.4   Key Hazard Recognition Factors

Some key recognition factors that can help the safety analyst visualize hazards, include the following:

1. Hazard sources
2. Hazard checklists
3. Lessons learned
4. Safety criteria
5. Key failure state questions
6. Good design practices
7. TLMs
8. SCFs
9. The SMM and its overall layout
10. Hazard triangle components

Hazard Sources   Identify the hazard sources in the system. Each hazard source will be responsible for one or more hazards. Use the system hierarchy table, list of equipment, and hazard checklists to identify hazard sources in the system. The hazard source is typically the item that causes damage during a mishap. Example hazard sources include

- fuel
- electricity
- radiation
- explosives
- rotating machinery

Hazard Checklists   Use hazard checklists to identify hazards in the system. The checklists are mental reminders of items, functions, and procedures that are hazardous. A good safety analyst should have many different hazard checklists at hand. Refer to Appendix C for some example checklists.

Lessons Learned   Use the past knowledge from experience and lessons learned. Review lessons learned documents and databases. Mishap and hazard information from a previously developed system that is similar in nature or design to the system under analysis will aid in the hazard recognition process. For example, by reviewing the mishaps of an earlier spacecraft system, it might be discovered that the use of a particular seal design on a specific part of the spacecraft resulted in several mishaps. Seal design on this part of the spacecraft should then be recognized as a potential hazardous area requiring special focus in new spacecraft design.

Safety Criteria   Review general design safety criteria, precepts, and principles applicable to the system under investigation. Use the criteria to identify specific hazards in the system design. By considering the reasoning and logic behind specific safety criteria, precepts, and principles, some types of hazards can be more easily recognized. For example, there is a good safety reason for the following safety criteria: "Do not supply primary and redundant system circuits with power from the same bus or circuit." This safety criterion is a clue that aids in recognizing hazards in systems involving redundancy and common cause failures. A hazard analysis should look for such hazards in a new design.

Key Failure State Questions   Use key failure state questions to identify hazards. This is a method involving a set of clue questions that must be answered, each of which can trigger the recognition of a hazard. The key states are potential states or ways the subsystem could fail, or operate incorrectly, and thereby result in creating a hazard. For example, when evaluating each subsystem, answering the question "what happens when the subsystem fails to operate?" may lead to the recognition of a hazard. For everything in the system (item, function, human, task, etc.), ask what happens if the item should do one of the following:

- Fails to operate
- Fails to function
- Malfunction
- Degraded function
- Inadvertent function
- Incorrect input to function
- Incorrect output from function
- Function timing error – early, late, and slow
- Unable to stop/control

- Operator confusion
- Operator misuse (reasonable misuse)
- Out of sequence
- Erratic function

Good Design Practices   Analyze known and recommended good design practices. If these practices are not in effect, then there may be hazards associated with these practices. By reverse engineering, good design practices found in various design documents and standards, hazardous design situations can be identified. For example, a good design practice for interrupting DC circuits is to place the switch or circuit breaker on the power branch, rather than the ground branch of the circuit. This prevents inadvertent operation should a fault to ground occur.

TLMs and SCFs   Review identified TLMs and SCFs to determine if any additional hazards have been overlooked in these categories.

SMM   Review the SMM to determine if all hazards have been identified for each of the undesired outcomes and TLMs.

Hazard Triangle   The hazard triangle concept provides a hazard recognition resource by evaluating individually each of the three hazard component categories from a systems context. For example, identify and evaluate all of the HS components in the unique system design as the first step. Next, evaluate all of the potential system failure modes to determine which failures are critical IMs for hazards. Then, identify and evaluate all potential targets and the possible threats to them.

When considering the HS component of the hazard triangle, focus on hazardous element checklists. The HS component involves items known to be a hazardous source, such as explosives, fuel, batteries, electricity, acceleration, chemicals, and the like. This means that generic hazard source checklists can be utilized to help recognize hazards within the unique system design. If a component in the system being analyzed is on one of the hazard source checklists, then this is a direct pointer to potential hazards that may be in the system. Speculate all of the possible ways that the hazardous element can be hazardous within the system design. There are many different types of hazard source checklists, such as those for energy sources, hazardous operations, chemicals, and so on.

Hazards can be recognized by focusing on known hazard triggering mechanisms (the IM hazard component). For example, in the design of aircraft, it is common knowledge that fuel ignition sources and fuel leakage sources are initiating mechanisms for fire/explosion hazards. Therefore, hazard recognition would benefit from detailed review of the design for ignition sources and leakage sources when fuel is involved. Component failure modes and human error are common triggering modes for hazards.

Hazards can be recognized by focusing on known or preestablished undesired outcomes or mishaps; review the potential targets and threats. This means considering and evaluating known undesired outcomes within the system. For example, a missile system has certain undesired outcomes that are known right from the conceptual stage of system development. By following these undesired outcomes backward, certain hazards can be more readily recognized. In the design of missile systems, it is well accepted that an inadvertent missile launch is an undesired mishap and, therefore, any conditions contributing to this event would formulate a hazard, such as auto-ignition, switch failures, and human error.

5.3.5 Hazard Recognition Basics

When identifying hazards, a key factor to keep in mind is the understanding of why hazards exist in the first place. Knowing why and how they exist helps in their identification. Hazards exist primarily for three reasons: (1) hazardous assets are used in the system, (2) design miscalculations and errors occur in the design process, and (3) failures, human errors, and environments impact the design during system operation.

Hazards are created because of the need for hazardous sources in the system, or they must interface with hazard sources, coupled with the fact that eventually everything fails, and these failures can unleash the undesired effects of the hazard source. Hazards also exist due to the need for safety-critical system functions, coupled with the potential for failures and human error within these safety-critical functions. Hazard creation can be summarized by the following factors, which can occur singularly or in combinations:

- The use of hazardous system elements (e.g., fuel, explosives, electricity, velocity, and stored energy).
- The system interfaces with hazard sources (e.g., fuel, electricity, etc.).
- Operation in hazardous environments (e.g., flood zones, ice, and heat).
- The need for hazardous functions (e.g., aircraft fueling, and welding).
- The use of safety-critical functions (e.g., flight control, arming, etc.).
- The inclusion of (unknown) design flaws, errors, and sneak paths.
- The potential for hardware wear, aging, and failure.
- Inadequacy in designing to tolerate critical failures.

5.3.6 Hazard Recognition Sources

Table 5.2 lists various sources that can be used to recognize and identify hazards.

5.4 DESCRIBING THE IDENTIFIED HAZARD

A very important aspect of hazard analysis that is often overlooked has to do with describing an identified hazard. A good hazard description depends upon the proper context; it describes a scenario that includes the hazard elements of HS, IM, and TTO. It should include only the basic hazard causal factors (HCFs) needed to cause the hazard and nothing more. If more is possible, then that is usually material for another similar hazard.

Describing a hazard is very similar to performing a mini-FTA of a hazard. The hazard description must be broken down into the three required components: HS, IMs, and TTO. An FTA begins with an undesired event and determines all of the contributing causal factors, and their methods of occurrence, which lead to the undesired event, down to the lowest level of detail. A hazard analysis identifies all of the hazard components and mechanisms of occurrence that can lead to the hazard's undesired mishap outcome. A hazard can contain several required mechanisms that must occur before the hazard becomes a mishap. For example, an uncontrolled office fire requires two mechanisms, a mechanism that starts the fire and a mechanism that prevents the automatic sprinkler system to fail.

Where a large fault tree (FT) usually has many unique cut sets that cause the undesired event, a hazard is typically a small FTA with a single cut set. A hazard is more like one cut set in

TABLE 5.2 **Hazard Recognition Methods**

	Hazard Recognition Subject Matter	Recognition Thoughts
1	Energy sources	Identify the various hazards that can spawn from energy sources used in the system. Consider different scenarios: for example, if the energy source fails to function, functions inadvertently, becomes uncontained, accidently contacted, and so on
2	Critical system functions	Determine which system functions are needed for safe system operation. Consider different scenarios: for example, if the function fails, functions inadvertently, functions out of tolerance, and so on
3	Redundant items	Identify the redundant subsystems and assemblies in the system. Determine which are safety-critical. Consider different scenarios: for example, if the function fails, functions inadvertently, functions out of tolerance, and so on
4	Expected environments	Identify all potential environments that could be possibly encountered. Identify systems/equipment that may be sensitive to certain environments
5	System hierarchy table	Evaluate every item on the hierarchy table. Identify items that are hazardous assets or hazardous if they malfunction
6	Hazardous assets	Determine which equipment in the system hierarchy table is hazardous in some way, and postulate the hazards that might be possible. For example, aircraft flight controls are hazardous if they fail
7	Existing design safety features	Look at the design of safety features claimed in the system design and evaluate the impact if they should fail or malfunction
8	TLMs/SMM	While looking at the TLMs and the SMM, consider the items in the hierarchy table and determine if any of them can contribute a hazard to one of the categories
9	Hazardous subsystems	Determine which of the subsystems in the system hierarchy table are hazardous in some way, and postulate the hazards that might be possible. For example, aircraft flight controls are hazardous if they fail
10	Critical software functions	Determine which software functions are needed for safe system operation. Consider different scenarios: for example, if the function fails, functions inadvertently, functions out of tolerance, and so on
11	System dependencies	Identify equipment that can cause hazards if they are caused to fail simultaneously by common cause faults
12	Hazard checklists	Review hazard checklists to see if they trigger ideas for any hazards in the system design
13	Lessons learned	Review lessons learned and mishap reports to see if they trigger ideas for any hazards in the system design
14	Design requirements	Review design requirements to identify hazardous situations created by the requirements and/or missing requirements affecting safety
15	Hazard organization	Organizing hazards below TLMs and TLHs helps in visualizing and writing hazards. Be sure to establish TLMs and TLHs during the process

an FT. Many different hazards leading to the same undesired event can be linked together by the FT structure. When a hazard is a minimal cut set of a larger group of hazards, they all fall under the same general TLM. For example, inadvertent missile launch can be caused by many different potential failures, each of which is a separate hazard. The FT combines them together for a single probability of occurrence for the sum of the individual hazards.

Sometimes, analysts try to put too much, or too little, information into a hazard. With too little information, the hazard cannot be fully comprehended or the risk calculated. Too little information produces pseudo-hazards, such as "aircraft collision." With too much information, there are usually multiple causal ORed factors included, which should be divided into separate hazards. For example, "inadvertent missile launch occurs due to switch fails closed OR human error closes switch." Note: the OR operator refers to the Boolean OR operator used in probabilistic mathematics; as opposed to the AND operator.

Write hazards to ensure they are characterized in complete system context:

- Describe the hazard source (HS), initiating mechanism (IM), and target/threat outcome (TTO).
- Do not abbreviate or assume readers understand program-special lingo and acronyms.
- Describe the hazard scenario in context (e.g., "fuel" is not a hazard, but, "fuel leak and an ignition source leading to fire and system loss" is a hazard)
- Make sure the hazard context includes the specific HFCs and the specific hazard–mishap effects
- Write the hazard in a complete sentence

5.5 HAZARD TYPES BY GENERAL CIRCUMSTANCES

Hazards typically fall into one of the several categories; understanding these categories helps in the hazard identification process. The following are some of these hazard categories based on general circumstances:

1. *Inherent Hazard*   These are hazards resulting from the inherently hazardous nature of the components, equipment, or processes in the system, such as hazardous materials, energy sources, or safety-critical functions. This probably accounts for the majority of system hazards.

2. *Software Hazard*   These are hazards resulting from software and its interface and control of system functions. The software associated with these system-level functions is safety related, and errors in these functions may be a hazard causal factor. For example, a missile launch is a safety-critical function that has serious safety implications, such as an inadvertent launch.

3. *Timing Hazard*   These are hazards resulting from errors or faults in areas where timing is safety-critical. These hazards can result from hardware, software, and/or human performance. This is often an overlooked area because timing is often taken for granted to be safe, until an accident or mishap occurs.

4. *Hardware-Induced Hazard*   These are hazards resulting from software errors caused by a computer hardware failure that causes a bit error, resulting in an erroneous instruction. For example, an intended word with bit pattern "1101" that means "add to register A" may be changed by an induced bit error to "1001" that means "subtract from register A." This type of hazard has safety implications in electronic data storage and transfer methods. Hardware failures can modify software and/or induce totally unpredictable results in the software and in system timing. For example, a hardware bit perturbation can result in an incorrect instruction or data, or a jump to an incorrect memory location. Hard hardware failures can be traced and corrected, but soft failures (e.g., alpha particle radiation) occurs at random and are usually not repeatable when debugging

5. *Latent Hazard*   These are hazards resulting from a hidden condition in the hardware/software design that is not hazardous until a particular unanticipated, unplanned, or untested set of circumstances occur. It could also be the result of a built-in unintended

function or the result of a sneak path in the software. This is somewhat of a misleading category, because in reality all hazards are essentially latent.

6. *Systemic Hazard*   These are hazards caused by the systemic system design. For example, an error in the hardware–software design, integration, or implementation that results in a system-level hazard. These types of hazards are generally the result of hardware–software interface flaws.

7. *Code Error Hazard*   These are hazards caused by a software coding error.

8. *Explosives Hazard* These are hazards caused by the use of explosives or explosive materials in the system.

9. *Common Cause Hazard*   These are hazards caused by faults and failures that cause redundant system elements to fail simultaneously. For example, flooding can cause multiple simultaneous failures.

10. *Sneak Circuit Hazard*   These are hazards caused by sneak, or unintended, electrical paths through system circuits that perform unwanted functions.

11. *HSI Hazard*   These are hazards caused by the user interface design. For example, the system operator can be forced into committing errors when he is confused by the poor layout of a multitude of switches, gauges, and displays.

12. *Organizational Hazard*   These are hazards caused by faults, failures, or errors in an organization. Typically, they deal with poor, inadequate, incorrect, or lack of performance in regard to safety issues.

5.6   HAZARD TYPES BY ANALYSIS CATEGORY

Hazards typically fall into one of the several categories, and understanding these categories helps in the hazard identification process. The following are some of these hazard categories based on system mode:

1. *Functional Hazard*   Hazards that are functional in nature; typically identified from an FHA. These hazards involve functions that must be performed by the system, and the functions are premature, fail, out of tolerance, unintended, and so on. For example, an inadvertent missile launch is a functional failure hazard. The functional error would have to be traced back to the hardware, software, or human root cause.

2. *System Hazard*   Hazards that are systemic in nature; typically identified from an SHA. These hazards involve synergistic system relationships. They can involve more than one subsystem and interfaces and interrelationships between subsystems.

3. *Subsystem Hazard*   Hazards involving or contained within a single subsystem; typically identified from an SSHA. They may pertain to individual subsystems or the system. Discovery methods include SSHA, FMEA, and PHA.

4. *Operational Hazard*   Hazards involving operating and support tasks; typically derived from an O&SHA. The effect of procedures as they are integrated with the system design.

5. *Health Hazard*   Hazards that are human health in nature; typically derived from an HHA. Hazards involving the health of system operating or support personnel. Typically, a subsystem problem but could also be a system hazard. Examples include operator injury from repetitive motion, operator injury from excessive noise, or operator injury from silica inhalation.

6. *Test Hazard*   Hazards involving testing. Includes test procedures, system equipment, software, and test support equipment. These hazards are typically derived from a test hazard analysis.

7. *Software Hazard*   Hazards that are strictly oriented to software-related HCFs. These hazards are typically derived from specially performed software safety analyses.

8. *Operator Hazard*   These are hazards caused by operator error, operator performance, and so on. These hazards are typically derived from PHAs, SSHAs, SHAs, and O&SHAs.

5.7   MODELLING HAZARD SPACE

A system's hazard space consists of all the things that can go wrong in a system causing the system to result in system failure or a mishap. It is the opposite of success space, which is the system's correct and successful operation. Hazard space is comprised of hazards and the causal factors to these hazards.

Hazard analysis is the process of identifying hazards within a system. It seems like an uncomplicated process; however, it actually requires a set of complex, methodical, and structured activities. One very important hazard analysis activity, which is commonly over-looked, is the structured organization of hazards and potential mishap outcomes. Quite often, hazard analyses are performed without any thought being given to understanding the overall hazard–mishap relationships for the complete system being analyzed (i.e., poor organization). Inexperienced analysts often just jump into a hazard analysis and start randomly listing perceived hazards. The situation is similar to driving to a distant unfamiliar location; for example, if you don't have a map showing how to get there, then any road will do, and you will likely become quickly lost. As Yogi Berra is purported to have said, "If you don't know where you are going, you will wind up somewhere else." Unfortunately, in hazard analysis, any road will not do, a roadmap of where you are going is necessary for an effective and successful hazard analysis.

During hazard analysis of a large system, the number of potential hazards can become so large and diverse that the problem becomes one of where to start the hazard analysis and how to proceed through the system without becoming lost in the maze. Also, large and complex systems often confuse the analyst between hazard sources, hazard casual factors, and hazard outcomes. This situation requires that a method of hazard organization be established, such as a hazard–mishap roadmap, in order to accurately identify the correct and complete set of hazards. A hazard analysis roadmap in the fashion of a system mishap model is a very effective hazard analysis tool.

It may seem simple to perform a hazard analysis . . . . merely look at all the system energy sources and list the bad things that can happen if they fail or malfunction. However, the energy sources provide only a partial segment of hazards. Or, look at system control laws and evaluate the effect of potential malfunctions. Unfortunately, there are many more different types of hazards that do not fall under the category of control laws. Or, you may be tempted to evaluate only system functions and the result of their failure or malfunction. However, functions are just part of the hazard analysis equation.

Although these approaches sound plausible and are taken by many, the problem is that effective hazard recognition is not quite that simple. Again, a map of the overall system hazard space is needed to ensure all types of hazards are covered and the various hazard analysis methods are funneled in the right direction.

When a hazard analysis is *not* properly organized and managed, the results can be quite disastrous and meaningless. An unorganized hazard analysis quite often results in incorrect, inadequate, misleading, missing, and/or conflicting hazards. For this reason, it is necessary to develop an SMM at the beginning of the hazard analysis process. The SMM is *not* the hazard analysis itself, but rather a guide for the analysis; it provides a layout of the relationships between major potential undesired outcomes and hazards, while developing the pathways between the two. An SMM helps in both hazard recognition and in ensuring that the correct hazards are identified and described; it provides the means for a focused hazard analysis.

One of the major problems encountered during hazard analysis is properly organizing the overall analysis such that the correct and proper hazards can be identified. Sometimes, systems are so large and complex that an analyst easily goes off in the wrong direction and misses or misidentifies hazards. This situation is analogous to the old adage that one cannot see the trees for the forest. The solution is to develop an SMM that aids in visualizing the trees in the forest. This is akin to visualizing the hazards in a large complex system and organizing them in a manageable form. The SMM depends significantly on the top-level mishap concept.

5.7.1   System Mishap Model

The SMM is a model of a system's hazard space, depicting the pathways to hazards by diagramming all of the system's TLMs and the system fault paths (SFPs) they generate. The SMM guides an analyst to ask the right questions when identifying hazards. In order to describe the SMM, several terms must first be defined; these terms serve as building blocks for the SMM concept and include the following: hazard, TLM, top-level hazard (TLH), HCF, and hazard causal pathways (HCPs).

> **Hazard**   A hazard is typically defined as "a condition that can result in danger, an undesired event, or a mishap." Although somewhat correct, this definition is incomplete and insufficient, thus has become misleading and a tool for distortion. In reality, a hazard is a complex entity, as well as a complex concept, which requires a clear and precise definition. A hazard is a specific set of dormant conditions that exist, within a system design, which can result in a mishap when the conditions are activated, where these conditions consist of three required components: a hazard source, an initiating mechanism, and a target–threat outcome. This is an expanded version of the source–mechanism–outcome (S-M-O) concept developed by Pat Clemens. A detailed explanation is contained in Refs. 1 and 2. A hazard is not an event, it is a dormant condition and thus a *potential event*.

> **Mishap**   A mishap is an actual undesired event that has occurred. It is a hazard scenario that has actually happened; it is the outcome consequences of an actuated hazard, in terms of outcome and severity. All hazards are unique with a unique outcome, but some hazards have partially common outcomes.

> **Top-Level Mishap**   A TLM is a generic mishap category for collecting and correlating related hazards that share the same general type of mishap outcome event. A common mishap outcome that can be caused by one or more hazards, the TLM's purpose is to serve as a collection point for all the potential hazards that can result in the same overall TLM outcome but have different causal factors. TLMs provide both a hazard focal point and a design safety focal point. TLMs help highlight and track major safety concerns. "Top level" implies an inherent level of safety importance, particularly for visibility at the system level. The TLM is derived by extracting the significant and common generic

outcome event portion of the contributing hazard–mishap description. A detailed explanation is contained in Ref. 3.

Top-Level Hazard A TLH is similar to a TLM (sometimes the terms are used interchangeably). A TLH is the top tier hazard category below a TLM. TLHs are an intermediate categorization level below the TLM to help refine the categorization of hazards below the TLM. This approach is used only when many different hazards are involved and further refinement of categories is needed. A TLH is a generic hazard category for collecting various hazards into different categories below a TLM category in order to help establish the hazard fault path.

Hazard Causal Factors HCFs are the particular and unique set of circumstances that form a hazard, including the initiating mechanisms that transform a hazard into a mishap. Causal factors may be the result of failures, malfunctions, external events, environmental effects, errors, poor design, or a combination thereof. Causal factors generally break down into categories of hardware, software, human action, procedures, and/or environmental factors.

Hazard Causal Pathways HCPs are the paths from the HCFs to the final mishap outcome. When identifying hazards, it is extremely useful to develop these pathways in order to correctly recognize the hazards. These pathways help in establishing the SMM roadmap. HCPs link the system fault relationships that lead to identifying specific hazards.

The TLM is a generic mishap category that is an important component in the SMM. TLMs are high-level undesired outcomes that provide a safety focus. They also provide a starting point for establishing HCPs. Working from the TLM back into the hazards is a very powerful approach. Quite often, the TLM outcomes are known even before the hazard analysis begins and are established early in the system safety program. TLMs are also identified during the preliminary hazard list (PHL) analysis or the preliminary hazard analysis (PHA) phases. Each unique system will have its own unique set of TLMs. As TLMs are established and compared with one another, it becomes clear where the major hazard analysis focus should be applied.

Figure 5.1 illustrates the TLM concept. In this example, there are five different top-level hazards resulting in an uncontrolled aircraft fire. Each hazard has different causal factors, but

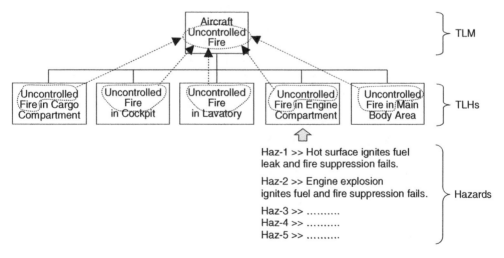

Figure 5.1 *Example TLM concept.*

a common *top-level* outcome, which is "Uncontrolled Fire" in the aircraft, for instance. This common outcome is extracted from the hazards and used as a common TLM to unite the hazards into a group. A different hazard such as "landing gear fails to lower" could not be placed directly under this TLM; it would fall under a different TLM such as "Loss of Landing Gear."

Systems typically have several different TLMs, depending on the size and safety criticality of the system and the desired safety focal points. Also, different types of systems have different types of TLMs, although there may be a few similar TLMs between some system types. The TLM "Uncontrolled Fire" is just one TLM for a typical aircraft system. Many other TLMs would also be developed, such as "Loss of Engines," "Loss of Landing Gear," "Controlled Flight into Terrain," Inadvertent Thrust Reverser Operation," and so on.

Hazards can be recognized by focusing on known or preestablished undesired outcomes or mishaps (the TTO hazard component). This means considering and evaluating known undesired outcomes within the system. For example, a missile system has certain undesired outcomes that are known right from the conceptual stage of system development. By following these undesired outcomes backward, certain hazards can be more readily recognized. In the design of missile systems, it is understood and well accepted that inadvertent missile launch is an undesired TLM and, therefore, any conditions contributing to this event would formulate a hazard, such as auto-ignition, switch failures, and human error.

The SMM is one of the most powerful organization and identification tools that can be applied when conducting a hazard analysis. When potential system mishap vulnerability is known (i.e., TLM), hazards leading to that mishap category can be more easily identified. An established methodology for an SMM involves utilizing a mishap tree linking system top-level mishaps to hazards, by identifying system pathway between the two; it is a combination of a hazard–mishap outcome model and a hazard pathway model. An SMM organizes the major potential mishaps that a system is susceptible to; it establishes and links the major mishap categories and hazards in a system. The SMM establishes hazard causal paths. The purpose of an SMM is to better understand a system, its associated hazards, and it potential mishaps. The SMM provides hazard analysis direction and visibility. The SMM is an a priori (before the fact) approach to understanding hazards and mishaps.

The SMM keeps the analyst focused on what are mishaps, what are hazards, and what are causal factors. Sometimes, when the hazard analysis is rushed, these variables become mixed and confusing. Development of the SMM is initiated prior to actual hazard analysis and finalized at hazard analysis completion. The SMM presents a hierarchical depiction of ways in which system events occur. A SMM shows the relationship of lower levels of assembly to higher levels of assembly and system function. The system hierarchies help clarify severity levels and risk levels. One of the most important steps in hazard analysis is the creation of an SMM before the actual identification of hazards. Hazards should not be developed in an arbitrary manner; they should be developed in an organized and consistent process established by developing an SMM.

The SMM can take many forms, such as a list, spreadsheet, or tree diagram. The important aspect of an SMM is that it presents a logical depiction of the ways in which system mishaps occur, establishing the path of relationships involving system function, hardware assemblies, system modes, and hazard sources. The SMM links the hazard source, initiating mechanisms, and target–threat outcome elements of a hazard.

A TLM is a generic mishap category for collecting together various hazards that share the same general outcome or type of mishap. A TLM is a common mishap outcome that can be caused by one or more hazards; its purpose is to serve as a collection point for all the

TABLE 5.3   Example TLMs for Different System Types

Missile System	Aircraft System	Spacecraft System
• Inadvertent missile launch	• Controlled flight into terrain	• Loss of oxygen
• Inadvertent warhead initiation	• Loss of all engines	• System Fire
• Incorrect missile target	• Loss of all flight controls	• Re-entry failure
• Self-destruct fails	• Loss of landing gear	• Temperature control fails
• Electrical injuries	• Inadvertent thrust reverser operation	• Communications fail
• Mechanical injuries	• Electrical injuries	• Electrical injuries
• RF radiation injuries	• Mechanical injuries	• Mechanical injuries
• Weapon ship fratricide		

potential hazards that can result in the same outcome but have different causal factors. TLMs provide a design safety focal point for a particular safety concern (i.e., the TLM outcome). Each contributing hazard has different initiating mechanisms or causal factors but a common TLM outcome event. This common outcome is extracted from the hazards and used as a common TLM to unite the hazards. A TLM is essentially a *top-level mishap outcome*.

During hazard analysis of a large system, the number of potential hazards can become so large and diverse that the problem becomes one of how to easily and accurately represent the safety risks of the system design. When several different hazards can result in the same mishap, that mishap is categorized as a TLM. The TLM becomes a generic mishap category for collecting various hazards contributing to it. It is referred to as a top-level mishap rather than a top-level hazard because it is a collection of several different hazards, each with the same overall mishap.

Systems typically have several different TLMs, depending on the size and safety criticality of the system and the desired safety focal points. Also, different types of systems have different types of TLMs, although there may be a few similar TLMs between some system types. Some example TLMs for three different system types are shown in Table 5.3.

5.7.2   System Mishap Model Examples

When systems are large and complex, it can be difficult to see the trees (i.e., hazards) for the forest (i.e., system). In these situations, it is important to utilize system mishap models that outline the forest by linking top-level mishaps with hazard paths and hazards. A system mishap model is a model organizing the major potential mishaps that a system is susceptible to; it establishes and links the major mishap categories comprising a system. The purpose of an SMM is to better understand a system, its associated hazards and potential mishaps. The SMM provides hazard analysis direction and visibility. It also provides a means for summing risk at the individual mishap level. The SMM is an a priori (before the fact) approach to understanding hazards and mishaps that should be performed prior to conducting the detailed hazard analysis.

The SMM helps the safety analyst to organize a hazard analysis and to visualize the trees within the forest; it maps the overall hazard space of a system. The SMM is not a hazard analysis but rather an aid. This chapter explains the SMM and its usage, along with several examples.

The SMM keeps the analyst focused on what are mishaps, what are hazards, and what are causal factors. Sometimes, when the hazard analysis is rushed, these variables get confused and mixed. Development of the SMM is initiated prior to an actual hazard analysis and finalized at hazard analysis completion. The SMM can take many forms, such as a list, spreadsheet, or tree diagram. An established methodology for an SMM involves utilizing a mishap tree linking system top-level mishaps.

An SMM can be developed using one of the several different approaches. Some common approaches include the following:

- Mishap tree
- Mind-map (see Refs. 1 and 2 for additional information)
- Fault tree
- Spreadsheet

A mishap tree is a structured diagram that organizes system mishaps into an easy-to-visualize-and- understand schema. The most effective approach for establishing a mishap tree is to utilize TLMs. A TLM is a generic mishap category for collecting and correlating related hazards that share the same general type of mishap event outcome. A TLM is a common mishap outcome caused by one or more hazards; its purpose is to serve as a collection point for the potential hazards that can result in the same overall TLM outcome but have different causal factors.

TLMs help highlight and track major safety concerns and provide a design safety focal point. "Top level" implies an inherent level of safety importance, particularly for risk visibility at the system level for a risk acceptance authority. The TLM severity will be the same as that of the highest contributing hazard's severity. Most hazards within the TLM will have the same severity level as the TLM; however, some may have a lesser severity. Figure 5.2 shows the generic concept of a mishap tree utilizing TLMs.

A mishap tree provides major safety focus. It identifies the potential mishap types that are safety-critical versus those that are serious and minor in severity. Safety-critical mishap types require more attention and hazard analysis focus. Mishap tree TLMs are derived by extracting the significant and common generic outcome event portion of the contributing hazard–mishap description. There are significant advantages in utilizing the TLM categorization process. It groups similar hazards, while also grouping similar risk categories. Although it is not realistic to sum the risk of all identified system hazards, it is feasible to sum the risk within TLM categories. The risk presented by all hazards cannot be summed because it is not meaningful to sum hazards with significantly different severity categories.

The mishap tree and TLMs focus on just the significant portion of the outcome described within a hazard. The TLM is restated as a simple generic outcome event when the TLM wording is shortened or slightly revised in order to make it a generic statement. TLM wording should focus on the particular safety issue of concern, such as inadvertent launch, aircraft controlled flight in terrain (CFIT), electrocution, and so on.

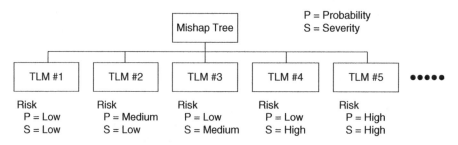

Figure 5.2   *Mishap tree concept.*

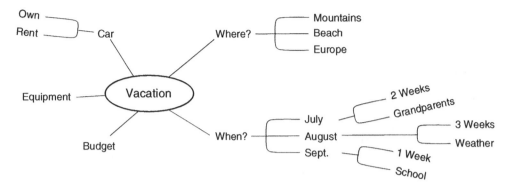

Figure 5.3 Example mind map.

The mind mapping (MM) approach to problem solving is useful for developing an SMM. MM is a methodology for visualizing and modeling an idea multidimensionally by linking the various associated thoughts and ideas that are relevant (see Refs. 4 and 5). It is essentially a brainstorming method for creating and organizing thought patterns. MM integrates the left and right spheres of the brain, which ultimately produces a better analysis result. MM stimulates ideas, provides for clustering of ideas and thoughts, and ties related ideas together. MM can be used for problem solving, outlines, taking notes, writing documents, preparing to-do lists, and so on. Reference 6 states that Boeing created a 25-foot-long mind map summarizing an aircraft engineering manual, which helped save millions of dollars of staff time.

Figure 5.3 shows an example of a mind map of brainstorming ideas for a family vacation.

A mind map is a diagram that is used to visually portray the relationships and connection between ideas, words, and other items around a central idea or keyword. The basic steps in the MM process include the following:

- Establish the central focus on a few short key words.
- Place the central focus of analysis in a bubble in the center of the page.
- Develop subtopics emanating from the central focus.
- Build the diagram by developing and connecting subtopics, sub-subtopics, and so on.
- Ideas are allowed to flow freely without judgment outward from the bubble.
- Key words are used to represent ideas and are expanded as desired.
- Key ideas are connected with lines and color can be utilized for emphasis.

Research shows that the mind's attention span is short, between 5 and 7 min. The mind works best in these short bursts and MM takes advantage of this tendency. MM trains the brain to see the whole picture along with the details, and to integrate logic and imagination. MM applies a nonlinear thought process that is not restrictive like other methods, such as outlining. It allows the mind to explore patterns and associations; it allows for more flexibility and thought. In some respects, the MM approach to developing an SMM is very much like fault tree analysis (FTA) in visually modeling logical SFPs, except without the rigid methodology requirements of FTA.

Figure 5.4 shows a mind map adapted to the SMM concept, utilizing TLMs as a major analysis focus. The TLMs are clustered around the system bubble. Each TLM is then evaluated for system aspects, such as mode, phase, failure modes, and so on, which will lead to hazards.

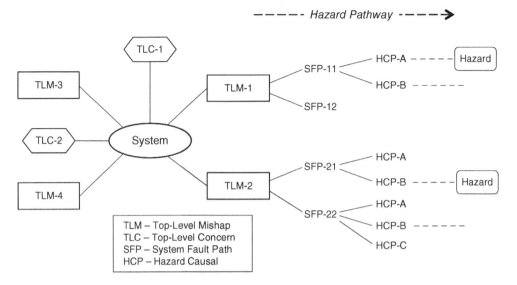

Figure 5.4 *SMM concept using the mind map modeling technique.*

Assuming the basic system design and operation are clearly understood, the basic steps to be taken in the SMM process include the following:

1. Identify the top undesired outcomes for the system; these include the top-level mishaps and the top-level concerns (TLCs).
2. Place the TLMs and TLCs on the SMM diagram and link them directly to the main bubble in mind map.
3. For each TLM and TLC, begin to identify the next item in the system design that would be of concern or a causal factor.
4. Continue developing the SFPs as far as possible.
5. Follow each path backwards going through functions, hardware, and components until discrete hazards can be clearly identified.
6. When the hazard paths are complete enough, the actual hazard analysis can proceed.

A TLM is a generic mishap category for collecting together various hazards that share the same general outcome or type of mishap. A TLM is a common mishap outcome that can be caused by one or more hazards; it serves as a collection point for all the hazards that can result in the same outcome but have different causal factors.

A TLC is a top-level safety concern that requires further investigation. It may eventually turn out to be a TLM or it may be revised into a different format after more is known about it and the SFPs. For the moment, it is a placeholder of something that is of significant safety concern.

The system mishap model is a model of the *hazard space* within a system. It is not an HA, but rather a map of the TLMs and TLC, along with their associated hazard paths drawn into the system design. This visual map of the hazard space aids in performing an HA.

Figure 5.5 shows an example SMM application using the mind mapping approach. This example involves the generic operation of an unmanned air vehicle (UAV).

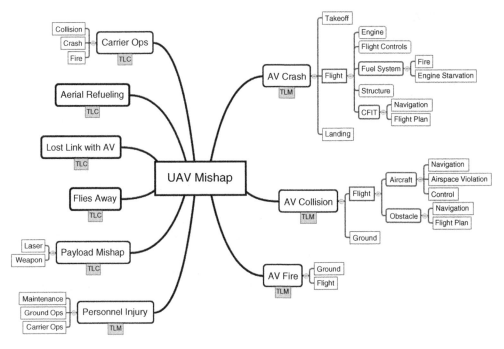

Figure 5.5 Example SMM of a UAV.

The overall generic hazard analysis process, utilizing an SMM, is depicted in Figure 5.6. This diagram shows the types of information that feed into, and is utilized by, the SMM and hazard analysis.

Another advantage of the SMM model is that the TLMs in the SMM can be used for developing a safety case for the system. Each TLM can be traced out to identify the hazards and any associated evidence mitigating the hazards to an acceptable level of risk.

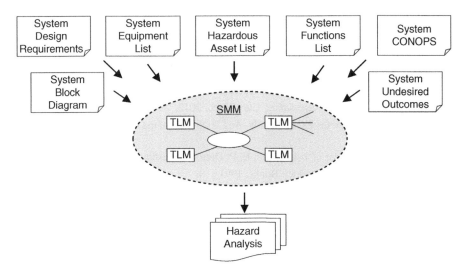

Figure 5.6 Hazard analysis using the mind mapping SMM.

5.8 SUMMARY

The following are the basic principles that help summarize the discussion in this chapter:

1. Most hazard analyses have a common set of steps.
2. Tools and aids should be used in hazard recognition.
3. An often overlooked step in hazard analysis is planning the overall analysis process prior to performing the analysis.
4. When systems are large and complex, it can be difficult to see the trees (i.e., hazards) for the forest (i.e., system). In these situations, it is important to utilize system mishap models that outline the forest by linking top-level mishaps with hazard paths and hazards.
5. When performing a hazard analysis, the use of hazard checklists is invaluable; however, a hazard checklist should never be considered a complete and final list, but merely a catalog for stimulating hazard ideas.
6. Do not exclude any thoughts, ideas, or concerns when postulating hazards. In addition to identifying real hazards, it is also important to show that certain hazards were suspected and considered, even though they were later found to be not possible for various design reasons. This provides evidence that all possibilities were considered.
7. Write a full, credible, and meaningful hazard description that is understandable to the reader and not just to the analyst. Do not assume the reader understands the hazard from a brief, abbreviated statement filled with project unique terms and acronyms. Avoid a hazard description that is a pseudo-hazard.
8. When writing a hazard description, remember to identify the three elements comprising a hazard within the description:
   - Hazard source (source)
   - Initiating mechanism (mechanism)
   - Target/threat outcome (outcome)

REFERENCES

The following are some references regarding hazard theory and hazard analysis:

1. C. Ericson, *Hazard Analysis Primer*, CreatSpace, 2012.
2. C. Ericson, *Clif's Notes on System Safety*, CreatSpace, 2012.
3. Top Level Mishap, Clif's Notes #12 in the July–Aug 2006 issue of the Journal of System Safety (JSS), Vol. 42 No. 4.
4. Tony Buzan, *Use Both Sides of Your Brain*, E. P. Dutton Publishing, 1974.
5. Joyce Wycoff, *Mindmapping*, Berkley Books, 1991.
6. T. Buzan and B. Buzan, *The Mind Map Book: How to use Radiant Thinking to Maximize Your Brain's Untapped Potential*, Plume Publishing, 1996.

Chapter *6*

Functional Hazard Analysis

6.1 FHA INTRODUCTION

Functional hazard analysis (FHA) is a tool for identifying hazards through the rigorous evaluation of system and/or subsystem functions including software. Systems are designed to perform a series of functions, which can be broken into subfunctions, sub-subfunctions, and so on. Functional objectives are usually well understood even when detailed design details are not available or understood. FHA is an inductive hazard analysis approach (inductively determines the effect of fault events) that evaluates the functional failure, corruption, and malfunction of functions.

6.2 FHA BACKGROUND

The primary focus of this technique is on the analysis of system functions. Either it could be performed during preliminary design in support of a preliminary design hazard analysis type (PD-HAT) or it could be performed during detailed design in support of a system design hazard analysis type (SD-HAT). Within the safety community, the acronym FHA has been used for both functional hazard analysis and fault hazard analysis.

The purpose of FHA is to identify system hazards by analyzing functions. Functions are the means by which a system operates to accomplish its mission or goals. System hazards are identified by evaluating the safety impact of a function failing to operate, operating incorrectly, or operating at the wrong time. When a function's failure can be determined as hazardous, the casual factors of the malfunction should be investigated in greater detail.

FHA is applicable to the analysis of all types of systems, equipment, and software. FHA can be implemented on a single subsystem, a complete functional system, or an integrated set of

Hazard Analysis Techniques for System Safety, Second Edition. Clifton A Ericson, II.
© 2016 John Wiley & Sons, Inc. Published 2016 by John Wiley & Sons, Inc.

systems. The level of analysis detail can vary, depending upon the level of functions being analyzed. For example, the analysis of high level system functions will result in a high-level hazard analysis, whereas the analysis of low-level (detailed design) subsystem functions will yield a more detailed functional analysis.

The technique, when methodically applied to a given system by experienced safety personnel, is thorough in identifying system functional hazards. Through a logical analysis of the way a system is functionally intended to operate, the FHA provides for the identification of hazards early in the design process. A basic understanding of system safety concepts and experience with the particular type of system is essential to create a correct list of potential hazards. The technique is uncomplicated and easily learned. Standard, easily followed FHA forms and instructions are provided in this chapter.

FHA is a powerful, efficient, and comprehensive system safety analysis technique for the discovery of hazards. It is especially powerful for the safety assessment of software. After a functional hazard is identified, further analysis of that hazard may be required to determine the specific causal factors of the functional hazards, such as hardware failures, human error, and so on. Since the FHA focuses on functions, it might overlook other types of hazards, such as those dealing with hazardous energy sources, sneak circuit paths, hazardous material, and so on. For this reason, the FHA should not be the sole hazard analysis performed but should be done in support of other types of hazard analysis, for example, PHA or SSHA.

6.3 FHA HISTORY

The exact history of this technique is unknown. The Technique seems to have naturally evolved over the years for the early analysis of systems when functional design information is available prior to the development of a detailed design. The FAA required FHA as part of the analysis of commercial aircraft and is documented in SAEARP4761, Guidelines and Methods for Conducting the Safety Assessment Process on Civil Airborne Systems and Equipment (Appendix A: Functional Hazard Assessment, 1996). In 2012, MIL-STD-882E formally incorporated FHA into the system safety process.

6.4 FHA THEORY

Figure 6.1 shows an overview of the basic FHA process and summarizes the important relationships involved. This process involves the evaluation of system functions for the identification and mitigation of hazards.

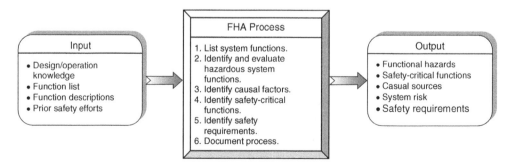

Figure 6.1 *FHA overview.*

Input information for the FHA consists of all design information relating to functional system operation. Typically, the following types of information are available and utilized in the FHA:

1. System design and operation information
2. A developed list of all system functions
3. Information from the PHL, PHA, and SSHA (if previously performed)
4. Functional flow diagrams of system operation
5. Hazard checklists (hazardous functions, tasks, etc.)

The primary purpose of the FHA is to identify and mitigate hazards resulting from the malfunction or incorrect operation of system functions. As such, the following information is typically output from the FHA:

1. Functional hazards
2. The identification of safety-critical functions
3. Hazard causal factors (failures, design errors, human errors, etc.)
4. Risk assessment
5. Safety requirements to mitigate the hazard

6.5 FHA METHODOLOGY

Table 6.1 lists and describes the basic steps in the FHA process. This process involves performing a detailed analysis focused on system functions.

Figure 6.2 depicts the overall FHA methodology. A key element of this methodology is to identify and understand all system functions. A function list must be created, and the use of

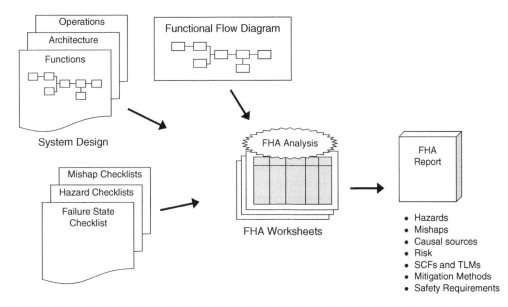

Figure 6.2 *FHA methodology.*

TABLE 6.1 FHA Process

Step	Task	Description
1	Define operation	Define the scope and boundaries of the operation to be performed. Understand the operation and its objective
2	Acquire data	Acquire all the necessary design and operational data needed for the analysis. This includes both schematics and manuals
3	List functions	Make a detailed list of all the functions to be considered in the FHA. This might be taken directly from a design document that is already written. It is important that all functions are considered
4	Conduct FHA	Perform the FHA on each item in the function list. This involves evaluating the effect of each functional failure mode. It is important that all functional failure modes be identified, along with the mission phase under analysis. Utilize existing hazard analyses results to assist in identifying hazards. Also, use hazard checklists to assist in hazard recognition. Identify hazards and any existing design features to eliminate or mitigate the hazard
5	Evaluate system risk	Identify the level of mishap risk presented by the identified hazards
6	Identify safety-critical functions	Based on level of risk criticality, identify those functions that are considered safety-critical
7	Recommend corrective action	Recommend corrective action for hazards with unacceptable risk. Develop derived safety requirements to mitigate identified hazards. Also, identify safety features already in the design or procedures for hazard mitigation
8	Monitor corrective action	Review design requirements to ensure that corrective action is being implemented
9	Track hazards	Transfer identified hazards into the hazard tracking system (HTS)
10	Document FHA	Document the entire FHA process on the worksheets. Update for new information and closure of assigned corrective actions

1.	Fails to operate
2.	Operates incorrectly/erroneously
3.	Operates inadvertently
4.	Operates at wrong time (early, late)
5.	Unable to stop operation
6.	Receives erroneous data
7.	Sends erroneous data
8.	Conflicting data or information

Figure 6.3 *Example of hazard checklist for failure states.*

functional flow diagrams is recommended because they provide an invaluable aid to the analysis. Checklists are applied against the function list to help identify hazards.

Figure 6.3 provides an example checklist of common failure states, which used for the analysis of functions. Each of the system functions should be evaluated for the effect of the failure state on the system. Since this is a partial list, the safety analyst should expand this list from experience and system knowledge.

6.6 FHA WORKSHEETS

It is desirable to perform the FHA analysis using a worksheet. The worksheet will help to add structure and rigor to the analysis, help record the process and data, and help support

System: ① Subsystem: ②			Functional Hazard Analysis				Analyst: ③ Date: ④		
Function	Hazard No.	Hazard	Effect	Causal Factors	IMRI	Recommended Action	FMRI	Comments	Status
⑤	⑥	⑦	⑧	⑨	⑩	⑪	⑫	⑬	⑭

Page: 1 of n

Figure 6.4 *Recommended FHA worksheet.*

justification for the identified hazards. The format of the analysis worksheet is not critical, and typically columnar-type worksheets are utilized.

The following basic information should be obtained from the FHA analysis worksheet:

1. Hazards
2. Hazard effects (mishaps)
3. Hazard causal factors (to subsystem identification)
4. Safety-critical factors or parameters
5. Risk assessment (before and after design safety features are implemented)
6. Derived safety requirements for eliminating or controlling the hazards

A recommended FHA analysis worksheet is shown in Figure 6.4. This particular FHA analysis worksheet utilizes a columnar-type format. Other worksheet formats may exist because different organizations often tailor their FHA analysis worksheet to fit their particular needs. The specific worksheet to be used may be determined by the system safety program (SSP), system safety working group, or the FHA customer.

The following instructions describe the information required under each column entry of the FHA worksheet:

1. *System* This column identifies the system under analysis.
2. *Subsystem* This column identifies the subsystem under analysis.
3. *Analyst* This column identifies the analyst performing the analysis.
4. *Date* This column identifies the date of the analysis.
5. *Function* This column identifies the design function. List and describe each of the system functions to be performed. If possible, include the purpose and the mode or phase of operation being performed.
6. *Hazard Number* This column identifies the number assigned to the identified hazard in the FHA (e.g., FHA-1, FHA-2, etc.). This hazard number is for future reference to the

particular hazard source and will be recorded in the associated hazard action record (HAR).

7. *Hazard*   This column identifies the specific hazard being postulated and evaluated for the stated functional failure. (Remember: Document all hazard considerations, even if they are later proven to be nonhazardous). Generally, a hazard is identified by considering the effect of function failure, erroneous function, incorrect function timing, and so on.

8. *Effect*   This column identifies the effect and consequences of the hazard, should it occur. Generally, the worst-case mishap result is the stated effect.

9. *Causal Factors*   This column identifies the causal factors involved in causing the functional failure and in causing the final effect resulting from the failure.

10. *Initial Mishap Risk Index (IMRI)*   This column provides a qualitative measure of mishap risk for the potential effect of the identified hazard, given that no mitigation techniques are applied to the hazard. Risk measures are a combination of mishap severity and probability, and the recommended values from MIL-STD-882 are shown below.

Severity	Probability
1: Catastrophic	A: Frequent
2: Critical	B: Probable
3: Marginal	C: Occasional
4: Negligible	D: Remote
	E: Improbable

11. *Recommended Action*   This column establishes recommended preventive measures to eliminate or control identified hazards. Safety requirements in this situation generally involve the addition of one or more barriers to keep the energy source away from the target. The preferred order of precedence for design safety requirements is as shown below.

Order of Precedence
1: Eliminate hazard through design selection
2: Control hazard through design methods
3: Control hazard through safety devices
4: Control hazard through warning devices
5: Control hazard through procedures and training

12. *Final Mishap Risk Index (FMRI)*   This column provides a qualitative measure of mishap risk significance for the potential effect of the identified hazard, given that mitigation techniques and safety requirements are applied to the hazard. The same metric definitions used in column 10 are also used here.

13. *Comments*   This column provides a space to record useful information regarding the hazard or the analysis process that is not noted elsewhere.

TABLE 6.2   Basic Aircraft Functions

No.	Function	Failure Condition
1	Control flight path	Inability to control flight path
2	Control touchdown and rollout	Inability to control touchdown and rollout
3	Control thrust	Inability to control thrust
4	Control cabin environment	Inability to control cabin environment
5	Provide spatial orientation	Inability to control to provide spatial orientation
6	Fire protection	Loss of fire protection

14. *Status*   This column states the current status of the hazard, as either being open or closed. This follows the hazard tracking methodology established for the program. A hazard can be closed only when it has been verified through analysis, inspection, and/or testing that the safety requirements are implemented in the design and successfully tested for effectiveness.

Note that in this analysis methodology, *every system function* is listed and analyzed. For this reason, not every entry in the FHA form will constitute a hazard, since not every function is hazardous. The analysis documents, however, that all functions were considered by the FHA. Note also that the analysis becomes a traceability matrix, tracing each function and its safety impact.

In filling out the columnar FHA form, the dynamic relationship between the entries should be kept in mind. The hazard, cause, and effect columns should completely describe the hazard. These columns should provide the three sides of the hazard triangle: source, mechanism, and outcome. Also, the FHA can become somewhat of a living document that is continually being updated as new information becomes available.

6.7   FHA EXAMPLE 1: AIRCRAFT FLIGHT FUNCTIONS

Table 6.2 provides a list of aircraft functions derived from the initial design concept for the aircraft system. This table also identifies the major failure condition of concern for each of the functions. These high-level functions are analyzed by FHA.

Tables 6.3 and 6.4 contain the FHA worksheets for this example.

6.8   FHA EXAMPLE 2: AIRCRAFT LANDING GEAR SOFTWARE

Figure 6.5 provides an example generic landing gear system that is computer driven using software. In this example, FHA analyzes the software functions.

The GDnB button is pressed to lower the landing gear and the GupB button is pressed to raise the gear. When the gear is up, switch S1 sends a true signal to the computer; otherwise, it sends a false signal. When the gear is down, switch S2 sends a true signal to the computer; otherwise, it sends a false signal. The purpose of the switches is for the system to know what position the landing gear is actually in, and prevent conflicting commands. S_{WOW} is the weight-on-wheels switch.

TABLE 6.3 FHA Example 1; Worksheet 1

System: Aircraft
Subsystem: Critical Functions

Functional Hazard Analysis

Analyst:
Date:

Function	Hazard No.	Hazard	Effect	Causal Factors	IMRI	Recommended Action	FMRI	Comments	Status
Control Flight Path (pitch and yaw)	F-1	Fails to occur, causing aircraft crash	Inability to control flight path (e.g., elevator hard over)	Loss of hydraulics Flight controls Software	1C			Safety-Critical Function	Open
	F-2	Occurs erroneously, causing aircraft crash	Elevator hard over	Software	1C				Open
Control Touchdown and Rollout	F-3	Fails to occur, causing aircraft crash	Inability to control flight path	Loss of hydraulics Flight controls Software	1C			Safety-Critical Function	Open
	---	Occurs erroneously, causing aircraft crash	Not applicable						Open
Control Thrust (engine speed and power)	F-4	Fails to occur, causing aircraft crash	Loss of aircraft thrust when needed	Engine hardware Software	1C			Safety-Critical Function	Open
	F-5	Occurs erroneously, causing aircraft crash	Incorrect aircraft thrust	Engine hardware Software	1C				Open

Page 1 of 2

100

TABLE 6.4 FHA Example 1; Worksheet 2

System: Aircraft
Subsystem: Critical Functions

Functional Hazard Analysis

Analyst:
Date:

Function	Hazard No.	Hazard	Effect	Causal Factors	IMRI	Recommended Action	FMRI	Comments	Status
Control Cabin Environment	F-6	Fails to occur, causing passenger becomes sick	Passenger comfort	Computer fault; Software	2D				Open
	F-7	Occurs erroneously, causing passenger becomes sick	Passenger comfort	Computer fault; Software	2D				Open
Provide Spatial Orientation	F-8	Fails to occur, causing aircraft crash	Pilot loses spatial orientation during critical flight	Computer fault; Software Displays fail	1C	Provide three independent displays		Safety-Critical Function	Open
	F-9	Occurs erroneously, causing aircraft crash	Pilot loses spatial orientation during critical flight	Computer fault; Software Displays fail	1C				Open
Fire Protection	F-10	Fails to occur, causing aircraft crash	Unable to extinguish onboard fire	Computer fault; Software;	1C			Safety-Critical Function	Open
	F-11	Occurs erroneously, causing equipment damage	Equipment damage	Computer fault; Software Displays fail	3C				Open

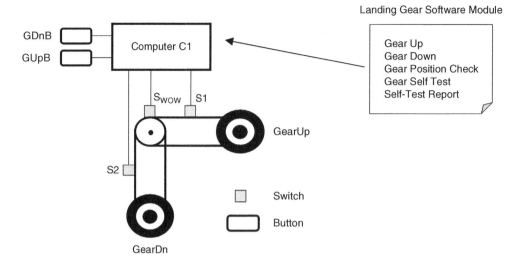

Figure 6.5 *Example aircraft landing gear system.*

The major software functions for the landing gear are listed in Figure 6.5. These functions have been identified early in the design, but have not yet been actually developed. Even though software code modules are yet to be developed to perform these functions, they can still be analyzed by FHA.

Tables 6.5 and 6.6 contain the FMEA worksheets evaluating the functions involved in the aircraft landing gear.

6.9 FHA EXAMPLE 3: ACE MISSILE SYSTEM

In order to demonstrate the FHA methodology, the same hypothetical small missile system from Chapters 7 and 8 will be used. The basic system design information provided in the PHA is shown again in Figure 6.6.

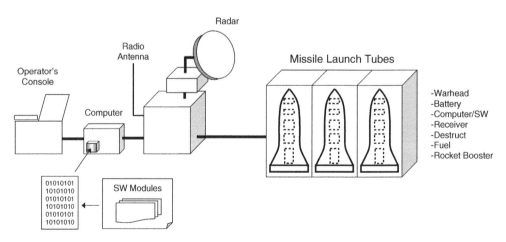

Figure 6.6 *Ace Missile System.*

TABLE 6.5   FHA Example 2; Worksheet 1

System: Aircraft
Subsystem: Landing Gear Software Functions

Functional Hazard Analysis

Analyst:
Date:

Function	Hazard No.	Hazard	Effect	Causal Factors	IMRI	Recommended Action	FMRI	Comments	Status
Gear Up	F-1	Fails to issue after takeoff, causing damage from drag	Aircraft flight with gear down	Computer; Software; Wiring	2C				Open
	F-2	Issues prematurely, causing aircraft damage during taxi	Gear raises during taxi or takeoff	Computer; Software	2C				Open
Gear Down	F-3	Fails to issue for landing, causing damage or injury during landing	Unable to lower landing gear and aircraft must land with landing gear up	Computer; Software; Wiring	1C	- Provide redundant design - Provide multiple sensors for software logic tests		Gear Down is Safety-Critical Function	Open
		Issues prematurely during flight, causing damage from drag	Aircraft flight with gear down	Computer; Software	2C				Open
Gear Position Check	F-4	Function fails, resulting in unknown gear position, causing system to not lower gear for landing	Unable to lower landing gear and aircraft must land with landing gear up	Computer; Software; Electronic faults	1C				Open
	F-5	Function error that reports incorrect gear position, causing system to not lower gear for landing	Unable to lower landing gear and aircraft must land with landing gear up	Computer; Software; Electronic faults	1C				Open

Page 1 of 2

TABLE 6.6   FHA Example 2; Worksheet 2

System: Aircraft
Subsystem: Landing Gear Software Functions

Analyst:
Date:

Functional Hazard Analysis

Function	Hazard No.	Hazard	Effect	Causal Factors	IMRI	Recommended Action	FMRI	Comments	Status
Gear Self Test	F-6	Self Test fails, resulting in gear fault not reported prior to takeoff, causing system to not lower gear for landing	Landing gear fault not detected. Unable to raise or lower gear when desired, no warning	Computer; Software; Electronic faults	1C				Open
	F-7	Self Test error, resulting in gear fault erroneously reported prior to takeoff, causing unnecessary maintenance delay	Incorrect landing gear status reported	Computer; Software; Electronic faults	3C				Open
Self Test Report	F-7	Self Test Report fails, resulting in gear fault not reported prior to takeoff, causing system to not lower gear for landing	BIT data not reported. No warning of fault state.	Computer; Software; Electronic faults; Memory fault	1C			System may still be correctly operational.	Open
	F-8	Self Test Report error, resulting in gear fault erroneously reported prior to takeoff, causing unnecessary maintenance delay	BIT data reported incorrectly. No warning of fault state, or incorrect warning	Computer; Software; Electronic faults; Memory fault	3C			System may still be correctly operational.	Open

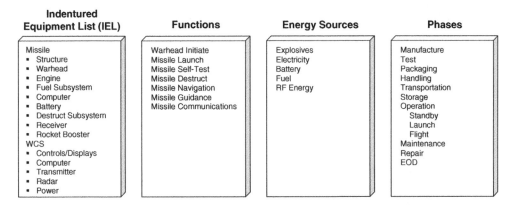

Figure 6.7   *Missile system component list and function list.*

Figure 6.7 lists the major system components, functions, phases, and energy sources that should be considered for any system analysis. The major segments of the system are the missile and the weapon control system (WCS). FHA will analyze the functions in this chart.

The FHA worksheets for the Ace Missile System are shown in Tables 6.7 and 6.8.

6.10   FHA ADVANTAGES AND DISADVANTAGES

The following are advantages of the FHA technique:

1. Is easily and quickly performed.
2. Does not require considerable expertise.
3. Is relatively inexpensive, yet provides meaningful results.
4. Provides rigor for focusing on hazards associated with system functions.
5. Is a good tool for software safety analysis.

The FHA technique has the following disadvantages:

1. Since the technique focuses on functions, it might overlook other types of hazards, such as those dealing with hazardous energy sources or sneak circuit paths.
2. After a functional hazard is identified, further analysis is required to identify the specific causal factors.

6.11   COMMON FHA MISTAKES TO AVOID

When first learning how to perform an FHA, it is commonplace to commit some typical errors. The following is a list of common errors made during the conduct of an FHA.

1. Each system function is not evaluated and documented.
2. The hazard description is incomplete, ambiguous, or too detailed.
3. Causal factors are not adequately identified or investigated.
4. The MRI is not stated or is incomplete.

TABLE 6.7 Ace Missile System FHA; Worksheet 1

System: Ace Missile System
Subsystem: System Functions

Analyst:
Date:

Functional Hazard Analysis

Function	Hazard No.	Hazard	Effect	Causal Factors	IMRI	Recommended Action	FMRI	Comments	Status
W/H Arm-1	F-1	Missile W/H Arm-1 function occurs Inadvertently	Inadvertent W/H initiation	Faults cause inadvertent missile W/H Arm-1 function	1C	Design for multiple events being required before initiation can occur (i.e., Arm-1 and Arm-2 and Power).	1E		Open
	F-2	Missile W/H Arm-1 function fails to occur	Unable to initiate W/H	Faults cause failure of missile W/H Arm-1 function	4E		4E	Dud weapon; not a safety concern	Closed
W/H Arm-2	F-3	Missile W/H Arm-2 function occurs Inadvertently	Inadvertent W/H initiation	Faults cause inadvertent missile W/H Arm-2 function	1C	Design for multiple events being required before initiation can occur (i.e., Arm-1 and Arm-2 and Power).	1E	Dud weapon; not a safety concern	Open
	F-4	Missile W/H Arm-2 function fails to occur	Unable to initiate W/H	Faults cause failure of missile W/H Arm-2 function	4E		4E		Closed
Missile Launch	F-5	Inadvertent missile launch function occurs.	Inadvertent missile launch	Faults cause inadvertent missile launch signal	1C	- Review software code - Design for safe HMI - Launch must require multiple design events	1E		Open
	F-6	Missile launch function fails to occur when intended	Missile battery may have been activated; unsafe missile state	Faults prevent missile launch when intended	2C	- Use redundant design	2E		Open
	F-7	Incorrect missile is launched	Inadvertent missile launch	Incorrect missile is selected and launched	1C	- Design for safe HMI - Review code for software safety	1E		Open

Page 1 of 2

106

TABLE 6.8 Ace Missile System FHA; Worksheet 2

System: Ace Missile System
Subsystem: System Functions

Functional Hazard Analysis

Analyst:
Date:

Function	Hazard No.	Hazard	Effect	Causal Factors	IMRI	Recommended Action	FMRI	Comments	Status
Missile Self Test	F-8	Missile launch Self-Test function fails, resulting in unknown missile status.	Unsafe missile state	Faults cause erroneous missile data to WCS operator and system	2C	- Use redundant design - Design for safe HMI	2E		Open
Missile Destruct	F-9	Missile destruct function inadvertently occurs.	Missile strikes undesired target	Faults cause inadvertent missile destruct signal	1C	- Launch must require multiple events - Review software code - Design for safe HMI	1E	Dud weapon; not a safety concern	Closed
	F-10	Missile destruct function fails to occur when requested	Missile fails to destruct when necessary to avoid undesired target	Faults prevent missile destruct when destruct has been elected	1C	- Use redundant design	1E		Open
Missile Navigation	F-11	Navigation function error occurs, causing undesired target strike, resulting in death/injury.	Missile strikes undesired target	Faults cause incorrect missile navigation, resulting in striking undesired target	1C	- Design for safe HMI - Review code for software safety	1E		Open
Missile Guidance	F-12	Guidance function error occurs, causing undesired target strike, resulting in death/injury.	Missile striking undesired target	Faults cause incorrect missile guidance, resulting in striking undesired target	1C	- Use redundant design - Design for safe HMI - Analyze guidance system	1E		Open

Page 2 of 2

107

5. The hazard mitigation method is insufficient for hazard risk.
6. The hazard is closed prematurely or incorrectly.
7. Modes other than operation are often overlooked, for example, maintenance, training, and test.

6.12 FHA SUMMARY

This chapter discussed the FHA technique. The following are the basic principles that help summarize the discussion in this chapter:

1. FHA is a qualitative analysis tool for the evaluation of system functions.
2. The primary purpose of FHA is to identify functions that can lead to the occurrence of an undesired event or hazard.
3. FHA is useful for software analysis.
4. The use of a functional block diagram greatly aids and simplifies the FHA process.

FURTHER READINGS

Ericson, C., *Hazard Analysis Primer*, II, CreatSpace, 2012.

SAE ARP4761, Guidelines and Methods for Conducting the Safety Assessment Process on Civil Airborne Systems and Equipment, Appendix A: Functional Hazard Assessment, 1996.

Chapter 7

Preliminary Hazard List Analysis

7.1 PHL INTRODUCTION

Preliminary hazard list (PHL) analysis technique is used for identifying and listing potential hazards and mishaps that may exist in a system. The PHL is performed during conceptual or preliminary design, and is the starting point for all subsequent hazard analyses. Once a hazard is identified in the PHL, the hazard will be used to launch in-depth hazard analyses and evaluations, as more system design details become available. The PHL is a means for management to focus on hazardous areas that may require more resources to eliminate the hazard or control risk to an acceptable level. Every hazard identified on the PHL will be analyzed with more detailed analysis techniques.

7.2 PHL BACKGROUND

This technique falls under the conceptual design hazard analysis type (CD-HAT). The PHL evaluates design at the conceptual level, without detailed information, and it provides a preliminary list of hazards. The analysis types are described in detail in Chapter 4.

This analysis technique falls under the preliminary design hazard analysis type (PD-HAT) because it evaluates design at the preliminary level without detailed information. The primary purpose of the PHL is to identify and list potential system hazards. A secondary purpose of the PHL is to identify safety-critical (SC) parameters and mishap categories. The PHL analysis is usually performed very early in the design development process and prior to performing any other hazard analysis. The PHL is used as a management tool to allocate resources to particularly hazardous areas within the design, and it becomes the foundation for all other

Hazard Analysis Techniques for System Safety, Second Edition. Clifton A Ericson, II.
© 2016 John Wiley & Sons, Inc. Published 2016 by John Wiley & Sons, Inc.

subsequent hazard analyses performed on the program. A follow-up on hazard analyses will evaluate these hazards in greater detail as the design detail progresses. The intent of the PHL is to affect the design for safety as early as possible in the development program.

The PHL can be applied to any type of system at the conceptual or preliminary stage of development. The PHL can be performed on a subsystem, a single system, or an integrated set of systems. The PHL is generally based on preliminary design concepts and is usually performed early in the development process, sometimes during the proposal phase or immediately after contract award in order to influence design and mishap risk decisions as the design is formulated and developed.

The technique, when applied to a given system by experienced system safety personnel, is thorough at identifying high-level system hazards and generic hazards that may exist in a system. A basic understanding of hazard theory is essential, so is the knowledge of system safety concepts. Experience with the particular type of system under investigation, and its basic components, is necessary in order to identify system hazards. The technique is uncomplicated and easily learned. Typical PHL forms and instructions are provided in this chapter.

The PHL technique is similar to a brainstorming session, whereby hazards are postulated and collated in a list. This list is then the starting point for subsequent hazard analyses, which will validate the hazard and begin the process of identifying causal factors, risk, and mitigation methods. Generating a PHL is a prerequisite to performing any other type of hazard analysis. The use of this technique is highly recommended. It is the starting point for a more detailed hazard analysis and safety tasks, and it is easily performed.

7.3 PHL HISTORY

The technique was established very early in the history of the system safety discipline. It was formally instituted and promulgated by the developers of MIL-STD-882.

7.4 PHL THEORY

The PHL analysis is a simple and straightforward analysis technique that provides a list of known and suspected hazards. A PHL analysis can be either as simple as conducting a hazard brainstorming session on a system or a slightly more structured process that helps ensure that all hazards are identified. The PHL method described here is a process with some structure and rigor, with the application of a few basic guideline rules.

This analysis should involve a group of engineers/analysts with expertise in a variety of specialized areas. The methodology described herein can be used by an individual analyst or a brainstorming group to help focus on the analysis. The recommended methodology also provides a vehicle for documenting the analysis results on a worksheet.

Figure 7.1 shows an overview of the basic PHL process and summarizes the important relationships involved in the PHL process. This process consists of combining design information with known hazard information to identify hazards. Known hazardous elements and mishap lessons learned are compared with the system design to determine if the design concept utilizes any of these potential hazard elements.

In order to perform the PHL analysis, the system safety analyst must have two things – design knowledge and hazard knowledge. Design knowledge means the analyst must possess a

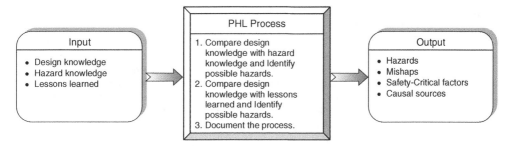

Figure 7.1 *Preliminary hazard list overview.*

basic understanding of the system design, including a list of major components. Hazard knowledge means the analyst needs to have a basic understanding of hazards, hazard sources, hazard components, and hazards in similar systems. Hazard knowledge is primarily derived from hazard checklists and from lessons learned on the same or similar systems and equipment.

In performing the PHL analysis, the analyst compares the design knowledge and information with hazard checklists. This allows the analyst to visualize or postulate possible hazards. For example, if the analyst discovers that the system design will be using jet fuel, he then compares jet fuel to a hazard checklist. From the hazard checklist, it will be obvious that jet fuel is a hazardous element and that a jet fuel fire/explosion is a potential mishap with many different ignition sources presenting many different hazards.

The primary output from the PHL is a list of hazards and the hazard sources that spawn them. It is also necessary and beneficial to collect and record additional information, such as the prime hazard causal factors (e.g., hardware failure, software error, human error, etc.), the major mishap category for the hazard (e.g., fire, inadvertent launch, physical injury, etc.), and any safety-critical factors that will be useful for subsequent analysis (e.g., SC function, SC hardware item, etc.).

7.5 PHL METHODOLOGY

Table 7.1 lists and describes the basic steps of the PHL process and summarizes the important relationships involved. A worksheet is utilized during this analysis process.

The PHL process begins by acquiring design information in the form of the design concept, the operational concept, major components planned for use in the system, major system functions, and software functions. Sources for this information could include statement of work (SOW), design specifications, sketches, drawings, or schematics. Additional design integration data can be utilized to better understand, analyze, and model the system. Typical design integration data includes functional block diagrams, indentured equipment lists (IEL) (e.g., work breakdown structure (WBS), reliability block diagrams, and concept of operations). If the design integration data is not available, the safety analyst may have to make assumptions in order to perform the PHL analysis. All assumptions should be documented.

The next step in the PHL analysis is to acquire the appropriate "hazard checklists." Hazard checklists are generic lists of items known to be hazardous or might create potentially hazardous designs or situations. The hazard checklist should not be considered complete

TABLE 7.1 PHL Analysis Process

Step	Task	Description
1	Define system	Define the scope and boundaries of the system. Define the mission, mission phases and mission environments. Understand the system design operational concepts and major system components
2	Plan PHL	Establish PHL goals, definitions, worksheets, schedule, and process. Identify system elements and functions to be analyzed
3	Select team	Select all team members to participate in PHL and establish responsibilities. Utilize team member expertise from several different disciplines (e.g., design, test, manufacturing, etc.)
4	Acquire data	Acquire all the necessary design, operational and process data needed for the analysis (e.g., equipment lists, functional diagrams, operational concepts, etc.). Acquire hazard checklists, lessons learned and other hazard data applicable to the system
5	Conduct PHL	a. Construct list of hardware components and system functions b. Evaluate conceptual system hardware; compare with hazard checklists c. Evaluate system operational functions; compare with hazard checklists d. Identify and evaluate system energy sources to be used; compare with energy hazard checklists e. Evaluate system/software functions; compare with hazard checklists f. Evaluate possible failure states
6	Build hazard list	Develop list of identified and suspected system hazards and potential system mishaps. Identify SCFs and TLMs if possible from information available
7	Recommend corrective action	Recommend safety guidelines and design safety methods that will eliminate or mitigate hazards
8	Document PHL	Document the entire PHL process and PHL worksheets in a PHL report. Include conclusions and recommendations

or all-inclusive. Hazard checklists help trigger the analyst's recognition of potential hazardous sources, from past lessons learned. Typical hazard checklists include the following:

a. Energy sources

b. Hazardous functions

c. Hazardous operations

d. Hazardous components

e. Hazardous materials

f. Lessons learned from similar type systems

g. Undesired mishaps

h. Failure mode and failure state considerations

When all the data is available, the analysis can begin. PHL analysis involves comparing the design and integration information with the hazard checklists. If the system design uses a known hazard component, hazardous function, hazardous operation, and the like, then a potential hazard exists. This potential hazard is recorded on the analysis form and then further evaluated with the level of design information that is available. Checklists also aid in the brainstorming process for new hazard possibilities brought about by the unique system design. The PHL output includes identified hazards, hazard causal factor areas (if possible), resulting mishap effect, and safety-critical factors (if any).

The PHL methodology is illustrated in Figure 7.2*a*. In this methodology, a system list is constructed, which identifies planned items in the hardware, energy sources, functions, and software categories. Items on the system list are then compared with items on the various safety checklists. Matches between the two lists trigger ideas for potential hazards, which are then compiled in the PHL.

The PHL methodology shown in Figure 7.2*a* is demonstrated by a brief example in Figure 7.2*b*. The system in this example involves the conceptual design for a new nuclear-powered aircraft carrier system. This example shows how to mentally link the system items to the checklist as a method for identifying hazards.

From the design and operational concept information, an IEL is constructed for the PHL (note that a master equipment list (MEL) can also be used). The equipment on the IEL is then compared with the hazard checklists to stimulate hazard identification. For example, *nuclear reactor* appears both on the IEL and on the hazardous energy source checklist. This match (1a) triggers the identification of one or more possible hazards, such as "reactor overtemperature." This hazard is then added to the PHL (1b) as hazard 1.

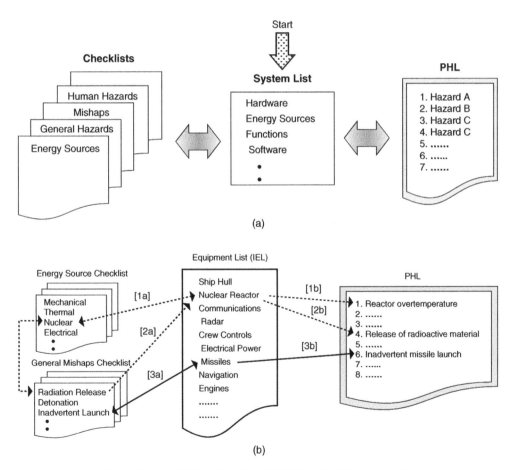

Figure 7.2 *(a) PHL methodology. (b) PHL methodology example.*

Nuclear reactor appears on the IEL and it also appears on the general mishaps checklist. This match (2a) triggers the identification of one or more possible hazards, "accidental release of radioactive material." This hazard is then added to the PHL (2b) as hazard 4.

Missiles appear on the IEL and inadvertent weapon launch appears on the general mishaps checklist. This match (3a) triggers the identification of "inadvertent missile launch" as a possible hazard, which is added to the PHL (3b) as hazard 6.

7.6 PHL WORKSHEET

It is desirable to perform the PHL analysis using a worksheet. The worksheet will help to add rigor to the analysis, help record the process and data, and help support justification for the identified hazards. The format of the analysis worksheet is not critical, and typically columnar-type worksheets are utilized.

The following basic information should be obtained from the PHL analysis worksheet:

- Hazard sources in the system
- Postulated hazards resulting from the hazard sources
- Top-level mishap (TLM) categories stemming from identified hazards
- Recommendations (such as safety requirements/guidelines that can be applied)

The primary purpose of a worksheet is to provide structure and documentation to the analysis process. The recommended PHL worksheet for system safety usage is shown in Figure 7.3.

In this PHL worksheet, the second column contains a list of system items, from which hazards can easily be recognized. For example, by listing all the system functions, hazards can be postulated by answering the questions "what if the function fails to occur" or "what if the function occurs inadvertently."

Preliminary Hazard List Analysis				
System Hazard Source Type: (1)				
No.	Hazard Source Item	Hazard	Hazard Effects	Comments
(2)	(3)	(4)	(5)	(6)

Figure 7.3 *PHL worksheet.*

The PHL worksheet columns are defined as follows:

1. *System Element Type*   This column identifies the type of system items under analysis, such as system hardware, system functions, system software, energy sources, and so on.
2. *Hazard No.*   This column identifies the hazard number for reference purposes.
3. *System Item*   This column is a subelement of data item 1 and identifies the major system items of interest in the identified category. In the example to follow, the items are first broken into categories of hardware, software, energy sources, and functions. Hazards are postulated through a close examination of each listed item under each category. For example, if explosives constitute an intended hardware element, then explosives would be listed under hardware and again under energy sources. There may be some duplication, but this allows for the identification of all explosives-related hazards.
4. *Hazard*   This column identifies the specific hazard that is created as a result of the indicated system item. (Remember: Document all potential hazards, even if they are later proven by other analyses to be nonhazardous in this application.)
5. *Hazard Effects*   This column identifies the effect of the identified hazard. The effect would be described in terms of resulting system operation, misoperation, death, injury, damage, and so on. Generally, the effect is the resulting mishap.
6. *Comments*   This column records any significant information, assumptions, recommendations, and so on resulting from the analysis. For example, safety-critical functions (SCFs), TLMs, or system safety design guidelines might be identified here.

7.7   HAZARD CHECKLISTS

Hazard checklists provide a common source for readily recognizing hazards. Since no single checklist is ever really adequate in itself, it becomes necessary to develop and utilize several different checklists. Utilizing several checklists may generate some repetition, but it will also result in improved coverage of hazardous elements.

Remember that a checklist should never be considered a complete and final list, but merely a mechanism or catalyst for stimulating hazard recognition. Refer to Appendix C for a more complete set of hazard checklists. To illustrate the hazard checklist concept, some example checklists are provided in Figures 7.4–7.8. These example checklists are not intended to

1. Fuels	12. Electrical generators
2. Propellants	13. R F energy sources
3. Initiators	14. Radioactive energy sources
4. Explosive charges	15. Falling objects
5. Charged electrical capacitors	16. Catapulted objects
6. Storage batteries	17. Heating devices
7. Static electrical charges	18. Pumps, blowers, and fans
8. Pressure containers	19. Rotating machinery
9. Spring-loaded devices	20. Actuating devices
10. Suspension systems	21. Nuclear
11. Gas generators	22. Cryogenics

Figure 7.4   *Example of hazard checklist for energy sources.*

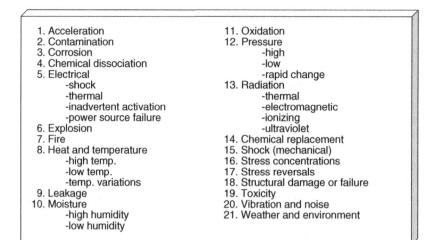

1. Acceleration	11. Oxidation
2. Contamination	12. Pressure
3. Corrosion	-high
4. Chemical dissociation	-low
5. Electrical	-rapid change
-shock	13. Radiation
-thermal	-thermal
-inadvertent activation	-electromagnetic
-power source failure	-ionizing
6. Explosion	-ultraviolet
7. Fire	14. Chemical replacement
8. Heat and temperature	15. Shock (mechanical)
-high temp.	16. Stress concentrations
-low temp.	17. Stress reversals
-temp. variations	18. Structural damage or failure
9. Leakage	19. Toxicity
10. Moisture	20. Vibration and noise
-high humidity	21. Weather and environment
-low humidity	

Figure 7.5 *Example of hazard checklist for general sources.*

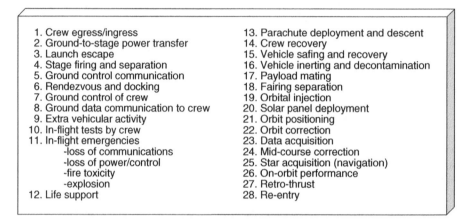

1. Crew egress/ingress	13. Parachute deployment and descent
2. Ground-to-stage power transfer	14. Crew recovery
3. Launch escape	15. Vehicle safing and recovery
4. Stage firing and separation	16. Vehicle inerting and decontamination
5. Ground control communication	17. Payload mating
6. Rendezvous and docking	18. Fairing separation
7. Ground control of crew	19. Orbital injection
8. Ground data communication to crew	20. Solar panel deployment
9. Extra vehicular activity	21. Orbit positioning
10. In-flight tests by crew	22. Orbit correction
11. In-flight emergencies	23. Data acquisition
-loss of communications	24. Mid-course correction
-loss of power/control	25. Star acquisition (navigation)
-fire toxicity	26. On-orbit performance
-explosion	27. Retro-thrust
12. Life support	28. Re-entry

Figure 7.6 *Example of hazard checklist for space functions.*

1. Welding
2. Cleaning
3. Extreme temperature operations
4. Extreme weight operations
5. Hoisting, handling, and assembly operations
6. Test chamber operations
7. Proof test of major components/subsystems/systems
8. Propellant loading/transfer/handling
9. High-energy pressurization/hydrostatic-pneumostatic testing
10. Nuclear component handling/checkout
11. Ordnance installation/checkout/test
12. Tank entry/confined space entry
13. Transport and handling of end item
14. Manned vehicle tests
15. Static firing

Figure 7.7 *Example of hazard checklist for general operations.*

1. Fails to operate.
2. Operates incorrectly/erroneously.
3. Operates inadvertently.
4. Operates at incorrect time (early, late).
5. Unable to stop operation.
6. Receives erroneous data.
7. Sends erroneous data.

Figure 7.8   *Example of hazard checklist for failure states.*

represent ultimate checklist sources, but some typical example checklists are used in recognizing hazards.

Figure 7.4 is a checklist of energy sources that are considered to be hazardous elements when used within a system. The hazard is generally from the various modes of energy release that are possible from hazardous energy sources. For example, electricity/voltage is a hazardous energy source. The various hazards that can result from undesired energy release include personnel electrocution, ignition source for fuels and/or materials, sneak path power for an unintended circuit, and so on.

Figure 7.5 contains a checklist of general sources that have been found to produce hazardous conditions and potential accidents, when the proper system conditions are present.

Figure 7.6 is a checklist of functions that are hazardous due to the critical nature of the mission. This checklist is an example particularly intended for space programs.

Figure 7.7 is a checklist of operations that are considered hazardous due to the materials used or due to the critical nature of the operation.

Figure 7.8 is a checklist of possible failure modes or failure states that are considered hazardous, depending on the critical nature of the operation or function involved.

This checklist is a set of key questions to ask regarding the state of the component, subsystem, or system functions. These are potential ways the subsystem could fail and thereby result in creating a hazard. For example, when evaluating each subsystem, answering the question "does *fail to operate* cause a hazard" may lead to the recognition of a hazard.

Note that when new hardware elements and functions are invented and used, new hazardous elements will be introduced requiring expanded and updated checklists.

7.8   PHL GUIDELINES

The following are some basic guidelines that should be followed when completing the PHL worksheet:

1. Remember that the objective of the PHL is to identify system hazards and/or mishaps.
2. The best approach is to start by investigating *system hardware items*, *system functions*, and *system energy sources*.
3. Utilize hazard checklists and lessons learned for hazard recognition.
4. A hazard write-up should be understandable but does not have to be detailed in description (i.e., the PHL hazard does not have to include all three elements of a hazard: hazardous element, initiating mechanisms, and outcome).

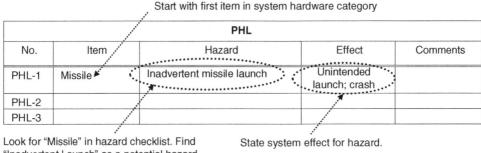

Start with first item in system hardware category

PHL				
No.	Item	Hazard	Effect	Comments
PHL-1	Missile	Inadvertent missile launch	Unintended launch; crash	
PHL-2				
PHL-3				

Look for "Missile" in hazard checklist. Find "Inadvertent Launch" as a potential hazard. Note simplified hazard write-up.

State system effect for hazard.

Figure 7.9   *PHL guidelines.*

Chapter 3 described the three components of a hazard: (1) hazardous element, (2) initiating mechanism, and (3) threat and target (outcome). Typically, when a hazard is identified and described, the hazard write-up description will identify and include all three components. However, in the PHL a complete and full hazard description is not always provided. This is primarily because of the preliminary nature of the analysis, and because all identified hazards are more fully investigated and described in the preliminary hazard analysis (PHA) and subsystem hazard analysis (SSHA).

Figure 7.9 shows how to apply the PHL guidelines when using the PHL worksheet.

7.9   PHL EXAMPLE: ACE MISSILE SYSTEM

In order to demonstrate the PHL methodology, a hypothetical small missile system will be analyzed. The basic system design is shown in Figure 7.10 for the Ace Missile System. The

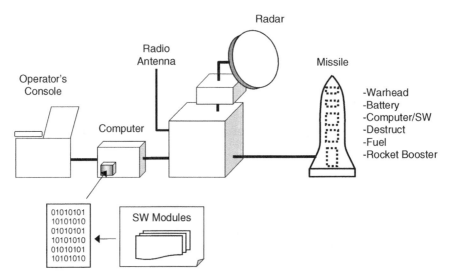

Figure 7.10   *Ace Missile System.*

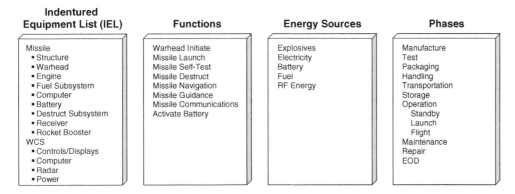

Figure 7.11   *Ace Missile System conceptual information.*

major segments of the system are the missile segment and the weapon control system (WCS) segment. The missile segment includes only those components specifically comprising the missile. The WCS segment includes those components involved in command and control over the missile, such as the operator's console, system computer, radar, system power, and so on.

The basic equipment and functions for this system are identified in Figure 7.11. During the conceptual design stage, this is the typical level of information that is available. Some basic design decisions may be necessary, such as the type of engine to be utilized, jet or solid rocket. A design safety trade study might be performed to evaluate the hazards of a jet system versus a rocket system. From this basic design information, a very credible list of hazards can easily be generated.

Figure 7.12 shows the basic planned operational phases for the missile system. As design development progresses, each of these phases will be expanded in greater detail.

The lists of components, functions, and phases are generated by the missile project designers or the safety analyst. The PHL begins by comparing each system component and function with hazard checklists, to stimulate ideas on potential hazards involved with this system design.

Tables 7.2–7.4 contain a PHL analysis of the system hardware, functions, and energy sources, respectively. For example, Table 7.2 evaluates system hardware starting with the first component in the IEL, missile body, the warhead, the engine, and so on. In this example, the PHL worksheet was developed as a single long table extending over several pages, but the worksheet could have been broken into many single pages.

The following results should be noted from the PHL analysis of the Ace Missile System:

1. A total of 40 hazards have been identified by the PHL analysis.
2. No recommended action resulted from the PHL analysis, only the identification of hazards. These hazards provide design guidance to the system areas that will present mishap risk and require further design attention for safety.

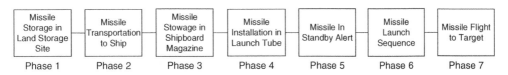

Figure 7.12   *Missile functional flow diagram of operational phases.*

TABLE 7.2 PHL Analysis of Ace Missile System: System Hardware Evaluation

	Preliminary Hazard List Analysis			
System Element Type: System Hardware				
No.	System Item	Hazard	Hazard Effects	Comments
PHL-1	Missile structure	Missile body breaks up resulting in fuel leakage; and ignition source causing fire	Missile fire	Ground operations
PHL-2	Missile structure	Missile body breaks up causing missile crash	Missile crash	Flight
PHL-3	Missile warhead (W/H)	Detonation of W/H explosives from fire, bullet, shock, and so on	W/H explosives detonation	Use insensitive munitions (IM)
PHL-4	Missile W/H	Initiation of W/H from inadvertent initiation commands	Inadvertent W/H initiation	Initiation requires both arm and fire signals
PHL-5	Missile W/H	Missile W/H fails to initiate	Dud	Unexploded ordnance (UXO) concern
PHL-6	Missile engine	Engine fails to start (missile crash)	Incorrect target	Unsafe missile state, fuel release
PHL-7	Missile engine	Engine fails during flight resulting in crash	Incorrect target	
PHL-8	Missile fuel subsystem	Engine fuel tank leakage and ignition source present resulting in fire	Missile fire	
PHL-9	Missile computer	Computer inadvertently generates W/H arm-1 and arm-2 commands, causing W/H initiation	Inadvertent W/H initiation	
PHL-10	Missile computer	Computer fails to generate W/H arm-1 or arm-2 commands	Inability to initiate W/H	Dud; not a safety concern
PHL-1	Missile computer	Computer inadvertently generates missile destruct command	Inadvertent destruct	Safe separation issue
PHL-12	Missile computer	Computer fails to generate missile destruct command	Inability to destruct missile	
PHL-13	Missile battery	Battery is inadvertently activated, providing power for W/H arm and fire commands	Inadvertent W/H initiation	Mishap also requires arm and fire signals
PHL-14	Missile battery	Battery electrolyte leakage occurs and ignition source present resulting in fire	Missile fire	
PHL-15	Missile destruct subsystem	Destruct system fails	Unable to destruct missile	Also requires fault necessitating destruct

TABLE 7.2. *(Continued)*

Preliminary Hazard List Analysis				
System Element Type: System Hardware				
No.	System Item	Hazard	Hazard Effects	Comments
PHL-16	Receiver	Receiver fails – no communication with missile	Unable to destruct missile	
PHL-17	Receiver	Receiver fails – creates erroneous destruct command	Inadvertent missile destruct	
PHL-18	Rocket booster	Inadvertent ignition of rocket	Inadvertent launch	Uncontrolled flight
PHL-19	WCS computer	Computer inadvertently generates missile launch commands	Inadvertent missile launch	
PHL-20	WCS radar	Electromagnetic radiation (EMR) injures exposed personnel	Personnel RF energy injury	
PHL-21	WCS radar	EMR causes ignition of explosives	Explosives detonation	
PHL-22	WCS radar	EMR causes ignition of fuel	Missile fuel fire	
PHL-23	WCS power	High-voltage electronics causes fire	Cabinet fire	System damage or personnel injury

3. Each of the 40 hazards identified in the PHL will be carried into the PHA for further analysis and investigation.
4. The TLMs and SCFs shown in Table 7.5 have been established from the entire list of PHL hazards. All of the identified hazards have been consolidated into these TLM categories. After establishing the TLMs, it was then possible to identify SCFs that are associated with certain TLMs, as shown in the table.

7.10 PHL ADVANTAGES AND DISADVANTAGES

The following are advantages of the PHL technique:

1. The PHL is easily and quickly performed.
2. The PHL does not require considerable expertise for technique application.
3. The PHL is comparatively inexpensive, yet provides meaningful results.
4. The PHL process provides rigor for focusing on hazards.
5. The PHL provides an indication of where major system hazards and mishap risk will exist.

TABLE 7.3   PHL Analysis of Ace Missile System: System Functions Evaluation

Preliminary Hazard List Analysis				
System Element Type: System Functions				
No.	System item	Hazard	Hazard effects	Comments
PHL-24	Warhead initiate	Warhead initiate function occurs inadvertently	Inadvertent W/H initiation	Initiation requires arm-1 and arm-2 functions
PHL-25	Warhead initiate	Warhead initiate function fails to occur	Dud warhead	Not a safety concern
PHL-26	Missile launch	Missile launch function occurs inadvertently	Inadvertent missile launch	
PHL-27	Missile self-test	Self-test function fails, resulting in unknown missile status	Unsafe missile state	
PHL-28	Missile destruct	Missile destruct function occurs inadvertently	Inadvertent missile destruct	
PHL-29	Missile navigation	Errors occur in missile navigation function	Incorrect target	
PHL-30	Missile guidance	Errors occur in missile guidance function	Incorrect target	
PHL-31	Communications with missile	Communication is lost, causing inability to initiate missile destruct system	Inability to destruct missile	

There are no real disadvantages in the PHL technique. There is a concern that it is sometimes strongly depended upon as the only hazard analysis technique that is applied. Also, due to the simplicity of the technique, it is sometime performed incorrectly.

7.11   COMMON PHL MISTAKES TO AVOID

When first learning how to perform a PHL, it is commonplace to commit some typical errors. The following is a list of errors often made during the conduct of a PHL:

1. Not listing all concerns or credible hazards. It is important to list all possible suspected or credible hazards and not leave any potential concerns out of the analysis.
2. Failure to document hazards identified but found to be not credible. The PHL is a historical document encompassing all hazard identification areas that were considered.
3. Not utilizing a structured approach of some type. Always use a worksheet and include all equipment, energy sources, functions, and so on.
4. Not collecting and utilizing common hazard source checklists.

TABLE 7.4   PHL Analysis of Ace Missile System: System Energy Sources Evaluation

\multicolumn{5}{c}{Preliminary Hazard List Analysis}

No.	System Item	Hazard	Hazard Effects	Comments
PHL-32	Explosives	Inadvertent detonation of W/H explosives	Inadvertent W/H initiation	
PHL-33	Explosives	Inadvertent detonation of missile destruct explosives	Inadvertent missile destruct	
PHL-34	Electricity	Personnel injury during maintenance of high-voltage electrical equipment	Personnel electrical injury	
PHL-35	Battery	Missile battery inadvertently activated	Premature battery power	Power to missile subsystems and W/H
PHL-36	Fuel	Missile fuel ignition causing fire	Missile fuel fire	
PHL-37	RF Energy	Radar RF energy injures personnel	Personnel injury from RF energy	
PHL-38	RF Energy	Radar RF energy detonates W/H explosives	Explosives detonation	
PHL-39	RF Energy	Radar RF energy detonates missile destruct explosives	Explosives detonation	
PHL-40	RF Energy	Radar RF Energy ignites fuel	Missile fuel fire	

System Element Type: System Energy Sources

TABLE 7.5   Missile System TLMs and SCFs from PHL Analysis

TLM No.	Top-Level Mishap	SCF
1	Inadvertent W/H initiation	Warhead initiation sequence
2	Inadvertent missile launch	Missile launch sequence
3	Inadvertent missile destruct	Destruct initiation sequence
4	Incorrect target	
5	Missile fire	
6	Missile destruct fails	Destruct initiation sequence
7	Personnel injury	
8	Unknown missile state	
9	Inadvertent explosives detonation	

5. Not researching similar systems or equipment for mishaps and lessons learned that can be applied.

6. Not establishing a correct list of hardware, functions, and mission phases.

7. Assuming the reader will understand the hazard description from a brief abbreviated statement filled with project unique terms and acronyms.

7.12 PHL SUMMARY

This chapter discussed the PHL analysis technique. The following are the basic principles that help summarize the discussion in this chapter:

1. The primary purpose of the PHL analysis is to identify potential hazards and mishaps existing in a system's conceptual design.
2. The PHL provides hazard information that serves as a starting point for the PHA.
3. The PHL is an aid for safety and design decision making early in the development program and provides information on where to apply safety and design resources during the development program.
4. The use of a functional flow diagram and an indentured equipment list greatly aids and simplifies the PHL process.
5. In performing the PHL analysis, hazard checklists are utilized; however, a hazard checklist should never be considered a complete and final list but merely a catalog for stimulating hazard ideas.
6. Do not exclude any thoughts, ideas, or concerns when postulating hazards. In addition to identifying real hazards, it is also important to show that certain hazards were suspected and considered, even though they were later ruled out for various design reasons. This provides evidence that all possibilities were considered.
7. Write a full, credible, and meaningful hazard description that is understandable to the reader and not just to the analyst. Do not assume the reader understands the hazard from a brief abbreviated statement filled with project unique terms and acronyms.
8. When possible, identify the three elements comprising a hazard:
   - Hazardous element (source)
   - Initiating mechanism (mechanism)
   - Target/threat (outcome)
9. Typically, when a hazard is identified and described, the hazard write-up description will identify and include all three elements of a hazard. However, in the PHL a complete and full hazard description is not always provided. This is primarily because of the preliminary nature of the analysis and because all identified hazards are more fully investigated and described in the PHA.

FURTHER READINGS

Layton, D., *System Safety: Including DOD Standards*, Weber Systems, Inc., Chesterland, OH, 1989.

Roland, H. E. and Moriarty, B., *System Safety Engineering and Management*, 2nd edition, John Wiley & Sons, 1990.

Stephans, R. A., *System Safety for the 21st Century*, John Wiley & Sons, 2004.

Stephenson, J., *System Safety 2000*, John Wiley & Sons, 1991.

System Safety Analysis Handbook, System Safety Society.

Vincoli, J. W., *A Basic Guide to System Safety*, Van Nostrand Reinhold, 1993.

Chapter 8

Preliminary Hazard Analysis

8.1 PHA INTRODUCTION

The preliminary hazard analysis (PHA) technique is a safety analysis tool for identifying hazards, their associated causal factors, effects, level of risk, and mitigating design measures when detailed design information is not available. The PHA provides a methodology for identifying and collating hazards in the system and establishing the initial system safety requirements (SSRs) for design from preliminary and limited design information. The intent of the PHA is to affect the design for safety as early as possible in the development program. The PHA normally does not continue beyond the subsystem hazard analysis (SSHA).

8.2 PHA BACKGROUND

This analysis technique falls under the preliminary design hazard analysis type (PD-HAT) because it evaluates design at the preliminary level without detailed information. The analysis types are described in Chapter 4. Gross hazard analysis (GHA) and potential hazard analysis are other names for this analysis technique. The name gross hazard analysis originated from MIL-SPEC-38130, which was the predecessor to MIL-STD-882.

The purpose of the PHA is to analyze identified hazards, usually provided by the preliminary hazard list (PHL), and to identify previously unrecognized hazards early in the system development. The PHA is performed at the preliminary design level, as its name implies. In addition, the PHA identifies hazard causal factors, consequences, and relative risk associated with the initial design concept. The PHA provides a mechanism for identifying initial design SSRs that assist in designing-in safety early in the design process. The PHA also identifies

Hazard Analysis Techniques for System Safety, Second Edition. Clifton A Ericson, II.
© 2016 John Wiley & Sons, Inc. Published 2016 by John Wiley & Sons, Inc.

safety-critical functions (SCFs) and top-level mishaps (TLMs) that provide a safety focus during the design process.

The PHA is applicable to the analysis of all types of systems, facilities, operations, and functions, and it can be performed on a unit, subsystem, system, or an integrated set of systems. The PHA is generally based on preliminary or baseline design concepts and is usually generated early in the system development process, in order to influence design and mishap risk decisions as the design is developed into detail.

The PHA technique, when methodically applied to a given system by experienced safety personnel, is thorough in identifying system hazards based on the limited design data available.

A basic understanding of hazard analysis theory is essential, so is the knowledge of system safety concepts. Experience with or a good working knowledge of the particular type of system and subsystem is necessary in order to identify and analyze all hazards. The PHA methodology is uncomplicated and easily learned. Standard PHA forms and instructions are provided in this chapter and standard hazard checklists are readily available.

The PHA is probably the most commonly performed hazard analysis technique. In most cases, the PHA identifies the majority of the system hazards. The remaining hazards are usually uncovered when subsequent hazard analyses are generated and more design details are available. Subsequent hazard analyses refine the hazard cause–effect relationship and uncover previously unidentified hazards and refine the design safety requirements.

There are no alternatives to a PHA. A PHL might be done in place of the PHA, but this is *not* recommended, since the PHL is only a list of hazards, is not as detailed as a PHA, and does not provide all of the required information. A SSHA might be done in place of the PHA, but this is *not* recommended since the SSHA is a more detailed analysis primarily of faults and failures that can create safety hazards. A modified failure mode and effects analysis (FMEA) could be used as a PHA, but this is *not* recommended since the FMEA primarily looks at failure modes only, while the PHA considers many more system aspects.

The use of the PHA technique is highly recommended for every program, regardless of size or cost, to support the goal of identifying and mitigating all system hazards early in the program. The PHA is the starting point for further hazard analysis and safety tasks, is easily performed, and identifies a majority of the system hazards. The PHA is a primary system safety hazard analysis technique for a system safety program.

8.3 PHA HISTORY

The PHA technique was established very early in the history of the system safety discipline. It was formally instituted and promulgated by the developers of MIL-STD-882. It was originally called a GHA because it was performed at a gross (preliminary) level of detail (see MIL-S-38130).

8.4 PHA THEORY

Figure 8.1 shows an overview of the basic PHA process and summarizes the important relationships involved in the PHA process. The PHA process consists of utilizing both design information and known hazard information to identify and evaluate hazards and to identify SC factors that are relevant to design safety. The PHA evaluates hazards identified by the PHL analysis in further detail.

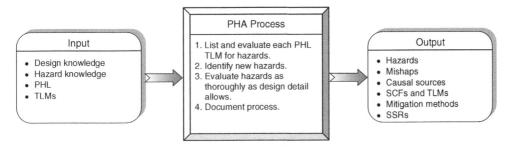

Figure 8.1 *PHA overview.*

   The purpose of the PHA is to identify hazards, hazard causal factors, hazard–mishap risk, and SSRs to mitigate hazards with unacceptable risk during the preliminary design phase of system development. To perform the PHA analysis, the system safety analyst must have three things – design knowledge, hazard knowledge, and the PHL. Design knowledge means the analyst must possess a basic understanding of the system design, including a list of major components. Hazard knowledge means the analyst needs a basic understanding about hazards, hazard sources, hazard components (hazard element, initiating mechanism, and target/threat), and hazards in similar systems. Hazard knowledge is primarily derived from hazard checklists and from lessons learned on the same or similar systems.

   The starting point for the PHA is the PHL collection of identified hazards. The PHA evaluates these hazards in more detail. In addition, the analyst compares the design knowledge and information with hazard checklists in order to identify previously unforeseen hazards. This allows the analyst to visualize or postulate possible hazards. For example, if the analyst discovers that the system design will be using jet fuel, he then compares jet fuel with a hazard checklist. From the hazard checklist, it will be obvious that jet fuel is a hazardous element (HE), and that a jet fuel fire/explosion is a potential mishap with many different ignition sources presenting many different hazards.

   Output from the PHA includes identified and suspected hazards, hazard causal factors, the resulting mishap effect, mishap risk, SCFs, and TLMs. PHA output also includes design methods and SSRs established to eliminate and/or mitigate identified hazards. It is important to identify SCFs because these are the areas that generally affect design safety and that are usually involved in major system hazards.

   Since the PHA is initiated very early in the design phase, the data available to the analyst may be incomplete and informal (i.e., preliminary). Therefore, the analysis process should be structured to permit continual revision and updating as the conceptual approach is modified and refined. When the subsystem design details are complete enough to allow the analyst to begin the SSHA in detail, the PHA is generally terminated.

8.5   PHA METHODOLOGY

The PHA methodology is shown in Figure 8.2. This process uses design and hazard information to stimulate hazard and causal factor identification. The PHA analysis begins with hazard identified from the PHL. The next step is to once again employ the use of hazard checklists (as done in the PHL analysis) and undesired mishap checklists. The basic inputs for the PHA include the functional flow diagram, the reliability block diagram, indentured

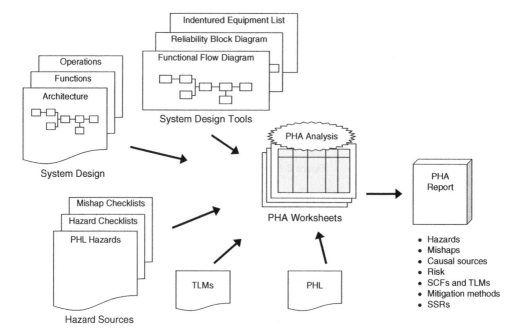

Figure 8.2   *Preliminary hazard analysis methodology.*

equipment list, system design, PHL hazards, hazard checklists, and mishap checklists – the first three of these are derived from the system design by the various system design organizations.

Hazard checklists are generic lists of known hazardous items and potentially hazardous designs, functions, or situations and are fully defined and discussed in Chapter 4. Hazard checklists should not be considered complete or all-inclusive, but merely a list of items to help trigger the analyst's recognition of potential hazard sources from past lesson learned. Typical hazard checklists include the following:

1. Energy sources
2. Hazardous functions
3. Hazardous operations
4. Hazardous components
5. Hazardous materials
6. Lessons learned from similar type of systems
7. Undesired mishaps
8. Failure mode and failure-state considerations

Refer to Chapter 7, on PHL analysis, for examples of each of these hazard checklists. Appendix C contains a more complete set of hazard checklists that can be used in a PHA.

Table 8.1 lists and describes the basic steps of the PHA process. This process involves analyzing PHL identified hazards in more detail and performing a detailed analysis of the system against hazard checklists.

When performing a PHA, the following factors should be considered, as a minimum:

a. Hazardous components (e.g., energy sources, fuels, propellants, explosives, and pressure systems).

TABLE 8.1   PHA Process

Step	Task	Description
1	Define system	Define the scope and boundaries of the system. Define the mission, mission phases, and mission environments. Understand the system design, operation, and major system components
2	Plan PHA	Establish PHA definitions, worksheets, schedule, and process. Identify system elements and functions to be analyzed
3	Establish safety criteria	Identify applicable design safety criteria, safety precepts/principles, safety guidelines, and safety-critical factors
4	Acquire data	Acquire all of the necessary design and operational and process data needed for the analysis (e.g., functional diagrams, drawings, operational concepts, etc.). Acquire hazard checklists, lessons learned, and other hazard data applicable to the system. Acquire all regulatory data and information that are applicable
5	Conduct PHA	a. Construct a list of equipment, functions, and energy sources for analysis b. Prepare a worksheet for each identified equipment item, function, and energy source c. Compare system hardware items with hazard checklists and TLMs d. Compare system operational functions with hazard checklists and TLMs e. Compare system energy sources with energy hazard checklists and TLMs f. Compare system software functions with hazard checklists and TLMs g. Expand the list of SCFs and TLMs, and use the same in the analysis h. Be cognizant of functional relationships, timing, and concurrent functions when identifying hazards i. Utilize hazard/mishap lessons learned from other systems
6	Evaluate risk	Identify the level of mishap risk presented for each identified hazard, both with and without hazard mitigation in the system design
7	Recommend corrective action	Recommend corrective action necessary to eliminate or mitigate identified hazards. Work with the design organization to translate the recommendations into SSRs. Also, identify safety features already in the design or procedures that are present for hazard mitigation
8	Monitor corrective action	Review test results to ensure that safety recommendations and SSRs are effective in mitigating hazards as anticipated
9	Track hazards	Transfer newly identified hazards into the HTS. Update the HTS as hazards, hazard causal factors, and risk are identified in the PHA
10	Document PHA	Document the entire PHA process and PHA worksheets in a PHA report. Include conclusions and recommendations

b. Subsystem interfaces (e.g., signals, voltages, timing, human interaction, hardware, etc.).

c. System compatibility constraints (e.g., material compatibility, electromagnetic interference, transient current, ionizing radiation, etc.).

d. Environmental constraints (e.g., drop, shock, extreme temperatures, noise and health hazards, fire, electrostatic discharge, lightning, X-ray, electromagnetic radiation, laser radiation, etc.).

e. Undesired states (e.g., inadvertent activation, fire/explosive initiation and propagation, failure to safe, etc.).

f. Malfunctions to the system, subsystems, or computing system.

g. Software errors (e.g., programming errors, programming omissions, logic errors, etc.).

h. Operating, test, maintenance, and emergency procedures.

i. Human error (e.g., operator functions, tasks, and requirements, etc.).

j. Crash and survival safety (e.g., egress, rescue, salvage, etc.).

k. Life cycle support (e.g., demilitarization/disposal, EOD, surveillance, handling, transportation, and storage).

l. Facilities, support equipment, and training.

m. Safety equipment and safeguards (e.g., interlocks, system redundancy, fail–safe design considerations, subsystem protection, fire suppression systems, personal protective equipment, warning labels, etc.).

n. Protective clothing, equipment, or devices.

o. Training and certification pertaining to safe operation and maintenance of the system.

p. System phases (test, manufacture, operations, maintenance, transportation, storage, disposal, etc.).

8.6   PHA WORKSHEET

The PHA is a detailed hazard analysis utilizing structure and rigor. It is desirable to perform the PHA using a specialized worksheet. Although the format of the PHA analysis worksheet is not critical, it is important that, as a minimum, the PHA generates the following information:

1. System hazards
2. Hazard effects (e.g., actions, outcomes, and mishaps)
3. Hazard causal factors (or potential causal factor areas)
4. Mishap risk assessment (before and after design safety features are implemented)
5. SCFs and TLMs
6. Recommendations for eliminating or mitigating the hazards

Figure 8.3 shows the columnar format PHA worksheet recommended for SSP usage. This particular worksheet format has proven to be useful and effective in many applications and it provides all of the information necessary from a PHA.

The following instructions describe the information required under each column entry of the PHA worksheet:

1. *System*   This entry identifies the system under analysis.
2. *Subsystem/function*   This entry identifies the subsystem or function under analysis.

Figure 8.3 *Recommended PHA worksheet.*

3. *Analyst* This entry identifies the name of the PHA analyst.

4. *Date* This entry identifies the date of the analysis.

5. *Hazard Number* This column identifies the number assigned to the identified hazard in the PHA (e.g., PHA-1, PHA-2, etc.). This is for future reference to the particular hazard source and may be used, for example, in the hazard action record (HAR) and the hazard tracking system (HTS).

6. *Hazard* This column identifies the specific hazard being postulated and evaluated. (Remember: Document all hazard considerations, even if they are later proven to be nonhazardous.)

7. *Causes* This column identifies conditions, events, or faults that could cause the hazard to exist and the events that can trigger the HEs to become a mishap or an accident.

8. *Effects* This column identifies the effects and consequences of the hazard, should it occur. Generally, the worst-case result is the stated effect. The effect ultimately identifies and describes the potential mishap involved.

9. *Mode* This entry identifies the system mode(s) of operation, or operational phases, where the identified hazard is of concern.

10. *Initial Mishap Risk Index (IMRI)* This column provides a qualitative measure of mishap risk significance for the potential effect of the identified hazard, given that no mitigation techniques are applied to the hazard. Risk measures are a combination of mishap severity and probability, and the recommended values from MIL-STD-882 are shown below.

Severity	Probability
I – Catastrophic	A – Frequent
II – Critical	B – Probable
III – Marginal	C – Occasional
IV – Negligible	D – Remote
	E – Improbable

11. *Recommended Action* This column establishes recommended preventive measures to eliminate or mitigate the identified hazards. Recommendations generally take the

form of guideline safety requirements from existing sources or a proposed mitigation method that is eventually translated into a new derived SSR intended to mitigate the hazard. SSRs are generated after coordination with the design and requirements organizations. Hazard mitigation methods should follow the preferred order of precedence established in MIL-STD-882 for invoking or developing safety requirements, which are shown below.

12. Order of Precedence

    1 – Eliminate hazard through design selection
    2 – Incorporate safety devices
    3 – Provide warning devices
    4 – Develop procedures and training

13. *Final Mishap Risk Index (FMRI)* This column provides a qualitative measure of mishap risk for the potential effect of the identified hazard, given that mitigation techniques and safety requirements are applied to the hazard. The same risk matrix table used to evaluate column 10 is also used here.

14. *Comments* This column provides a place to record useful information regarding the hazard or the analysis process that are not noted elsewhere. This column can be used to record the final SSR number for the developed SSR, which will later be used for traceability.

15. *Status* This column states the current status of the hazard, as being either open or closed.

8.7 PHA GUIDELINES

The following are some basic guidelines that should be followed when completing the PHA worksheet:

1. Remember that the objective of the PHA is to identify system hazards, effects, causal factor areas, and risk. Another by-product of the PHA is the identification of TLMs and SCFs.

2. Start by listing and systematically evaluating system hardware subsystems, system functions, and system energy sources on separate worksheet pages. For each of these categories, identify hazards that may cause the TLMs identified from the PHL. Also, use hazard checklists to identify new TLMs and hazards.

3. PHL hazards must be converted to TLMs for the PHA. Use TLMs, along with hazard checklists and lessons learned for hazard recognition, to identify hazards.

4. Do not worry about reidentifying the same hazard when evaluating system hardware, system functions, and system energy sources. The idea is to provide a thorough coverage in order to identify all hazards.

5. Expand each identified hazard in more detail to identify causal factors, effects, and risk.

6. As causal factors and effects are identified, hazard risk can be determined or estimated.

7. Continue to establish TLMs and SCFs as the PHA progresses and use them in the analysis.

8. A hazard write-up in the PHA worksheet should be clear and understandable with as much information necessary to understand the hazard.

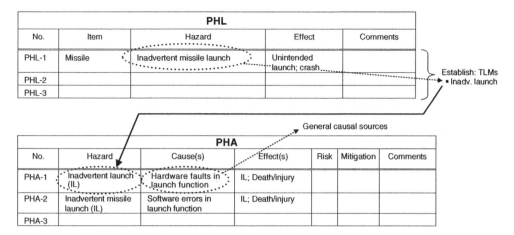

Figure 8.4 *PHA guidelines.*

9. The PHL hazard column does not have to contain all three elements of a hazard: Hazardous element, initiating mechanisms (IMs), and outcome (O). The combined columns of the PHA worksheet can contain all three components of a hazard. For example, it is acceptable to place the HE in the hazard section, the IMs in the cause section, and the O in the effect section. The hazard, causes, and effects columns should together completely describe the hazard.

10. Use analysis aids to help recognize and identify new hazards, such as hazard checklists, lesson learned from hazard databases and libraries, mishap investigations, and so on. Also, use applicable hazards from the PHA of other similar systems.

11. After performing the PHA, review the PHL hazards, to ensure all have been covered via the TLM process. This is because the PHL hazards were not incorporated one-for-one.

Figure 8.4 shows how to apply the PHA guidelines when using the PHA worksheet.

8.8   PHA EXAMPLE: ACE MISSILE SYSTEM

In order to demonstrate the PHA methodology, the same hypothetical Ace Missile System from Chapter 7 will be used. The basic preliminary design is shown in Figure 8.5; however, note that the conceptual design changed slightly from the concept phase to the preliminary design phase (as happens in many development programs). The design concept has now expanded from a single missile system to multiple missiles in launch tubes. These changes will be reflected in the PHA. The major segments of the system are the missile segment and the weapon control system (WCS) segment.

During preliminary design development, the system design has been modified and improved to include the following:

1. Multiple missiles instead of a single missile.
2. The missiles are now contained in launch tubes.
3. A radio transmitter was added to WCS design for missile destruct subsystem.

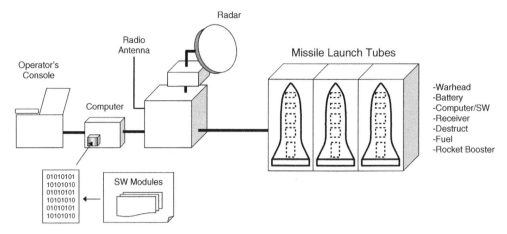

Figure 8.5 *Ace Missile System.*

Figure 8.6 lists the major system components, functions, phases, and energy sources that should be considered for the PHA. This is the typical level of information available for the PHA.

Figure 8.6 contains a preliminary indentured equipment list (IEL) for this system that will be used for the conduct of the PHA. This is the level of information typically available during preliminary design. As the design development progresses, the IEL will be expanded in breadth and depth for the SSHA. The IEL is basically a hierarchy of equipment that establishes relationships, interfaces, and equipment types. A new PHA worksheet page will be started for each IEL item.

Sometimes, a more detailed hierarchy is available at the time of the PHA, but usually it is not. The basic ground rule is that the higher level of detail goes into the PHA, and the more detailed breakdown goes into the SSHA. Sometimes, the decision is quite obvious, while at other times, the decision is somewhat arbitrary. In this example, the computer software would be included in the PHA only as a general category, and it would also be included in the SSHA when the indentured list is continued to a lower level of detail for the software (e.g., module level).

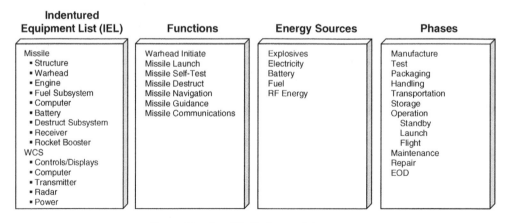

Figure 8.6 *Ace Missile System information.*

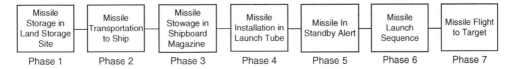

Figure 8.7 *Ace missile functional flow diagram of operational phases.*

The PHA will analyze the system at the subsystem level because that is the level of design details available. The SSHA will utilize the PHA hazards and carry the analysis a level lower as more design detail becomes available.

It is also helpful when performing the PHA to utilize functional flow diagrams (FFDs) of the system if they are available. The FFD shows the steps that must take place in order to perform a particular system function. The FFD helps identify subsystem interfaces and relationships that can be used in the analysis. Sometimes, it is necessary for the system safety analyst to develop both IELs and FFDs if the project design team has not developed them.

Figure 8.7 is an FFD showing the basic planned operational phases for the missile system. As design development progresses, each of these phases will be expanded in greater detail.

Figure 8.8 is an FFD showing the elements and sequence of events required to generate the missile launch signal.

Figure 8.9 is an FFD showing the elements and sequence of events required to generate the missile warhead launch signal.

The hazards identified by the PHL analysis form the initial set of hazards for the PHA. Since the PHL hazards are generalized, brief, and mixed, it is better to condense the PHL hazards to TLMs and then use the TLMs for the hazard categories that the PHA should be considering for all aspects of the system design and operation. Table 8.2 contains the list of TLM resulting from the PHL analysis in Chapter 7.

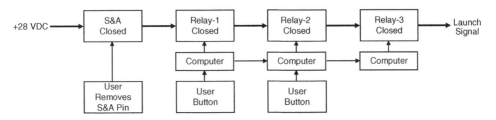

Figure 8.8 *Ace missile launch signal functional flow diagram.*

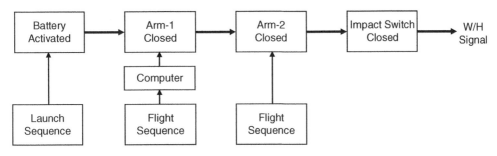

Figure 8.9 *Ace missile warhead initiate signal functional flow diagram.*

TABLE 8.2   Missile System TLMs from PHL Analysis

TLM No.	Top-Level Mishap
1	Inadvertent initiation of W/H explosives
2	Inadvertent launch
3	Inadvertent missile destruct
4	Incorrect target
5	Missile fire
6	Missile destruct fails
7	Personnel injury
8	Unknown missile state
9	Inadvertent detonation of explosives

If a new PHA worksheet were started for every IEL item, system function, and system energy source, there would be a minimum of 26 worksheets (14 IEL items + 7 function + 5 energy sources). In order to not overwhelm the reader only six worksheets are provided (five IEL items and one function). These samples should provide sufficient information on how to perform the PHA. These same examples will be carried into the SSHA.

Tables 8.3–8.8 contain five example worksheets from the example missile system PHA.

The following should be noted from the PHA of the Ace Missile System:

The recommended action is not always in the form of a direct SSR. Additional research may be necessary to convert the recommendation into a meaningful design requirement.

As a result of the PHA, TLM 10 was added to the list of TLMs. The new TLM list is shown in Table 8.9.

8.9   PHA ADVANTAGES AND DISADVANTAGES

The PHA technique affords the following advantages:

1. Is easily and quickly performed.
2. Is comparatively inexpensive, yet provides meaningful results.
3. Provides rigor for focusing for the identification and evaluation of hazards.
4. Is a methodical analysis technique.
5. Identifies majority of system hazards and provides an indication of system risk.
6. Commercial software is available to assist in the PHA process.

There are no real disadvantages to the PHA technique. There is a concern that it is sometimes improperly depended upon as the only hazard analysis technique that is applied. Also, due to the simplicity of the technique, it is sometimes performed incorrectly.

8.10   COMMON PHA MISTAKES TO AVOID

When first learning how to perform a PHA, it is commonplace to commit some typical errors. The following is a list of common errors made during the conduct of a PHA.

1. Not listing all concerns or credible hazards. It is important to list all possible suspected or credible hazards and not leave any potential concerns out of the analysis.

TABLE 8.3 Ace Missile System PHA; Worksheet 1

System: Ace Missile System
Subsystem: Missile Structure Subsystem

Preliminary Hazard Analysis

Analyst:
Date:

No.	Hazard	Causes	Effects	Mode	IMRI	Recommended Action	FMRI	Comments	Status
PHA-1	Missile structure fails resulting in unstable missile flight and missile crash	Manufacturing defect; design error	Unstable flight resulting in crash causing death/injury, incorrect target	Flight	1D	- Use 5× safety factor on structure design	1E		Open
PHA-2	Missile body breaks up resulting in fuel leakage, ignition source causing fire	Manufacturing defect, design error	Missile fire causing death/injury	Ground operations	1D	- Use 5× safety factor on structure design	1E		Open
PHA-3	Missile structure fails during handling, resulting in personnel injury	Manufacturing defect, design error, handling equipment failure	Personnel injury	PHS&T	2D	- Use 5× safety factor on structure design	2E		Open
						- Establish SSRs for handling equipment			

Teaching notes:

• Focus of this PHA this worksheet is on the missile structure subsystem
• PHA-1 was identified from missile structure contributions to TLM-4 (incorrect target).
• PHA-2 was identified from missile structure contributions to TLM-5 (missile fire).
• PHA-3 was identified from missile structure contributions to TLM-7 (personnel injury).

TABLE 8.4 Ace Missile System PHA; Worksheet 2

| | | | | Preliminary Hazard Analysis | | | | Analyst:
Date: | | |
No.	Hazard	Causes	Effects	Mode	IMRI	Recommended Action	FMRI	Comments	Status
System: Ace Missile System Subsystem: Missile Warhead **Subsystem**									
PHA-4	Inadvertent initiation of W/H explosives due to erroneous initiate commands	Erroneous commands from hardware faults, software faults, human error	Personnel death/injury		1D	- Use multiple independent switches in fuze design	1E		Open
PHA-5	Inadvertent initiation W/H explosives due to external environment	Bullet strike, shrapnel, heat	Personnel death/injury		1D	- Conduct FTA of fuze design	1E		Open
PHA-6	Failure of W/H explosives to initiate when commanded	Hardware faults, software faults	Dud missile, not a safety concern		—	- Use insensitive munitions	—	Not a safety concern	Closed
						- Provide protective covering when possible			

Teaching notes:

- Focus of this PHA worksheet is on the missile warhead subsystem.
- PHA-4 was identified by considering W/H contributions to TLM-1 and failure-state hazard checklist.
- PHA-5 was identified by considering W/H contributions to TLM-1 and explosives insensitive munitions considerations.
- PHA-6 was identified by considering failure-state hazard checklist as applied to the W/H.

TABLE 8.5   Ace Missile System PHA; Worksheet 3

System: Ace Missile System
Subsystem: Missile Destruct Subsystem

Preliminary Hazard Analysis

Analyst:
Date:

No.	Hazard	Causes	Effects	Mode	IMRI	Recommended Action	FMRI	Comments	Status
PHA-7	Inadvertent missile destruct occurs during flight	Erroneous commands from hardware faults, software faults, human error	Missile debris lands on occupied area resulting in death/injury, incorrect target	Flight	1D	- Provide safety interlock inhibiting signal until required	1E		Open
PHA-8	Inadvertent missile destruct occurs during ground operations	Wire short, explosives detonation	Explosion and debris cause personnel death/injury	Standby PHS&T	1D	- Ensure high reliability of S&A - Isolate critical pins in connector designs	1E		Open
PHA-9	Missile destruct fails to occur when commanded	Command error, radio transmission fault	Missile strikes undesired area resulting in death/injury, incorrect target	Flight	1D	- Ensure high reliability of S&A - Provide redundant design	1E		Open

Teaching notes:
- Focus of this PHA worksheet is on the missile destruct subsystem.
- PHA-7 was identified by considering destruct subsystem contributions to TLM-3.
- PHA-8 was identified by considering destruct subsystem contributions to TLM-3.
- PHA-9 was identified by considering failure-state hazard checklist as applied to the destruct subsystem.

TABLE 8.6   Ace Missile System PHA; Worksheet 4

System: Ace Missile System
Subsystem: Missile Rocket Booster Subsystem

Preliminary Hazard Analysis

Analyst:
Date:

No.	Hazard	Causes	Effects	Mode	IMRI	Recommended Action	FMRI	Comments	Status
PHA-10	Inadvertent ignition	Erroneous commands from hardware faults, software faults, human error, igniter failure	Inadvertent missile launch resulting in death/injury when missile hits ground	Flight	1D	- Provide safety interlock inhibiting signal until required - Verify software design	1E		Open
PHA-11	Explosion of rocket propellant	Manufacturing defect, bullet strike, fire	Explosives ignition resulting in personnel death/injury	PHS&T	1D	- Provide protective covering - Use safe propellant	1E		Open
PHA-12	Failure of rocket booster to start	Failure of commands, igniter failure	Unable to launch missile, not a safety concern	Launch	–		–	Not a safety concern	Closed
PHA-13	Failure of booster in flight	Manufacturing defect, installation error	Unstable flight resulting in crash causing death/injury, incorrect target	Flight	1D	- QA of manufacturing and installation	1E		Open

Teaching Notes:

• Focus of this PHA worksheet is on the missile rocket booster subsystem.
• PHA-10 was identified by considering rocket booster contributions to TLM-2.
• PHA-11 was identified by considering rocket booster contributions and energy sources and TLM-9.
• PHA-12 was identified by considering the failure-state hazard checklist as applied to the rocket booster.
• PHA-13 was identified by considering the failure-state hazard checklist as applied to the rocket booster.

140

TABLE 8.7 Ace Missile System PHA; Worksheet 5

System: Ace Missile System
Subsystem: WCS Radar Subsystem

Preliminary Hazard Analysis

Analyst:
Date:

No.	Hazard	Causes	Effects	Mode	IMRI	Recommended Action	FMRI	Comments	Status
PHA-14	Electromagnetic radiation (EMR) injures exposed personnel	Personnel in excessive RF energy zone	Personnel injury from RF energy	Ground operations	1D	- Establish safe personnel distances and place warning in all procedures	1E		Open
PHA-15	EMR causes ignition of explosives	Fuel in excessive RF energy zone	Explosives ignition resulting in personnel death/injury	Ground operations	1D	- Establish safe explosives distances and place warning in all procedures	1E		Open
PHA-16	EMR causes ignition of fuel	Explosives in excessive RF energy zone	Fuel fire resulting in personnel death/injury	Ground operations	1D	- Establish safe fuel distances and place warning in all procedures	1E		Open
PHA-17	High-voltage radar electronics causes fire	Overheating of high-voltage radar electronics causes fire	Fire causing system damage and/ or death/injury	Ground operations	1D	- Provide temperature warning	1E		Open
PHA-18	Personnel injury from high-voltage electronics during maintenance	Personnel touches exposed contacts with high-voltage present	Personnel injury from electrical hazard	Maintenance	2D	- Design unit to prevent personnel contact with voltage	2E		Open

Teaching notes:

• Focus of this PHA worksheet is on the WCS radar subsystem.
• PHA-14 was identified by considering radar contributions to TLM-7 and RF energy source.
• PHA-15 was identified by considering radar and RF energy sources and TLM-9.
• PHA-16 was identified by considering radar contributions to TLM-5 and RF energy source.
• PHA-17 was identified by considering radar contributions to TLM-7 and electrical energy source.
• PHA-18 was identified by considering radar contributions to TLM7 and electrical energy source.

TABLE 8.8   Ace Missile System PHA; Worksheet 6

System: Ace Missile System
Subsystem: Warhead Initiate Function

Preliminary Hazard Analysis

Analyst:
Date:

No.	Hazard	Causes	Effects	Mode	IMRI	Recommended Action	FMRI	Comments	Status
PHA-19	Warhead initiate function occurs inadvertently	Erroneous commands from hardware faults, software faults, human error	Premature warhead initiation resulting in death/injury	PHS&T Flight	1D	- Provide safety interlock inhibiting signal until required	1E		Open
PHA-20	Warhead initiate function fails to occur	Hardware faults, software faults	Warhead fails to initiate when desired, dud warhead	Flight	—		—	Not a safety concern	Closed
PHA-21	Unable to safe warhead after initiate command	Hardware faults, software faults	Warhead explodes when missile strikes ground, resulting in death/injury	Flight	1D	- Design for high reliability	1E	Although not in design data, safing function is needed New TLM	Open

Teaching notes:

• Focus of this PHA worksheet is on the warhead initiate function.

• PHA-119 was identified by considering warhead initiate function to TLM-1.

• PHA-20 was identified by considering warhead initiate function to the failure-state hazard checklist.

• PHA-21 was identified by considering warhead initiate safing function to the failure-state hazard checklist.

TABLE 8.9   Expanded List of TLMs from PHA

No.	TLM
1	Inadvertent initiation of W/H explosives
2	Inadvertent launch
3	Inadvertent missile destruct
4	Incorrect target
5	Missile fire
6	Missile destruct fails
7	Personnel injury
8	Unknown missile state
9	Inadvertent detonation of explosives
10	Unable to safe warhead

2. Failure to document hazards identified but found to be not credible. The PHA is a historical document encompassing all hazard identification areas that were considered.

3. Not utilizing a structured approach of some type. Always use a worksheet and include all equipment, energy sources, functions, and so on.

4. Not collecting and utilizing common hazard source checklists.

5. Not researching similar systems or equipment for mishaps and lessons learned that can be applied.

6. Not establishing a correct list of hardware, functions, and mission phases.

7. Assuming the reader will understand the hazard description from a brief abbreviated statement filled with project unique terms and acronyms.

8. Inadequately describing the identified hazard (insufficient detail, too much detail, incorrect hazard effect, wrong equipment indenture level, not identifying all three elements of a hazard, etc.).

9. Inadequately describing the causal factors (the identified causal factor does not support the hazard, the causal factor is not detailed enough, not all of the causal factors are identified, etc.).

10. Inadequately describing the hazard–mishap risk index (MRI). For example, the MRI is not stated or is incomplete, the hazard severity level does not support actual hazardous effects, the final MRI is a higher risk than the initial MRI, the final severity level in the risk is less than the initial severity level (sometimes possible, but not usually), and the hazard probability is not supported by the causal factors.

11. Providing recommended hazard mitigation methods that do not address the actual causal factor(s).

12. Incorrectly closing the hazard.

13. The PHA is initiated beyond the preliminary design stage or the PHA is continued beyond the preliminary design stage.

14. Not establishing and utilizing a list of TLMs and SCFs.

8.11   PHA SUMMARY

This chapter discussed the PHA technique. The following are the basic principles that help summarize the discussion in this chapter:

1. The PHA is an analysis tool for identifying system hazards, causal factors, mishap risk, and safety recommendations for mitigating risk. It is based upon preliminary design information.

2. The primary purpose of the PHA is to identify and mitigate hazards early in the design development process in order to influence the design when the cost impact is minimal.

3. The use of a specialized worksheet provides structure and rigor to the PHA process.

4. The use of a functional flow diagram, reliability block diagram, and an indentured equipment list greatly aids and simplifies the PHA process.

5. Do not exclude any thoughts, ideas, or concerns when postulating hazards. In addition to identifying real hazards, it is important to show that certain hazards were suspected and considered, even though they were later found not to be possible for various design reasons. This provides evidence that all possibilities were considered.

6. Write a full, credible, and meaningful hazard description that is understandable to the reader and not just to the analyst. Do not assume the reader understands the hazard from a brief abbreviated statement filled with project unique terms and acronyms.

7. Identify the three elements comprising a hazard using the PHA worksheet columns:
   - *Hazard Column* Hazard source (source)
   - *Causes Column* Initiating mechanism (mechanism)
   - *Effects Column* Target/threat outcome (outcome)

FURTHER READINGS

Ericson, C. A., Boeing document D2-113072-1, System Safety Analytical Technology: Preliminary Hazard Analysis, 1969.

Layton, D., *System Safety: Including DOD Standards*, Weber Systems, Inc., 1989.

Roland, H. E. and Moriarty, B., *System Safety Engineering and Management*, John Wiley & Sons, 2nd edition, 1990.

Stephans, R. A., *System Safety for the 21st Century*, John Wiley & Sons, 2004.

Stephenson, J., *System Safety 2000*, John Wiley & Sons, 1991.

System Safety Analysis Handbook, System Safety Society.

Vincoli, J. W., *A Basic Guide to System Safety*, Van Nostrand Reinhold, 1993.

Chapter 9

Subsystem Hazard Analysis

9.1   SSHA INTRODUCTION

The subsystem hazard analysis (SSHA) technique is a safety analysis tool for identifying hazards, their associated causal factors, effects, level of risk, and mitigating design measures. The SSHA is performed when detailed design information is available as it provides a methodology for analyzing in greater depth the causal factors for hazards previously identified by earlier analyses, such as the preliminary hazard analysis (PHA). The SSHA helps derive detailed system safety requirements (SSRs) for incorporating design safety methods into the system design.

9.2   SSHA BACKGROUND

This analysis technique falls under the detailed design hazard analysis type (DD-HAT) because it evaluates design at the detailed subsystem level of design information. The hazard analysis types are described in Chapter 4.

The purpose of the SSHA is to expand upon the analysis of previously identified hazards, and to identify new hazards, from detailed design information. The SSHA provides for the identification of detailed causal factors of known and newly identified hazards and, in turn, provides for the identification of detailed SSRs for design. The SSHA provides a safety focus from a detailed subsystem viewpoint, through analysis of the subsystem structure and components. The SSHA helps verify subsystem compliance with safety requirements contained in subsystem specifications.

Hazard Analysis Techniques for System Safety, Second Edition. Clifton A Ericson, II.
© 2016 John Wiley & Sons, Inc. Published 2016 by John Wiley & Sons, Inc.

The SSHA is applicable to the analysis of all types of systems and subsystems, and is typically performed at the detailed component level of a subsystem. The SSHA is usually performed during detailed design development and helps to guide the detailed design for safety.

The technique provides sufficient thoroughness to identify hazards and detailed hazard causal factors when applied to a given system/subsystem by experienced safety personnel. An understanding of hazard analysis theory, as well as the knowledge of system safety concepts, is essential. Experience with and/or a good working knowledge of the particular type of system and subsystem is necessary in order to identify and analyze all hazards. The methodology is uncomplicated and easily learned. Standard SSHA forms and instructions have been developed that are included as part of this chapter.

The SSHA is an in-depth and detailed analysis of hazards previously identified by the PHA. The SSHA also identifies new hazards. It requires detailed design information and a good understanding of the system design and operation.

The use of the SSHA technique is recommended for identification of subsystem-level hazards and further investigation of detailed causal factors of previously identified hazards. A fault hazard analysis (FHA) or failure mode and effects analysis (FMEA) may be done in place of the SSHA. Using these alternative techniques is *not* recommended. Both the FHA and FMEA techniques focus on failure modes and can, therefore, miss or overlook certain hazards.

9.3   SSHA HISTORY

The technique was established very early in the history of the system safety discipline. It was formally instituted and promulgated by the developers of MIL-STD-882. It was developed to replace the FHA technique, which was previously used for the hazard analysis of subsystems.

9.4   SSHA THEORY

The SSHA is performed to evaluate previously identified hazards at the subsystem level, identify new subsystem level hazards, and determine mishap risk. The SSHA refines the hazard causal factors to the detailed root cause level. Through this refining process, the SSHA ensures design SSRs adequately control hazards at the subsystem design level. The SSHA is a robust, rigorous, and structured methodology that consists of utilizing both detailed design information and known hazard information to identify hazards and their detailed causal factors.

Hazards from the PHA are imported into the SSHA and the causal factors are identified from the detailed component design information. The SSHA can be structured to follow an indentured equipment list (IEL) that has been expanded in detail from the IEL used in the PHA.

Output from the SSHA includes identified and suspected hazards, hazard causal factors, the resulting mishap effect, and safety-critical factors. SSHA output also includes design methods and SSRs established to eliminate and/or mitigate identified hazards with unacceptable risk. The SSHA also identifies safety-critical functions (SCFs) and top-level mishaps (TLMs) that provide a safety focus during the design process.

Figure 9.1 shows an overview of the SSHA process and summarizes the important relationships involved.

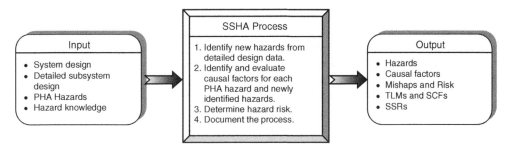

Figure 9.1 *Subsystem hazard analysis overview.*

9.5 SSHA METHODOLOGY

The SSHA methodology is shown in Figure 9.2. The SSHA process uses different kinds of design information to stimulate hazard identification. The analysis begins with hazards identified from the PHA. Hazard checklists (or catalogs) and undesired mishap checklists are employed to help identify new hazards previously unseen. Three of the best tools for aiding the SSHA analyst are the functional flow diagrams (FFDs), the reliability block diagram, and an IEL, all of which are derived from the system design. Using the detailed design information that is available during the SSHA, hazard causal factors can be evaluated in greater detail.

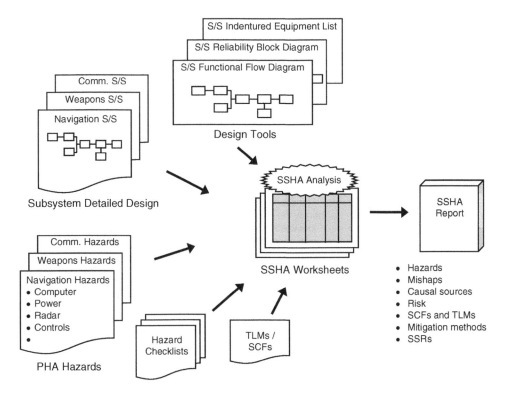

Figure 9.2 *SSHA methodology.*

The functional flow diagram simplifies system design and operation for clarity and understanding. It shows the relationships and interfaces between subsystems. It also shows the functions that must be performed by the system. The IEL identifies all of the specific hardware and software in the system design. By comparing the detailed equipment with known hazard checklists, hazards are easily identified.

Table 9.1 lists and describes the basic steps of the SSHA process. The SSHA process involves performing a detailed analysis of each subsystem in the system under investigation.

As a minimum, when performing the SSHA, consideration should be given to the following:

1. Performance of the subsystem hardware
2. Performance degradation of the subsystem hardware
3. Inadvertent functioning of the subsystem hardware

TABLE 9.1 SSHA Process

Step	Task	Description
1	Define system	Define scope and boundaries of the operation to be performed. Understand the system design and operation. Understand the detailed subsystem designs
2	Plan SSHA	Establish SSHA goals, definitions, worksheets, schedule, and process. Identify the subsystems to be analyzed
3	Establish safety criteria	Identify applicable design safety criteria, safety precepts/principles, safety guidelines, and safety-critical factors
4	Acquire data	Acquire all of the necessary detailed design and operational data needed for the analysis. This includes both schematics, operation manuals, functional flow diagrams, reliability block diagrams, and so on. Acquire hazard checklists, lessons learned, and other hazard data applicable to the system
5	Conduct SSHA	1. Construct list of discrete subsystem components for SSHA worksheets 2. Begin by populating the SSHA worksheet with the hazards identified in the PHA 3. Evaluate each discrete component in the list and identify hazard causal factors from the subsystem components 4. Utilize TLMs and SCFs to identify new hazards 5. Utilize hazard checklists to identify new hazards 6. Utilize hazard/mishap lessons learned to identify new hazards 7. Be cognizant of functional relationships, timing, and concurrent functions when identifying hazards
6	Evaluate risk	Identify the level of mishap risk presented to the system by each identified hazard, both before and after recommended hazard mitigations have been established for the system design
7	Recommend corrective action	Recommend corrective action necessary to eliminate or mitigate identified hazards. Work with the design organization to translate the recommendations into SSRs. Also, identify safety features already in the design or procedures that are present for hazard mitigation
8	Monitor corrective action	Review test results to ensure that safety recommendations and SSRs are effective in mitigating hazards as anticipated
9	Track hazards	Transfer newly identified hazards into the hazard tracking system (HTS). Update hazards in the HTS as causal factors and risk are identified in the SSHA
10	Document SSHA	Document the entire SSHA process on the worksheets. Update for new information and closure of assigned corrective actions

4. Functional failure of the subsystem hardware

5. Common mode failures

6. Timing errors

7. Design errors or defects

8. Human error and the human–system interface design

9. Software errors and the software–machine interface

10. Functional relationships or interfaces between components and equipment comprising each subsystem

9.6   SSHA WORKSHEET

It is desirable to perform the SSHA analysis using a worksheet. The worksheet will help to add rigor to the analysis, help record the process and data, and help support justification for the identified hazards and safety recommendations. The format of the analysis worksheet is not critical, and typically columnar-type worksheets are utilized.

As a minimum, the following basic information should be obtained from the SSHA analysis worksheet:

1. Hazards

2. Hazard effects (mishaps)

3. Detailed hazard causal factors (materials, processes, excessive exposures, failures, etc.)

4. Risk assessment (before and after design safety features are implemented)

5. Recommendations for eliminating or mitigating the hazards (which can be converted into derived SSRs)

The recommended SSHA worksheet for SSP usage is the columnar format shown in Figure 9.3. This particular worksheet format has proven to be useful and effective in many applied situations and it provides all of the information required from a SSHA.

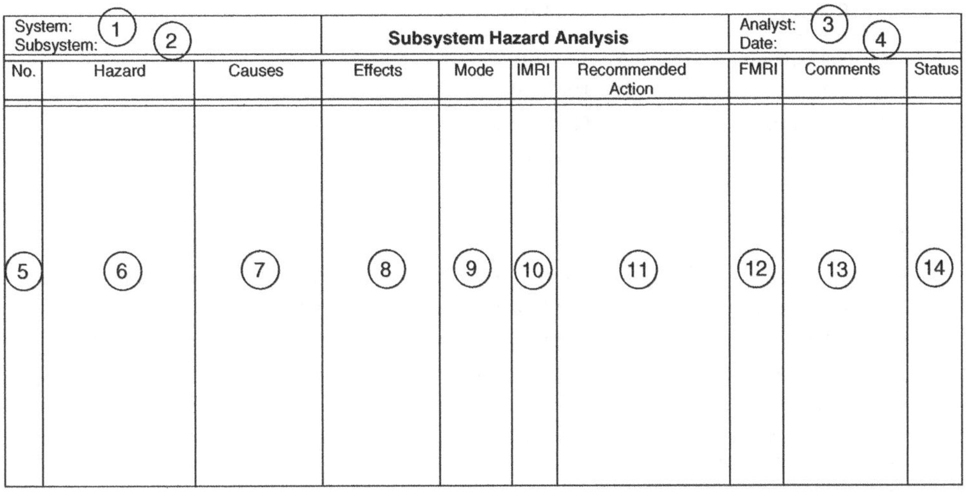

System: (1) Subsystem: (2)			Subsystem Hazard Analysis				Analyst: (3) Date: (4)		
No.	Hazard	Causes	Effects	Mode	IMRI	Recommended Action	FMRI	Comments	Status
(5)	(6)	(7)	(8)	(9)	(10)	(11)	(12)	(13)	(14)

Figure 9.3 *Recommended SSHA worksheet.*

The following instructions describe the information required under each column entry of the SSHA worksheet:

1. *System* This entry identifies the system under analysis.
2. *Subsystem* This entry identifies the subsystem under analysis.
3. *Analyst* This entry identifies the name of the SSHA analyst.
4. *Date* This entry identifies the date of the analysis.
5. *Hazard Number* This column identifies the number assigned to the identified hazard in the SSHA (e.g., SSHA-1, SSHA-2, etc.). This is for future reference to the particular hazard source and may be used, for example, in the hazard action record (HAR).
6. *Hazard* This column identifies the specific hazard being postulated. (Remember, document all hazard considerations, even if they are proven to be nonhazardous.) The SSHA is started by transferring all hazards from the PHA into the SSHA, for a more thorough detailed analysis.
7. *Causes* This column identifies conditions, events, or faults that could cause the hazard to exist and the events that can trigger the hazardous elements (HE) to become a mishap or accident. The detailed design information that is available during the SSHA provides for the identification of the specific hazard causal factors.
8. *Effects* This column identifies the effect and consequences of the hazard, assuming it occurs. Generally, the worst-case result is the stated effect.
9. *Mode* This entry identifies the system mode(s) of operation, or operational phases, where the identified hazard is of concern.
10. *Initial mishap risk index (IMRI)* This column provides a qualitative measure of mishap risk significance for the potential effect of the identified hazard, given that no mitigation techniques are applied to the hazard. Risk measures are a combination of mishap severity and probability, and the recommended values from MIL-STD-882 are shown below.

Severity	Probability
1 – Catastrophic	A – Frequent
2 – Critical	B – Probable
3 – Marginal	C – Occasional
4 – Negligible	D – Remote
	E – Improbable

11. *Recommended Action* This column establishes recommended preventive measures to eliminate or mitigate the identified hazards. Recommendations generally take the form of guideline safety requirements from existing sources, or a proposed mitigation method that is eventually translated into a new derived SSR intended to mitigate the hazard. SSRs are generated after coordination with the design and requirements organizations. Hazard mitigation methods should follow the preferred order of precedence established in MIL-STD-882 for invoking or developing safety requirements, which are shown below.

Order of Precedence
1 – Eliminate hazard through design selection
2 – Control hazard through design methods
3 – Control hazard through safety devices
4 – Control hazard through warning devices
5 – Control hazard through procedures and training

12. *Final mishap risk index (FMRI)*   This column provides a qualitative measure of mishap risk significance for the potential effect of the identified hazard, given that mitigation techniques and safety requirements are applied to the hazard. The same values used in column 10 are also used here.
13. *Comments*   This column provides a place to record useful information regarding the hazard or the analysis process.
14. *Status*   This column states the current status of the hazard, as being either open or closed.

9.7   SSHA GUIDELINES

The following are some basic guidelines that should be followed when performing the SSHA:

1. Remember that the objective of the SSHA is to identify detailed subsystem causes of identified hazards, plus previously undiscovered hazards. It refines risk estimates and mitigation methods.
2. Isolate the subsystem and look only within that subsystem for hazards. The effect of a SSHA hazard goes to the subsystem boundary only. The SHA identifies hazards at the SSHA interface and includes interface boundary causal factors.
3. Start the SSHA by populating the SSHA worksheet with hazards identified from the PHA. Evaluate the subsystem components to identify the specific causal factors to these hazards. In effect, the PHA functional hazards and energy source hazards are transferred to the SSHA subsystem responsible for those areas.
4. Identify new hazards and their causal factors by evaluating the subsystem hardware components and software modules. Use analysis aids to help recognize and identify new hazards, such as TLMs, hazard checklists, lesson learned, mishap investigations and hazards from similar systems.
5. Most hazards will be inherent type hazards (contact with high voltage, excessive weight, fire, etc.). Some hazards may contribute to system hazards (e.g., inadvertent missile launch), but generally several subsystems will be required for this type of system hazard (thus need for SHA).
6. Consider erroneous input to subsystem as the cause of a subsystem hazard (command fault).
7. The PHA and SSHA hazards establish the TLMs. The TLMs are used in the SHA for hazard identification. Continue to establish TLMs and SCFs as the SSHA progresses, and utilize them in the analysis.
8. A hazard write-up in the SSHA worksheet should be clear and understandable with as much information as necessary to understand the hazard.

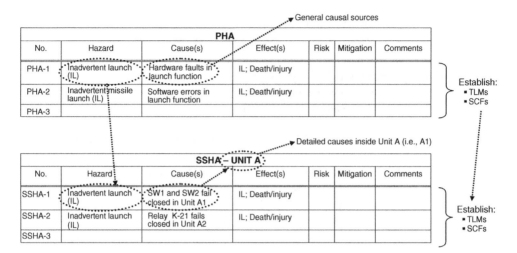

Figure 9.4 SSHA guidelines.

9. The SSHA hazard column does not have to contain all three elements of a hazard: HE, initiating mechanisms (IMs), and outcome (O). The combined columns of the SSHA worksheet can contain all three components of a hazard. For example, it is acceptable to place the HE in the hazard section, the IMs in the cause section and the O in the effect section. The hazard, causes, and effects columns should together completely describe the hazard. These columns should provide the three sides of the hazard triangle.

10. The SSHA does not evaluate system functions but only functions that reside entirely within the subsystem. Functions tend to cross subsystem boundaries and are, therefore, evaluated in the SHA.

Figure 9.4 shows how to apply the SSHA guidelines when using the SSHA worksheet.

9.8 SSHA EXAMPLE: ACE MISSILE SYSTEM

In order to demonstrate the SSHA methodology, the same hypothetical small missile system from Chapters 7 and 8 will be used. The basic system design information provided is shown in Figure 9.5.

Figure 9.6 lists the major system components, functions, phases, and energy sources that should be considered for the SSHA. The major segments of the system are the missile and the weapon control system (WCS).

Figure 9.7 shows the basic planned operational phases for the missile system.

Table 9.2 shows the indentured equipment list for the Ace Missile System, which will be used for the SSHA. Note that this IEL has been expanded in detail from the PHA.

The SSHA is initiated when sufficient subsystem detailed design data becomes available. The IEL given in Table 9.2 shows the information available for the PHA and for the SSHA. The IEL shows all of the system equipment as it is "indentured" by system, subsystem, assembly, unit, and component. It is a hierarchy of equipment that establishes relationships, interfaces, and equipment types.

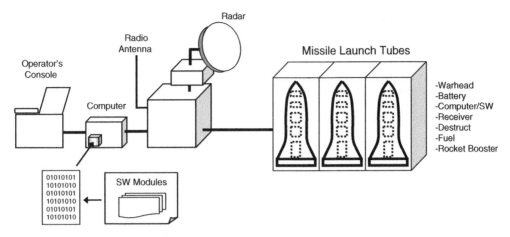

Figure 9.5   *Ace Missile System.*

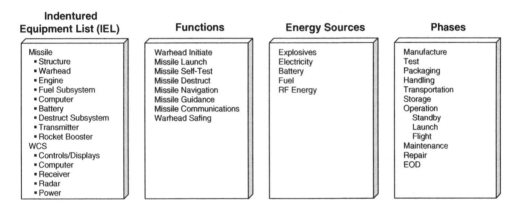

Figure 9.6   *Ace Missile System component list and function list.*

Figure 9.7   *Ace missile functional flow diagram of operational phases.*

When establishing which equipment level should go into the PHA and which should go into the SSHA, the basic ground rule is that the higher level of detail goes into the PHA, and the more detailed breakdown goes into the SSHA. Sometimes, the decision is quite obvious, while at other times the decision is somewhat arbitrary. In this example, the computer software would be included in the PHA as a general category, and it would also be included in the SSHA when the indentured list is continued to a lower level of detail.

It is also helpful when performing the SSHA to utilize functional flow diagrams of the system if they are available. The FFD shows the steps that must take place in order to perform a particular system function. The FFD helps identify subsystem interfaces and relationships that

TABLE 9.2   Ace Missile System IEL

No.	Indentured Equipment List	PHA Level	SSHA Level
1.0	Missile structure subsystem	X	X
	Structural frame		X
	skin		X
	Deployable fins		X
	Flight controls		X
2.0	Missile warhead subsystem	X	X
	2.1 Electronics		X
	2.2 Fuze		X
	2.3 Safe and arm (S&A) device		X
	2.4 High explosives		X
3.0	Missile engine subsystem	X	X
	3.1 Jet engine		X
	3.2 Engine controls		X
4.0	Missile fuel subsystem	X	X
	4.1 Fuel tank		X
	4.2 Fuel		X
5.0	Missile computer subsystem	X	X
	5.1 Electronics		X
	5.2 Embedded software		X
6.0	Missile battery subsystem	X	X
	6.1 Battery structure		X
	6.2 Battery electrolyte		X
	6.3 Battery squib		X
7.0	Missile destruct subsystem	X	X
	7.1 Safe and arm (S&A)		X
	7.2 Initiator		X
	7.3 Explosives		X
8.0	Missile transmitter subsystem	X	X
	8.1 Electronics unit		X
	8.2 Power supply		X
9.0	Missile rocket booster subsystem	X	X
	9.1 Initiator		X
	9.2 Solid propellant		X
10.0	WCS controls and displays subsystem	X	X
	8.1 Displays		X
	8.2 Controls		X
11.0	WCS computer subsystem	X	X
	9.1 Electronics		X
	9.2 Embedded software		X
12.0	WCS receiver subsystem	X	X
	10.1 Electronics unit		X
	10.2 Power Supply		X
13.0	WCS radar subsystem	X	X
	11.1 Electronics		X
	11.2 Wave guides		X
14.0	WCS power subsystem	X	X
	12.1 Electrical power		X
	12.1 Circuit breakers		X

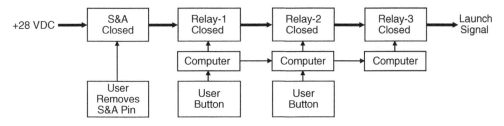

Figure 9.8 *Missile launch signal functional flow diagram.*

can be used in the analysis. Sometimes, it is necessary for the system safety analyst to develop both IELs and FFDs if none has been developed by the project.

Figure 9.8 is an FFD showing the elements and sequence of events required to generate the missile launch signal.

Figure 9.9 is an FFD showing the elements and sequence of events required to generate the missile warhead initiate signal.

During detailed design development, the system design has been modified and improved to include the following:

1. Spring-loaded wings have been added to the missile body.
2. The missile wings are normally closed within the missile body and spring open after launch.
3. As a result of the PHA, a warhead safing function has been added to the system design and the list of functions.

The SSHA does not evaluate functions, unless the function resides entirely within the subsystem. Functions tend to cross subsystem boundaries and are, therefore, evaluated in the SHA that looks for system interface hazards.

The SSHA is started by transferring the PHA hazards to the hazard column of the SSHA. These hazards are then evaluated in detail for subsystem causal factors. Also, with the new detailed design information, additional hazards may be identified. Note that the PHA hazards are repeated for each unit with the subsystem for the SSHA. The causal factors for that unit are then evaluated and applied against the hazard. In some cases, the unit cannot possibly contribute to the hazard, while in other cases the detailed unit causes are identified. Demonstration that the unit cannot contribute to a hazard is a beneficial by-product of the SSHA.

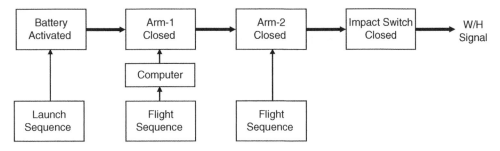

Figure 9.9 *Ace missile warhead initiate signal functional flow diagram.*

Note that the lower level items in the system hierarchy table (i.e., IEL) are addressed by the SSHA, and the higher level information was addressed by the PHA.

If a new SSHA worksheet were started for every IEL item, there would be a minimum of 14 worksheets (14 IEL items). In order to not overwhelm the reader only five worksheets are provided (five IEL items). These samples should provide sufficient information on how to perform the SSHA.

Tables 9.3–9.7 contain the SSHA worksheets for the example missile system.

The following should be noted from the SSHA of the Ace Missile System:

1. The recommended action is not always in the form of an SSR. Additional research may be necessary to convert the recommendation into a meaningful design requirement.
2. Each of the PHA hazards has been carried into the SSHA for further analysis and investigation.
3. Three new hazards were identified by the SSHA that were not in the PHA.
4. The original 10 TLMs from the PHA have not changed and still apply.

9.9   SSHA ADVANTAGES AND DISADVANTAGES

The SSHA technique has the following are advantages:

1. Provides rigor for focusing on hazards and detailed causal factors.
2. Focuses on hazards and not just failure modes as is done in an FMEA.
3. Is cost-effective in providing meaningful results.

There are no disadvantages of the SSHA technique.

9.10   COMMON SSHA MISTAKES TO AVOID

When first learning how to perform a SSHA, it is commonplace to commit some typical errors. The following is a list of common errors made during the conduct of an SSHA.

1. Not listing all concerns or credible hazards. It is important to list all possible suspected or credible hazards and not leave any potential concerns out of the analysis.
2. Failure to document hazards identified but found to be not credible. The SSHA is a historical document encompassing all hazard identification areas that were considered.
3. Not utilizing a structured approach of some type. Always use a worksheet and include all equipment, energy sources, functions, and so on.
4. Not collecting and utilizing common hazard source checklists.
5. Not researching similar systems or equipment for mishaps and lessons learned that can be applied.
6. Not establishing a correct list of hardware, functions, and mission phases.
7. Assuming the reader will understand the hazard description from a brief abbreviated statement filled with project unique terms and acronyms.

TABLE 9.3   Ace Missile System SSHA; Worksheet 1

System:
Subsystem: *Missile Structure Subsystem*

Subsystem Hazard Analysis

Analyst:
Date:

No.	Hazard	Causes	Effects	Mode	IMRI	Recommended Action	FMRI	Comments	Status
SSHA-1	Missile structure fails resulting in unstable missile flight and missile crash	Frame collapses or skin tears open from flight stresses	Unstable flight resulting in crash causing death/injury; incorrect target	Flight	1D	- Use 5x safety factor on structure design	1E	From PHA-1	Open
SSHA-2	Missile body breaks up resulting in fuel leakage; ignition source causing fire	Missile is dropped, frame collapses from loading stresses	Missile fire causing death/injury	PHS&T	1D	- Use 5x safety factor on structure design	1E	From PHA-2	Open
SSHA-3	Missile structure fails during handling resulting in personnel injury	Handling equipment fails, frame collapses from loading stresses	Personnel injury	PHS&T	2D	- Use 5x safety factor on structure design	2E	From PHA-3	Open
SSHA-4	Flight controls fail resulting in errant uncontrolled flight	Hydraulic failure, jam in mechanical flight controls	Unstable flight resulting in crash causing death/injury; incorrect target	Flight	1D	- Design flight controls to prevent jamming	1E	New hazard (not in PHA)	Open
SSHA-5	Fins accidentally deploy during handling	Deploy switch fails, release mechanism fails	Personnel injury	PHS&T	2D	- Design fin deploy removal locking pin for PHS&T	2E	New hazard (not in PHA)	Open

Teaching notes:
• In the SSHA for the missile structure subsystem, the structure components comprising the subsystem will be evaluated – frame, skin, deployable fins, and flight controls.
• The same PHA hazards are analyzed; however, the detailed causal factors are now visible from the component information. In addition, new hazards may become visible.

TABLE 9.4 Ace Missile System SSHA; Worksheet 2

System:
Subsystem: ***Missile Warhead Subsystem***

Analyst:
Date:

No.	Hazard	Causes	Effects	Subsystem Hazard Analysis						
				Mode	IMRI	Recommended Action	FMRI	Comments	Status
SSHA-6	Inadvertent initiation of W/H explosives due to erroneous initiate commands	Fuze failure, software errors	Personnel death/ injury	Flight PHS&T	1D	- Conduct FTA of fuze design - Use 3 independent switches in fuze design	1E	From PHA-4 (human error in controls subsystem)	Open
SSHA-7	Inadvertent detonation of W/H explosives due to external environment	Bullet strike, shrapnel, heat	Personnel death/ injury	PHS&T	1D	- Use insensitive munitions - Provide protective covering when possible	1E	From PHA-5	Open
SSHA-8	Unable to safe warhead after initiate command	Failure in electronics, software error, fuze switches cannot be reversed	Warhead explodes when missile impacts ground, resulting in death/injury	Flight	1D	- Use redundancy in design - Verify software code - Use reversible fuze switches	1E	From PHA-21	Open

Teaching notes:
• In the SSHA for the warhead subsystem, the warhead components comprising the subsystem will be evaluated – electronics, fuze, S&A, and high explosives.
• The same PHA hazards are analyzed; however, the detailed causal factors are now visible from the component information. In addition, new hazards may become visible.

TABLE 9.5   Ace Missile System SSHA; Worksheet 3

System:

Subsystem: *Missile Destruct Subsystem*

Subsystem Hazard Analysis

Analyst:

Date:

No.	Hazard	Causes	Effects	Mode	IMRI	Recommended Action	FMRI	Comments	Status
SSHA-9	Inadvertent missile destruct occurs during flight	Initiator fails, wire short providing voltage to initiator	Missile debris lands on occupied area resulting in death/injury, incorrect target	Flight	1D	- Provide safety interlock inhibiting signal until required	1E	From PHA-7 (command faults from computer)	Open
SSHA-10	Inadvertent missile destruct occurs during ground operations	S&A device fails and power present, initiator faults, wire short	Explosion and debris cause personnel death/injury	Standby PHS&T	1D	- Ensure high reliability of S&A - Isolate critical pins in connector designs	1E	From PHA-8	Open
SSHA-11	Missile destruct fails to occur when commanded	S&A pin not removed, S&A fails in safe mode	Missile strikes undesired area resulting in death/injury, incorrect target	Flight	1D	- Ensure high reliability of S&A	1E	From PHA-9 (command error; radio transmission fault)	Open
SSHA-12	Inadvertent detonation of high explosives	Bullet strike, shrapnel, heat	Explosion and debris cause personnel death/injury	Ground operations	1D	- Use IM explosives	1E	New hazard (not in PHA)	Open

Teaching notes:

• In the SSHA for the missile destruct subsystem, the destruct components comprising the subsystem will be evaluated – S&A, initiator, and high explosives.

• The same PHA hazards are analyzed; however, the detailed causal factors are now visible from the component information. In addition, new hazards may become visible.

159

TABLE 9.6   Ace Missile System SSHA; Worksheet 4

				Subsystem Hazard Analysis					Analyst: Date:	

System:
Subsystem: ***Missile Rocket Booster Subsystem***

No.	Hazard	Causes	Effects	Mode	IMRI	Recommended Action	FMRI	Comments	Status
SSHA-13	Inadvertent ignition	Erroneous commands from hardware faults, software faults, human error, and igniter failure	Inadvertent missile launch resulting in death/injury when missile hits ground	Flight	1D	- Provide safety interlock inhibiting signal until required - Verify software design	1E	From PHA-10	Open
SSHA-14	Explosion of rocket propellant	Manufacturing defect, bullet strike, fire	Explosives ignition resulting in personnel death/injury	PHS&T	1D	- Provide protective covering - Use safe propellant	1E	From PHA-11	Open
SSHA-15	Failure of booster in flight	Manufacturing defect, installation error	Unstable flight resulting in crash causing death/injury, incorrect target	Flight	1D	- QA of manufacturing and installation	1E	From PHA-13	Open

Teaching notes:

• In the SSHA for the missile rocket booster subsystem, the rocket booster components comprising the subsystem will be evaluated – initiator and solid propellant.
• The same PHA hazards are analyzed; however, the detailed causal factors are now visible from the component information. In addition, new hazards may become visible.

TABLE 9.7   Ace Missile System SSHA; Worksheet 5

System:
Subsystem: *WCS Radar Subsystem*

Subsystem Hazard Analysis

Analyst:
Date:

No.	Hazard	Causes	Effects	Mode	IMRI	Recommended Action	FMRI	Comments	Status
SSHA-16	Electromagnetic radiation (EMR) injures exposed personnel	Personnel in excessive RF energy zone	Personnel injury from RF energy	Ground operations	1D	Establish safe personnel distances and place warning in all procedures	1E	From PHA-14	Open
SSHA-17	EMR causes ignition of explosives	Fuel in excessive RF energy zone	Explosives ignition resulting in personnel death/injury	Ground operations	1D	Establish safe explosives distances and place warning in all procedures	1E	From PHA-15	Open
SSHA-18	EMR causes ignition of fuel	Explosives in excessive RF energy zone	Fuel fire resulting in personnel death/injury	Ground operations	1D	Establish safe fuel distances and place warning in all procedures	1E	From PHA-16	Open
SSHA-19	High-voltage radar electronics causes fire	Overheating of high-voltage radar electronics causes fire	Fire causing system damage and/ or death/ injury	Ground operations	1D	Provide temperature warning	1E	From PHA-17	Open
SSHA-20	Personnel injury from high-voltage electronics during maintenance	Personnel touches exposed contacts with high-voltage present	Personnel injury from electrical hazard	Maintenance	2D	Design unit to prevent personnel contact with voltage	2E	From PHA-18	Open

Teaching notes:
• In the SSHA for the WCS radar subsystem, the radar components comprising the subsystem will be evaluated – electronics, wave guide, electrical power, and mechanical controls.
• The same PHA hazards are analyzed; however, the detailed causal factors are now visible from the component information. In addition, new hazards may become visible.

8. Inadequately describing the identified hazard (insufficient detail, too much detail, incorrect hazard effect, wrong equipment indenture level, not identifying all three elements of a hazard, etc.).

9. Inadequately describing the causal factors (the identified causal factor does not support the hazard, the causal factor is not detailed enough, not all of the causal factors are identified, etc.).

10. Inadequately describing the hazard mishap risk index (MRI). For example, the MRI is not stated or is incomplete, the hazard severity level does not support actual hazardous effects, the final MRI is a higher risk than the initial MRI, the final severity level in the risk is less than the initial severity level (sometimes possible, but not usually), the hazard probability is not supported by the causal factors.

11. Providing recommended hazard mitigation methods that do not address the actual causal factor(s).

12. Incorrectly closing the hazard.

13. Not establishing and utilizing a list of TLMs and SCFs.

9.11   SSHA SUMMARY

This chapter discussed the SSHA technique. The following are the basic principles that help summarize the discussion in this chapter:

1. The SSHA is an analysis tool for identifying system hazards, causal factors, mishap risk, and safety recommendations for mitigating risk. It is based upon detailed design information.

2. The primary purpose of the SSHA is to identify and mitigate hazards in the design development process in order to influence the design when the cost impact is minimal.

3. The use of a specialized worksheet provides structure and rigor to the SSHA process.

4. The use of a functional flow diagram, reliability block diagram, and an indentured equipment list greatly aids and simplifies the SSHA process.

5. Do not exclude any thoughts, ideas, or concerns when postulating hazards. In addition to identifying real hazards, it is important to show that certain hazards were suspected and considered, even though they were later found to be not possible for various design reasons. This provides evidence that all possibilities were considered.

6. Write a full, credible, and meaningful hazard description that is understandable to the reader and not just to the analyst. Do not assume the reader understands the hazard from a brief abbreviated statement filled with project unique terms and acronyms.

7. Identify the three elements comprising a hazard using the SSHA worksheet columns:
   - *Hazard Column*   Hazardous element (source)
   - *Causes Column*   Initiating mechanism (mechanism)
   - *Effects Column*   Target/threat (outcome)

FURTHER READINGS

Layton D., *System Safety: Including DOD Standards*, Weber Systems, Inc., 1989.

Roland H. E. and Moriarty B., *System Safety Engineering and Management*, 2nd edition, John Wiley & Sons, Inc., 1990.

Stephans R. A., *System Safety for the 21st Century*, John Wiley & Sons, Inc., 2004.

Stephenson J., *System Safety 2000*, John Wiley & Sons, 1991.

System Safety Society, *System Safety Analysis Handbook*, System Safety Society

Vincoli J. W., *A Basic Guide to System Safety*, Van Nostrand Reinhold, 1993.

Chapter 10

System Hazard Analysis

10.1 SHA INTRODUCTION

The system hazard analysis (SHA) is an analysis methodology for identifying hazards, evaluating risk, and safety compliance at the system level, with a focus on interfaces and safety-critical functions (SCFs). The SHA ensures that identified hazards are understood at the system level, that all causal factors are identified and mitigated, and that the overall system risk is known and accepted. SHA also provides a mechanism for identifying previously unforeseen interface hazards and evaluating causal factors in greater depth.

The SHA is a detailed study of hazards resulting from system integration. This means evaluating all identified hazards and hazard causal factors across subsystem interfaces. The SHA expands upon the subsystem hazard analysis (SSHA) and may use techniques, such as fault tree analysis (FTA), to assess the impact of certain hazards at the system level. The system-level evaluation should include analysis of all possible causal factors from sources such as design errors, hardware failures, human errors, software errors, and so on.

Overall, the SHA performs the following functions:

a. Verifies system compliance with safety requirements contained in the system specifications and other applicable documents.

b. Identifies hazards associated with the subsystem interfaces and system functional faults.

c. Assesses the risk associated with the total system design including software, and specifically of the subsystem interfaces.

d. Recommends actions necessary to eliminate identified hazards and/or control their associated risk to acceptable levels.

Hazard Analysis Techniques for System Safety, Second Edition. Clifton A Ericson, II.
© 2016 John Wiley & Sons, Inc. Published 2016 by John Wiley & Sons, Inc.

10.2   SHA BACKGROUND

This analysis technique falls under the system design hazard analysis type (SD-HAT). The analysis types are described in Chapter 4. There are no alternative names for this technique, although there are other safety analysis techniques that are sometimes used in place of the SHA, such as FTA.

The SHA assesses the safety of the total system design by evaluating the integrated system. The primary emphasis of the SHA, inclusive of hardware, software, and human systems integration (HSI), is to verify that the product is in compliance with the specified and derived system safety requirements (SSRs) at the system level. This includes compliance with acceptable mishap risk levels. The SHA examines the entire system as a whole by integrating the essential outputs from the SSHAs. Emphasis is placed on the interactions and the interfaces of all the subsystems as they operate together.

The SHA evaluates subsystem interrelationships for the following:

a. Compliance with specified safety design criteria

b. Possible independent, dependent, and simultaneous hazardous events including system failures, failures of safety devices, and system interactions that could create a hazard or result in an increase in mishap risk

c. Degradation in the safety of a subsystem or the total system from normal operation of another subsystem

d. Design changes that affect subsystems

e. Effects of human errors

f. Degradation in the safety of the total system from commercial off-the-shelf (COTS) hardware or software

g. Assurance that SCFs are adequately safe from a total system viewpoint and that all interface and common cause failure (CCF) considerations have been evaluated

The SHA can be applied to any system; it is applied during and after detailed design to identify and resolve subsystem interface problems. The SHA technique, when applied to a given system by experienced safety personnel, is thorough in evaluating system-level hazards and causal factors, and ensuring safe system integration. The success of the SHA highly depends on the completion of other system safety analyses, such as the preliminary hazard analysis (PHA), SSHA, safety requirements/criteria analysis (SRCA), and operations and support hazard analysis (O&SHA).

A basic understanding of hazard analysis theory is essential, so is the knowledge of system safety concepts. Experience with and/or a good working knowledge of the particular type of system and subsystems is necessary in order to identify and analyze all hazards. The SHA methodology is uncomplicated and easily learned. Standard SHA forms and instructions have been developed and are included as part of this chapter.

The overall purpose of the SHA is to ensure safety at the integrated system level. The preferred approach utilizes the SHA worksheet presented in this chapter. Particular interface concerns and/or top-level mishaps (TLMs) may require a separate analysis, such as FTA, to identify all the unique details and causal factors of the TLMs. High-consequence TLMs may require a common cause failure analysis (CCFA) to ensure that all design redundancy is truly independent and effective.

The use of this technique is recommended for the identification of system-level interface hazards. There is no alternative hazard analysis technique for the SHA. Other hazard analysis techniques, such as the FTA or CCFA, may be used to supplement the SHA but are *not* recommended as a substitute for the SHA.

10.3   SHA HISTORY

The SHA technique was established very early in the history of the system safety discipline. It was formally instituted and promulgated by the developers of MIL-STD-882. It was developed to ensure safety at the integrated system level.

10.4   SHA THEORY

The intent of SHA is to ensure complete system-level hazard mitigation and demonstrate safety compliance with system-level safety requirements. Two key concepts involved with an SHA are safety-critical function and safety-critical function thread.

Figure 10.1 shows an overview of the basic SHA process and summarizes the important relationships involved. The SHA provides a mechanism to identify all hazard causal factors and their mitigation. It also provides the means to evaluate all subsystem interfaces for hazard causal factors.

System design, design criteria, and previously identified hazards are starting considerations in the SHA. Hazard action records (HARs), from the hazard tracking system (HTS), contain the previously identified hazards, causal factors, and actions resulting from the PHA, SSHA, O&SHA, and HHA. Through a review of the hazards with possible interface concerns, the SHA either identifies interface-related hazards that were previously undiscovered or interface causal factors to existing hazards.

As part of the SHA, it is beneficial if all identified hazards are combined under TLMs. The SHA then evaluates each TLM to determine if all causal factors are identified and adequately mitigated to an acceptable level of system risk. A review of the TLMs in the SHA will indicate if additional in-depth analysis of any sort is necessary, such as for a safety-critical hazard or an interface concern.

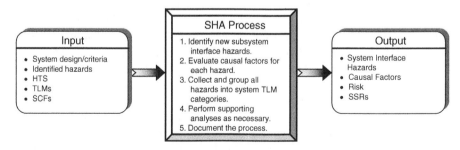

Figure 10.1   *SHA overview.*

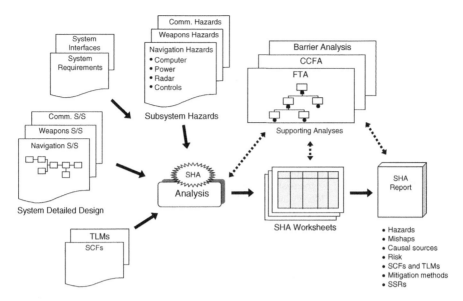

Figure 10.2   *SHA methodology.*

10.5   SHA METHODOLOGY

The SHA is initiated when detailed design data is available. The SHA is usually initiated when the SSHA is essentially complete and when the O&SHA has been initiated but not necessarily completed. The O&SHA information is utilized as it is developed.

An overview of the SHA methodology is shown in Figure 10.2.

The SHA process involves reviewing and utilizing the results of previously identified hazards. This review is primarily focused on the evaluation of subsystem interfaces for potential hazards not yet identified. System and subsystem design safety requirements are utilized by the SHA to evaluate system compliance. Subsystem interface information, primarily from the SSHA, and interface specifications are also utilized by the SHA to assist in the identification of interface-related hazards.

The TLMs must be established during the SHA, if not previously done during the PHA or SSHA. This is accomplished by reviewing all identified hazards and their resulting mishap consequences. Design requirement and guideline documents help establish TLMs, such as the "inadvertent missile launch" TLM.

Table 10.1 lists and describes the basic steps in the SHA process.

Most hazards will have most likely already been identified by the time the PHL, PHA, SSHA, and O&SHA are complete. Since the SHA focuses on interface hazards and causal factors, the SHA sometimes does not result in a large quantity of hazards.

10.6   SHA WORKSHEET

It is desirable to perform the SHA analysis using a worksheet. The worksheet will help to add rigor to the analysis, help record the process and data, and help support justification for the

TABLE 10.1   SHA Process

Step	Task	Description
1	Define system	Define the system, scope, and system boundaries. Define the mission, mission phases, and mission environments. Understand the system design and operation.
2	Plan SHA	Establish SHA goals, definitions, worksheets, schedule, and process. Identify the subsystems and interfaces to be analyzed. It is very likely that SCFs and TLMs will provide the interfaces of most interest
3	Establish safety criteria	Identify applicable design safety criteria, safety precepts/principles, safety guidelines, and safety-critical factors
4	Establish TLMs and SCFs	If not previously done, establish the TLMs from the identified hazards and the system SCFs
5	Identify system interface hazards	Identify system SCFs. Build SCF threads identifying the component/function within the thread
		Evaluate each SCF thread and identify interface-type hazards from the thread component
		Utilize hazard checklists and mishap lessons learned to identify hazards
		Conduct supporting analyses, such as FTA, as found necessary during the SHA.
6	Perform supporting analyses	Certain safety-critical hazards may require a more detailed and/or quantitative analysis, such as an FTA or CCFA. Perform these analyses as found necessary during the SHA
7	Evaluate risk	Identify the level of mishap risk presented to the system by each identified hazard, both before and after recommended hazard mitigations have been established for the system design
8	Recommend corrective action	Recommend any corrective action as found necessary during the SHA
9	Monitor corrective action	Review test results to ensure that safety recommendations and SSRs are effective in mitigating hazards as anticipated
10	Track hazards	Transfer newly identified hazards into the HTS. Update the HTS as hazards, hazard causal factors, and risk are identified in the SHA
11	Document SHA	Document the entire SHA process on the worksheets. Update for new information and closure of assigned corrective actions

identified hazards and safety recommendations. The format of the analysis worksheet is not critical, and typically columnar-type worksheets are utilized.

As a minimum, the following basic information should be obtained from the SHA analysis worksheet:

1. Interface or system hazards
2. Hazard causal factors
3. Hazard risk
4. Information to determine if further causal factor analysis is necessary for a particular hazard

The recommended SHA columnar-type worksheet is shown in Figure 10.3.

The following instructions describe the information required under each column entry of the SHA worksheet:

1. *System*   This entry identifies the system under analysis.
2. *Analyst*   This entry identifies the name of the SHA analyst.

System: (1)			System Hazard Analysis		Analyst: (2) (3) Date:				
No.	TLM / SCF	Hazard	Causes	Effects	IMRI	Recommended Action	FMRI	Status
(4)	(5)	(6)	(7)	(8)	(9)	(10)	(11) (12)	

Figure 10.3 *Recommended SHA worksheet.*

3. *Date*   This entry identifies the date of the SHA analysis.

4. *Hazard Number*   This column identifies the number assigned to the identified hazard in the SHA (e.g., SHA-1, SHA-2, etc.). This is for future reference to the particular hazard source and may be used, for example, in the HAR and the HTS.

5. *TLM/SCF*   This column identifies the TLM or the SCF that is being investigated for possible interface hazards.

6. *Hazard*   This column identifies the specific hazard being postulated. (Remember: Document all hazard considerations, even if they are proven to be nonhazardous.)

7. *Causes*   This column identifies conditions, events, or faults that could cause the hazard to exist, and the events that can trigger the hazardous elements to become a mishap or accident.

8. *Effects*   This column identifies the effects and consequences of the hazard, should it occur. Generally, the worst-case result is the stated effect.

9. *Initial Mishap Risk Index (IMRI)*   This column provides a qualitative measure of mishap risk for the potential effect of the identified hazard, given that no mitigation techniques are applied to the hazard. Risk measures are a combination of mishap severity and probability, and the recommended values from MIL-STD-882 as shown below.

Severity	Probability
I – Catastrophic	A – Frequent
II – Critical	B – Probable
III – Marginal	C – Occasional
IV – Negligible	D – Remote
	E – Improbable

10. *Recommended Action*   This column establishes recommended preventive measures to eliminate or mitigate the identified hazards. Recommendations generally take the form of guideline safety requirements from existing sources, or a proposed mitigation

method that is eventually translated into a new derived SSR intended to mitigate the hazard. SSRs are generated after coordination with the design and requirements organizations. Hazard mitigation methods should follow the preferred order of precedence established in MIL-STD-882 for invoking or developing safety requirements, which are shown below.

Order of Precedence

1 – Eliminate hazard through design selection
2 – Incorporate safety devices
3 – Provide warning devices
4 – Develop procedures and training

Recommended action may also include actions such as further and more detailed analysis using other techniques to determine if and where other mitigation is required.

11. *Final Mishap Risk Index (FMRI)*   This column provides a qualitative measure of the mishap risk for the potential effect of the identified hazard, given that mitigation techniques and safety requirements are applied to the hazard. The same risk matrix table used to evaluate column 9 are also used here. The implementation of recommended mitigation measures should be verified and validated prior to accepting the final mishap residual risk.

12. *Status*   This column states the current status of the hazard, as being either open or closed.

10.7   SHA GUIDELINES

The following are some basic guidelines that should be followed when completing the SHA worksheet:

1. The SHA identifies hazards caused by subsystem interface factors, environmental factors, or common cause factors.

2. The SHA should not be a continuation of subsystem hazards (e.g., personnel contacts high voltage in Unit B) because these types of hazards have been adequately treated by the SSHA. Do not place all SSHA hazards into the SHA.

3. Start the SHA by considering TLMs and SCFs for interface hazards. These are the significant safety area that can provide the sources for identifying system interface hazards.

4. For each SCF, determine the hazardous undesired event(s) for that function. For example, the SCF "missile launch function" creates an undesired event of "inadvertent launch," which may consist of several different interface hazards.

5. For each SC function, create an SCF thread. This thread consists of the items necessary for the safe operation of the SCF.

6. Evaluate and analyze the items in each SCF thread for interface causal factors contributing to the undesired event for the thread.

7. For TLMs that do not directly relate to an SCF, it may be necessary to establish a pseudo-SCF for them. The pseudo-SCF thread then can be evaluated in a similar manner.

8. The SSHA does not evaluate functions, unless the function resides entirely within the subsystem. Functions tend to cross subsystem boundaries and are, therefore, evaluated in the SHA.

9. Perform supporting analyses as determined necessary (e.g., FTA, bent pin analysis, and CCFA).

SCFs and TLMs emphasize safety-critical areas that should receive special attention from the analyst. In many instances, there is a direct relationship between SCFs and TLMs, as some TLMs are the inverse of an SCF. This is the reason for the SCF/TLM column in the SHA worksheet.

Figure 10.4 depicts an SCF thread for a hypothetical missile launch function under normal expected conditions. Note that HSI stands for human system integration (i.e., operator interfaces).

Figure 10.5 shows an example SCF thread for the hypothetical missile launch function. This thread can be used to understand the hazard causal factors within a thread when evaluating the undesired event for that thread.

Figure 10.6 depicts how an SCF thread can be utilized in the SHA.

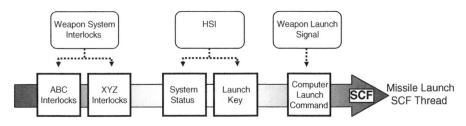

Figure 10.4 *SCF thread with intended results.*

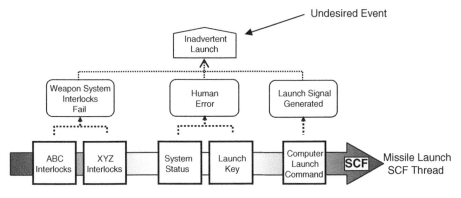

Figure 10.5 *SCF thread with unintended results.*

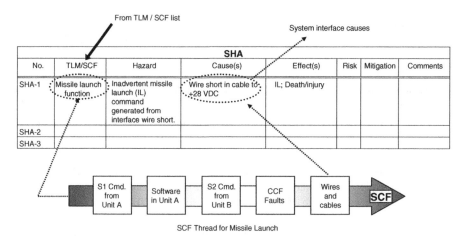

Figure 10.6   *SCF thread used in SHA.*

10.8   SHA EXAMPLE

In order to demonstrate the SHA methodology, the same hypothetical small missile system from Chapters 7 and 8 will be used. The basic system design information provided is shown in Figure 10.7.

Figure 10.8 lists the major system components, functions, phases, and energy sources that should be considered for the SSHA. The major segments of the system are the missile and the weapon control system (WCS).

Figure 10.9 shows the basic planned operational phases for the missile system.

Table 10.2 contains the list of top-level mishaps resulting from previous hazard analyses. These are the TLMs that will be used in the SHA worksheets.

Table 10.3 contains the SHA worksheet for the entire missile weapon system.

Since TLM 1 and TLM 2 are safety-critical mishaps, it has been recommended in SHA-1 and SHA-2 that an FTA is performed on these TLMs to ensure all causal factors have been identified and that no interface problems or common cause faults have been overlooked.

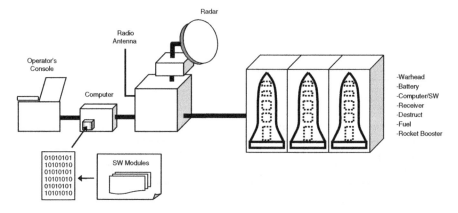

Figure 10.7   *Ace Missile System.*

TABLE 10.2   Ace Missile System TLMs

No.	TLM
1	Inadvertent W/H explosives initiation
2	Inadvertent launch
3	Inadvertent missile destruct
4	Incorrect target
5	Missile fire
6	Missile destruct fails
7	Personnel injury
8	Unknown missile state
9	Inadvertent detonation of explosives
10	Unable to safe warhead

Figure 10.10 is a functional block diagram (FBD) for the missile warhead initiation function. This FBD shows what tasks and functions must transpire in order for missile initiation to occur.

As part of the SHA, FTAs would be performed on TLM 1 and TLM 2 as recommended in the SHA worksheets. To continue this example, a top-level fault tree (FT) will be constructed for TLM 2, "inadvertent warhead initiation." In performing the FTA, the safety analyst will use the FBD, detailed schematics, software code, and so on.

Figure 10.11 contains the top-level FTA for TLM 2. The FTA cannot be completed in this example, because not enough design information has been provided. The top-level FTA is

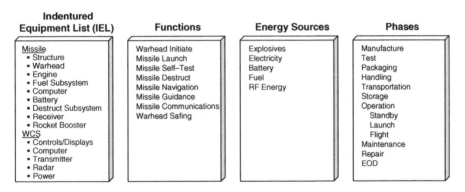

Figure 10.8   Missile system component list and function list.

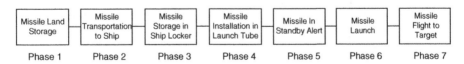

Figure 10.9   Ace missile functional flow diagram of operational phases.

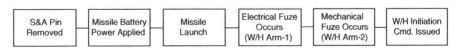

Figure 10.10   Functional block diagram for missile warhead initiation.

TABLE 10.3　Missile System SHA Worksheet

System:

Analyst:
Date:

Hazard No.	TLM/SCF	Hazard	System Hazard Analysis Causes	Effects	IMRI	Recommended Action	FMRI	Status
SHA-1	Missile launch function	Inadvertent missile launch signal is generated by a short circuit in missile interface cable	After missile is installed in launch tube, the S&A is armed. A short in missile interface cable between WCS and missile provides +28 V to missile launch initiator	Inadvertent launch of missile, resulting in death/injury	1D	Design connector such that missile launch signal pin is isolated from pins with voltage Perform FTA on IL to study in detail	1E	Open
SHA-2	Warhead initiation function	Premature warhead initiation signal is generated by damaged fuze and impact switch due to common cause shock environment	The impact switch and three fuze switches are failed, closed from missile being dropped. After the missile is launched and warhead power is applied, the initiation signal goes immediately to warhead	Inadvertent warhead initiation, resulting in death/injury	1D	Design to prevent shock sensitivity Do not allow the use of dropped missiles Perform FTA on W/H initiation to study in detail	1E	Open
SHA-3	Missile destruct function	Inadvertent missile destruct signal is generated by erroneous radio link	Radio link with missile is faulty due to weather, and missile computer erroneously interprets radio transmission as requesting a missile destruct	Inadvertent initiation of missile destruct, resulting in death/injury	1D	Design destruct command to be a unique word that cannot easily be created Design such that two different destruct commands must be sent and interpreted by computer	1E	Open
SHA-4	Missile destruct function	Missile destruct fails when commanded by operator due to RF interference or jamming	RF interference or jamming in area prevents correct radio transmission to missile resulting in loss of destruct command	Inability to destroy errant missile, resulting in death/injury upon impact	1D	Design such that multiple RF channels can be used	1E	Open

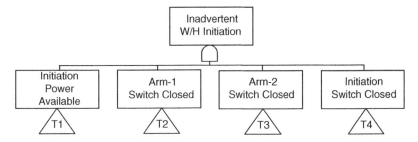

Figure 10.11   *Inadvertent warhead ignition FT.*

shown here to demonstrate the SHA process. A completed FTA may uncover CCFs that bypass the designed-in redundancy. Further, more detailed, information on FTA can be found in Chapter 15.

The following results should be noted from the SHA of the Ace Missile System:

1. All of the identified hazards fall under one of the 10 TLM categories.
2. The recommended action is not always in the form of a direct SSR. Additional research may be necessary to convert the recommendation into a meaningful design requirement.
3. The SHA often recommends further detailed analysis of a particular hazard, TLM, or safety concern. In this case, the SHA recommended that FTAs be performed on TLM 1 and TLM 2. When complete, these FTAs should show that these undesired events are within acceptable levels of probability of occurrence and that no critical common cause faults exist. If the FTAs do not show acceptable risk, then new interface hazard causal factors may have been identified that have not been mitigated.

10.9   SHA ADVANTAGES AND DISADVANTAGES

The SHA technique affords the following advantages:

1. Identifies system interface type hazards.
2. Consolidates hazards to ensure that all causal factors are thoroughly investigated and mitigated.
3. Identifies critical system-level hazards that must be evaluated in more detail through the use of other analysis techniques.
4. Provides the basis for making an assessment of overall system risk.

There are no disadvantages of the SHA technique.

10.10   COMMON SHA MISTAKES TO AVOID

When first learning how to perform an SHA, it is commonplace to commit some typical mistakes. The following is a list of errors often encountered during the conduct of a SHA:

1. The causal factors for a hazard are not thoroughly investigated.

2. The mishap risk index (MRI) risk severity level does not appropriately support the identified hazardous effects.

3. Hazards are closed prematurely without a complete causal factor analysis and test verification.

4. Failure to consider common cause events and dependent events.

5. A series of FTAs are used in place of the SHA worksheets. The FTA should be a supporting analysis only.

10.11   SHA SUMMARY

This chapter discussed the SHA technique. The following are the basic principles that help summarize the discussion in this chapter:

1. The primary purpose of the SHA is to ensure safety of the total system, which involves ensuring that system risk is acceptable. The SHA assesses system compliance with safety requirements and criteria, through traceability of hazards and verification of SSRs.

2. The SHA identifies system hazards overlooked by other analyses, particularly those types of hazards perpetuated through subsystem interface incompatibilities.

3. SCFs and safety-critical TLMs may require more detailed analysis by other techniques (e.g., FTA, CCFA, etc.) to ensure that all causal factors are identified and mitigated.

4. The use of worksheet forms provides structure and rigor to the SHA process.

FURTHER READINGS

Layton, D., *System Safety: Including DOD Standards*, Weber Systems, Inc., 1989.

Roland, H. E. and Moriarty, B. *System Safety Engineering and Management*, John Wiley & Sons, Inc., 2nd edition, 1990.

Stephans, R. A., *System Safety for the 21st Century*, John Wiley & Sons, Inc., 2004.

Stephenson, J., *System Safety 2000*, John Wiley & Sons, Inc., 1991.

System Safety Society, *System Safety Analysis Handbook*, System Safety Society 1999, Virginia, www.system-safety.org.

Vincoli, J. W., *A Basic Guide to System Safety*, Van Nostrand Reinhold, 1993.

Chapter *11*

Operating and Support Hazard Analysis

11.1 O&SHA INTRODUCTION

The Operating and support hazard analysis (O&SHA) is an analysis technique for identifying hazards in system operational tasks, along with the hazard causal factors, effects, risk, and mitigating methods. The O&SHA is an analysis technique for specifically assessing the safety of operations by integrally evaluating operational procedures, the system design, and the human system integration (HSI) interface. The scope of the O&SHA includes normal operation, test, installation, maintenance, repair, training, storage, handling, transportation, and emergency/rescue operations. Consideration is given to system design, operational design, hardware failure modes, human error, and task design. Human factors and HSI design considerations are a large factor both in system operation and, therefore, in the O&SHA. The O&SHA is conducted during system development in order to affect the design for future safe operations.

11.2 O&SHA BACKGROUND

This analysis technique falls under the operations design hazard analysis type (OD-HAT) because it evaluates procedures and tasks performed by humans. The basic analysis types are described in Chapter 4. An alternative name for this analysis technique is the operating hazard analysis (OHA).

The purpose of the O&SHA is to ensure the safety of the system and personnel in the performance of system operation. Operational hazards can be introduced by the system design,

Hazard Analysis Techniques for System Safety, Second Edition. Clifton A Ericson, II.
© 2016 John Wiley & Sons, Inc. Published 2016 by John Wiley & Sons, Inc.

procedure design, human error, and/or the environment. The overall O&SHA goal are as follows:

1. Provide safety focus from an operations and operational task viewpoint.
2. Identify task or operational oriented hazards caused by design, hardware failures, software errors, human error, timing, and the like.
3. Assess the operations mishap risk.
4. Identify design system safety requirements (SSRs) to mitigate operational task hazards.
5. Ensure all operational procedures are safe.

The O&SHA is conducted during system development and is directed toward developing safe design and procedures to enhance safety during operation and maintenance. The O&SHA identifies the functions and procedures that could be hazardous to personnel or, through personnel errors, could create hazards to equipment, personnel, or both. Corrective action resulting from this analysis is usually in the form of design requirements and procedural inputs to operating, maintenance, and training manuals. Many of the procedural inputs from system safety are in the form of caution and warning notes.

The O&SHA is applicable to the analysis of all types of operations, procedures, tasks, and functions. It can be performed on draft procedural instructions or detailed instruction manuals. The O&SHA is specifically oriented toward the hazard analysis of tasks for system operation, maintenance, repair, test, and troubleshooting.

The O&SHA technique provides sufficient thoroughness in identifying and mitigating operation- and support-type hazards when applied to a given system/subsystem by experienced safety personnel. A basic understanding of hazard analysis theory is essential, so is the knowledge of system safety concepts. Experience with or a good working knowledge of the particular type of system and subsystem is necessary in order to identify and analyze hazards that may exist within procedures and instructions. The methodology is uncomplicated and easily learned. Standard O&SHA forms and instructions have been developed that are included as part of this chapter.

The O&SHA evaluates the system design and operational procedures to identify hazards and to eliminate or mitigate operational task hazards. The O&SHA can also provide insight into design changes that might adversely affect operational tasks and procedures. The O&SHA effort should start early enough during system development to provide inputs to the design, and prior to system test and operation. The O&SHA worksheet provides a format for entering the sequence of operations, procedures, tasks, and steps necessary for task accomplishment. The worksheet also provides a format for analyzing this sequence in a structured process that produces a consistent and logically reasoned evaluation of hazards and controls.

Although some system safety programs (SSPs) may attempt to replace the O&SHA with a preliminary hazard analysis (PHA), this is *not* recommended since the PHA is not oriented specifically for the analysis of operational tasks. The use of the O&SHA technique is recommended for identification and mitigation of operational and procedural hazards.

11.3   O&SHA HISTORY

The O&SHA technique was established very early in the history of the system safety discipline. It was formally instituted and promulgated by the developers of MIL-STD-882. It was developed to ensure the safe operation of an integrated system. It was originally called OHA, but was later expanded in scope and renamed O&SHA to more accurately reflect all operational support activities.

11.4 O&SHA DEFINITIONS

To facilitate a better understanding of O&SHA, the following definitions of specific terms are provided:

11.4.1 Operation

An operation is the performance of procedures to meet an overall objective. For example, a missile maintenance operation may be "replacing missile battery." The objective is to perform all the necessary procedures and tasks to replace the battery.

11.4.2 Procedure

A procedure is a set of tasks that must be performed to accomplish an operation. Tasks within a procedure are designed to be followed sequentially to properly and safely accomplish the operation. For example, the above battery replacement operation may be comprised of two primary procedures: (1) battery removal and (2) battery replacement. Each of these procedures contains a specific set of tasks that must be performed.

11.4.3 Task

A task is an element of work, which together with other elements of work comprises a procedure. For example, battery removal may consist of a series of sequential elements of work, such as power shutdown, compartment cover removal, removal of electrical terminals, unbolting of battery hold down bolts, and battery removal.

Figure 11.1 portrays these definitions and their interrelationships. It should be noted that tasks might be further broken down into subtasks, sub-subtasks, and so on.

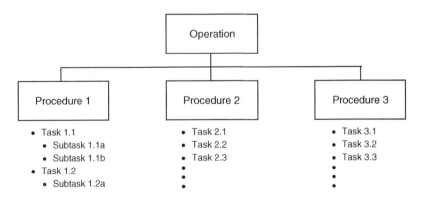

Figure 11.1 *Operation definitions.*

11.5 O&SHA THEORY

Figure 11.2 shows an overview of the basic O&SHA process and summarizes the important relationships involved. The intent of the O&SHA is to identify and mitigate hazards associated with the operational phases of the system, such as deployment, maintenance, calibration, test, training, and so on. This process consists of utilizing both design information and known hazard information to verify complete safety coverage and control of hazards. Operational task hazards are identified through the meticulous analysis of each detailed procedure that is to be performed during system operation or support.

Input information for the O&SHA consists of all system design and operation information, operation and support manuals, and hazards identified by other program hazard analyses. Typically, the following types of information are available and utilized in the O&SHA:

1. Hazards and top-level mishaps (TLMs) identified from the preliminary hazard list (PHL), PHA, subsystem hazard analysis (SSHA), system hazard analysis (SHA), and health hazard assessment (HHA).
2. Engineering descriptions of the system, support equipment, and facilities.
3. Written procedures and manuals for operational tasks to be performed.
4. Chemicals, materials, and compounds used in the system production, operation, and support.
5. Human factors engineering data and reports.
6. Lessons learned, including human error mishaps.
7. Hazard checklists.

The primary purpose of the O&SHA is to identify and mitigate hazards resulting from the system fabrication, operation, and maintenance. As such, the following information is typically output from the O&SHA:

1. Task hazards
2. Hazard causal factors (materials, processes, excessive exposures, errors, etc.)
3. Risk assessment
4. Safety design requirements to mitigate the hazard
5. The identification of caution and warning notes for procedures and manuals
6. The identification of special HSI design methods to counteract human error-related hazards

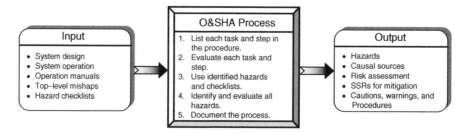

Figure 11.2 O&SHA overview.

Generally, the O&SHA evaluates manuals and procedural documentation that are in the draft stage. The output of the O&SHA will add cautions and warnings, and possibly new procedures to the final documentation.

11.6 O&SHA METHODOLOGY

The O&SHA process methodology is shown in Figure 11.3. The idea behind this process is that different types of information are used to stimulate hazard identification. The analyst employs hazard checklists, mishap checklists, and system tools. Typical system tools might include functional flow diagrams (FFDs), operational sequence diagrams (OSDs), and indentured task lists (ITLs).

Table 11.1 lists and describes the basic steps of the O&SHA process. The O&SHA process involves performing a detailed analysis of each step or task in the operational procedure under investigation.

The objective of the O&SHA is to identify and mitigate hazards that might occur during the operation and support of the system. The humans should be considered an element of the total system, both receiving inputs and initiating outputs during the conduct of this analysis. Hazards may result due to system design, support equipment design, test equipment, human error, HSI errors, and/or incorrect procedures. O&SHA consideration includes the environment, personnel, procedures, and equipment involved throughout the operation of a system. The O&SHA may be performed on such activities as testing, installation, modification, maintenance, support, transportation, ground servicing, storage, operations, emergency escape, egress, rescue, postaccident responses, and training. The O&SHA also ensures operation and maintenance manuals properly address safety and health requirements. The O&SHA may

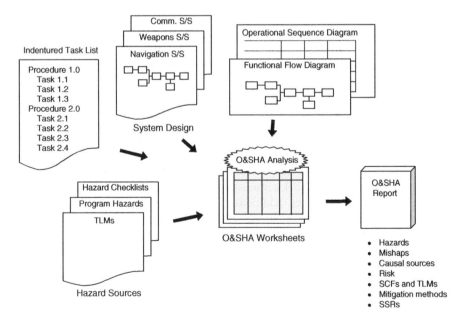

Figure 11.3 *O&SHA methodology.*

TABLE 11.1 O&SHA Process

Step	Task	Description
1	Define system operation	Define scope and boundaries of the operation to be performed. Understand the operation and its objective
2	Acquire data	Acquire all necessary design and operational data needed for the analysis. These data include both schematics and operation manuals
3	List procedures and detailed tasks	Make a detailed list of all procedures and tasks to be considered in the O&SHA. This list can be taken directly from manuals, procedures, or operational plans that are already written or in draft form
4	Conduct O&SHA	a. Input task list into the O&SHA worksheets b. Evaluate each item in the task list and identify hazards for the task c. Compare procedures and tasks with hazard checklists d. Compare procedures and tasks with lessons learned e. Be cognizant of task relationships, timing, and concurrent tasks when identifying hazards
5	Evaluate risk	Identify the level of mishap risk presented by the hazard with, and without, mitigations in the system design
6	Recommend corrective action	Recommend corrective action necessary to eliminate or mitigate identified hazards. Work with the design organization to translate the recommendations into SSRs. Also, identify safety features already in the design or procedures that are present for hazard mitigation
7	Ensure caution and warnings are implemented	Review documented procedures to ensure that corrective action is being implemented. Ensure that all caution and warning notes are inputted in manuals and/or posted on equipment appropriately, as recommended in the O&SHA
8	Monitor corrective action	Participate in verification and validation of procedures, and review the results to ensure that SSRs effectively mitigate hazards
9	Track hazards	Transfer identified hazards into the hazard tracking system (HTS). Update hazards in the HTS as causal factors and risk are identified in the O&SHA
10	Document O&SHA	Document the entire O&SHA process on the worksheets. Update for new information and closure of assigned corrective actions

also evaluate adequacy of operational and support procedures used to eliminate, control, or abate identified hazards or risks.

The O&SHA effort should start early enough to provide inputs to the design and prior to system test and operation. The O&SHA is most effective as a continuing closed loop iterative process, whereby proposed changes, additions, and formulation of functional activities are evaluated for safety considerations, prior to formal acceptance.

O&SHA considerations should include the following:

1. Potentially hazardous system states under operator control
2. Operator hazards resulting from system design (hardware aging and wear, distractions, confusion factors, worker overload, operational tempo, exposed hot surfaces, environmental stimuli, etc.)
3. Operator hazards resulting from potential human error

4. Errors in procedures and instructions

5. Activities that occur under hazardous conditions, their time periods, and the actions required to minimize risk during these activities/time periods

6. Changes needed in functional or design requirements for system hardware/software, facilities, tooling, or support/test equipment to eliminate or control hazards or reduce associated risks

7. Requirements for safety devices and equipment, including personnel safety and life support equipment

8. Warnings, cautions, and special emergency procedures (e.g., egress, rescue, escape, render safe, explosive ordnance disposal, back-out, etc.), including those necessitated by failure of a computer software-controlled operation to produce the expected and required safe result or indication

9. Requirements for packaging, handling, storage, transportation, maintenance, and disposal of hazardous materials

10. Requirements for safety training and personnel certification

11. The safety effect of nondevelopmental items (NDI) and commercial off-the-shelf (COTS) items, both in hardware and in software, during system operation

12. The safety effect of concurrent tasks and/or procedures

11.7 O&SHA WORKSHEET

The O&SHA is a detailed hazard analysis utilizing structure and rigor. It is desirable to perform the O&SHA using a specialized worksheet. Although the specific format of the analysis worksheet is not critical, as a minimum, the following basic information is required from the O&SHA:

1. Specific tasks under analysis

2. Identified hazard

3. Effect of hazard

4. Hazard causal factors (varying levels of detail)

5. Recommended mitigating action (design requirement, safety devices, warning devices, special procedures and training, caution and warning notes, etc.)

6. Risk assessment (initial and final)

Figure 11.4 shows the columnar format O&SHA worksheet recommended for SSP usage. This particular worksheet format has proven to be useful and effective in many applications and it provides all the information necessary from an O&SHA.

The following instructions describe the information required under each column entry of the O&SHA worksheet:

1. *System*   This entry identifies the system under analysis.

2. *Operation*   This entry identifies the system operation under analysis.

3. *Analyst*   This entry identifies the name of the O&SHA analyst.

4. *Date*   This entry identifies the date of the O&SHA analysis.

System: ① Operation: ②		Operating and Support Hazard Analysis				Analyst: ③ Date: ④			
Task	Hazard No.	Hazard	Causes	Effects	IMRI	Recommended Action	FMRI	Comments	Status
⑤	⑥	⑦	⑧	⑨	⑩	⑪	⑫	⑬	⑭

Figure 11.4 *Recommended O&SHA worksheet.*

5. *Task*   This column identifies the operational task being analyzed. List and describe each of the steps or tasks to be performed. If possible, include the purpose and the mode or phase of operation being performed.

6. *Hazard Number*   This is the number assigned to the identified hazard in the O&SHA (e.g., O&SHA-1, O&SHA-2, etc.). This is for future reference to the particular hazard source and may be used, for example, in the hazard action record (HAR). The hazard number is at the end of the worksheet because not all tasks listed will have hazards associated with them, and this column could be confusing at the front of the worksheet.

7. *Hazard*   This column identifies the specific hazard, or hazards, that could possibly result from the task. (Remember: document all hazard considerations, even if they are later proven to be nonhazardous).

8. *Causes*   This column identifies conditions, events, or faults that could cause the hazard to exist, and the events that can trigger the hazardous elements to become a mishap or accident.

9. *Effects* This column identifies the effect and consequences of the hazard, should it occur. The worst-case result should be the stated effect.

10. *Initial Mishap (IMRI) Risk*   This column provides a qualitative measure of mishap risk significance for the potential effect of the identified hazard, given that no mitigation techniques are applied to the hazard. Risk measures are a combination of mishap severity and probability, and the recommended values from MIL-STD-882 are shown in the following table.

Severity	Probability
1 – Catastrophic	A – Frequent
2 – Critical	B – Probable
3 – Marginal	C – Occasional
4 – Negligible	D – Remote
	E – Improbable

11. *Recommended Action*   This column establishes recommended preventive measures to eliminate or mitigate the identified hazards. Recommendations generally take the form of guideline safety requirements from existing sources, or a proposed mitigation method that is eventually translated into a new derived SSR intended to mitigate the hazard. SSRs are generated after coordination with the design and requirements organizations. Hazard mitigation methods should follow the preferred order of precedence established in MIL-STD-882 for invoking or developing safety requirements, which are shown in the following table.

Order of Precedence

1 – Eliminate hazard through design selection
2 – Control hazard through design methods
3 – Control hazard through safety devices
4 – Control hazard through warning devices
5 – Control hazard through procedures and training

12. *Final Mishap Risk (FMRI)*   This column provides a qualitative measure of mishap risk significance for the potential effect of the identified hazard, given that mitigation techniques and safety requirements are applied to the hazard. The same values used in column 10 are also used here.
13. *Comments*   This column provides a place to record useful information regarding the hazard or the analysis process that are not noted elsewhere.
14. *Status*   This column states the current status of the hazard, as being either open or closed.

Note that in this analysis methodology, each and every procedural task is listed and analyzed. For this reason, not every entry in the O&SHA form will constitute a hazard, since not all tasks will be hazardous. This process documents that the O&SHA considered all tasks and identifies which tasks are hazardous and which are not.

11.8   O&SHA HAZARD CHECKLISTS

Hazard checklists provide a common source for readily recognizing hazards. Since no single checklist is ever really adequate in itself, it becomes necessary to develop and utilize several different checklists. Utilizing several checklists may result in some repetition, but complete coverage of all hazardous elements will be more certain. If a hazard is duplicated, it should be recognized and condensed into one hazard. Remember that a checklist should never be considered a complete and final list but merely a catalyst for stimulating hazard recognition.

Chapter 7 on PHL analysis provided some example general-purpose hazard checklists applicable to system design. Figure 11.5 provides an example hazard checklist applicable to operational tasks. This example checklist is not intended to represent all hazard sources but some typical considerations for an O&SHA.

1. Work Area	4. Machines
Tripping, slipping, and corners	Cutting, punching, and forming
Illumination	Rotating shafts
Floor load, piling	Pinch points
Ventilation	Flying pieces
Moving objects	Projections
Exposed surfaces – hot, electric	Protective equipment
Cramped quarters	5. Tools
Emergency exits	No tools
2. Materials Handling	Incorrect tools
Heavy, rough, and sharp	Damaged tools
Explosives	Out of tolerance tools
Flammable	6. Emergency
Awkward, fragile	Plans, procedures, and numbers
3. Clothing	Equipment
Loose, ragged, and soiled	Personnel
Necktie, jewelry	Training
Shoes, high heels	7. Safety Devices
Protective	Fails to function
	Inadequate

Figure 11.5   *Operational hazard checklist.*

11.9   O&SHA SUPPORT TOOLS

The functional flow diagram (or functional block diagram) simplifies system design and operation for clarity and understanding. The use of the FFD for O&SHA evaluation of procedures and tasks is recommended.

Indentured equipment lists were defined in Chapter 2 as a valuable aid in understanding systems and performing hazard analyses. Indentured tasks lists are also developed to assist in the design and development of operations.

Operational sequence diagrams are a special type of diagrams that are used to define and describe a series of operations and tasks using a graphical format. The OSD plots a flow of information, data, or energy relative to time (actual or sequential) through an operationally

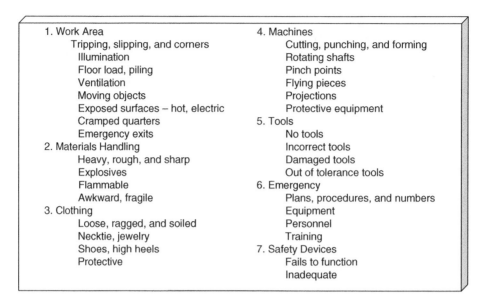

Figure 11.6   *Operational sequence diagram symbols.*

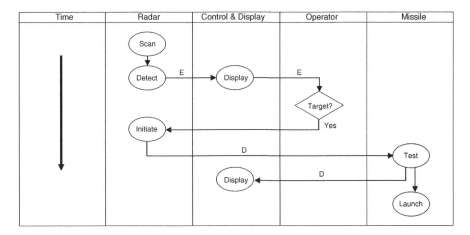

Figure 11.7 *Example operational sequence diagram.*

defined system using standard symbols to relate actions taken. Actions in the OSD may include inspections, data transmittal/receipt, storage, repair, decision points, and so on. The OSD helps to display and simplify activities in a highly complex system and identify procedural hazards.

Symbols used in the OSD are adapted from the ASME flow chart standards, as shown in Figure 11.6. The OSD methodology was originally defined in MIL-H-46855, Human Engineering Requirements for Military Systems, Equipment and Facilities.

An example OSD is shown in Figure 11.7 for a missile system. Note that the subsystems are denoted along the top, and time is denoted in the left-hand side column.

11.10 O&SHA GUIDELINES

The following are some basic guidelines that should be followed when completing the O&SHA worksheet:

1. Remember that the objective of the O&SHA is to evaluate the system design and operational procedures to identify hazards and to eliminate or mitigate operational task hazards.
2. Start the O&SHA by populating the O&SHA worksheet with the specific tasks under investigation.
3. A hazard write-up in the O&SHA worksheet should be clear and understandable with as much information as necessary to understand the hazard.
4. The O&SHA hazard column does not have to contain all three elements of a hazard: hazardous element (HE), initiating mechanisms (IMs), and outcome (O). The combined columns of the SSHA worksheet can contain all three components of a hazard. For example, it is acceptable to place the HE in the hazard section, the IMs in the cause section, and the O in the effect section. The hazard, causes, and effects columns should together completely describe the hazard. These columns should provide the three sides of the hazard triangle (see Chapter 3).

TABLE 11.2   Example Electrical Outlet Replacement Procedure

Step	Description of Task
1.0	Locate circuit breaker
2.0	Open circuit breaker
3.0	Tag circuit breaker
4.0	Remove receptacle wall plate – 2 screws
5.0	Remove old receptacle – 2 screws
6.0	Unwire old receptacle – disconnect 3 wires
7.0	Wire new receptacle – connect 3 wires
8.0	Install new receptacle – 2 screws
9.0	Install old wall plate – 2 screws
10.0	Close circuit breaker
11.0	Remove circuit breaker tag
12.0	Test circuit

11.11   O&SHA EXAMPLES

11.11.1   Example 1

To demonstrate the O&SHA methodology, a hypothetical procedure will be analyzed. The selected example procedure is to replace an electrical outlet receptacle in a weapons maintenance facility. The receptacle contains 220 VAC, so the procedure is a hazardous operation. The detailed set of tasks to accomplish this procedure is provided in Table 11.2.

Tables 11.3–11.5 contain the O&SHA worksheets for this example. The following should be noted from this example analysis:

1. Every procedural task is listed and evaluated on the worksheet.
2. Every task may not have an associated hazard.
3. Even though a task may not have an identified hazard, the task is still documented in the analysis to indicate that it has been reviewed.

11.11.2   O&SHA Example 2

In order to further demonstrate the O&SHA methodology, the same hypothetical Ace Missile System from Chapters 7–9 will be used. The system design is shown again in Figure 11.8.

Figure 11.9 shows the basic planned operational phases for the Ace Missile System. Phase 4 has been selected for O&SHA in this example.

The detailed set of tasks to accomplish the phase 4 procedure is provided in Table 11.6.

It should be noted that in a real-world system the steps in Table 11.3 would likely be more refined and consist of many more discrete and detailed steps. The steps have been kept simple here for the purpose of demonstrating the O&SHA technique.

Tables 11.7–11.10 contain the O&SHA worksheets for the Ace Missile System example.

TABLE 11.3   O&SHA Example 1; Worksheet 1

System: Missile Maintenance Facility
Operation: Replace 220 V Electrical Outlet

Operating and Support Hazard

Analyst:
Date:

Task	Hazard No.	Hazard	Causes	Analysis	IMRI	Recommended Action	FMRI	Comments	Status
1.0 Locate CB Locate panel and correct circuit breaker (CB) inside the panel	OHA-1	Wrong CB is selected	Human error	Circuit is not de-energized, live contacts are touched later in procedure resulting in electrocution	1D	- Warning note to test contacts prior to touching wires in Task 6	1E		Open
2.0 Open CB Manually open the CB handle	OHA-2	CD is not actually opened	Internal CB contacts are failed closed; human error	Circuit is not de-energized, live contacts are touched later in procedure resulting in electrocution	1D	- Warning note to test contacts prior to touching wires in Task 6	1E		Open
3.0 Tag CB Place tag on CB Indicating that it is not to be touched during maintenance.	OHA-3	Wrong CB is tagged and untagged CB is erroneously closed	Another person closes unmarked CB	Circuit is not de-energized, resulting in electrocution	1D	- Warning note to test contacts prior to touching wires in Task 6	1E		Open
4.0 Remove wall plate. Remove two screws from outlet wall plate; remove wall plate	—	None							

TABLE 11.4 O&SHA Example 1; Worksheet 2

System: Missile Maintenance Facility
Operation: Replace 220 V Electrical Outlet

Operating and Support Hazard Analysis

Analyst:
Date:

Task	Hazard No.	Hazard	Causes	Effects	IMRI	Recommended Action	FMRI	Comments	Status
5.0 Remove receptacle Remove two screws from old outlet receptacle; pull receptacle out from wall	—	None							
6.0 Unwire old receptacle. Disconnect the three wires from old receptacle wire mounts	OSHA-4	High voltage (220 VAC) is still on circuit	CB not opened; wires are energized	Electrocution	1D	- Warning note to test contacts prior to touching	1E		Open
7.0 Wire new receptacle Connect the three wires to the new receptacle wire mounts	—	Wires are incorrectly connected to wrong wire mounts	Human error; error in manual	Does not matter since AC current					
8.0 Install receptacle Install new receptacle into wall and install two receptacle screws	—	None							

TABLE 11.5   O&SHA Example 1; Worksheet 3

System: Missile Maintenance Facility Operation: Replace 220 V Electrical Outlet							Operating and Support Hazard Analysis			Analyst: Date:	
Task	Hazard No.	Hazard	Causes	Effects	IMRI	Recommended Action	FMRI	Comments	Status
9.0 Install wall plate Install old wall plate to new receptacle by installing two wall plate screws	—	None							
10.0 Close CB Close circuit breaker	OSHA-5	Arcing/sparking	Wires incorrectly installed and partially touching ground	Personnel injury	2D	- Warning note to visually inspect installation - Require installation QA check prior to applying power	2E		Open
11.0 Remove tag Remove CB tag	—	None							
12.0 Test circuit Place meter in new receptacle and test voltages	OSHA-6	Meter settings are incorrect, causing damage to meter	Human error	Damaged test equipment	3D	- Warning note in manual regarding potential meter damage	3E		Open

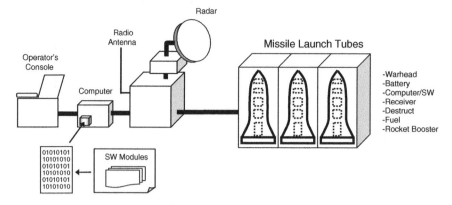

Figure 11.8 *Ace Missile System.*

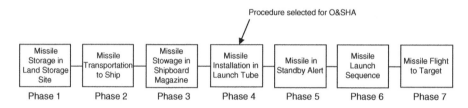

Figure 11.9 *Ace functional flow diagram of missile operational phases.*

TABLE 11.6   Missile Installation in Launch Tube
Procedure

Step	Description of Task
4.1	Remove missile from ship storage locker
4.2	Load missile onto handcart transporter
4.3	Transport missile to launch tube
4.4	Hoist missile into launch tube
4.5	Run missile tests
4.6	Install missile cables
4.7	Remove S&A pins
4.8	Place missile in standby alert

The following should be noted from the O&SHA of the Ace Missile System:

1. A total of 12 operational hazards were identified in the procedural steps for the Missile Installation in Launch Tube procedure.
2. All of the identified hazards fit within the previously established TLMs (see Chapter 8 on PHA).

TABLE 11.7 O&SHA Example 2; Worksheet 1

System: Ace Missile System
Operation: Missile Installation in Launch Tube

Operating and Support Hazard Analysis

Analyst:
Date:

Task	Hazard No.	Hazard	Causes	Effects	IMRI	Recommended Action	FMRI	Comments	Status
Task 4.1 Remove missile from ship magazine	OHA-1	Missile is dropped resulting in explosion	Human handling error and significant shock on missile	Shock causes ignition of W/H explosives	1D	- Require trained and qualified personnel - Conduct 40-Foot Drop Test - Warning note in manual on explosives hazard	1E		Open
	OHA-2	Missile is dropped resulting in damaged missile	Human handling error and missile hits a sharp surface when dropped	Missile skin is dented, resulting in unusable missile	2D	- Develop missile handling procedures - Ensure adequate missile handling equipment is used - Return all dropped missiles to depot, do not use - Warning note in manual on missile damage - Require trained and qualified personnel	2E		Open

(continued)

193

TABLE 11.7. (*Continued*)

				Operating and Support Hazard Analysis						

System: Ace Missile System
Operation: Missile Installation in Launch Tube

Analyst:
Date:

Task	Hazard No.	Hazard	Causes	Effects	IMRI	Recommended Action	FMRI	Comments	Status
Task 4.2 Load missile onto handcart transporter	OHA-3	Missile is dropped resulting in personnel injury	Human handling error and missile falls on personnel	Personnel injury	2D	- Require trained and qualified personnel - Warning note in manual on personnel hazard	2E		Open
	OHA-4	Missile is dropped resulting in damaged missile	Cart is overloaded causing axle failure resulting in dumping load of missiles	Missile skin is dented, resulting in several unusable missiles	2D	- Require trained and qualified personnel - Warning note in manual to not overload cart	2E		Open

TABLE 11.8 O&SHA Example 2; Worksheet 2

System: Ace Missile System
Operation: Missile Installation in Launch Tube

Operating and Support Hazard Analysis

Analyst:
Date:

Task	Hazard No.	Hazard	Causes	Effects	IMRI	Recommended Action	FMRI	Comments	Status
Task 4.3 Transport missile to launch tube	OHA-5	Missile falls off the cart resulting in damaged missile	Human handling error and missile hits sharp surface when dropped	Missile skin is dented, resulting in unusable missile	2D	- Develop missile handling procedures - Ensure adequate missile handling equipment is used - Return all dropped missiles to depot, do not use - Warning note in manual on missile damage - Require trained and qualified personnel	2E		Open
	OHA-6	Missile falls off the cart resulting in personnel injury	Human handling error and missile falls on personnel	Personnel injury	2D	- Require trained and qualified personnel - Warning note in manual on personnel hazard	2E		Open
	OHA-7	Missile is struck by bullet or shrapnel resulting in explosion	Terrorist or wartime activities	Initiation of W/H explosives resulting in death/injury	1D	- Use insensitive munitions (IM)	1E		Open
Task 4.4 Hoist missile into launch tube	OHA-8	Missile is dropped resulting in a fuel fire	Human handling error; hoist failure	Fuel tank is ruptured with ignition source present causing fire, resulting in death/injury	1D	- Develop procedures for fire-fighting equipment (available and personnel training) - Warning note in manual on fire hazard and responses - Inspect hoist equipment prior to use - Require trained and qualified personnel	1E		Open

TABLE 11.9 O&SHA Example 2; Worksheet 3

System: Ace Missile System
Operation: Missile Installation in Launch Tube

Operating and Support Hazard Analysis

Analyst:
Date:

Task	Hazard No.	Hazard	Causes	Effects	IMRI	Recommended Action	FMRI	Comments	Status
Task 4.5 Run missile tests	OHA-9	Missile test causes missile launch	Test equipment fault; stray voltage on test lines	Inadvertent missile launch resulting in personnel injury	1D	- Develop test equipment procedures and inspections - Require trained and qualified personnel - Caution note to ensure S&A pins are installed	1E		Open
	OHA-10	Missile test causes destruct system initiation	Test equipment fault; stray voltage on test lines	Inadvertent missile destruct initiation resulting in personnel injury	1D	- Develop test equipment procedures and inspections - Require trained and qualified personnel - Caution note to ensure S&A pins are installed	1E		Open
Task 4.6 Install missile cables	OHA-11	Cables incorrectly installed resulting in mismated connectors that cause wrong voltages on missile launch wire	Human error results in incorrect connector mating that places wrong voltages on critical connector pins	Inadvertent missile launch resulting in personnel death/ injury	1D	- Require trained and qualified personnel - Caution note to ensure S&A pins are installed - Design connectors to prevent mismating	1E		Open

TABLE 11.10   O&SHA Example 2; Worksheet 4

				Operating and Support Hazard Analysis				Analyst: Date:		
System: Ace Missile System Operation: Missile Installation in Launch Tube										
Task	Hazard No.	Hazard	Causes	Effects	IMRI	Recommended Action	FMRI	Comments	Status	
Task 4.7 Remove S&A pins	—	Missile launch S&A pin not removed	Human error	Unable to launch missile; not a safety concern, dud missile	—		—			
	OHA-12	Missile destruct S&A pin not removed	Human error	Unable to destruct errant missile, resulting in death/injury	1D	- Require trained and qualified personnel - Caution note to ensure S&A pins are removed	1E		Open	
Task 4.8 Place missile in standby alert	OHA-13	Missile is erroneously placed in training mode and system launches against false target	Human error	Missile strikes erroneous target resulting in death/injury	1D	- Require trained and qualified personnel - Warning note in manual on system mode hazard	1E		Open	

11.12 O&SHA ADVANTAGES AND DISADVANTAGES

The O&SHA technique has the following advantages:

1. Provides rigor for focusing on operational and procedural hazards.
2. Is cost-effective in providing meaningful safety results for operational hazards.

There are no disadvantages of the O&SHA technique.

11.13 COMMON O&SHA MISTAKES TO AVOID

The following common errors are made during the conduct of an O&SHA:

1. Some procedural tasks have not been identified, are incomplete, or have been omitted.
2. The hazard description is incomplete, ambiguous, or too detailed.
3. Causal factors are not adequately identified, investigated, or described.
4. The mishap risk index (MRI) is not stated, is incomplete, or is not supported by the hazard information provided.
5. The hazard mitigation does not support the final MRI.

11.14 SUMMARY

This chapter discussed the O&SHA technique. The basic principles that help summarize the discussion in this chapter are as follows:

1. The O&SHA is an analysis tool for identifying system operational hazards, causal factors, mishap risk, and system safety design requirements for mitigating risk.
2. The primary purpose of the O&SHA is to identify hazardous procedures, design conditions, failure modes, and human error that can lead to the occurrence of an undesired event or hazard during the performance of operational and support tasks.
3. The use of a specialized worksheet provides structure and rigor to the O&SHA process.
4. The use of functional flow diagrams, operational sequence diagrams, and indentured task lists greatly aids and simplifies the O&SHA process.

FURTHER READINGS

Ericson, C. A., Boeing document D2-113072-4, System Safety Analytical Technology: Operations and Support Hazard Analysis, 1971.

Layton, D., *System Safety: Including DOD Standards*, Weber Systems, Inc., 1989.

MIL-H-46855, Human Engineering Requirements for Military Systems, Equipment and Facilities.

Roland, H. E. and Moriarty, B., *System Safety Engineering and Management*, John Wiley & Sons, 2nd edition, 1990.

Stephans, R. A., *System Safety for the 21st Century*, John Wiley & Sons, Inc., 2004.

Stephenson, J., *System Safety 2000*, John Wiley & Sons, 1991.

System Safety Society, *System Safety Analysis Handbook*, System Safety Society 1999, Virginia, www.system-safety.org.

Vincoli, J. W., *A Basic Guide to System Safety*, Van Nostrand Reinhold, New York, 1993.

Chapter *12*

Health Hazard Analysis

12.1 HHA INTRODUCTION

The Health hazard analysis (HHA) is an analysis technique for evaluating the human health aspects of a system's design. These aspects include considerations for ergonomics, noise, vibration, temperature, chemicals, hazardous materials, and so on. The intent is to identify human health hazards during design and eliminate or mitigate them through design safety features. If health hazards cannot be eliminated, then protective measures must be used to reduce the associated risk to an acceptable level. Health hazards must be considered during manufacture, operation, test, maintenance, and disposal.

On the surface, the HHA appears to be very similar in nature to the operating and support hazard analysis (O&SHA), and the question often arises as to whether they both accomplish the same objectives. The O&SHA evaluates operator tasks and operations in conjunction with the system design for the identification of hazards, whereas the HHA focuses strictly on human health issues. There may occasionally be some overlap, but they each serve different interests.

12.2 HHA BACKGROUND

Note that the HHA was previously called the health hazard assessment. MIL-STD-882E changed the name from health hazard assessment to health hazard analysis; however, it is the same analysis methodology.

This analysis technique falls under the health design hazard analysis type (HD-HAT). The basic analysis types are described in Chapter 4. The HHA is performed over a period of time, continually being updated and enhanced as more design information becomes available.

Hazard Analysis Techniques for System Safety, Second Edition. Clifton A Ericson, II.
© 2016 John Wiley & Sons, Inc. Published 2016 by John Wiley & Sons, Inc.

The purpose of the HHA is to

1. provide a design safety focus from the human health viewpoint.
2. identify hazards directly affecting the human operator from a health standpoint.

The intent of the HHA is to identify human health hazards and propose design changes and/ or protective measures to reduce the associated risk to an acceptable level. Human health hazards can be the result of exposure to ergonomic stress, chemicals, physical stress, biological agents, hazardous materials, and the like. As previously stated, phases where human operators can be exposed to health hazards occur during manufacture, operation, test, maintenance, and disposal of the system.

The HHA is applicable to analysis of all types of systems, equipment, and facilities that include human operators. The HHA evaluates operator health safety during production, operation, maintenance, and disposal. The HHA technique, when applied to a given system by experienced safety personnel, should provide a thorough and comprehensive identification of the human health hazards that exist in a given system. A basic understanding of both the hazard analysis theory and the system safety concepts is essential. Experience with a particular type of system is helpful in generating a complete list of potential hazards. The technique is uncomplicated and easily learned. Standard, easily followed HHA worksheets and instructions are provided in this chapter.

The HHA concentrates on human health hazards during the production, test, and operational phases of the system in order to eliminate or mitigate human health hazards through the system design. The HHA should be completed and system risk known prior to the conduct of any of the production or operational phases. Although some of the hazards identified through the HHA may have already been identified by the preliminary hazard list (PHL), preliminary hazard analysis (PHA), or subsystem hazard analysis (SSHA) techniques, the HHA should not be omitted since it may identify hazards overlooked by these other analyses. The use of this technique by a system safety program (SSP) is highly recommended.

12.3   HHA HISTORY

The HHA technique was established very early in the history of the system safety discipline. It was formally instituted and promulgated by the developers of MIL-STD-882.

12.4   HHA THEORY

The HHA focuses on the identification of potential human health hazards resulting from a system operator's exposure to known human health hazard sources. In general terms, these hazard sources stem from system tasks, processes, environments, chemicals, and materials. Specific health hazards and their impact on the humans are assessed during the HHA.

Figure 12.1 shows an overview of the basic HHA process and summarizes the important relationships involved in the HHA process. This process consists of utilizing both design information and known hazard information to identify hazards. Known hazardous elements and mishap lessons learned are compared with the system design to determine if the design concept contains any of these potential hazard elements.

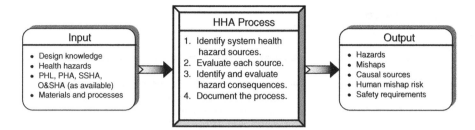

Figure 12.1 *HHA overview.*

The HHA process involves the following:

1. Identifying the human hazard source agents (noise, radiation, heat stress, cold stress, etc.) involved with the system and its logistical support.
2. Determining the critical quantities or exposure levels involved, based on the use, quantity, and type of substance/agent used.
3. Establishing design mitigation methods to eliminate or reduce exposures to acceptable levels of risk.

12.5 HHA METHODOLOGY

Table 12.1 lists and describes the basic steps of the HHA process, which involves performing a detailed analysis of all potentially hazardous human health hazard sources.

The thought process behind the HHA methodology is shown in Figure 12.2. The idea supporting this process is that different kinds of design information are used to facilitate human health hazard identification. The analysis begins with hazards identified from the PHL and PHA, which is the starting point for the HHA. The next step is to once again employ the use of hazard checklists and undesired mishap checklists. Of particular interest are hazard checklists dealing with human health issues. Also, data on human limitations and regulatory requirements are used to identify human health hazards.

The HHA is strongly dependent upon the use of health hazard checklists. Hazard checklists are generic lists of known hazardous items and potentially hazardous designs or situations and should not be considered complete or all-inclusive. Checklists are intended as a starting point to help the analyst recognize the potential hazard sources from the past lessons learned.

Typical health hazard checklist categories include the following:

a. Ergonomic
b. Noise
c. Vibration
d. Temperature
e. Chemicals
f. Biological
g. Hazardous materials
h. Physical stress

TABLE 12.1   HHA Process

Step	Task	Description
1	Acquire design information	Acquire all the design, operational, and manufacturing data for the system
2	Acquire health hazard checklists	Acquire checklists of known health hazard sources, such as chemicals, materials, processes, and so on. Also, acquire checklists of known human limitations in the operation of systems, such as noise, vibration, heat, and so on
3	Acquire regulatory information	Acquire all regulatory data and information that are applicable to human health hazards
4	Identify health hazard sources	Examine the system and identify all potentially hazardous health sources and processes within the system. Include quantity and location when possible. Utilize the checklists
5	Identify hazards	Identify and list potential hazards created in the system design resulting from the health hazard sources
6	Identify safety barriers	Identify design mitigation methods or barriers in the path of the health hazard source. Also, identify existing design features to eliminate or mitigate the hazard
7	Evaluate system risk	Identify the level of mishap risk presented to personnel, both with and without design controls in the system design
8	Recommend corrective action	Determine if the design controls present are adequate and if not, recommend controls that should be added to reduce the mishap risk
9	Track hazards	Transfer newly identified hazards into the HTS. Update the HTS as hazards, hazard causal factors, and risk are identified in the HHA
10	Document HHA	Document the entire HHA process on the worksheets. Update for new information as necessary

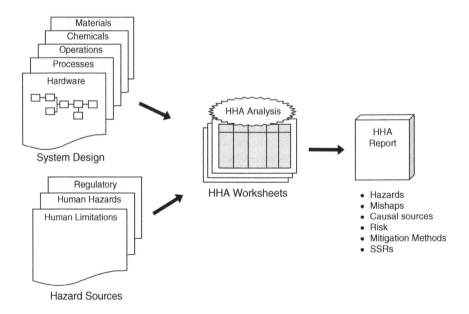

Figure 12.2   HHA methodology.

When performing the HHA, the following factors should be given consideration:

a. Toxicity, quantity, and physical state of materials.

b. Routine or planned uses and releases of hazardous materials or physical agents.

c. Accidental exposure potentials.

d. Hazardous waste generated.

e. Hazardous material handling, transfer, and transportation requirements.

f. Protective clothing/equipment needs.

g. Detection and measurement devices required to quantify exposure levels.

h. Number of personnel potentially at risk.

i. Design controls that could be used, such as isolation, enclosure, ventilation, noise or radiation barriers, and so on.

j. Potential alternative materials to reduce the associated risk to users/operators.

k. The degree of personnel exposure to the health hazard.

l. System, facility, and personnel protective equipment design requirements (e.g., ventilation, noise attenuation, radiation barriers, etc.) to allow safe operation and maintenance.

m. Hazardous material and long-term effects (such as potential for personnel and environmental exposure, handling and disposal issues/requirements, protection/control measures, and life cycle costs).

n. Means for identifying and tracking information for each hazardous material.

o. Environmental factors that effect exposure (wind, temperature, humidity, etc.)

When hazardous materials must be used in the system, the following considerations must be evaluated and documented:

1. Identify hazardous materials data:
   a. Name
   b. Stock number
   c. Affected system components and processes
   d. Quantity, characteristics, and concentrations of the materials in the system
   e. Source documents relating to the materials
2. Determine under which conditions the hazardous materials can pose a health threat.
3. Characterize material hazards and determine reference quantities and hazard ratings (e.g., acute health, chronic health, carcinogenic, contact, flammability, reactivity, and environmental hazards).
4. Estimate the expected usage rate of each hazardous material.
5. Recommend the disposition (disposal, recycle, etc.) of each hazardous material identified.

When feasible engineering designs are not available to reduce hazards to acceptable levels, alternative protective measures must be specified (e.g., protective clothing, specific operation, or maintenance practices to reduce risk to an acceptable level). Identify potential nonhazardous or less-hazardous alternatives to hazardous materials if they exist or provide a justification why an alternative cannot be used.

12.6 HHA WORKSHEET

The HHA is a detailed hazard analysis utilizing structure and rigor. It is desirable to perform the HHA using a specialized worksheet. Although the format of the analysis worksheet is not critical, typically matrix- or columnar-type worksheets are used to help maintain focus and structure in the analysis. Sometimes, a textual document layout worksheet is utilized. As a minimum, the following basic information is required from the HHA:

1. Personnel health hazards
2. Hazard effects (mishaps)
3. Hazard causal factors (materials, processes, excessive exposures, etc.)
4. Risk assessment (before and after design safety features are implemented)
5. Derived safety requirements for eliminating or mitigating the hazards.

The recommended HHA worksheet is shown in Figures 12.3. This particular HHA worksheet utilizes a columnar-type format and has been proven effective in many applications. The worksheet can be modified as found necessary by the system safety program.

The information required under each column entry of the worksheet is described as follows:

1. *System* This column identifies the system under analysis.
2. *Subsystem* This column identifies the subsystem under analysis.
3. *Operation* This column identifies the operation under analysis.
4. *Mode* This column identifies the system mode under analysis.
5. *Analyst* This column identifies the name of the HHA analyst.
6. *Date* This column identifies the date of the analysis.
7. *Hazard Type* This column identifies the type of human health concerns being analyzed, such as vibration, noise, thermal, chemical, and so on.
8. *Hazard No.* This column identifies the number assigned to the identified hazard in the HHA (e.g., HHA-1, HHA-2, etc.). This is for future reference to the particular hazard

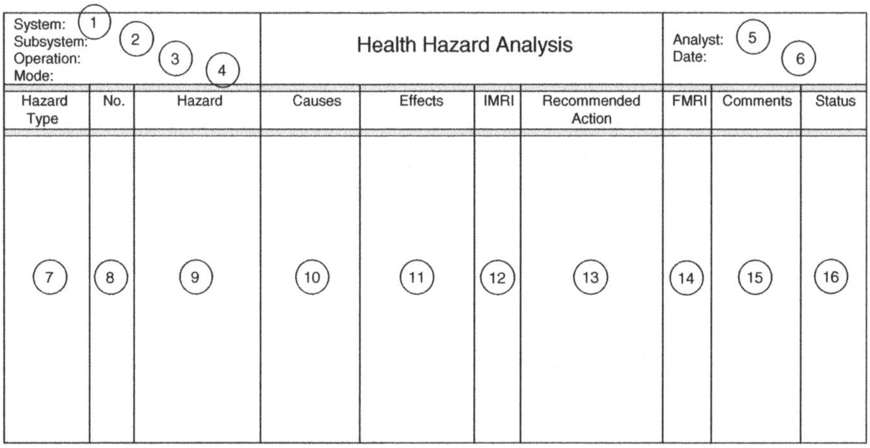

Figure 12.3 *Recommended HHA worksheet.*

source and may be used, for example, in the hazard action record (HAR) and the hazard tracking system (HTS).

9. *Hazard* This column identifies the particular human health hazard. It should describe the hazard source, mechanism, and outcome. The specific system mode or phase of concern should also be identified.

10. *Cause* This column identifies conditions, events, or faults that could cause the hazard to exist and the events that can trigger the hazardous elements to become a mishap or accident.

11. *Effect/Mishap* This column identifies the effect and consequences of the hazard, should it occur. Generally, the worst-case result is the stated effect.

12. *Initial Mishap Risk Index (IMRI)* This column provides a qualitative measure of mishap risk for the potential effect of the identified hazard, given that no mitigation techniques are applied to the hazard. Risk measures are a combination of mishap severity and probability, and the recommended values are listed in the following table.

Severity	Probability
1 – Catastrophic	A – Frequent
2 – Critical	B – Probable
3 – Marginal	C – Occasional
4 – Negligible	D – Remote
	E – Improbable

13. *Recommended Action* This column establishes recommended preventive measures to eliminate or control identified hazards. Safety requirements in this situation generally involve the addition of one or more barriers to keep the energy source away from the target. The preferred order of precedence for design safety requirements is as shown in the following table:

Order of Precedence
1 – Eliminate hazard through design selection
2 – Control hazard through design methods
3 – Control hazard through safety devices
4 – Control hazard through warning devices
5 – Control hazard through procedures and training

14. *Final mishap risk index (FMRI)* This column provides a qualitative measure of mishap risk significance for the potential effect of the identified hazard, given that mitigation techniques and safety requirements are applied to the hazard. The same values used in column 12 are also used here.

15. *Comments* This column provides a place to record useful information regarding the hazard or the analysis process that are not noted elsewhere.

16. *Status* This column states the current status of the hazard, as either being open or closed. This follows the hazard tracking methodology established for the program. A hazard can be closed only when it has been verified through analysis, inspection, and/or testing that the safety requirements are implemented in the design and successfully tested for effectiveness.

12.7   HUMAN HEALTH HAZARD CHECKLIST

Table 12.2 provides a list of typical health hazard sources that should be considered when performing an HHA.

TABLE 12.2   Typical Human Health Hazard Sources

HHA Category	Examples
Acoustic Energy	
Potential energy existing in a pressure wave transmitted through the air may interact with the body to cause loss of hearing or internal organ damage	• Steady-state noise from engines • Impulse noise from shoulder-fired weapons
Biological Substances	
Exposures to microorganisms, their toxins, and enzymes	• Sanitation concerns related to waste disposal
Chemical Substances	
Exposure to toxic liquids, mists, gases, vapors, fumes, or dusts	• Combustion products from weapon firing • Engine exhaust products • Degreasing solvents
Oxygen Deficiency	
Hazard may occur when atmospheric oxygen is displaced in a confined/enclosed space and falls below 21% by volume; also used to describe the hazard associated with the lack of adequate ventilation in crew spaces	• Enclosed or confined spaces associated with shelters, storage tanks, and armored vehicles • Lack of sufficient oxygen and pressure in aircraft cockpit cabins • Carbon monoxide in armored tracked vehicles
Ionizing Radiation Energy	
Any form of radiation sufficiently energetic to cause ionization when interacting with living matter	• Radioactive chemicals used in light sources for optical sights and instrumented panels
Nonionizing Radiation	
Emissions from the electromagnetic spectrum that has insufficient energy to produce ionization, such as lasers, ultraviolet, and radio frequency radiation sources	• Laser range finders used in weapons systems; microwaves used with radar and communication equipment
Shock	
Delivery of a mechanical impulse or impact to the body. Expressed as a rapid acceleration or deceleration	• Opening forces of a parachute harness • Back kick of firing a handheld weapon
Temperature Extremes	
Human health effects associated with hot or cold temperatures	• Increase in the body's heat burden from wearing total encapsulating protective chemical garments • Heat stress from insufficient ventilation to aircraft or armored vehicle crew spaces
Trauma	
Injury to the eyes or body from impact or strain	• Physical injury cause by blunt or sharp impacts • Musculoskeletal trauma caused by excessive lifting

TABLE 12.2 *(Continued)*

HHA Category	Examples
Vibration	
Adverse health effects (e.g., back pain, hand-arm vibration syndrome (HAVS), carpel tunnel syndrome, etc.) caused by contact of a mechanically oscillating surface with the human body	• Riding in and/or driving/piloting armored vehicles or aircraft • Power hand tools • Heavy industrial equipment
Human–Machine Interface	
Various injuries such as musculoskeletal strain, disk hernia, carpel tunnel syndrome, and so on resulting from physical interaction with system components	• Repetitive ergonomic motion • Manual material handling – lifting assemblies or subassemblies • Acceleration, pressure, velocity, and force
Hazardous Materials Exposures to toxic materials hazardous to humans	• Lead • Mercury

12.8 HHA EXAMPLE

In order to demonstrate the HHA methodology, an HHA is performed on an example system. The particular system for this example is a hypothetical diesel submarine system. An HHA is performed on the diesel engine room of this system.

Tables 12.3–12.5 contain the HHA worksheets for this example.

12.9 HHA ADVANTAGES AND DISADVANTAGES

The HHA technique has the following advantages:

1. The HHA is easily and quickly performed.
2. The HHA does not require considerable expertise for technique application.
3. The HHA is relatively inexpensive, yet provides meaningful results.
4. The HHA provides rigor for focusing on system health hazards.
5. The HHA quickly provides an indication of where major system health hazards will exist.

There are no notable disadvantages of the HHA technique.

12.10 COMMON HHA MISTAKES TO AVOID

When first learning how to perform an HHA, it is commonplace to commit some traditional errors. The following is a list of typical errors made during the conduct of an HHA.

1. The hazard identified is other than a human health hazard.
2. The hazard description is not detailed enough.
3. Design mitigation factors not stated or provided.

TABLE 12.3   Example HHA; Worksheet 1

System:
Subsystem:
Operation:
Mode:

Health Hazard Analysis

Analyst:
Date:

HH Type	No.	Hazard	Causes	Effects	IMRI	Recommended Action	FMRI	Comments	Status
Noise	HH-1	Excessive exposure to engine noise causes operator ear damage	Constant engine noise above xx dB	Ear damage; loss of hearing	3C	Ear protection; limit exposure time	3E		Open
Vibration	--	No hazard; within limits	Engine vibration	None	4E	None	4E		Closed
Temperature	--	No hazard; within limits	Engine room temperature	None	4E	None	4E		Closed
Oxygen Deficiency	HH-2	Loss of oxygen in engine room, causing operator death	Closed compartment and faults cause oxygen loss	Operator death	1C	Sensors and warning devices	1E		Open

TABLE 12.4   Example HHA; Worksheet 2

System:
Subsystem:
Operation:
Mode:

Health Hazard Analysis

Analyst:
Date:

HH Type	No.	Hazard	Causes	Effects	IMRI	Recommended Action	FMRI	Comments	Status
Biological Substance	--	None present; no hazard	None	None	4E	None	4E		Closed
Chemical Substance	HH-3	Exposure to diesel fuel fumes	Operator sickness	Operator sickness	3B	Sensors and air purge	3E		Open
Ergonomic	--	None	None	None	4E	None	4E		Closed
Physical stress	HH-4	Operator injury from lifting heavy objects	Operator injury	Operator injury	3C	All items must not exceed one-man weight limit	3E		Open

Page: 2 of 3

209

TABLE 12.5   Example HHA; Worksheet 3

System:
Subsystem:
Operation:
Mode:

Health Hazard Analysis

Analyst:
Date:

HH Type	No.	Hazard	Causes	Effects	IMRI	Recommended Action	FMRI	Comments	Status
Ionizing Radiation	---	No hazard; none present	None	None	4E	None	4E		Closed
Non-Ionizing Radiation	---	No hazard; none present	None	None	4E	None	4E		Closed
Hazardous Material	HH-5	Exposure to asbestos causes operator death	Exposure to asbestos particles	Operator death	1C	Prohibit use of asbestos in system	1E		Open

4. Design mitigation factors do not address the actual causal factor(s).
5. Overuse of project specific terms and abbreviations.

12.11 SUMMARY

This chapter discussed the HHA technique. The following are the basic principles that help summarize the discussion in this chapter:

1. The primary purpose of HHA is to identify human health hazards.
2. The use of a human health hazard checklist greatly aids and simplifies the HHA process.
3. The use of the recommended HHA worksheet aids the analysis process and provides documentation of the analysis.

FURTHER READINGS

DoD Data Item Description DI-SAFT-80106, Occupational Health Hazard Assessment Report.

System Safety Society, *System Safety Analysis Handbook*, System Safety Society 1999, Virginia, www.system-safety.org.

Chapter *13*

Requirements Hazard Analysis

13.1 RHA INTRODUCTION

Requirement hazard analysis (RHA) is an analysis for evaluating design requirements and system safety requirements (SSRs). As the name implies, RHA evaluates SSRs and their criteria. The RHA has a twofold purpose: (1) to ensure that every identified hazard has at least one corresponding safety requirement and (2) to verify that all safety requirements are implemented and are validated successfully. The RHA is essentially a traceability analysis to ensure that there are no holes or gaps in the safety requirements and that all identified hazards have adequate and proven design mitigation coverage. The RHA applies to hardware, software, firmware, and test requirements.

RHA also applies to the traceability of design criteria and guidelines to SSRs, such as those specified in the DoD Joint Software System Safety Handbook (DoDJSSSH) and SEB6-A for software, and MIL-STD-1316 for fuze systems. Guideline traceability ensures that the appropriate safety guidelines and criteria are incorporated into the SSRs.

13.2 RHA BACKGROUND

Note that the RHA was previously called the safety requirement/criteria analysis (SRCA). MIL-STD-882E (2012) changed the name; however, it is essentially the same analysis.

This analysis technique falls under the requirements design hazard analysis type (RD-HAT). The basic analysis types are described in Chapter 4. Alternative names for this technique are requirements hazard analysis (RHA) and requirements traceability analysis.

The purpose of the RHA is to ensure that all identified hazards have corresponding design safety requirements to eliminate or mitigate the hazard, and that the safety requirements are verified and

Hazard Analysis Techniques for System Safety, Second Edition. Clifton A Ericson, II.
© 2016 John Wiley & Sons, Inc. Published 2016 by John Wiley & Sons, Inc.

validated as being successful in the system design and operation. The intent is to ensure that the system design requirements have no "safety" gaps (i.e., no hazard has been left unmitigated) and to verify that all safety requirements are adequately implemented or created as necessary.

RHA is applicable to analysis of all types of systems, facilities, and software where hazards and safety requirements are involved during development. RHA is particularly useful when used in a software safety program. The RHA technique, when applied to a given system by experienced safety personnel, is very thorough in providing an accurate traceability of safety design requirement verification and safety test requirement validation.

A basic understanding of system safety and the design requirement process is essential. Experience with the particular type of system is helpful. The technique is uncomplicated and easily learned. Standard easily followed RHA worksheets and instructions are provided in this chapter.

The RHA is very useful for ensuring that design safety requirements exist for all hazards and that the safety requirements are in the design and test specifications. The RHA provides assurance that safety requirements exist for all identified hazards and ensures that all safety requirements are incorporated into design and test specifications. The application of the RHA to software development has proven to be very effective for successfully evaluating software to ensure compliance with safety requirements.

13.3 RHA HISTORY

The RHA technique was established very early in the history of the system safety discipline, when it was known as SRCA. It was formally instituted and promulgated by the developers of MIL-STD-882; the name was changed in version E to more accurately reflect the actual task being performed.

13.4 RHA THEORY

As previously stated, the purpose of the RHA is to ensure that safety requirements exist for all identified hazards, that all safety requirements are incorporated into the design and test specifications, and that all safety requirements are successfully tested. Figure 13.1 shows an overview of the basic RHA process and summarizes the important relationships involved in the RHA process. This process consists of comparing the SSRs with design requirements and identified hazards. In this way any missing safety requirements will be identified. In addition, SSRs are traced into the test requirements to ensure that all SSRs are tested.

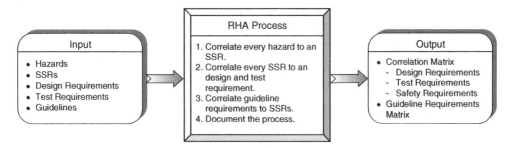

Figure 13.1 *RHA overview.*

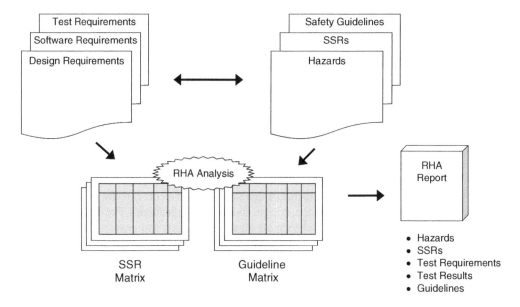

Figure 13.2 *RHA methodology.*

13.5 RHA METHODOLOGY

The idea behind this thought process is that a matrix worksheet is used to correlate safety requirements with design requirements, test requirements, and identified hazards. If a hazard does not have a corresponding safety requirement, then there is an obvious gap in the safety requirements. If a safety requirement is not included in the design requirements, then there is a gap in the design requirements. If a safety requirement is missing from the test requirements, then that requirement cannot be verified and validated. If an SSR cannot be shown to have passed testing, then the associated hazard cannot be closed.

Figure 13.2 shows an overview of the RHA thought process for the RHA technique.

The RHA is a detailed correlation analysis, utilizing structure and rigor to provide traceability for all SSRs. The RHA begins by acquiring the system hazards, SSRs, design requirements and test requirements. A traceability matrix is then constructed that correlates the hazards, SSRs, design requirements, and test requirements together. The completed traceability matrix ensures that every hazard has a corresponding safety requirement and that every safety requirement has a corresponding design and test requirement.

The RHA consists of two separate correlation traceability analyses: (1) an SSRs correlation and (2) a guideline compliance correlation. The guideline correlation applies only to systems where guidelines exist and are applied to the system design. For example, the safety guideline requirements from the DoDJSSSH are generally applied to the design of software, and the guideline requirements from MIL-STD-1316 are applied to fuze system designs.

Table 13.1 lists and describes the basic steps of the RHA process.

13.6 RHA WORKSHEETS

The RHA is a detailed hazard analysis utilizing structure and rigor. It is desirable to perform the RHA using specialized worksheets. Although the format of the analysis worksheet is not

TABLE 13.1   RHA Process

Step	Task	Description
1	Acquire requirements	Acquire all of the design (hardware and software) and test requirements for the system
2	Acquire safety data	Acquire all of the hazards and SSRs for the system
3	Acquire safety guidelines	Acquire all safety guidelines that are applicable to the system
4	Establish an SSR traceability matrix	Correlate SSRs with hazards, design requirements, and test requirements
5	Establish guideline traceability matrix	Correlate safety guidelines and criteria with SSRs
6	Identify requirement gaps	Identify hazards that have no corresponding safety requirement
7	Recommend corrective action	Determine if the design controls present are adequate, and if not, recommend controls that should be added to reduce the mishap risk
8	Track hazards	Transfer identified hazards into the hazard tracking system (HTS)
9	Document RHA	Document the entire RHA process on the worksheets. Update for new information as necessary

critical, typically, matrix- or columnar-type worksheets are used to help maintain focus and structure in the analysis. Software packages are available to aid the analyst in preparing these worksheets.

The purpose of the RHA is to establish traceability of SSRs and to assist in the closure of mitigated hazards. As a minimum, the following basic information is required from the RHA:

1. Traceability matrix of all identified hazards to corresponding SSRs.
2. Traceability matrix of all safety design requirements to test requirements and test results.
3. Identification of new safety design requirements and tests necessary to cover gaps discovered by 1 and 2 above.
4. Traceability matrix of all safety guidelines and criteria to SSRs.
5. Data from 1, 2, 3, and 4 above supporting closure of hazards.

The specific worksheet to be used may be determined by the managing authority, the safety working group, the safety team, or the safety analyst performing the analysis.

Figure 13.3 contains an example RHA requirements' correlation worksheet for the traceability of SSRs.

The information required under each column of the requirements correlation matrix worksheet is described as follows:

1. *System*   This column identifies the system under analysis.
2. *Subsystem*   This column identifies the subsystem under analysis.
3. *SSR Number*   This column identifies the SSR number.
4. *SSR*   This column states the actual verbiage of the SSR.
5. *Safety-Critical (SC)*   Place a Yes in this column to indicate that the SSR is an SC requirement.
6. *Hazard Action Record (HAR) Number*   This column identifies the HAR or HARs associated with the SSR. The SSR may be providing mitigation for one or more hazards.

System: (1) (2) Subsystem:		SSR Traceability Matrix					RHA			
SSR No.	System Safety Requirement (SSR)	SC	HAR No.	TLM No.	Design Req. No.		Test Req. No.	Test		
								M	C	R
(3)	(4)	(5)	(6)	(7)	(8)		(9)	(10)		

Figure 13.3 *RHA requirements correlation matrix worksheet.*

7. *Top-Level Mishap (TLM) Number*   This column identifies the TLM associated with the SSR.
8. *Design Requirement Number*   This column identifies the specific design requirement that implements the SSR.
9. *Test Requirement Number*   This column identifies the specific test requirement or requirements that test the SSR.
10. *Test*   This column provides information regarding testing of the SSR. Three different types of information are provided:
    a. *M*   This column identifies the test method used: *T*est, *A*nalysis, *I*nspection, and *N*ot done.
    b. *C*   This column identifies the test coverage: *E*xplicitly tested by a specific test or *I*mplicitly tested through another test.
    c. *R*   This column identifies the test results: *P*ass or *F*ail.

Figure 13.4 contains an example RHA correlation worksheet for the traceability of SSR compliance with safety guideline and criteria.

The information required under each column of the guideline correlation matrix worksheet is described as follows:

1. *System*   This column identifies the system under analysis.
2. *Subsystem*   This column identifies the subsystem under analysis.
3. *Guideline Number*   This column identifies the requirement number from the guideline or criteria document.
4. *Guideline Requirement*   This column contains the actual text of the guideline requirement.
5. *SSR Number*   This column identifies the specific SSR that implements the design guideline requirement.

System: ①②		Guideline Compliance Matrix		RHA			
Subsystem:							
Guideline Number	Guideline Requirement		SSR Number	Comments	Implement		
					F	P	N
③	④		⑤	⑥	⑦		

Figure 13.4 *Guideline correlation matrix worksheet.*

6. *Comments* This column provides for any necessary discussion associated with the guideline requirement. For example, if the guideline is only partially implemented or not implemented at all, sufficient rationale and justification must be provided.

7. *Implement* This column provides information regarding implementation of the design guideline. Check the particular column that is applicable:

a. *F*ull implementation by the SSR.

b. *P*artial implementation by the SSR.

c. *N*ot implemented (not applicable or not possible).

13.7 RHA EXAMPLE

This example involves a fuze subsystem for a missile weapon system, which includes both hardware and software. In these examples, not all of the SSRs or design requirements are included, only a small example portion of the overall requirements to demonstrate the technique.

Note from this example that more than one safety requirement may exist for a particular hazard. Also, note that the HAR number is provided for each hazard. The identified design requirement is derived from the program design specification document, and the test requirement is derived from the program test requirements document.

Table 13.2 provides an RHA requirements traceability matrix for this example system. Note that only a single page is shown for this example. A typical RHA would consist of many pages.

Table 13.3 provides a compliance matrix for the DoDJSSSH software guidelines (Appendix E). A software RHA correlation compliance analysis includes the review of each of these guidelines, determines if it is applicable to the system design, provides rationale for those not applicable and provides the corresponding safety requirement references.

Table 13.4 provides a MIL-STD-1316 (series) compliance matrix for this example system.

Table 13.5 provides a MIL-HDBK-454 (series) compliance matrix for this example system.

TABLE 13.2 RHA Requirements Traceability Matrix Example

SSR Traceability Matrix

System: Lark Missile
Subsystem: Fire Control System

Requirements Hazard Analysis

SSR Number	System Safety Requirement (SSR)	SC	HAR Number	TLM Number	Design Req't Number	Test Req't Number	Test T/A/I/N	Test E/I	Test P/F
SSR 31	No single point failure can result in missile launch.	Y	21	1	SS 7.7.21	TS 4.7.21	T	E	P
SSR 32	Three separate and independent means are required for missile arming.	Y	81, 95	1	SS 7.7.22	TS 4.7.22	T	E	P
SSR 33									
SSR 34									

Page:

TABLE 13.3 DoDJSSSH Compliance Matrix Example

System: Subsystem:	**DODJSSSH Software Compliance Matrix**		Requirements Hazard Analysis			
DoDJSSSH Number	**DoDJSSSH Requirement**	SSR Number	Comments	Compliance		
				F	P	N
E.4.3	Primary computer failure. The system shall be designed such that a failure of the primary control computer will be detected and the system returned to a safe state.	SSR 71		X		
E.5.2	CPU selection. CPUs, microprocessors and computers that can be fully represented mathematically are preferred to those that cannot		Not possible for program to obtain a CPU meeting this requirement. Intel Pentium II has been selected and approved. This requirement is too stringent for program.			X
E.6.4	Operational checks. Operational checks of testable safety-critical system elements shall be made immediately prior to performance of a related safety-critical operation.	SSR 72		X		
E.9.4	Safety-critical displays. Safety-critical operator displays, legends and other interface functions shall be clear, concise and unambiguous and, where possible, be duplicated using separate display devices.	SSR 73		X		
E.11.2	Modular code. Software design and code shall be modular. Modules shall have one entry and one exit point.	SSR 74		X		
E.11.18	Variable declaration. Variables or constants used by a safety-critical function will be decla red/initialized at the lowest possible level	SSR 75		X		
E.11.19	Unused executable code. Operational program loads shall not contain unused executable code.	SSR 76		X		
				Page:		

TABLE 13.4 MIL-STD-1316E Compliance Matrix Example

MIL-STD-1316E Compliance Matrix

Requirements Hazard Analysis

System:
Subsystem:

1316E Number	MIL-STD-1316E Requirement	SSR Number	Comments	Compliance F	Compliance P	Compliance N
4.2.2	Arming Delay. A safety feature of the fuze shall provide and arming delay, which assures that a safe separation distance, can be achieved for all defined operational conditions.	SSR 47		X		
4.2.3	Manual Arming. An assembled fuze shall not be capable of being armed manually.	SSR 48		X		
4.3	Safety System Failure Rate. The fuze safety system failure rate shall be calculated for all logistical and tactical phases from manufacture to safe separation. The safety system failure rate shall be verified to the extent practical by test and analysis during fuze evaluation and shall not exceed the rates given for the following phases:	SSR 49	Verified by FTA.	X		
4.3a	Prior to intentional initiation of the arming sequence: 1×10^{-6} to prevent arming or functioning.					
4.3b	Prior to the exit (for tubed launch munitions): 1×10^{-4} to prevent arming 1×10^{-6} to prevent functioning					
4.3c	Between initiation of arming sequence and safe separation: 1×10^{-3} to prevent arming ALARP with established risk to prevent functioning		ALARP – As Low As Reasonably Practical			
			Page:			

TABLE 13.5   MIL-HDBK-454 Compliance Matrix Example

System: Subsystem:		MIL-HDBK-454 Compliance Matrix			Requirements Hazard Analysis			
						Compliance		
454 Number	MIL-STD-454 Requirement		SSR Number	Comments		F	P	N
4.1	The equipment shall provide fail-safe features for safety of personnel during installation, operation, maintenance, and repair or interchanging of a complete assembly or component parts.		SSR 101			X		
4.2	Electric equipment shall be bonded in accordance with MIL-B-5087, Class R/L/H		SSR 102			X		
4.3a	At an ambient temperature of 25 degrees C, the operating temperature of control panels and operating controls shall not exceed 49 degrees C (120 degrees F).		SSR 103			X		
4.3b	At an ambient temperature of 25 degrees C (77 degrees F), exposed parts subject to contact by personnel (other than control panels and operating controls) shall not exceed 60 degrees C (140 degrees F).		SSR 104			X		
				MIL- HDBK -454M, General Guide lines for Electronic Equipment, Guideline 1 - Safety Design Criteria for Personnel Hazards				
								Page:

221

13.8   RHA ADVANTAGES AND DISADVANTAGES

The RHA technique has the following advantages:

1. RHA is easily and quickly performed.
2. RHA correlates SSRs with hazards and design requirements and specifications.
3. RHA verifies and validates SSRs through correlation of test results.
4. Software packages are available to aid the analyst in preparing RHA worksheets.

The RHA technique has no major disadvantages.

13.9   COMMON RHA MISTAKES TO AVOID

When first learning how to perform an RHA, it is commonplace to commit some typical errors. The following is a list of common errors made during the conduct of an RHA:

1. Failing to put all safety guidelines and requirements into SSRs.
2. Failing to perform a traceability compliance analysis on all safety guidelines and requirements SSRs.

13.10   SUMMARY

This chapter discussed the RHA technique. The following basic principles help summarize the discussion in this chapter:

1. The primary purpose of the RHA is to assess the SSRs and
   - identify hazards without associated SSRs (gaps in the design SSRs).
   - identify requirements that are not in the system design requirements.
   - identify requirements that are not tested and validated for effectiveness.
   - identify safety guidelines and requirements that are not implemented in the SSRs.
2. The RHA relates the identified hazards to the design SSRs to ensure that all identified hazards have at least one safety requirement, whose implementation will mitigate the hazard.
3. The RHA relates the design SSRs to the test requirements to ensure all safety requirements are verified and validated by test.
4. The RHA is also used to incorporate safety guidelines and requirements that are specifically safety related but not tied to a specific hazard.
5. The RHA process ensures that the design safety requirements are properly developed and translated into the system hardware and software requirement documents.
6. The use of the recommended RHA worksheets simplifies the process and provides documentation of the analysis.
7. An RHA is really a secondary-type analysis technique; it supports the primary hazard analyses. A safety program should not rely solely on an RHA, other hazard analyses must also be performed, such as the PHA, SSHA, and SHA.

FURTHER READINGS

DoD Joint Software System Safety Handbook (DoDJSSSH), Appendix E: Generic Requirements and Guidelines, 1999.

EIA SEB6-A, System Safety Engineering in Software Development, Appendix A: Software System Safety Checklist, 1990. (EIA stands for Electronic Industries Association).

MIL-HDBK-454M (Series), General Guidelines for Electronic Equipment, Guideline 1: Safety Design Criteria for Personnel Hazards.

MIL-STD-1316 (Series), Safety Criteria for Fuze Design: DoD Design Criteria Standard.

MIL-STD-882E, Standard Practice for System Safety, 2012.

System Safety Society, *System Safety Analysis Handbook*, System Safety Society 1999, Virginia, www.system-safety.org.

Chapter *14*

Environmental Hazard Analysis (EHA)

14.1 EHA INTRODUCTION

Environmental hazard analysis (EHA) is an analysis technique for evaluating the adverse effects of a system on the environment. The purpose is to ensure that the system under analysis does not damage the environment during the life cycle of the system, where the system life cycle includes design, development, test, manufacture, operation, and disposal. The EHA is a method by which a safe balance is maintained between the human activity and the environment.

The EHA is performed during system design in order to appropriately influence design decisions in order to prevent decisions that may have an adverse impact on the environment, during both normal and abnormal system operations. It is also intended to help modify the design when necessary to ensure that the risk presented to the environment is minimal and acceptable.

The intent of the EHA is to identify environment-related hazards during design and eliminate them through design features. If these hazards cannot be eliminated, then protective measures must be used to reduce the associated risk to an acceptable level. Environmental hazards must be specifically considered for system manufacture, operation, test, maintenance, modification, and disposal, under the conditions of normal operation and abnormal operation (e.g., failures, errors, etc.). An EHA does not stop when the system design is analyzed for environmental safety, it must also include other system-related aspects. For example, it must cover manufacturing processes, repair processes, transportation, storage, disposal, by-products, residue, and the hazardous ingredients involved (i.e., hazard sources).

The EHA may provide system-specific data to support National Environmental Policy Act (NEPA) and Executive Order (EO) 12114 requirements that may be applicable. The EHA

Hazard Analysis Techniques for System Safety, Second Edition. Clifton A Ericson, II.
© 2016 John Wiley & Sons, Inc. Published 2016 by John Wiley & Sons, Inc.

could be construed as being similar to an Environmental Impact Statement (EIS) that is often required by federal regulations.

14.2   EHA BACKGROUND

This analysis technique falls under the detailed design hazard analysis type (DD-HAT). The basic analysis types are described in Chapter 4. The EHA is performed over a period of time, continually being updated and enhanced as more design information becomes available. The EHA is applicable to analysis of all types of systems, equipment and facilities, and their potential impact on the environment. The safety concern is for environmental hazards that may arise during normal system operation, and environmental hazards that may result from abnormal operation caused by failures, errors, common cause events, and so on.

The EHA has the following purposes:

a. Provide a design safety focus on potential damage to the environment.
b. Identify system hazards directly affecting the environment.
c. Determine the hazard risk involved.
d. Establish design safety measures to eliminate or reduce the risk.

When performing an EHA two questions must be answered: (1) what environments does the system touch upon and (2) how can the system cause damage to those environments. The EHA should address these questions in detail. This requires looking at all aspects of system operation, both normal and abnormal.

Since the term *environment* is not well defined and somewhat vague, another, even larger question that must also be addressed is, what is an environment and how many are there. Environments would seem to include both natural environments and man-made environments, as some man-made environments may cause damage to the natural environment. For example, RF radiation is a man-made environment that can cause damage to animals.

Typical natural environments include, but are not limited to, the following:

- Air
- Land
- Water – streams, rivers, lakes, and oceans
- Animals
- Trees
- Plants
- Atmosphere
- Food
- Biological environments affecting plants, animal, and humans
- Marine life

Typical man-made environments include, but are not limited to, the following:

- Electromagnetic radiation
- Nuclear radiation

- Pollution
- Hazardous material exposure
- Noise
- Biological agent exposure

The EHA technique, when applied to a given system by experienced safety personnel, should provide a thorough and comprehensive identification of hazards that exist in a given system design that can cause adverse damage to the environment. A basic understanding of the hazard analysis theory and the system safety concepts is essential. Experience with the particular type of system is helpful in generating a complete list of potential hazards. Also, it is necessary to fully understand the environments involved and the mechanisms by which they can cause damage. The technique is uncomplicated and easily learned. Standard, easily followed EHA worksheets and instructions are provided in this chapter.

14.3   EHA HISTORY

Safety of the environment has been a concern for many years. Some projects performed an EHA, or EHA-like analysis, automatically, while other projects did so only when specifically required. The EHA technique was formally instituted and required in version E of MIL-STD-882, which was released in 2012.

14.4   EHA THEORY

The EHA focuses on the identification of potential hazards to the environment resulting from a systems manufacture, test, operation, maintenance, modification, and disposal. The scope of the EHA should consider the entire system life cycle and address hazards associated with, but not limited to, the following:

- Environmental impacts on sea, air, space, and land resources and ecosystems.
- Hazardous materials use and generation.
- Demilitarization and disposal requirements
- Exposure to chemical, biological, and other hazards impacting public health.
- Noise generation resulting from operation of the system.
- Pollutant emissions generation (e.g., air, water, and solid waste).
- Release of hazardous substances incidental to the routine maintenance and operation of the system.
- Inadvertent releases of hazardous/toxic materials.

Programs should begin the process of identifying environmental hazards and safety requirements using sources such as the following:

- Identification of environments that the system will interface with.
- Environmental hazard analysis data and information, risk assessments, mishaps, and lessons learned from legacy and similar systems.

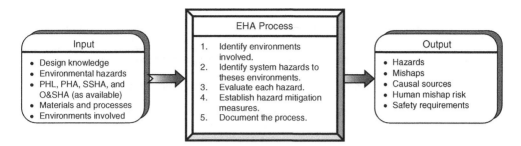

Figure 14.1 *EHA overview.*

- User requirements documents (e.g., joint capabilities integration and development system, concept of operations, etc.).
- System design data and information (e.g., design specifications).
- Demilitarization and disposal of legacy and similar systems.
- Environmental issues at legacy and similar system locations and potential locations throughout the life cycle.
- Programmatic Environment, Safety, and Occupational Health Evaluation (PESHE) and NEPA documents from legacy and similar systems.
- Preliminary hazard list (PHL)/preliminary hazard analysis (PHA) for the system under development.
- Life cycle sustainment plan(s) for legacy or similar systems.

Figure 14.1 shows an overview of the basic EHA process and summarizes the important relationships involved. This process consists of utilizing both design information and known hazard information to identify hazards. Known hazardous elements and mishap lessons learned are compared with the system design to determine if the design concept contains any of these potential hazard elements.

14.5 EHA METHODOLOGY

Table 14.1 lists and describes the basic steps of the EHA process, which involves performing a detailed analysis of the system design and operations as they affect the environment from a safety standpoint.

The thought process behind the EHA methodology is shown in Figure 14.2. The idea supporting this process is that different kinds of design information are used to facilitate hazard identification. The analysis begins with hazards identified from the PHL and PHA, which is the starting point for the EHA. The next step is to once again employ the use of hazard checklists and undesired mishap checklists. Of particular interest are hazard checklists dealing with hazard sources that can cause environmental damage.

The EHA process is best addressed via two different viewpoints. Viewpoint 1 is the environment view – it covers all environments that could be adversely affected by the systems operation or malfunction. Viewpoint 2 is the hazard source view – it covers all hazard sources within the system that could adversely affect the environment if they were purposely or inadvertently released into the environment. This dual-pronged approach helps ensure complete hazard coverage.

TABLE 14.1 EHA Process

Step	Task	Description
1	Acquire design information	Acquire all of the design, operational, and manufacturing data for the system
2	Establish the environments involved	Determine all of the environments the system will be involved with
3	Acquire environmental hazard checklists	Acquire checklists of known hazard sources, such as chemicals, materials, processes, and so on. Also, acquire checklists of known human limitations in the operation of systems, such as noise, vibration, heat, and so on
4	Acquire regulatory information	Acquire all regulatory data and information that are applicable to environmental hazards
5	Identify hazard sources	Examine the system and identify all potentially hazardous sources and processes within the system. Include quantity and location when possible. Utilize the checklists Hazard sources are typically the elements that actually damage the environment
6	Identify hazards	Identify and list potential hazards created in the system design resulting from the hazard sources
7	Identify hazard barriers	Identify design mitigation methods or barriers in the path of the hazard source. Also, identify existing design features to eliminate or mitigate the hazard
8	Evaluate system risk	Identify the level of mishap risk presented to personnel, both with and without design controls in the system design
9	Recommend corrective action	Determine if the design controls present are adequate, and if not, recommend controls that should be added to reduce the mishap risk
10	Track hazards	Transfer newly identified hazards into the HTS. Update the HTS as hazards, hazard causal factors, and risk are identified in the EHA
11	Document EHA	Document the entire EHA process on the worksheets. Update for new information as necessary

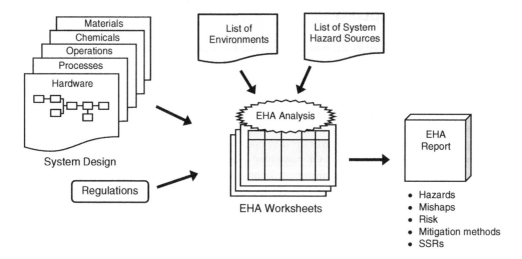

Figure 14.2 EHA methodology.

The EHA strongly depends upon the use of hazard checklists. Hazard checklists are generic lists of known hazardous items and potentially hazardous designs or situations and should not be considered complete or all-inclusive. Checklists are intended as a starting point to help trigger the analyst's recognition of potential hazard sources from past lessons learned.

When performing the EHA, the following factors should be given consideration:

a. Toxicity, quantity, and physical state of materials.

b. Routine or planned uses and releases of hazardous materials or physical agents.

c. Accidental exposure potentials.

d. Hazardous waste generated.

e. Hazardous material handling, transfer, and transportation requirements.

f. Protective clothing/equipment needs.

g. Detection and measurement devices required to quantify exposure levels.

h. Number of personnel potentially at risk.

i. Design controls that could be used, such as isolation, enclosure, ventilation, noise or radiation barriers, and so on.

j. Potential alternative materials to reduce the associated risk to users/operators.

k. The degree of personnel exposure to the health hazard.

l. System, facility, and personnel protective equipment design requirements (e.g., ventilation, noise attenuation, radiation barriers, etc.) to allow safe operation and maintenance.

m. Hazardous material and long-term effects (such as potential for personnel and environmental exposure, handling and disposal issues/requirements, protection/control measures, and life cycle costs).

n. Means for identifying and tracking information for each hazardous material.

o. Environmental factors that affect exposure (wind, temperature, humidity, etc.)

When hazardous materials must be used in the system, the following considerations must be evaluated and documented:

1. Identify hazardous materials data:
   a. Name
   b. Stock number
   c. Affected system components and processes
   d. Quantity, characteristics, and concentrations of the materials in the system
   e. Source documents relating to the materials
2. Determine under which conditions the hazardous materials can pose a health threat.
3. Characterize material hazards and determine reference quantities and hazard ratings (e.g., acute health, chronic health, carcinogenic, contact, flammability, reactivity, and environmental hazards).
4. Estimate the expected usage rate of each hazardous material.
5. Recommend the disposition (disposal, recycle, etc.) of each hazardous material identified.

When feasible engineering designs are not available to reduce hazards to acceptable levels, alternative protective measures must be specified (e.g., protective clothing, specific operation, or maintenance practices to reduce risk to an acceptable level). Identify potential nonhazardous or less-hazardous alternatives to hazardous materials if they exist, or provide a justification why an alternative cannot be used.

14.6 EHA WORKSHEET

In conducting an EHA, two worksheets are recommended to ensure full system and hazard coverage. The two worksheets provide two different viewpoints: one view is of all the environments that could be adversely affected by the systems operation or malfunction, while the second view evaluates the hazard sources within the system that could adversely affect the environment if they were purposely or inadvertently released. When both worksheets are applied, there is greater assurance that all of the system and environmental aspects have been adequately covered by the analysis.

Worksheet 1, shown in Figure 14.3, is utilized to ensure all environments are covered by the analysis. The analysis is initiated by listing all the environments in the left-hand column that the system is expected to encounter or interface with. This particular EHA worksheet utilizes a columnar-type format and has been proven effective in many applications. The worksheet can be modified as found necessary by the system safety program.

The information required under each column entry of the worksheet is described as follows:

1. *Environment* This column lists all the normally expected environments that the system would encounter.
2. *Hazard* This column identifies what can go wrong and the resulting undesirable consequences and damage to the environment. If a hazard cannot be postulated, then it should be so stated on the worksheet; this helps document that everything was considered and not left out of the analysis.
3. *Cause* This column identifies conditions, events, or faults that could cause the hazard to exist, and the events that can trigger the hazardous elements to become a mishap or accident.
4. *Effect/Mishap* This column identifies the effect and consequences of the hazard, should it occur. Generally, the worst-case result is the stated effect.

Environmental Hazard Analysis							
Environment	Hazard	Causes	Effects	IMRI	Recommended Action	FMRI	Comments
①	②	③	④	⑤	⑥	⑦	⑧

Figure 14.3 *Recommended EHA worksheet 1.*

5. *Initial Mishap Risk Index (IMRI)* This column provides a qualitative measure of mishap risk for the potential effect of the identified hazard, given that no mitigation techniques are applied to the hazard. Risk measures are a combination of mishap severity and probability, and the recommended values are shown below. Risk knowledge helps in the prioritization of mitigation measures.

Severity	Probability
1 – Catastrophic	A – Frequent
2 – Critical	B – Probable
3 – Marginal	C – Occasional
4 – Negligible	D – Remote
	E – Improbable

6. *Recommended Action* This column establishes recommended preventive measures to eliminate or control identified hazards. Safety requirements in this situation generally involve the addition of one or more barriers to keep the energy source away from the target. The preferred order of precedence for design safety requirements are as shown in the following table.

Order of Precedence
1 – Eliminate hazard through design selection
2 – Control hazard through design methods
3 – Control hazard through safety devices
4 – Control hazard through warning devices
5 – Control hazard through procedures and training

7. *Final Mishap Risk Index (FMRI)* This column provides a qualitative measure of mishap risk significance for the potential effect of the identified hazard, given that mitigation techniques and safety requirements are applied to the hazard. The same values used in column 5 are also used here. The initial and final risk values are important because they show the amount of risk improvement.

8. *Comments* This column provides a place to record useful information regarding the hazard or the analysis process that are not noted elsewhere.

Worksheet 2, shown in Figure 14.4, is utilized to ensure all hazard sources involved with the system are covered by the analysis. The analysis is initiated by listing all of the hazard sources in the left-hand column that the system utilizes. The reason is that typically it is the release of energy sources into the environment that causes environmental damage. This particular EHA worksheet utilizes a columnar-type format and has been proven effective in many applications. The worksheet can be modified as found necessary by the system safety program.

The information required under each column entry of the worksheet is described as follows:

1. *Hazard Source* This column lists all hazard sources that exist within the system and system operations such as fuel, oil, explosive materials, RF energy, biological material, nuclear material, and so on.

2. *Hazard* This column identifies what can go wrong and the resulting undesirable consequences and damage to the environment. If a hazard cannot be postulated, then it

Environmental Hazard Analysis							
Hazard Source	Hazard	Causes	Effects	IMRI	Recommended Action	FMRI	Comments
①	②	③	④	⑤	⑥	⑦	⑧

Figure 14.4   *Recommended EHA worksheet 2.*

should be so stated on the worksheet; this helps document that everything was considered and not left out of the analysis.

3. *Cause*   This column identifies conditions, events, or faults that could cause the hazard to exist, and the events that can trigger the hazardous elements to become a mishap or accident.

4. *Effect/Mishap*   This column identifies the effect and consequences of the hazard, should it occur. Generally, the worst-case result is the stated effect.

5. *IMRI*   This column provides a qualitative measure of mishap risk for the potential effect of the identified hazard, given that no mitigation techniques are applied to the hazard. Risk measures are a combination of mishap severity and probability, and the recommended values are shown below. Risk knowledge helps in the prioritization of mitigation measures.

6. *Recommended Action*   This column establishes recommended preventive measures to eliminate or control identified hazards. Safety requirements in this situation generally involve the addition of one or more barriers to keep the energy source away from the target.

7. *FMRI*   This column provides a qualitative measure of mishap risk significance for the potential effect of the identified hazard, given that mitigation techniques and safety requirements are applied to the hazard. The same values used in column 5 are also used here. The initial and final risk values are important because they show the amount of risk improvement.

8. *Comments*   This column provides a place to record the useful information regarding the hazard or the analysis process that is not noted elsewhere.

14.7   EXAMPLE CHECKLISTS

In order to aid the EHA process, several checklists are provided to help stimulate the analysis thought process. These following checklists are not to be considered complete but for examples only:

- Table 14.2 provides example environments to consider.

TABLE 14.2 Typical Environments to Consider

Environment	Safety Concerns
Air • Indoor; outdoor	Pollution, chemicals, toxins, viruses, and poor air quality
Atmosphere	Pollution; damage to ozone layer
Water • streams, rivers, lakes, and oceans	Pollution, chemicals, toxins, viruses, and poison
Land • Ground, soil, mountains, and glaciers	Pollution, chemicals, toxins, and contamination
Trees	Pollution, chemical damage, and toxins
Plants	Pollution, chemical damage, toxins, biological damage, and genetic modification
Food	Pollution, chemicals, toxins, poison, and genetic modification
Animals	Chemical damage, biological damage, air pollution, and spread of disease
Insects	Spread of disease
Marine life	Chemical damage, biological damage, and water pollution

- Table 14.3 provides example environmental safety concerns.
- Table 14.4 provides example hazard sources and their impact on the environment.

It should be noted that these lists are examples and should not be considered complete; the safety analyst should be responsible for developing a complete list for the project involved.

14.8 EHA EXAMPLE

In order to demonstrate the EHA methodology, an EHA is performed on two hypothetical example systems. Table 14.5 uses the *environment* EHA worksheet to analyze and underwater oil drilling system. Table 14.6 uses the *hazard source* EHA worksheet to analyze an oil transportation ship system. These are rather oversimplified examples to demonstrate the basic EHA concept and methodology.

14.9 EHA ADVANTAGES AND DISADVANTAGES

The EHA technique offers the following advantages:

1. The EHA is easily and quickly performed.
2. The EHA does not require considerable expertise for technique application.
3. The EHA is relatively inexpensive, yet provides meaningful results.
4. The EHA provides rigor for focusing on system environmental hazards.

There are no disadvantages of the EHA technique.

TABLE 14.3   Typical Environment Safety Concerns to Consider

Environmental Safety Concerns

Indoor air pollutants
• Asbestos
• Formaldehyde
• Radon
• Environmental tobacco smoke
• Volatile organic compounds

Indoor air quality
• Sick building syndrome
• Ventilation
• Legionnaire's disease

Biological water pollutants
• Pathogens
• Overgrowth of aquatic plants

Toxic water pollutants
• Inorganic chemicals
• Radioactive materials
• Synthetic organic chemicals

Soil damage
• Oil spills/leaks

Marine life damage
• Oil spills/leaks
• Ship biowaste leaks

Outdoor air poisoning
• Nuclear radiation leaks/releases
• Toxic chemical leaks/releases

TABLE 14.4   Typical Environmental Hazard Sources to Consider

Hazard Sources	Environmental Concerns
Electromagnetic radiation	Injury to humans/animals
Nuclear radiation	Injury to humans/animals
Biological agents	Injury to humans/animals
Hazardous materials	Lead poisoning; mercury poisoning
Noise	Hearing damage to humans/animals
Chemical substances	Air/water pollution
Ionizing radiation energy (nuclear)	Radiation damage
Nonionizing radiation (laser)	Eye sight damage
Temperature extremes	Injury to humans/animals
Vibration	Injury to humans/animals
Oil	Pollution; contamination
Fuel – auto, aviation	Pollution; contamination

TABLE 14.5   Example EHA; Worksheet 1

	Environmental Hazard Analysis								
System: Subsystem:					Analyst: Date:				
Environment	Hazard	Causes	Effects	IMRI	Recommended Action	FMRI	Comments	Status
Ocean water	Water pollution due to oil well leak	Valve failure; Pipe failure	Water pollution	2C	xxxxx	---	No SSR	Open
	Damage to ocean plants from oil resulting from oil well leak	Valve failure; Pipe failure	Oil contamination damages plants	2C	xxxxx	2E	Mitigated by SSR #14	Closed
	Injury/death to ocean animal life(mammals) from oil resulting from oil well leak	Valve failure; Pipe failure	Oil contamination causes injury death to exposed animals	2C	xxxxx	2E	Mitigated by SSR #21	Closed
	Injury/death to ocean birds from oil resulting from oil well leak	Valve failure; Pipe failure	Oil contamination causes injury death to exposed birds	2C	xxxxx	---	No SSR	Open

Page: 1

235

TABLE 14.6   Example EHA; Worksheet 2

			Environmental Hazard Analysis			Analyst: Date:		
System: Subsystem:								
Hazard Source	Hazard	Causes	Effects	IMRI	Recommended Action	FMRI	Comments	Status
Oil	Water pollution and contamination, injures fish, mammals, fauna	Leakage from oil container ship	Injury/death of fish/animals	2C	None	2E	Mitigated by SSR #10	Closed
Gasoline	Water pollution and contamination, injures fish, mammals, fauna	Leakage from oil container ship	Injury/death of fish/animals	2C	Sensors and air purge	---	No SSR	Open
Bio waste	Water pollution and contamination, injures fish, mammals, fauna	Leakage from oil container ship	Injury/death of fish/animals	2C	None	2E	Mitigated by SSR #99	Closed
Ship propellers	Injury/death to fish and animals in water	Rotating propeller blades strike fish or animals	Injury/death of fish/animals	2C	All items must not exceed one-man weight limit	---	No SSR	Open

Page: 1

236

14.10 COMMON EHA MISTAKES TO AVOID

When first learning how to perform an EHA, it is commonplace to commit some traditional errors. The following is a list of typical errors made during the conduct of an EHA.

1. The hazard identified is other than an environmental hazard.
2. The hazard description is not detailed enough.
3. Design mitigation factors not stated or provided.
4. Design mitigation factors do not address the actual causal factor(s).
5. Overuse of project specific terms and abbreviations.
6. Failing to identify and evaluate all environments actually involved.

14.11 SUMMARY

This chapter discussed the EHA technique. The following basic principles help summarize the discussion in this chapter:

1. The primary purpose of EHA is to identify and mitigate environmental hazards.
2. The use of an environmental checklist greatly aids and simplifies the EHA process.
3. The use of the recommended EHA worksheets aids the analysis process and provides for documentation of the analysis.

14.12 REFERENCES

There are no significant references for the EHA technique that describe it in detail. Although many textbooks on system safety discuss the methodology in general, they do not provide detailed explanation with examples.

14.13 NATIONAL ENVIRONMENTAL POLICY ACT

The NEPA was established in 1969 to develop, for government agencies and interests, formal pollution prevention guidelines that minimize emissions and pollutants to acceptable levels. NEPA is a public law that requires the identification and analysis of potential environmental impacts of certain industry and Federal actions before those actions are initiated. The process for implementing the law is codified in the Council on Environmental Quality (CEQ) Regulations. NEPA was one of the first laws ever written that establishes a broad national framework for protecting the environment. The Environmental Protection Agency (EPA) serves as an independent agency that ensures that the NEPA is followed.

The NEPA, as implement by EOs 11514 and 11991 and the CEQ, requires that all agencies of the Federal government implement the provisions of NEPA. NEPA mandates that Federal agencies "utilize a systematic, multidisciplinary approach that will ensure the integrated use of natural and social sciences and the environmental design arts in planning and decision making, which may have an impact on the environment."

NEPA's basic policy is to assure that industry and government give proper consideration to the environment prior to undertaking any major action that may affect the environment. NEPA procedures ensure that environmental information is available to public officials and citizens before decisions are made and before proposed actions can be taken.

Section 102 of the NEPA created the CEQ in the Executive Office of the President. The CEQ's purpose is to tell federal agencies what they must do to comply with the procedures and achieve the goals of the NEPA. The President, the federal agencies, and the courts share responsibility for enforcing NEPA so as to achieve the substantive requirements of NEPA, Section 101. The CEQ promulgates regulations that implement Section 102(2) of the NEPA and CEQ's regulations are binding on the Navy. The CEQ also provides guidance documents, which aid Federal agencies in their implementation of NEPA's myriad of procedural requirements.

The five basic tenets of the NEPA and CEQ Regulations are as follows:

1. Procedures must be in place to ensure that environmental information is available to decision makers and citizens before decisions are made and before proposed actions are taken.
2. The NEPA process should identify and assess reasonable alternatives to proposed actions that would avoid or minimize adverse environmental effects.
3. The NEPA analysis should identify and discuss the potential impacts of health and safety issues caused by environmental factors. The Manpower and Personnel Integration (MANPRINT) process is a technique for evaluating human factors.
4. NEPA's purpose is to help agency officials make decisions based on an understanding of environmental effects, enabling them to take actions that protect, restore, and enhance the environment.
5. Agencies must integrate the NEPA process with other planning agencies at the earliest possible time to ensure that planning and decisions reflect environmental values, to avoid delays later in the process, and to head off potential conflicts.

US Congress has directed that, to the fullest extent possible, the policies, regulations, and public laws of the United States shall be interpreted and administered in accordance with the policies set forth in the NEPA, Chapter 55 Section 4332. Congress also directed that all agencies of the federal government should

a. utilize a systematic, interdisciplinary approach that will ensure the integrated use of the natural and social sciences and the environmental design arts in planning and in decision making which may have an impact on man's environment.
b. identify and develop methods and procedures, in consultation with the CEQ, which will insure that environmental amenities and values, which presently are not quantified may be given appropriate consideration in decision making along with economic and technical considerations.

14.14   ENVIRONMENTAL PROTECTION AGENCY

The EPA serves as an independent agency under Section 309 of the Clean Air Act (42 U.S.C. 7609) whose purpose is to superintend NEPA policies and laws. A major function of the EPA is to ensure that agencies preparing EIS documentation under NEPA have the benefit of a review by a federal agency whose primary mission is the protection of the environment. The EPA must comment in writing, and make these comments available for public review. Section 309 also provides authority for the EPA to determine independently that a proposed action is a major action (or threat) that would significantly affect the environment, even if the proposing agency has determined otherwise. EPA's review is primarily concerned with identifying and recommending measures for the mitigation of any significant environmental effects associated with the proposed action.

Chapter 15

Fault Tree Analysis

15.1 FTA INTRODUCTION

Fault tree analysis (FTA) is a systems analysis technique used to determine the root causes and probability of occurrence of a specified undesired event. FTA is employed to evaluate large complex dynamic systems, in order to understand and prevent potential problems. Using a rigorous and structured methodology, FTA allows the systems analyst to model the unique combinations of fault events that can cause an undesired event to occur. The undesired event may be a system hazard of concern or a mishap that is under accident investigation.

A Fault tree (FT) is a model that logically and graphically represents the various combinations of possible events, faulty and normal, occurring in a system that can lead to an undesired event or state. The analysis is deductive, in that it transverses from the general problem to the specific causes. The FT develops the logical fault paths from a single undesired event at the top to all of the possible root causes at the bottom. The strength of FTA is that it is easy to perform, easy to understand, provides useful system insight, and shows all the possible causes for a problem under investigation.

FTs are graphical models using logic gates and fault events to model the cause–effect relationships involved in causing the undesired event. The graphical model can be translated into a mathematical model to compute failure probabilities and system importance measures. FT development is an iterative process, where the initial structure is continually updated to coincide with design development.

In the analysis of systems, there are two applications of FTA. The most commonly used application is the *proactive FTA*, performed during system development to influence design by predicting and preventing future problems. The other application is the *reactive FTA*, performed after an accident or mishap has occurred. The techniques used for both applications

Hazard Analysis Techniques for System Safety, Second Edition. Clifton A Ericson, II.
© 2016 John Wiley & Sons, Inc. Published 2016 by John Wiley & Sons, Inc.

are identical except that the reactive FTA includes the use of mishap evidence and the evidence event gate (EEG).

When used as a system safety analysis tool, the FT results in a graphic and logical representation of the various combinations of possible events, both faulty and normal, occurring within a system, which can cause a predefined undesired event. An undesired event is any event, which is identified as objectionable and unwanted, such as a potential accident, hazardous condition, or undesired failure mode. This graphic presentation exposes the inter-relationships of system events, and their interdependence upon each other, which results in the occurrence of the undesired event.

The completed FT structure can be used to determine the significance of fault events and their probability of occurrence. The validity of action taken to eliminate or control fault events can be enhanced in certain circumstances by quantifying the FT and performing a numerical evaluation. The quantification and numerical evaluation generates three basic measurements for decision making relative to risk acceptability and required preventive measures:

1. The probability of occurrence of the undesired event,
2. The probability and significance of fault events (cut sets) causing the undesired event, and
3. The risk significance or importance of components.

In most circumstances, a qualitative evaluation of the fault tree will yield effective results at a reduced cost. Careful thought must be given in determining whether to perform a qualitative or a quantitative FTA. The quantitative approach provides more useful results; however, it requires more time and experienced personnel. The quantitative approach also requires the gathering of component failure rate data for input to the FT.

Since an FT is both a graphic and a logical representation of the causes or system faults leading to the undesired event, it can be used in communicating and supporting decisions to expend resources to mitigate hazards. As such, it provides the required validity in a simple and highly visible form to support decisions of risk acceptability, and preventive measure requirements.

The FT process can be applied during any life cycle phase of a system – from concept to usage. However, FTA should be used as early in the design process as possible since the earlier necessary design changes are made, the less they cost.

An important time- and cost-saving feature of the FT technique is that only those system elements that contribute to the occurrence of the undesired event need to be analyzed. During the analysis, noncontributing elements are ruled out and are, thus, not included in the analysis. This means that a majority of the effort is directed toward the elimination or control of the source or sources of the problem area. However, system elements not involved with the occurrence of one undesired event may be involved with occurrence of another undesired event.

In summary, the FT is used to investigate the system of concern, in an orderly and concise manner, to identify and depict the relationships and causes of the undesired event. A quantitative evaluation may be performed in addition to a qualitative evaluation to provide a measure of the probability of the occurrence of the top-level event and the major faults contributing to the top-level event. The analyst may use the results of an FTA as follows:

1. Verification of design compliance with established safety requirements.
2. Identification of design safety deficiencies (subtle or obvious) that have developed in spite of existing requirements.

3. Identification of common mode failures.
4. Establishment of preventive measures to eliminate or mitigate identified design safety deficiencies.
5. Evaluation of the adequacy of the established preventive measures.
6. Establishment or modification of safety requirements suitable for the next design phase.

15.2   FTA BACKGROUND

This analysis technique falls under the system design hazard analysis type (SD-HAT). Refer to Chapter 4 for a description of the analysis types. The FTA technique has been referred to as logic tree analysis and logic diagram analysis.

FTA has several basic purposes, which include the following:

1. To find the root causes of a hazard or undesired event during design development in order that they can be eliminated or mitigated.
2. To establish the root causes of a mishap that has occurred and prevent them from recurring.
3. To identify the undesired event causal factor combinations and their relative probability.
4. To determine high-risk fault paths and their mechanisms.
5. To identify risk importance measures for components and fault events.
6. To support a probabilistic risk assessment (PRA) of system designs.

FTA can be used to model an entire system, with analysis coverage given to subsystems, assemblies, components, software, procedures, environment, and human error. FTA can be conducted at different abstraction levels, such as conceptual design, top-level design, and detailed component design. FTA has been successfully applied to a wide range of systems, such as missiles, ships, spacecraft, trains, nuclear power plants, aircraft, torpedoes, medical equipment, and chemical plants. The technique can be applied to a system very early in design development and thereby identify safety issues early in the design process. Early application helps system developers to design-in safety of a system during early development rather than having to take corrective action after a test failure or a mishap.

A basic understanding of the FTA theory is essential to developing FTs of small and noncomplex systems. In addition, it is crucial for the analyst to have a detailed understanding of the system regardless of complexity. As system complexity increases, increased knowledge and experience in FTA is also required. Overall, FTA is very easy to learn and understand. Proper application depends on the complexity of the system and the skill of the analyst.

Applying FTA to the analysis of a system design is not a difficult process. It is more difficult than an analysis technique such as a PHA, primarily because it requires a logical thought process, an understanding of FTA construction methodology, and a detailed knowledge of system design and operation. FTA does not require knowledge of high-level mathematics compared to Markov or Petri net analyses.

FTA enjoys a favorable reputation among system safety analysts in all industries utilizing the technique. In some industries, it is the only tool that can provide the necessary probability calculations for verification that numerical requirements are being met. Many commercial computer programs are available to assist the analyst in building, editing, and mathematically evaluating FTs.

Some analysts criticize the FTA tool because it does not always provide probabilities to six-decimal place accuracy when modeling certain designs. However, comparison of FT model results with those of other tools, such as Markov analysis (MA) show that FTA provides very comparable results with much greater simplicity in modeling difficulty.

Although FTA is classified as a hazard analysis, it is primarily used as a root cause analysis tool to identify and evaluate the causal factors of a hazard. In addition, it can provide a probability risk assessment.

Markov analysis could be utilized in place of FTA for probability calculations; however MA has limitations that FTA does not (refer to Chapter 26). For example, it is difficult to model large complex systems, the mathematics are more cumbersome, it is difficult to visualize fault paths in an MA model, and an MA model does not produce cut sets.

FTA is highly recommended for detailed analysis of an undesired event to determine and understand all the possible fault combinations that can cause the undesired event to occur. Since it is a highly intensive analysis, it is usually recommended for the evaluation of hazards that are safety-critical or high consequence.

15.3   FTA HISTORY

The FTA technique was invented and developed by H. Watson and Allison B. Mearns of Bell Labs for use on the Minuteman Guidance System. Dave Haasl of the Boeing Company recognized the power of FTA and applied it for quantitative safety analysis of the entire Minuteman Weapon System. The analytical power and success of the technique was recognized by the commercial aircraft industry and the nuclear power industry and they then began using it for safety evaluations. Many individuals in each of these industries contributed to enhancing the state of the art in fault tree mathematics, graphics, and computer algorithms.

15.4   FTA THEORY

FTA is a robust, rigorous, and structured methodology requiring the application of certain rules of Boolean algebra, logic, and probability theory. The FT itself is a logic diagram of all the events (failure modes, human error, and normal conditions) that can cause the top undesired event to occur.

When the FT is complete, it is evaluated to determine the critical cut sets (CSs) and probability of failure. The cut sets are the combination of failure events that can cause the top event to occur. The FT evaluation provides the necessary information to support risk management decisions.

As shown in Figure 15.1, the theory behind FTA is to start with a top undesired event (UE) (e.g., hazard) and model all of the system faults that can contribute to this top event. The FT model is a reflection of the system design, from a failure state viewpoint. In this example, the UE might be "inadvertent warhead initiation due to system faults."

FTs are developed in layers, levels, and branches using a repetitive analysis process. Figure 15.2 demonstrates an FT developed in layers, with each major layer representing significant aspects of the system. For example, the top FT structure usually models the system functions and phases, the intermediate FT structure models subsystem fault flows, and the bottom FT structure models assembly and component fault flows.

244 FAULT TREE ANALYSIS

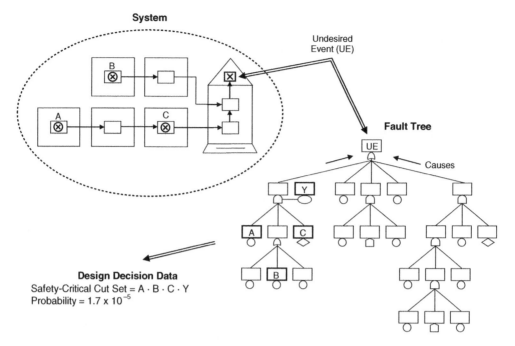

Figure 15.1 FTA overview.

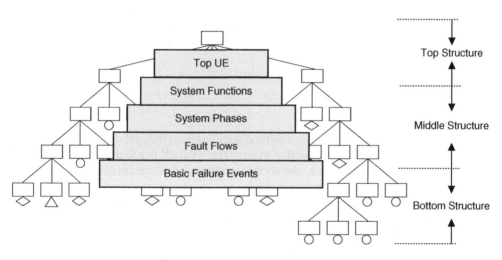

Figure 15.2 Major levels of a fault tree.

15.5 FTA METHODOLOGY

There are eight basic steps in the FTA process, as shown in Figure 15.3. These are the steps
required to perform a complete and accurate FTA. Some analysts may combine or expand some
of the steps, but these are the basic procedures that must be followed.

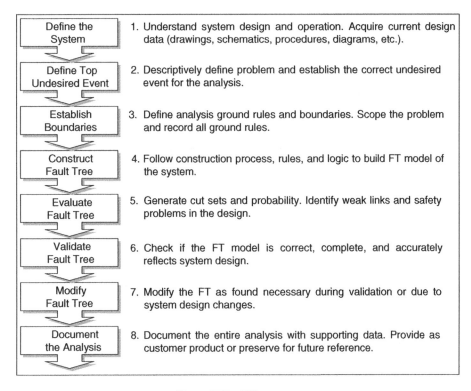

Define the System	1. Understand system design and operation. Acquire current design data (drawings, schematics, procedures, diagrams, etc.).
Define Top Undesired Event	2. Descriptively define problem and establish the correct undesired event for the analysis.
Establish Boundaries	3. Define analysis ground rules and boundaries. Scope the problem and record all ground rules.
Construct Fault Tree	4. Follow construction process, rules, and logic to build FT model of the system.
Evaluate Fault Tree	5. Generate cut sets and probability. Identify weak links and safety problems in the design.
Validate Fault Tree	6. Check if the FT model is correct, complete, and accurately reflects system design.
Modify Fault Tree	7. Modify the FT as found necessary during validation or due to system design changes.
Document the Analysis	8. Document the entire analysis with supporting data. Provide as customer product or preserve for future reference.

Figure 15.3 *FTA process.*

15.5.1 FT Building Blocks

FTs consist of nodes interlinked together in a tree-like structure. The nodes represent fault/failure paths and are linked together by Boolean logic and symbols. The FT symbols form the basic building blocks of FTA and consist of four categories:

1. Basic events
2. Gate events
3. Conditional events
4. Transfer events

Figure 15.4 shows the standard symbols for basic events (BE), condition events (CE), and transfer events (TE) as they would appear on an FT, and their associated definitions. Note that the rectangle is nothing more than a placeholder for text. When FTA was first developed, the text was placed directly in the BE symbols and the rectangle was used only for gate nodes; but with the advent of computer graphics this became cumbersome, so the rectangle was adopted for all nodes.

Figure 15.5 shows the gate event symbols, definitions, and probability calculation formulas. It is through the gates that the FT logic is constructed and the tree grows in width and depth.

The symbols shown in Figures 15.4 and 15.5 are generally considered the standard FT symbols; however, some FT software programs do utilize slightly different symbols. Figure 15.6 shows some alternative and additional symbols that might be encountered.

Symbol	Type	Description
	Node Text Box	Contains the text for all FT nodes. Text goes in the box, and the node symbol goes below the box.
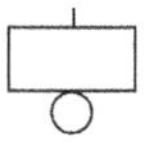	Primary Failure (BE)	A basic component failure; the primary, inherent, failure mode of a component. A random failure event.
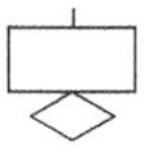	Secondary Failure (BE)	An externally induced failure or a failure mode that could be developed in more detail if desired.
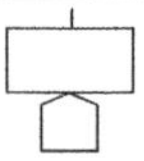	Normal Event (BE)	An event that is expected to occur as part of normal system operation.
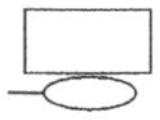	Condition (CE)	A conditional restriction or probability.
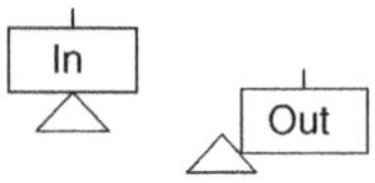	Transfer (TE)	Indicates where a branch or subtree is marked for the same usage elsewhere in the tree. In and Out or To/From symbols.

Figure 15.4 *FT symbols for basic events, conditions, and transfers.*

Symbol	Gate Type	Description
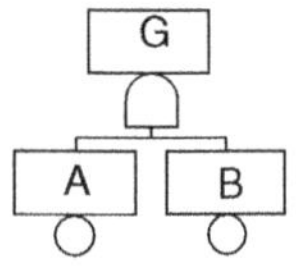	AND Gate	The output occurs only if all of the inputs occur together. $P = P_A \cdot P_B = P_A P_B$ (2 input gate) $P = P_A \cdot P_B \cdot P_C = P_A P_B P_C$ (3 input gate)
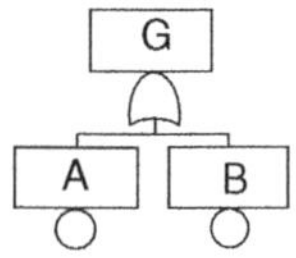	OR Gate	The output occurs only if at least one of the inputs occurs. $P = P_A + P_B - P_A P_B$ (2 input gate) $P = (P_A + P_B + P_C) - (P_{AB} + P_{AC} + P_{BC}) + (P_{ABC})$ (3 input gate)
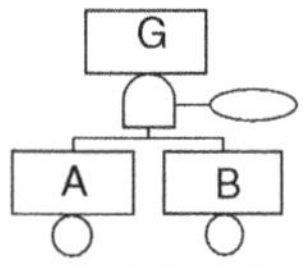	Priority AND Gate	The output occurs only if all of the inputs occur together, and A must occur before B. The priority statement is contained in the Condition symbol. $P = (P_A P_B) / N!$ Given $\lambda_A \approx \lambda_B$ and $N =$ number of inputs to gate
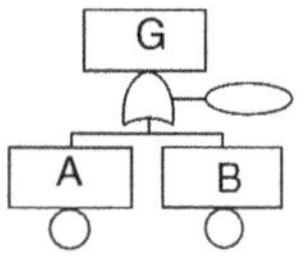	Exclusive OR Gate	The output occurs if either of the inputs occurs, but not both. The exclusivity statement is contained in the condition symbol. $P = P_A + P_B - 2(P_A P_B)$
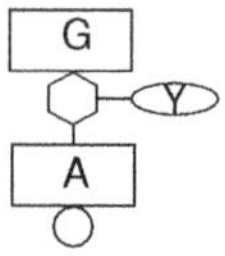	Inhibit Gate	The output occurs only if the input event occurs and the attached condition is satisfied. $P = P_A \cdot P_Y = P_A P_Y$

Figure 15.5 *FT symbols for gate events.*

Typical Symbol	Action	Description	Alternative Symbol
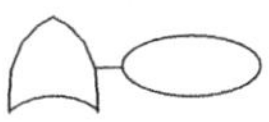	Exclusive OR Gate	Only one of the inputs can occur, not both. Disjoint events.	
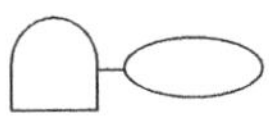	Priority AND Gate	All inputs must occur, but in given order, from left to right.	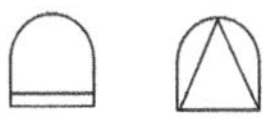
	M of N Gate	M of N combination of inputs causes output to occur. Voting gate.	
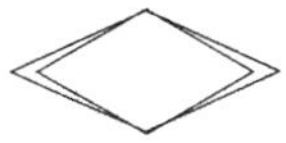	Double Diamond	User defined event for special uses.	

Figure 15.6 *Alternative FT symbols.*

15.5.2 FT Definitions

In addition to the FT symbol definitions, the following definitions define important concepts utilized in FTA:

Cut Set   A set of events that together cause the top UE to occur. Also referred to as a fault path.

Min Cut Set (MinCS or MCS)   A CS that has been reduced to the minimum number of events that cause the top UE to occur. The CS cannot be further reduced and still guarantee the occurrence of the top UE.

CS Order   The CS order is the number of items in a CS. A one-order CS is a single-point failure (SPF). A two-order CS has two items AND'd together.

Multiple Occurring Event (MOE)   The FT basic Event that occurs in more than one place in the FT.

Multiple Occurring Branch (MOB)   An FT branch that is used in more than one place in the FT. This is one place in the FT where the transfer symbol is used. All BEs below the MOB are automatically MOEs.

Failure   The occurrence of a basic inherent component failure, for example, "Resistor Fails Open."

Fault   The occurrence or existence of an undesired state of a component, subsystem, or system. For example, "light off" is an undesired fault state that may be due to light bulb failure, loss of power, or operator action. (*Note*: all failures are faults, but not all faults are failures.)

Primary Fault/Failure   An independent *component failure that cannot be further defined* at a lower level. For example, "diode inside a computer fails (due to materiel flaw)."

Secondary Fault/Failure   An independent component *failure that is caused by an external force* on the system. For example, "diode fails due to excessive RF/EMI energy in system." Failure due to out-of-tolerance operational or environmental conditions.

Command Fault/Failure   An item that is "commanded" to fail or forced into a fault state by system design. For example, "light off" is the command fault for the light, that is, it is commanded to fail off if certain system faults cause the loss of power. A command fault can be the normal operational state but, at the wrong time, and sometimes, it is lack of the desired normal state when desired or intended. (This is the "transition" to look for in the analysis.)

Exposure Time (ET)   The length of time a component is effectively exposed to failure during system operation. ET has a large effect on FT probability calculations ($P = 1.0 - e^{-\lambda T}$). ET can be controlled by design, repair, circumvention, testing, and monitoring.

Critical Path   The highest probability CS that drives the top UE probability. The most dramatic system improvement is usually made by reducing the probability of this CS.

Importance Measure   A measure of the relative importance (sensitivity) of a BE or CS in the overall FT.

Figure 15.7 demonstrates the usage of FT transfer symbols and the MOE/MOB concepts. This figure shows an example of three FT pages. On page 1, the node with a triangle at the bottom with the name A represents a transfer-in. This means that a duplicate of branch A should also go here, but it is drawn somewhere else, page 2 in this case. In this case, A is not an MOB but merely the transfer of tree to start on a new page, due to lack of space on page 1. Transfer C represents an MOE, as it is intended to be repeated at two different places in the FT.

15.5.3   FT Construction: Basics

FT construction is an iterative process that begins at the treetop and continues down through all of the tree branches. The same set of questions and logic is applied on every gate, moving down the tree. After identifying the top UE, subundesired events are identified and structured into what is referred to as the "top fault tree" layer. The actual deductive analysis begins with the development of the "fault flow" or cause and effect relationship of fault and normal events through the system. This deductive reasoning involves determining the type of gate and the particular inputs to this gate at each gate level of the FT. The fault flow links the flow of events from the system level, through the subsystem level, to the component level.

The FT development proceeds through the identification and combination of the system normal and fault events, until all events are defined in terms of basic identifiable hardware faults, software faults, and human error. This is the level of basic events in the FT structure and is the end point for construction of the FT.

In developing the structure of the FT, certain procedures must consistently be followed in a repetitive manner. These procedures are necessary at each gate level to determine the type of gate to be used and the specific inputs to the gate. The established procedure revolves around three principal concepts:

1. The I-N-S concept
2. The SS–SC concept
3. The P-S-C concept

I-N-S Concept   This concept involves answering the question "What is Immediate, Necessary, and Sufficient to cause the event?" The "I-N-S" question identifies the most immediate cause(s) of the event, the causes that are absolutely necessary, and includes only the causes that are absolutely necessary and sufficient. For example, water is necessary to maintain a green lawn and rain is sufficient to provide it, or a sprinkler system is sufficient.

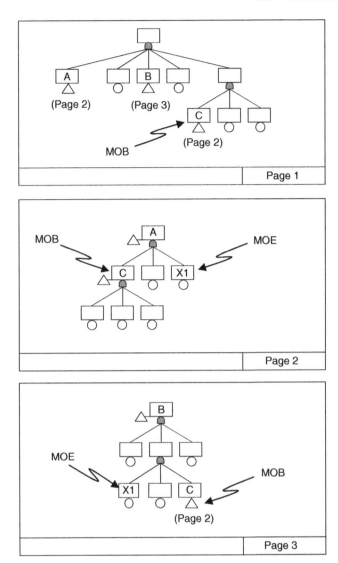

Figure 15.7 *FT transfers and MOE/MOB.*

This seems like an obvious question to ask, but too often it is forgotten in the turmoil of analysis. There are several reasons for stressing this question:

1. It helps keep the analyst from jumping ahead.
2. It helps focus on identifying the next element in the cause–effect chain.
3. It is a reminder to only include the minimum sufficient causes necessary and nothing extraneous

SS–SC Concept The SS–SC concept differentiates between the failure being "state of the system" (SS) and "state of the component" (SC). If a fault in the event box can be caused by a

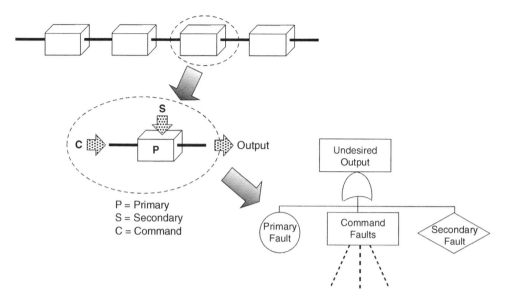

Figure 15.8   *The P-S-C concept.*

component failure, classify the event as an SC fault. If the fault cannot be caused by a component failure, classify the fault as an SS fault. If the fault event is classified as an SC, then the event will have an OR gate with P-S-C inputs. If the fault event is classified as an SS, then the event will be further developed using I-N-S logic to determine the inputs and gate type.

P-S-C Concept   This concept involves answering the question "What are the Primary, Secondary, and Command causes of the event?" The "P-S-C" question forces the analyst to focus on specific causal factors. The rationale behind this question is that every component fault event has only three ways of failing: a primary failure mode, a secondary failure mode, or a command path fault. Figure 15.8 demonstrates this concept. An added benefit of this concept is that if more than two of the three elements of P-S-C are present, then an OR gate is automatically indicated.

Figure 15.8 depicts how a system element is subdivided into primary, secondary, and command events for the FT structure. Two types of system events exist – those, which are intended, and those, which *are not intended*. The intended events follow the desired intended mode of system operation, while the command path faults follow the undesired modes of operation.

A primary failure is the inherent failure of a system element (e.g., a resistor fails open). The primary failure is developed only to the point where identifiable internal component failures will directly cause the fault event. The failure of one component is presumed to be unrelated to the failure of any other component (i.e., independent).

A secondary failure is the result of external forces on the component (e.g., a resistor fails open due to excessive external heat exposure). Development of the secondary failure event requires a thorough knowledge of all external influences affecting system components (e.g., excessive heat, vibration, EMI, etc.). The failure of one component may be related to the failure of other components (i.e., dependent). This type of component failure is due to any cause other than its own primary failure.

A command failure is an expected, or intended, event that occurs at an undesired time due to specific failures. For example, missile launch is an intended event at a certain point in the mission. However, this event can be "commanded" to occur prematurely by certain failures in the missile arm and fire functions. Failures and faults in this chain of events are referred to as the "command path" faults.

The command path is a chain of events delineating the path of command failure events through the system. Analysis of command path events creates an orderly and logical manner of fault identification at each level of the FT. A path of command events through the FT corresponds to the signal flow through the system. In developing command events, the question "what downstream event *commands* the event to occur" is asked for each event being analyzed. At the finish of each FT branch, the command path will terminate in primary and/or secondary events.

Note that the command path is primarily a guideline for analysis of fault event development through a system. Once an analysis is completed, comparison between the FT and the system signal flow diagram will show that the FT command path represents the signal flow through the system along a single thread.

For another example of a command path fault, consider a relay. When the relay coil is energized, the relay contacts will automatically close, as designed and intended. If a failure downstream of the relay provides inadvertent power to the relay coil, then the closing of the relay contacts is considered as a "command" failure. The relay operates as normally intended, except at the wrong time.

15.5.4   FT Construction: Advanced

As previously mentioned, FT building is a repetitive process. Figure 15.9 displays this iterative process; for every logic gate on the FT, the same set of three questions is asked: I-N-S, P-S-C, and SS/SC.

Answering these questions provides the gate input events and the gate logic involved. As can be seen from this diagram, as the iterative analysis proceeds downward, the cause–effect relationships are linked in an upward manner.

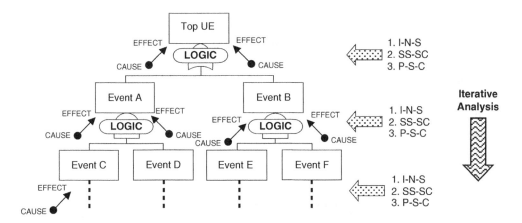

Figure 15.9   *FT building steps.*

The basic steps to follow when constructing the FT include the following:

1. Review and understand the fault event under investigation.
2. Identify all the possible causes of this event by asking the following questions:
   a. Immediate, necessary, and sufficient?
   b. State of component or state of system?
   c. Primary, secondary, and command?
3. Identify the relationship or logic of the cause–effect events.
4. Structure the tree with the identified gate input events and gate logic.
5. Double check logic to ensure that a jump in logic has not occurred.
6. Keep looking back to ensure identified events are not repeated.
7. Repeat for next fault event (i.e., gate).

Some critical items to remember while performing this process are as follows:

1. When possible, start analyzing in the design at the point where the undesired event occurs.
2. Work backward (through system) along signal or logic flow.
3. Keep node wording clear, precise, and complete.
4. Check to ensure all text boxes have unique text, no repeated text.
5. Ensure you do not jump ahead of a possible fault event.
6. Look for component or fault event transition states (e.g., "*no output* signal from component A" and "*no input* fluid to valve V1").

15.5.5 FT Construction Rules

Some basic rules for FT construction and development include the following:

1. Complete basic required data for each FT node (node type, node name, and text).
2. Give every node a unique identifying name.
3. No gate-to-gate connections are allowed (always have text box).
4. Always place relevant text in text box, never leave it blank.
5. State event fault state exactly and precisely, use state transition wording.
6. Complete the definition of all inputs to a gate before proceeding.
7. Keep events on their relative level for clarity.
8. Use meaningful naming convention.
9. Do not draw lines from two gates to a single input (use the MOE methodology).
10. Assume no miracles (i.e., miraculous component failure blocks other failures from causing UE).
11. I-N-S, P-S-C, and SS–SC are analysis concepts, do not use these words in text boxes.

Figure 15.10 demonstrates some typical violations of the FT construction rules. Violation of these rules creates many problems. For example, if a text box is missing, or has no text in it, no one reading the FT will be able to understand the logic involved.

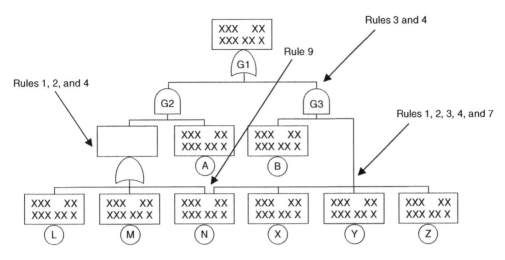

Figure 15.10   *FT construction errors.*

15.6   FUNCTIONAL BLOCK DIAGRAMS

When constructing FTs, an important concept to remember is the use of functional block diagrams (FBDs). The FBD gives a simplified representation of the system design and operation for clarity and understanding. It shows the subsystem interfaces and the component relationships. The FBD shows the functions that must be performed by the system for successful operation, thereby also indicating potential modes of faulty operation. When constructing an FT, it is often much easier to work from an FBD than from a large complex electrical schematic. A general rule of thumb is "*if an analyst cannot draw an FBD of the system being analyzed, then the analyst may not fully understand the system design and operation.*" As shown in Figure 15.11, in many cases the FBD forms the levels and events directly for the FTA.

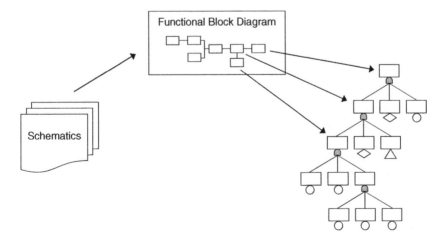

Figure 15.11   *Use of functional block diagrams.*

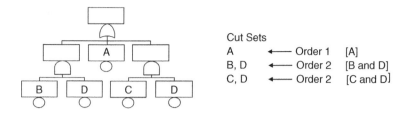

Figure 15.12 *Example FT cut sets.*

15.7 FT CUT SETS

CS's are one of the key products of FTA. They identify the component failures and/or event combinations that can cause the top UE to occur. CS's provide one mechanism for probability calculations. Essentially, CS reveal the critical and weak links in a system design by identifying safety problem components, high-probability CS, and where intended safety or redundancy features have been bypassed.

Figure 15.12 shows an example FT with its resulting CS's listed on the right. Per the definition of a CS, each of these CS's can cause the top UE to occur. CS's are generated through the rules of Boolean algebra, and many different algorithms exist for generating CS's.

In general, the following observations regarding CS tend to hold true:

1. A low-order CS indicates high safety vulnerability. A single-order CS (i.e., a single-point failure) tends to cause the greatest risk.
2. A high-order CS indicates low safety vulnerability. A high-order CS (e.g., a 5 input AND gate) tends to have a comparatively small probability and, therefore, presents less system risk.
3. For a large total number of CS, the analyst needs to evaluate the collective risk on the top UE. This is because all of the CS added together might reach an unacceptable value.

15.8 MOCUS ALGORITHM

One of the most common FT algorithms for generating CS's is the MOCUS (method of obtaining cut sets) algorithm, developed by J. Fussell and W. Vesely [1]. This is an effective top-down gate substitution methodology for generating CS's from an FT. MOCUS is based on the observation that AND gate increases the number of elements in CS and that OR gate increases the number of CS.

The basic steps in the MOCUS algorithm are as follows:

1. Name or number all gates and events.
2. Place the uppermost gate name in the first row of a matrix.
3. Replace top gate with its inputs, using the following notations:
   a. Replace an AND gate with its inputs, each input separated by a comma.
   b. Replace an OR gate by vertical arrangement, creating a new line for each input.
4. Reiteratively substitute and replace each gate with its inputs, moving down the FT.

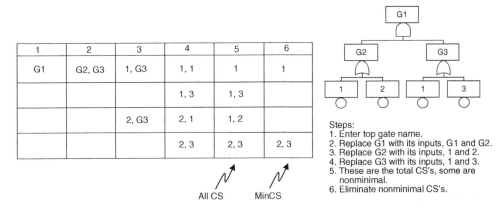

1	2	3	4	5	6
G1	G2, G3	1, G3	1, 1	1	1
			1, 3	1, 3	
		2, G3	2, 1	1, 2	
			2, 3	2, 3	2, 3

All CS   MinCS

Steps:
1. Enter top gate name.
2. Replace G1 with its inputs, G1 and G2.
3. Replace G2 with its inputs, 1 and 2.
4. Replace G3 with its inputs, 1 and 3.
5. These are the total CS's, some are nonminimal.
6. Eliminate nonminimal CS's.

Figure 15.13 *MOCUS example 1.*

5. When only basic inputs remain in the matrix, the substitution process is complete and the list of all cut sets has been established.

6. Remove all nonminimal CS and duplicate CS from the list using the laws of Boolean algebra.

7. The final list contains the Min cut sets.

Figure 15.13 provides an example of applying the MOCUS algorithm to an FT.

Figure 15.14 provides an example of applying the MOCUS algorithm to an FT using a modified technique.

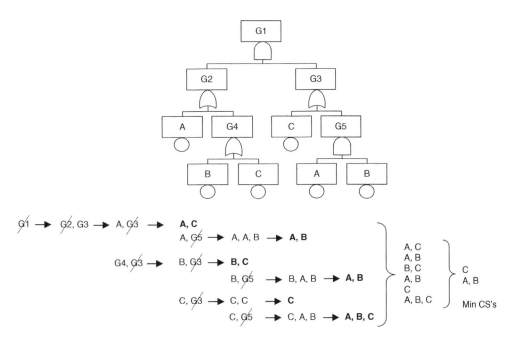

Figure 15.14 *MOCUS example 2.*

```
G5 = A · B = AB
G3 = C + G5 = C + AB
G4 = B + C
G2 = A + G4 = A + B + C
G1 = G2 · G3
   = (A + B + C) (C + AB)
   = AC + AAB + BC + BAB + CC + CAB
   = AC + AB + BC + AB + C + ABC
   = C + AC + BC + AB + ABC
   = C + AB
```

Figure 15.15 *Example using bottom-up method.*

15.9 BOTTOM-UP ALGORITHM

Another CS generation algorithm is the bottom-up algorithm, which is just the reverse of the MOCUS algorithm. Figure 15.15 shows how the algorithm works on the same FT used in Figure 15.14. Note that the results are identical for the two methods.

15.10 FT MATHEMATICS

FT mathematics is based on Boolean algebra, probability theory, and reliability theory. The following are some definitions for mathematical terms frequently encountered in FTA.

15.10.1 Probability of Success

The probability of success is the reliability (R) of a component, which is calculated by $R = e^{-\lambda T}$, where $\lambda =$ component failure rate and $T =$ component exposure time. Also, $\lambda = 1/\text{MTBF}$, where MTBF is the mean time between failure.

15.10.2 Probability of Failure

Unreliability (Q) is the probability of failure of a component, where

$$R + Q = 1 \quad \text{and} \quad Q = 1 - R = 1 - e^{-\lambda T}$$

When $\lambda T < 0.001$, then $Q \approx \lambda T$, which is a useful approximation for hand calculations. In safety work, Q is referred to as P, for probability of failure. Note that the longer the mission (or ET), the higher the probability of failure, and the smaller the failure rate, the lower the probability of failure.

15.10.3 Boolean Rules for FTA

The following Boolean laws apply directly to FTA for the reduction of CS to their minimum components. These rules are required for reducing trees with MOEs in them.

$$a \cdot a = a$$
$$a + a = a$$
$$a + ab = a$$
$$a(a + b) = a$$

15.10.4 AND Gate Probability Expansion

The probability for an AND gate is

$$P = P_A P_B P_C P_D P_E \ldots P_N$$

where N is equal to the number of inputs to the gate.

15.10.5 OR Gate Probability Expansion

The probability for an OR gate is

$$P = \left(\sum \text{1st term}\right) - \left(\sum \text{2nd term}\right) + \left(\sum \text{3rd term}\right)$$

$$- \left(\sum \text{4th term}\right) + \left(\sum \text{5th term}\right) - \left(\sum \text{6th term}\right) \cdots$$

$$P = (P_A + P_B + P_C) - (P_{AB} + P_{AC} + P_{BC}) + (P_{ABC})$$

$$\gg \text{example for 3 input AND gate}$$

15.10.6 FT Probability Expansion

The Boolean equation for an entire FT is all of the cut sets OR'd together. This means that the probability calculation is the OR expansion formula for all of the CS.

$$CS = \{CS1; CS2; CS3; CS4; CS5; CS6; CS7; CS8; CS9; CS10;$$

$$CS11; CS12; CS13; CS14; \cdots\}$$

$$P = \left(\sum \text{1st term}\right) - \left(\sum \text{2nd term}\right) + \left(\sum \text{3rd term}\right)$$

$$- \left(\sum \text{4th term}\right) + \left(\sum \text{5th term}\right) - \left(\sum \text{6th term}\right) \cdots$$

$$P = (PCS1 + PCS2 + \cdots) - (PCS1 \cdot PCS2 + PCS1 \cdot PCS3 + \cdots)$$

$$+ (PCS1 \cdot PCS2 \cdot PCS3 + PCS1 \cdot PCS2 \cdot PCS4 + \cdots) - \cdots$$

15.10.7 Inclusion–Exclusion Approximation

Most FT have a large number of CS. Formulating the exact equation for a large number of cut sets would result in an unwieldy equation, even for a computer. The inclusion–exclusion approximation method has been developed to resolve this numerical problem. This approximation says that the first term in the OR gate expansion is the upper bound probability for the tree. This means the true probability will be no worse than this value. The first and second terms together compute the lower bound tree probability. This means the true probability will be no better than this value. And, as successive terms are added to the computation, the tree probability approaches the exact calculation.

$$P = P_A + P_B + P_C + P_D - (P_{AB} + P_{AC} + P_{AD} + P_{BC} + P_{BD} + P_{CD}) + (P_{ABC} + P_{ABD} + P_{ACD} + P_{BCD}) - (P_{ABCD})$$

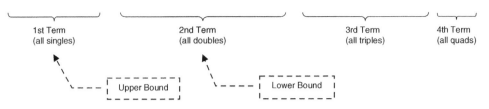

Figure 15.16 depicts the OR gate expansion formula along with the "terms" in the formula for just four CS's.

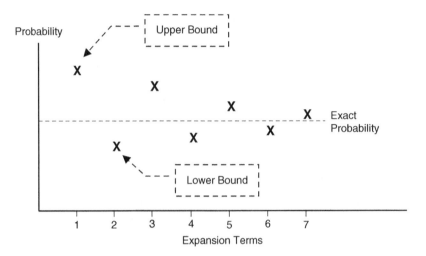

Figure 15.16   *OR gate expansion formula.*

Figure 15.17   *The first and second terms bound the tree probability.*

Figure 15.16 depicts the OR gate expansion formula along with the "terms" in the formula for just four CS's.

Figure 15.17 illustrates how by including each successive term in the calculation, the probability will approach the exact probability.

15.11   PROBABILITY

The top FT probability is a general term for the probability calculated for the top undesired event. The top event probability is calculated from the FT using the probabilities that are input for the basic events, in terms of a failure rate and exposure time or a straight probability. Depending on the specific top event definition, the top event probability can be the probability of the top event occurring during a mission, the probability of the top event occurring in a given period of time, a pure probability number for the top event, or the top event unavailability.

A gate probability is sometimes called an intermediate event probability since it is calculated for an intermediate event below the FT top event. The gate acts like a top event for the branch of FT below it. Analysis of intermediate gate event probabilities is sometimes useful during FTA. Whatever method is used to calculate gate probabilities (top-down or bottom-up), if MOEs are not correctly mathematically accounted for (i.e., resolved), the final results can be erroneous.

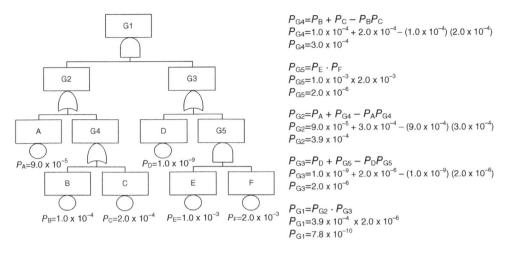

$P_{G4} = P_B + P_C - P_B P_C$
$P_{G4} = 1.0 \times 10^{-4} + 2.0 \times 10^{-4} - (1.0 \times 10^{-4})(2.0 \times 10^{-4})$
$P_{G4} = 3.0 \times 10^{-4}$

$P_{G5} = P_E \cdot P_F$
$P_{G5} = 1.0 \times 10^{-3} \times 2.0 \times 10^{-3}$
$P_{G5} = 2.0 \times 10^{-6}$

$P_{G2} = P_A + P_{G4} - P_A P_{G4}$
$P_{G2} = 9.0 \times 10^{-5} + 3.0 \times 10^{-4} - (9.0 \times 10^{-4})(3.0 \times 10^{-4})$
$P_{G2} = 3.9 \times 10^{-4}$

$P_{G3} = P_D + P_{G5} - P_D P_{G5}$
$P_{G3} = 1.0 \times 10^{-9} + 2.0 \times 10^{-6} - (1.0 \times 10^{-9})(2.0 \times 10^{-6})$
$P_{G3} = 2.0 \times 10^{-6}$

$P_{G1} = P_{G2} \cdot P_{G3}$
$P_{G1} = 3.9 \times 10^{-4} \times 2.0 \times 10^{-6}$
$P_{G1} = 7.8 \times 10^{-10}$

Figure 15.18   *Example bottom-up gate-to-gate calculation.*

There are several different methods to compute the top FT probability. The most common approaches are the following:

1. Direct analytical calculation using the FT CS's
2. Bottom-up gate-to-gate calculation
3. Simulation

The direct analytical calculation using the FT CS's approach merely sums all of the CS's using the OR gate expansion explained above. When the number of CS's becomes very large, it becomes too time-consuming and unwieldy to make an exact calculation, and the inclusion–exclusion approximation method explained above is then utilized.

The simulation method employees Monte Carlo techniques to simulate the random failure of events in the FT. Millions of trials are generally run, and then statistical calculations are made to compute the top FT probability.

The bottom-up gate-to-gate calculation method starts at the bottom of the FT and calculates each gate up the tree in a step-by-step process. Each gate is calculated using the appropriate gate probability formula. A lower level gate calculation is used as an input value to a higher level gate. There is one important caution with this technique: if the FT contains MOEs or MOBs, the calculation will be incorrect unless the MOEs and MOBs are correctly accounted for (i.e., Boolean reduction), which usually means falling back on the CS calculation.

An example bottom-up gate-to-gate calculation is shown in Figure 15.18.

15.12   IMPORTANCE MEASURES

One of the most important outputs of an FTA is the set of importance measures that are calculated for the FT events and CS. Importance measures help to identify weak links in the system design, and the components that will provide the most cost-effective mitigation. The FT importance measures establish the significance for all the events in the fault tree in terms of their contributions to the FT top event probability. Both intermediate gate events and basic events can be prioritized according to their importance. Top event importance measures can also be calculated that give the sensitivity of the top event probability to an increase or decrease in the probability of any event in the fault tree. Both absolute and relative importance measures can be calculated.

What is often useful about the top event importance measures is that they generally show that relatively few events are important contributors to the top event probability. In many FTs, less than 20% of the basic events in the fault tree are important contributors, contributing more than 80–90% of the top event probability. Moreover, the importance measures of events in the FT generally cluster in groups that differ by orders of magnitude from one another. In these cases, the importance measures are so dramatically different that they are generally not dependent on the preciseness of the data used in the FTA.

The FT top importance measures can be used to allocate program resources by identifying the significant areas for design improvement. Trade studies can be performed to show how much probability improvement can be achieved for various design change costs.

The basic importance measures that can be calculated for each event in the FT are described next.

15.12.1   Cut Set Importance

The CS importance measure evaluates the contribution of each min CS to the FT top event probability. This importance measure provides a method for ranking the impact of each CS. The CS importance is calculated by calculating the ratio of the CS probability to the overall FT top probability. The calculation is performed as follows:

Given the following min cut sets:

$$
\begin{aligned}
&2\\
&2,3\\
&2,4 \quad \ggg \quad I_{2,4} = (P_2 \cdot P_4)/P_{\text{TOP}} \quad \text{[for cut set } 2,4]\\
&7\\
&8,9
\end{aligned}
$$

15.12.2   Fussell–Vesely Importance

The Fussell–Vesely (F–V) importance measure evaluates the contribution of each event to the FT top event probability. This importance measure is sometimes called the *top contribution importance*. Both the absolute and the relative F–V importance are determinable for every event modeled in the fault tree, not only for the basic events but also for every higher level event and contributor. This provides a numerical significance of all the fault tree elements and allows them to be prioritized. The F–V importance is calculated by summing all the minimal cut sets (causes) of the top event involving the particular event and calculating the ratio to the top FT probability. The calculation is performed as follows:

Given the following min cut sets:

$$
\begin{aligned}
&2 \quad \ggg\\
&2,3 \quad \ggg \quad I_2 = [(P_2) + (P_2 \cdot P_3)(P_2 \cdot P_4)]/P_{\text{TOP}} \quad \text{[for event 2]}\\
&2,4 \quad \ggg\\
&7\\
&8,9
\end{aligned}
$$

15.12.3 Risk Reduction Worth

The risk reduction worth (RRW) importance measure evaluates the decrease in the probability of the top event if a given event is assured not to occur. This importance measure can also be called the *top decrease sensitivity*. This measure is related to the previous F–V importance. The risk reduction worth for a basic event shows the decrease in the probability of the top event that would be obtained if the lower level event (i.e., the failure) did not occur. It thus gives the maximum reduction in the top probability for the upgrade of an item. Both the absolute value and the relative value of the RRW are determinable for every event and contributor modeled in the fault tree. The RRW is normally calculated by requantifying the fault tree or the minimum cut sets with the probability of the given event set to 0.0. This calculation and that for risk achievement worth and Birnbaum's importance measure (below) is similar to a partial derivative, in that all other event probabilities are held constant.

15.12.4 Risk Achievement Worth

The risk achievement worth (RAW) importance measure evaluates the increase in the top event probability if a given event occurs. This importance measure can also be called the *top increase sensitivity*. The RAW shows where prevention activities should be focused to assure failures do not occur. Since the failures with the largest RAW have the largest system impacts, these are the failures that should be prevented. The RAW also shows the most significant events for contingency planning. Both the absolute and the relative RAW are obtainable for every event and contributor modeled in the fault tree. The RAW is normally calculated by requantifying the fault tree or the minimum cut sets with the probability of the given event set to 1.0.

15.12.5 Birnbaum's Importance Measure

The BB importance measure evaluates the rate of change in the top event probability as a result of change in the probability of a given event. The BB measure is equivalent to a sensitivity analysis and can be calculated by first calculating the top event probability with the probability of the given event set to 1.0 and then subtracting the top event probability with the probability of the given event set to 0.0. Because of the way BB measure is formulated, it does not account for the probability of an event. BB is related to RAW and RRW; when these are expressed on an interval scale (absolute value), BB = RAW + RRW.

The above importance and sensitivity measures can be calculated, not only for the FT but also for its equivalent success tree. When applied to the success tree, the measures give the importance of an event not occurring. The top event is now the nonoccurrence of the undesired event and each event is the event nonoccurrence. Therefore, when applied to an event in the success tree, the F–V importance gives the contribution of the nonoccurrence of the event to the nonoccurrence of the top event. The RRW gives the decrease in the nonoccurrence probability of the top event if the event nonoccurrence probability were zero, that is, if the event did occur. The RAW gives the increase in the nonoccurrence probability of the top if the nonoccurrence probability of the event were 1.0, that is, if the event were assured not to occur. The importance measures for the success tree give equivalent information as for the FT but from a non-occurrence, or success, standpoint.

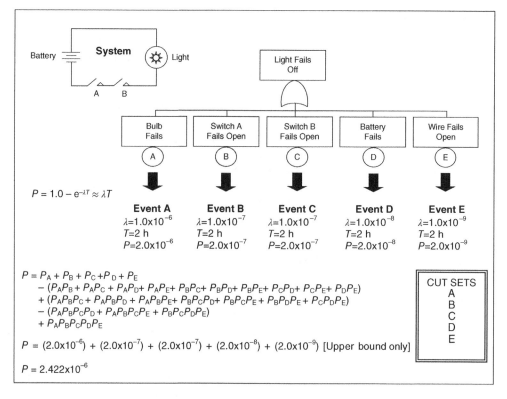

Figure 15.19   *Example FT with CS and probability calculation.*

15.13   FT EXAMPLE 1

Figure 15.19 is a summary example showing a sample system, with its corresponding FT, CS, and probability calculation.

15.14   FT EXAMPLE 2

Figure 15.20 is an example fire and arming circuit for a missile system, simplified for the purpose of demonstrating FT construction. This figure shows a basic electrical diagram of the functions for this system. Note that this example is a contrived and simplified system for the purpose of demonstrating the FT construction. The purpose of this example is to demonstrate construction and evaluation a proactive type FTA that is performed during system design.

Simplifying detailed circuits into block diagrams aids the FTA construction process significantly. Figure 15.21 shows the functional block diagram for the example missile arm-fire system circuit.

Figure 15.22 demonstrates the FT construction process by developing the first three levels of the FT from the FBD. The FTA begins at the warhead and progresses backward down the command path of the system.

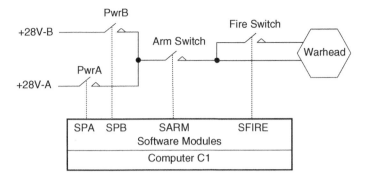

Figure 15.20 *Example missile arm fire system.*

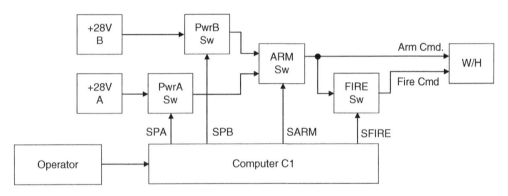

Figure 15.21 *System FBD.*

In Figure 15.22, the top-level construction of the FT is depicted through steps 1, 2, and 3. These steps are explained as follows:

Step 1   In step 1, the FT analysis begins where the UE begins, at the missile warhead. When asking what could cause inadvertent warhead ignition, examine the inputs to the warhead (and ignoring the internals to the warhead itself). In this case, there are two input signals, warhead arm and warhead fire commands. The I-N-S conditions that cause warhead ignition are: (a) warhead arm signal present AND (b) warhead fire signal present. This is a state of the system fault event requiring an AND gate.

Step 2   Step 2 evaluates the fault event "arm signal at warhead input." The question that must be answered is . . . what are the I-N-S causes that provide an arm signal to the warhead? This step involves tracing the arm signal backward to its source. The arm signal could be inadvertently provided (command fault) from the arm switch, or there could be a wire short to +28 VDC in the wiring. This is a state of the system fault event requiring an OR gate.

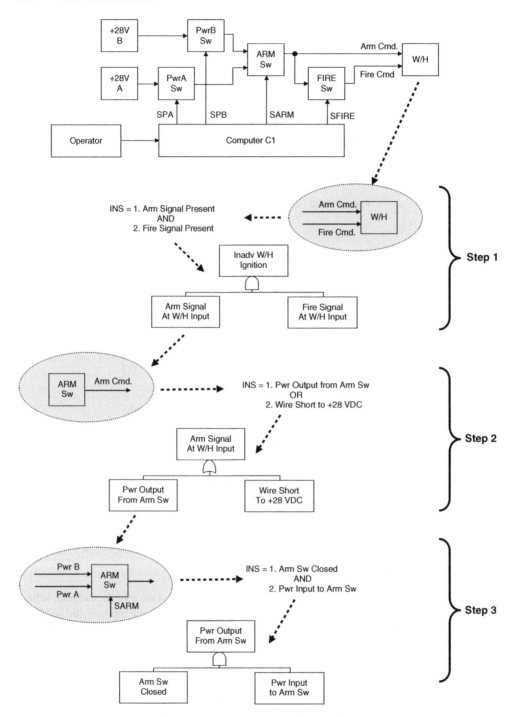

Figure 15.22 *System FBD and start of FT.*

Step 3   Step 3 evaluates the fault event "power is output from the arm switch." The question that must be answered is . . . what are I-N-S causes that provide power output from the arm switch? Examination of the arm switch diagram shows that is has two power inputs, and an input command that opens or closes the switch. This means that power can be output only when (a) power is present at the input of the arm switch and (b) when the arm switch is closed. This is a state of the system fault event requiring an AND gate.

Step 4   Step 4 (not shown in the figure) evaluates the fault event "arm switch is closed." The question that must be answered is . . . what are the I-N-S conditions that cause the arm switch to close? This is a state of the component fault event requiring an OR gate because the analysis is at a component level. There are three ways a component can fail, primary, secondary, and command (P-S-C). If each of these ways is feasible, they are enumerated under the OR gate. The command fault would be expanded as the next step in the analysis.

The FT for this system is shown in Figures 15.23–15.25. The three FT sections constructed in Figure 15.22 are combined together, and the rest of the FT is then completed.

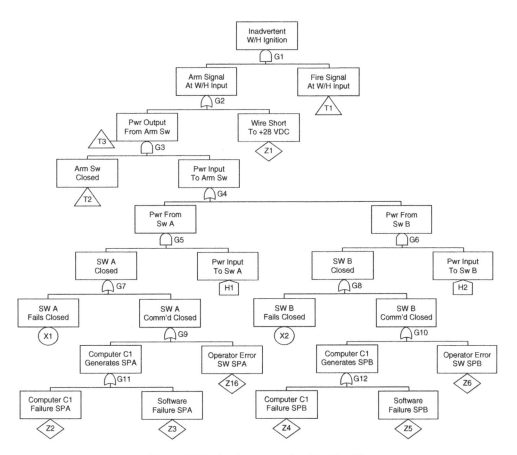

Figure 15.23   *Inadvertent warhead ignition FT.*

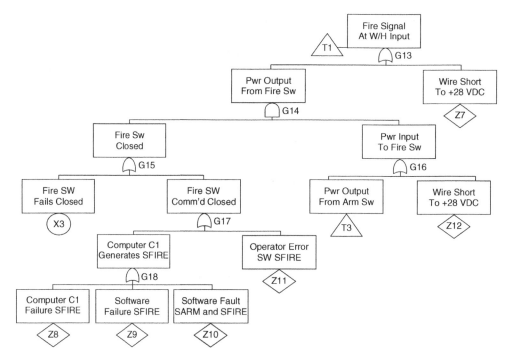

Figure 15.24   *FT branch T1.*

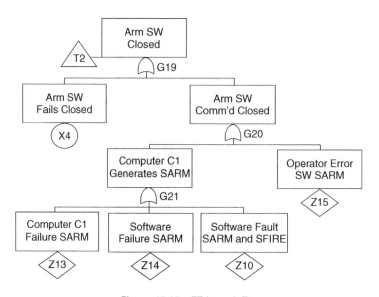

Figure 15.25   *FT branch T2.*

TABLE 15.1   Basic Event Data

NAME	LAMBDA	ET[a]	PROB.	TEXT	NOTES
X1	1.100E−005	10		Switch A fails closed	
X2	1.100E−005	10		Switch B fails closed	
X3	1.100E−005	10		FIRE switch fails closed	
X4	1.100E−005	10		ARM switch fails closed	
Z1	1.100E−009	10		Wire short-arm wire to +28 VDC	
Z2	1.100E−007	10		Computer C1 failure SPA	
Z3	1.100E−007	10		Software failure SPA	
Z4	1.100E−007	10		Computer C1 failure SPB	
Z5	1.100E−007	10		Software failure SPB	
Z6	1.100E−004	10		Operator error switch SPB	
Z7	1.100E−009	10		Wire short to +28 VDC	
Z8	1.100E−007	10		Computer C1 failure SFIRE	
Z9	1.100E−007	10		Software failure SFIRE	
Z10	1.100E−007	10		Software fault SARM and SFIRE	MOE
Z11	1.100E−004	10		Operator error switch SFIRE	
Z12	1.100E−009	10		Wire short to +28 VDC	
Z13	1.100E−007	10		Computer C1 failure SARM	
Z14	1.100E−007	10		Software failure SARM	
Z15	1.100E−004	10		Operator error switch SARM	
Z16	1.100E−004	10		Operator error switch SPA	
H1			1.0	Power input to switch A	
H2			1.0	Power input to switch B	

[a]ET: Exposure time.

After the FT is constructed, failure data must be collected for the components in the FT. Table 15.1 contains the failure data for the example missile system. This data is used in the FT probability calculation. Note that since this is a small and simple FT, a very simple node naming convention is used, whereby the first letter represents a specific node type, as follows:

X – represents primary failures (circles).

Z – represents secondary failures (diamonds).

H – represents normal events (houses).

G – represents a gate event.

Table 15.2 contains the list of minimum CS generated from the FT. Each CS is shown with the probability just for that CS. In this table the CSs are ordered by probability. Note that this FT yields 216 CSs for just 22 components in the FT. The CS list shows that there are no CSs of order 1 (i.e., SPF). It also shows that there are 2 CSs of order 2 and 205 CSs of order 3.

The computed top probability for this FT is $P = 5.142 \times 10^{-6}$. As a quick check, by mentally summing the probability of the top probability CS's and comparing that value with the FT probability, this overall FT probability looks reasonable.

Note that some three order CS's have a higher probability than some of the two order CS's. But, closer examination shows that many of these three order CS's have a house in them with $P = 1.0$, thus effectively making them two order CS's.

Figure 15.26 is an FT of the CS with the highest probability (CS 1). This CS-1 FT displays the basic events involved and the path taken to the top of the FT. The important item to note from this CS-1 FT is that contrary to the CS list, there is an SPF in the FT. CS-1 involves events

TABLE 15.2   FT Cut Sets

1	1.10E−06	H1	Z10		50	1.21E−12	H1	Z14	Z8
2	1.10E−06	H2	Z10		51	1.21E−12	H1	Z14	Z9
3	1.21E−06	H1	Z11	Z15	52	1.21E−12	H2	X4	Z7
4	1.21E−06	H2	Z11	Z15	53	1.21E−12	H2	Z13	Z8
5	1.21E−07	H1	X3	Z15	54	1.21E−12	H2	Z13	Z9
6	1.21E−07	H1	X4	Z11	55	1.21E−12	H2	Z14	Z8
7	1.21E−07	H2	X3	Z15	56	1.21E−12	H2	Z14	Z9
8	1.21E−07	H2	X4	Z11	57	1.21E−12	Z10	Z2	
9	1.21E−08	H1	X3	X4	58	1.21E−12	Z10	Z3	
10	1.21E−08	H2	X3	X4	59	1.21E−12	Z10	Z4	
11	1.21E−09	H1	Z11	Z13	60	1.21E−12	Z10	Z5	
12	1.21E−09	H1	Z11	Z14	61	1.33E−12	X1	X3	X4
13	1.21E−09	H1	Z15	Z8	62	1.33E−12	X2	X3	X4
14	1.21E−09	H1	Z15	Z9	63	1.33E−12	Z11	Z13	Z16
15	1.21E−09	H2	Z11	Z13	64	1.33E−12	Z11	Z13	Z6
16	1.21E−09	H2	Z11	Z14	65	1.33E−12	Z11	Z14	Z16
17	1.21E−09	H2	Z15	Z8	66	1.33E−12	Z11	Z14	Z6
18	1.21E−09	H2	Z15	Z9	67	1.33E−12	Z11	Z15	Z2
19	1.21E−09	Z10	Z16		68	1.33E−12	Z11	Z15	Z3
20	1.21E−09	Z10	Z6		69	1.33E−12	Z11	Z15	Z4
21	1.33E−09	Z11	Z15	Z16	70	1.33E−12	Z11	Z15	Z5
22	1.33E−09	Z11	Z15	Z6	71	1.33E−12	Z15	Z16	Z8
23	1.21E−10	H1	X3	Z13	72	1.33E−12	Z15	Z16	Z9
24	1.21E−10	H1	X3	Z14	73	1.33E−12	Z15	Z6	Z8
25	1.21E−10	H1	X4	Z8	74	1.33E−12	Z15	Z6	Z9
26	1.21E−10	H1	X4	Z9	75	1.33E−13	X1	Z11	Z13
27	1.21E−10	H2	X3	Z13	76	1.33E−13	X1	Z11	Z14
28	1.21E−10	H2	X3	Z14	77	1.33E−13	X1	Z15	Z8
29	1.21E−10	H2	X4	Z8	78	1.33E−13	X1	Z15	Z9
30	1.21E−10	H2	X4	Z9	79	1.33E−13	X2	Z11	Z13
31	1.21E−10	X1	Z10		80	1.33E−13	X2	Z11	Z14
32	1.21E−10	X2	Z10		81	1.33E−13	X2	Z15	Z8
33	1.33E−10	X1	Z11	Z15	82	1.33E−13	X2	Z15	Z9
34	1.33E−10	X2	Z11	Z15	83	1.33E−13	X3	Z13	Z16
35	1.33E−10	X3	Z15	Z16	84	1.33E−13	X3	Z13	Z6
36	1.33E−10	X3	Z15	Z6	85	1.33E−13	X3	Z14	Z16
37	1.33E−10	X4	Z11	Z16	86	1.33E−13	X3	Z14	Z6
38	1.33E−10	X4	Z11	Z6	87	1.33E−13	X3	Z15	Z2
39	1.21E−11	H1	Z15	Z7	88	1.33E−13	X3	Z15	Z3
40	1.21E−11	H2	Z15	Z7	89	1.33E−13	X3	Z15	Z4
41	1.33E−11	X1	X3	Z15	90	1.33E−13	X3	Z15	Z5
42	1.33E−11	X1	X4	Z11	91	1.33E−13	X4	Z11	Z2
43	1.33E−11	X2	X3	Z15	92	1.33E−13	X4	Z11	Z3
44	1.33E−11	X2	X4	Z11	93	1.33E−13	X4	Z11	Z4
45	1.33E−11	X3	X4	Z16	94	1.33E−13	X4	Z11	Z5
46	1.33E−11	X3	X4	Z6	95	1.33E−13	X4	Z16	Z8
47	1.21E−12	H1	X4	Z7	96	1.33E−13	X4	Z16	Z9
48	1.21E−12	H1	Z13	Z8	97	1.33E−13	X4	Z6	Z8
49	1.21E−12	H1	Z13	Z9	98	1.33E−13	X4	Z6	Z9

(continued)

TABLE 15.2 *Continued*

99	1.21E−14	H1	Z13	Z7	148	1.33E−16	X1	Z13	Z9
100	1.21E−14	H1	Z14	Z7	149	1.33E−16	X1	Z14	Z8
101	1.21E−14	H2	Z13	Z7	150	1.33E−16	X1	Z14	Z9
102	1.21E−14	H2	Z14	Z7	151	1.33E−16	X2	X4	Z7
103	1.33E−14	X1	X3	Z13	152	1.33E−16	X2	Z13	Z8
104	1.33E−14	X1	X3	Z14	153	1.33E−16	X2	Z13	Z9
105	1.33E−14	X1	X4	Z8	154	1.33E−16	X2	Z14	Z8
106	1.33E−14	X1	X4	Z9	155	1.33E−16	X2	Z14	Z9
107	1.33E−14	X2	X3	Z13	156	1.33E−16	X3	Z13	Z2
108	1.33E−14	X2	X3	Z14	157	1.33E−16	X3	Z13	Z3
109	1.33E−14	X2	X4	Z8	158	1.33E−16	X3	Z13	Z4
110	1.33E−14	X2	X4	Z9	159	1.33E−16	X3	Z13	Z5
111	1.33E−14	X3	X4	Z2	160	1.33E−16	X3	Z14	Z2
112	1.33E−14	X3	X4	Z3	161	1.33E−16	X3	Z14	Z3
113	1.33E−14	X3	X4	Z4	162	1.33E−16	X3	Z14	Z4
114	1.33E−14	X3	X4	Z5	163	1.33E−16	X3	Z14	Z5
115	1.33E−14	Z15	Z16	Z7	164	1.33E−16	X4	Z2	Z8
116	1.33E−14	Z15	Z6	Z7	165	1.33E−16	X4	Z2	Z9
117	1.33E−15	X1	Z15	Z7	166	1.33E−16	X4	Z3	Z8
118	1.33E−15	X2	Z15	Z7	167	1.33E−16	X4	Z3	Z9
119	1.33E−15	X4	Z16	Z7	168	1.33E−16	X4	Z4	Z8
120	1.33E−15	X4	Z6	Z7	169	1.33E−16	X4	Z4	Z9
121	1.33E−15	Z11	Z13	Z2	170	1.33E−16	X4	Z5	Z8
122	1.33E−15	Z11	Z13	Z3	171	1.33E−16	X4	Z5	Z9
123	1.33E−15	Z11	Z13	Z4	172	1.33E−17	Z13	Z16	Z7
124	1.33E−15	Z11	Z13	Z5	173	1.33E−17	Z13	Z6	Z7
125	1.33E−15	Z11	Z14	Z2	174	1.33E−17	Z14	Z16	Z7
126	1.33E−15	Z11	Z14	Z3	175	1.33E−17	Z14	Z6	Z7
127	1.33E−15	Z11	Z14	Z4	176	1.33E−17	Z15	Z2	Z7
128	1.33E−15	Z11	Z14	Z5	177	1.33E−17	Z15	Z3	Z7
129	1.33E−15	Z13	Z16	Z8	178	1.33E−17	Z15	Z4	Z7
130	1.33E−15	Z13	Z16	Z9	179	1.33E−17	Z15	Z5	Z7
131	1.33E−15	Z13	Z6	Z8	180	1.33E−18	X1	Z13	Z7
132	1.33E−15	Z13	Z6	Z9	181	1.33E−18	X1	Z14	Z7
133	1.33E−15	Z14	Z16	Z8	182	1.33E−18	X2	Z13	Z7
134	1.33E−15	Z14	Z16	Z9	183	1.33E−18	X2	Z14	Z7
135	1.33E−15	Z14	Z6	Z8	184	1.33E−18	X4	Z2	Z7
136	1.33E−15	Z14	Z6	Z9	185	1.33E−18	X4	Z3	Z7
137	1.33E−15	Z15	Z2	Z8	186	1.33E−18	X4	Z4	Z7
138	1.33E−15	Z15	Z2	Z9	187	1.33E−18	X4	Z5	Z7
139	1.33E−15	Z15	Z3	Z8	188	1.33E−18	Z13	Z2	Z8
140	1.33E−15	Z15	Z3	Z9	189	1.33E−18	Z13	Z2	Z9
141	1.33E−15	Z15	Z4	Z8	190	1.33E−18	Z13	Z3	Z8
142	1.33E−15	Z15	Z4	Z9	191	1.33E−18	Z13	Z3	Z9
143	1.33E−15	Z15	Z5	Z8	192	1.33E−18	Z13	Z4	Z8
144	1.33E−15	Z15	Z5	Z9	193	1.33E−18	Z13	Z4	Z9
145	1.21E−16	Z1	Z7		194	1.33E−18	Z13	Z5	Z8
146	1.33E−16	X1	X4	Z7	195	1.33E−18	Z13	Z5	Z9
147	1.33E−16	X1	Z13	Z8	196	1.33E−18	Z14	Z2	Z8

(*continued*)

TABLE 15.2 *Continued*

197	1.33E−18	Z14	Z2	Z9	207	1.33E−20	Z13	Z3	Z7
198	1.33E−18	Z14	Z3	Z8	208	1.33E−20	Z13	Z4	Z7
199	1.33E−18	Z14	Z3	Z9	209	1.33E−20	Z13	Z5	Z7
200	1.33E−18	Z14	Z4	Z8	210	1.33E−20	Z14	Z2	Z7
201	1.33E−18	Z14	Z4	Z9	211	1.33E−20	Z14	Z3	Z7
202	1.33E−18	Z14	Z5	Z8	212	1.33E−20	Z14	Z4	Z7
203	1.33E−18	Z14	Z5	Z9	213	1.33E−20	Z14	Z5	Z7
204	1.33E−19	Z1	Z11	Z12	214	1.33E−22	Z1	Z10	Z12
205	1.33E−20	X3	Z1	Z12	215	1.33E−22	Z1	Z12	Z8
206	1.33E−20	Z13	Z2	Z7	216	1.33E−22	Z1	Z12	Z9

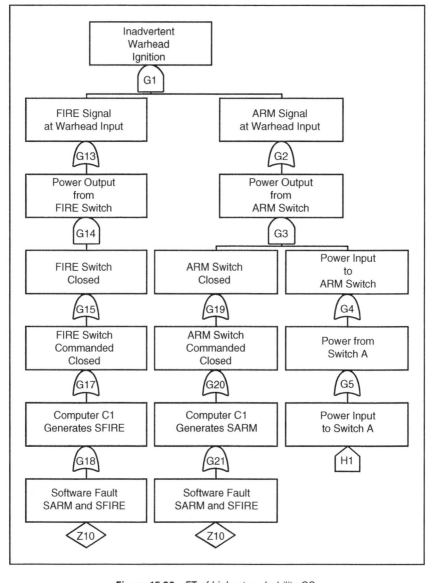

Figure 15.26 FT of highest probability CS.

Symbol	Gate Type	Description
A E B	Evidence Event	This gate does not show a logical combination, it is used to indicate that evidence is present to prove or disprove a particular path. Event B is developed only if evidence event E is true. Not used for a quantitative analysis.

Figure 15.27 *Evidence event gate definition.*

Z10 AND H1. But since the house event H1 is a normal event with probability equal to 1.0, it does not count as a failure. Therefore, only event Z10 is necessary and sufficient to cause the top undesired event. Another important item to note form this FT is that event Z10 is an MOE that occurs on both sides of an AND gate, thereby effectively bypassing intended system redundancy. Note also from the CS list that some three order CS's have a higher probability than two order CS's. This short CS analysis demonstrates that it is important to fully evaluate each CS.

15.15   FT EXAMPLE 3

This example involves a reactive-type FTA that is performed after the undesired event has already occurred. This example shows how to use the evidence event gate when performing an FTA for an accident investigation. Figure 15.27 shows the EEG definition. As the FT is developed downward, fault event B is hypothesized as a possible cause of the accident. If there is data from the accident investigation indicating that event B did not happen, that evidence is stated in E, and event B is not developed further because it has been proven as a noncause. If evidence indicates that B did occur, then event B is analyzed further because it is a positive cause. This is a method for quickly following productive causal paths. Figure 15.28 shows an FTA of the Titanic sinking. This FT uses the EEG and shows where collected evidence can be utilized in the FTA. It should be noted that this type of FT with the EEG is a qualitative-type analysis, not a quantitative-type analysis. Note in this FT that transfer events A and B would be further developed as a result of this analysis; however, the information is not available.

15.16   PHASE- AND TIME-DEPENDENT FTA

Typical FT quantification provides a single value for the probability of the top event for a system mission. This top event probability is not partitioned into contributions over different mission phases or time intervals. If the mission under consideration has different phases, which are reflected in the FT, then the top event probability obtained is the total probability for the mission. Similarly, if a system fault event is modeled over a time interval, then the top probability obtained is the total system failure probability over the time interval. In this case, individual probabilities for different segments of the time interval are not obtainable.

Most FT software cannot produce phase-dependent or time-dependent results. This is not the limitation of the fault tree model itself, but a limitation of the available software. Different mission phases can be modeled in an FT. Also, individual failure rates and time intervals can be provided for each component. However, typical FT software calculates the total probability

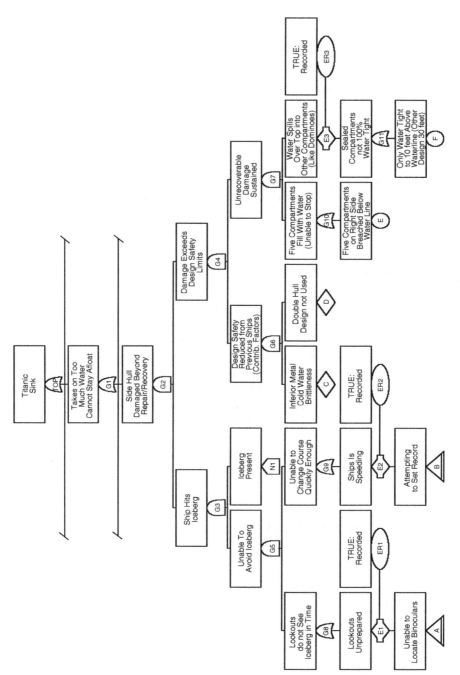

Figure 15.28 *FTA of titanic sinking.*

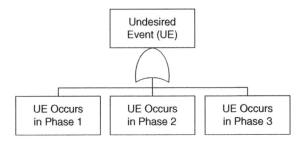

Figure 15.29   *Mission phased FT.*

only and does not have the capability of breaking the probability into more detailed contributions.

This limitation of the FT software is not generally a problem because most applications require only a single-mission probability. If phase-dependent or time-dependent results are desired, then there are two options:

1. Use a specialized software that has the capability to perform these calculations.
2. Break the FT model into phase-dependent or time-dependent segments and then perform the calculations with a standard FT software.

The second method involves modeling time dependence into the FT by dividing (partitioning) the top undesired event into smaller time segments. The top undesired event is divided into time interval events using an OR gate. This modeling technique is illustrated in Figure 15.29, whereby the top undesired event is divided into three mission phases, with each phase representing a different time interval.

In the mission-phased FT model, the basic event occurrences, such as component failures, have been separated into more specific phased events. The OR gate is more correctly a mutually exclusive OR gate since the event cannot occur in intervals 1, 2, and 3. If the FT software cannot handle mutually exclusive OR gates, then the simple OR gate can be used provided the minimal cut sets can be scanned to remove any minimal cut sets that contain both of the events.

This approach is both tedious and tricky. It is tedious because the number of basic events in the FT is expanded and hence can greatly expand the number of minimal CS's that are generated. It is tricky because OR'ing the three phases together is not entirely correct, and provides only an approximation.

The reason it is not entirely correct is because cross-phase CS's are not accounted for. For example, say that a two-phase FT had only one CS comprising A and B, as shown in Figure 15.30. Equation 3.1 shows the probability calculation for the single-phase FT. Equation 3.2 shows the calculation for the FT when the FT is broken into two phases that are then OR'd together, which is not correct. Equation 3.3 shows the correct probability calculation, which must include the two additional cross-phase CS's.

The way to read Equation 3.3 is as follows:

$$P = A_1 \cdot B_1 + A_2 \cdot B_2 + A_1 \cdot B_2 + A_2 \cdot B_1 \tag{3.3}$$

Rearrange the failure sequence as they could occur during the phased operation.

$$P = A_1 \cdot B_1 + A_2 \cdot B_2 + A_1 \cdot B_2 + B_1 \cdot A_2 \qquad \text{(Equation 3.3 rearranged)}$$

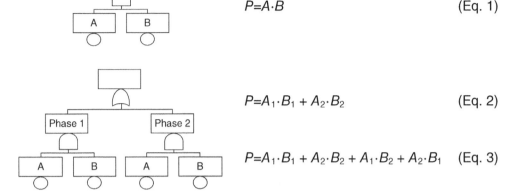

$$P = A \cdot B \qquad \text{(Eq. 1)}$$

$$P = A_1 \cdot B_1 + A_2 \cdot B_2 \qquad \text{(Eq. 2)}$$

$$P = A_1 \cdot B_1 + A_2 \cdot B_2 + A_1 \cdot B_2 + A_2 \cdot B_1 \qquad \text{(Eq. 3)}$$

Figure 15.30 *Example of phased FT cut sets.*

A fails in phase 1 AND B fails in phase 1

A fails in phase 2 AND B fails in phase 2

A fails in phase 1 AND B fails in phase 2

B fails in phase 1 AND A fails in phase 2

This approach considers all possible combinations of failure for the two components in the two phases in order to account for the cross-phase failures.

15.17 DYNAMIC FTA

Dynamic fault tree (DFT) is an extension of the standard FT analysis methodology and was developed specifically for the analysis of computer-based systems. Two special gates are part of the DFT methodology, the functional dependency (FDEP) gate and the spares gate. The DFT methodology was developed to provide a means for combining FTA with Markov analysis for sequence-dependent problems. Markov chains are commonly used to assess the reliability and performance of fault-tolerant computer-based systems. Markov models have the advantage of easily modeling the sequence-dependent behavior that is typically associated with fault-tolerant systems. However, Markov models have the disadvantage of being large and cumbersome and the generation of a Markov model for many systems can be tedious and error-prone. Thus, DFT combines the best of both techniques.

DFT works well with sequence-dependent failures. For example, suppose a system uses two redundant components, one of which operates as the primary unit and the other serves as a standby backup in case the first fails. Also, suppose the design uses a switch to change to the backup when the primary unit fails. If the switch controller fails after the primary unit fails (and thus the standby spare is already in use), the system can continue to operate. However, if the switch controller fails before the primary unit fails, then the standby spare unit cannot be switched to active operation and the system fails when the primary unit fails. The order in which both the primary unit and the switch fail determines if the system continues to operate, and thus the FT model is sequence dependent. The standard FT approach would be to model this situation using a priority AND gate. In standard FTA, the priority AND gate mathematical calculation provides a very good approximation, but not an exact answer.

If the inputs to a priority AND gate are not simple basic events (i.e., they are gate events), this calculation is more difficult and, in general, requires the solution of a set of integral equations or some approximations. The DFT methodology can relieve the analyst of the need to perform the calculation of the probability of the order of occurrence of the inputs to a priority AND gate. The DFT methodology automatically generates and solves the set of integral equations needed to solve the priority AND system via a built-in Markov algorithm.

Standard FT methodology will model everything that the DFT approach can model. The benefit of the DFT approach is that it provides increased accuracy in the final probability calculation. The disadvantage of DFT is that it is sometimes more difficult to understand the DFT model, and special software is required that can handle the special DFT gates. For further information on this approach, see Refs. [2–4].

15.18 FTA ADVANTAGES AND DISADVANTAGES

A proven technique with many years of successful use, the FTA has the following advantages:

1. It is a structured, rigorous, and methodical approach.
2. A large portion of the work can be computerized.
3. Can be effectively performed on varying levels of design detail.
4. Visual model displaying cause/effect relationships.
5. Relatively easy to learn, do, and follow.
6. Models complex system relationships in an understandable manner.
7. Follows fault paths across system boundaries.
8. Combines hardware, software, environment, and human interaction.
9. Permits probability assessment.
10. Scientifically sound; based on logic theory, probability theory, Boolean algebra, and reliability theory.
11. Commercial software is available.
12. FT's can provide value despite incomplete information.
13. FT approximations can provide excellent decision-making information.

Although a strong and powerful technique, FTA does have the following disadvantages:

1. Can easily become time-consuming if the analyst is not careful.
2. Can become the goal rather than the tool.
3. Modeling sequential timing and repair is more difficult.
4. Modeling multiple phases is more difficult.
5. Requires an analyst with some training and practical experience.

15.19   COMMON FTA MISTAKES TO AVOID

When first learning how to perform an FTA, it is commonplace to commit some traditional errors. The following is a list of typical errors made during the conduct of an FTA that an analyst should avoid:

1. Not including human error in the FTA.
2. Not fully understanding the system design and operation.
3. Jumping ahead in the system design further than the fault logic warrants.
4. Not placing text in every tree node.
5. Not placing sufficient descriptive text in every tree node.
6. Forgetting the correct FT definitions (incorrect event usage).
7. Incorrectly accounting for MOEs in FT mathematics.

15.20   SUMMARY

This chapter discussed the FTA technique. The following are the basic principles that help summarize the discussion in this chapter:

1. The primary purpose of FTA is to identify all events (failures, errors, environments, etc.) that can lead to the occurrence of a UE and show how they logically occur and relate to each other.
2. FTA is an analysis tool that provides the following:
   - An evaluation of complex systems and system relationships
   - A graphical model
   - A probability model
3. FTA is for system evaluation:
   - *Safety* Hazardous and catastrophic events
   - *Reliability* System unavailability
   - *Performance* Unintended functions
4. FTA is for decision making:
   - Root cause analysis
   - Risk assessment
   - Design assessment
5. There are two types of FTA:
   - Design evaluation (proactive: prevents accident)
   - Accident investigation (reactive: postaccident)
6. The established FT construction procedure revolves around three principal concepts:
   - The I-N-S concept
   - The SS-SC concept
   - The P-S-C concept
7. The use of a functional block diagram greatly aids and simplifies the FTA process.

REFERENCES

Although there are many papers and references on FTA, the following textbooks are probably the most relevant and useful:

1. J. B. Fussell and W. E. Vesely, A New Method for Obtaining Cutsets for Fault Trees, *Trans. ANS*, (15):262–263 (1972).
2. J. Dugan, S. Bavuso, and M. Boyd, Dynamic Fault Tree Models for Fault Tolerant Computer Systems, *IEEE Trans. Reliab.*, **41**(3):363–377 (1992).
3. J. D. Andrews and J. B. Dugan, Dependency Modeling Using Fault Tree Analysis, Proceedings of the 17th International System Safety Conference, pp. 67–76, 1999.
4. L. Meshkat, J. B. Dugan, and J. D. Andrews, Dependability Analysis of Systems with On-Demand and Active Failure Modes, Using Dynamic Fault Trees, *IEEE Trans. Reliability*, **51**(2):240–251 (2002).

FURTHER READINGS

Andrews, J. D. and Moss, T. R., *Reliability and Risk Assessment*, The American Society of Mechanical Engineers, New York, 2nd edition 2002, 1993.

Ericson, C. A., *Fault Tree Analysis Primer*, CreateSpace, 2011.

Fussell, J. B., et al., MOCUS: Λ Computer Program to Obtain Minimal Cutsets, 1974, Aerojet Nuclear ANCR-1156.

Henley, E. J. and Kumamoto, H., *Probabilistic Risk Assessment and Management for Engineers and Scientists*, 2nd edition, IEEE Press, 1996.

NASA, *Fault Tree Handbook with Aerospace Applications*, NASA. version 1.1, 2002.

Roberts, N. H., Vesely, W. E., Haasl, D. F., and Goldberg, F. F., *Fault Tree Handbook, NUREG-0492, 1981*, U.S. Government Printing Office Wash DC. 1981.

Schneeweiss, W. G., *The Fault Tree Method*, LiLoLe-Verlag, 1999.

Chapter *16*

Failure Mode and Effects Analysis

16.1 FMEA INTRODUCTION

Failure mode and effects analysis (FMEA) is a tool for evaluating the effect(s) of potential failure modes of subsystems, assemblies, components, or functions. It is primarily a reliability tool to identify failure modes that would adversely affect the overall system reliability. FMEA has the capability to include failure rates for each failure mode, in order to achieve a quantitative probabilistic analysis. Additionally, the FMEA can be extended to evaluate failure modes that may result in an undesired system state, such as a system hazard, and thereby also be used for hazard analysis.

A more detailed version of the FMEA is known as failure mode, effects, and criticality analysis (FMECA). The FMECA requires that more information be obtained from the analysis, particularly information dealing with the criticality and detection of the potential failure modes.

FMEA is a disciplined bottom-up evaluation technique that focuses on the design or function of products and processes, in order to prioritize actions to reduce the risk of product or process failures. In addition, the FMEA is a tool for documenting the analysis and capturing recommended design changes. Time and resources for a comprehensive FMEA must be allotted during design and process development, when design and process changes can most easily and inexpensively be implemented.

16.2 FMEA BACKGROUND

FMEA falls under the detailed design hazard analysis type (DD-HAT) because it is an analysis done at the component or functional level based on detailed system design. The basic hazard analysis types are described in Chapter 4.

Hazard Analysis Techniques for System Safety, Second Edition. Clifton A Ericson, II.
© 2016 John Wiley & Sons, Inc. Published 2016 by John Wiley & Sons, Inc.

An alternate name for this technique is FMECA. FMECA is basically the same as FMEA, except that it adds criticality evaluation to each failure mode, as well as the evaluation of possible failure mode detection methods.

The purpose of FMEA is to evaluate the effect of failure modes to determine if design changes are necessary due to unacceptable reliability, safety, or operation resulting from potential failure modes. When component failure rates are attached to the identified potential failure modes, a probability of subsystem or component failure can be derived. FMEA was originally developed to determine the reliability effect of failure modes, but it can also be used to identify mishap hazards resulting from potential failure modes.

The FMEA is applicable to any system or equipment, at any desired level of design detail – subsystem, assembly, unit, or component. FMEA is generally performed at the assembly or unit level, because failure rates are more readily available for the individual embedded components. The FMEA can provide a quantitative reliability prediction for the assembly or unit that can be used in a quantitative safety analysis (e.g., fault tree). FMEA tends to be more hardware and process oriented but can be used for software analysis when evaluating the failure of software functions.

The technique is thorough for evaluating potential individual failure modes and providing reliability information. However, for safety purposes, an FMEA is limited because it considers only single-item failures and not the combination of items failing together; generally, mishaps result from failure combinations. Also, an FMEA does not identify hazards arising from events other than failures (e.g., timing errors, radiation, high voltage, etc.).

The technique can be easily performed and mastered; however, a basic understanding of failures and the failure mode theory and the hazard analysis theory is necessary as well as knowledge of system safety concepts. Additionally, a detailed understanding of the system design and operation is required.

The methodology is uncomplicated and easily learned. Standard FMEA forms and instructions are included in this chapter.

FMEA is a valuable reliability tool for analyzing potential failure modes and calculating subsystem, assembly, or unit failure rates. Severity and probability evaluation of failure modes provides a prioritized list for corrective actions. FMEA can also be extended to identify hazards resulting from potential failure modes and evaluating the resulting mishap risk. Note, however, that an FMEA will likely not identify all system hazards, because it only looks at single-component failure modes, while hazards can be the result of multiple hazards and events other than failure modes. For this reason, FMEA is *not* recommended as the *sole tool* for hazard identification. FMEA should be used only for hazard analysis when conducted in conjunction with other hazard analysis techniques.

A modified FMEA for hazard identification is recommended for the evaluation of failure modes, when done in support of other hazard analyses. However, the FMEA is *not* recommended as the sole hazard analysis to be performed, since the FMEA primarily looks at single-failure modes only, while a hazard analysis considers many additional system aspects.

16.3   FMEA HISTORY

The FMEA was developed for the United States Military as a formal analysis technique. Military Procedure MIL-P-1629 (now MIL-STD-1629A), titled "Procedures for Performing a Failure Mode, Effects and Criticality Analysis," is originally dated November 9, 1949. It was used as a reliability evaluation technique to determine the effect of system and equipment failures. Failures were classified according to their impact on mission success and personnel/ equipment safety. The term *personnel/equipment*, taken directly from an abstract of Military Standard MIL-P-1629, is notable because of the significance given to personnel. Used for aerospace/rocket development, the FMEA and the more detailed FMECA were helpful in avoiding errors in small sample sizes of costly rocket technology.

The use of FMEA was encouraged in the 1960s for space product development and served well in getting man on the moon. Ford Motor Company reintroduced FMEA in the late 1970s for safety and regulatory consideration after multiple Pinto automobile exploding gas tank accidents. Ford Motor Company has also used FMEAs effectively for both production improvement and design improvement.

The Automotive Industry Action Group (AIAG) and the American Society for Quality Control (ASQC) copyrighted industry-wide FMEA standards, in February of 1993, are the technical equivalent of the Society of Automotive Engineers procedure SAE J-1739. The standards are presented in an FMEA Manual approved and supported by all three US auto makers, which provides general guidelines for preparing an FMEA.

16.4 FMEA DEFINITIONS

In order to facilitate a better understanding of FMEA, some definitions for specific terms are in order. The following are the basic FMEA terms:

Failure A failure is the departure of an item from its required or intended operation, function, or behavior; problems that users encounter. It is the inability of a system, subsystem, or component to perform its required function. The inability of an item to perform within previously prescribed limits is failure.

Failure Mode A failure mode is the manner by which an item fails; the mode or state the item is in after it fails. It is the way in which the failure of an item occurs.

Failure Cause The failure cause is the process or mechanism responsible for initiating the failure mode. The possible processes that can cause component failure include physical failure, design defects, manufacturing defects, environmental forces, and so on.

Failure Effect The consequence(s) a failure mode has on the operation, function, or status of an item and on the system.

Fault An undesired anomaly in the functional operation of an equipment or system. The occurrence of an undesired state, which may be the result of a failure.

Critical Item List (CIL) A list of items that are considered critical for reliable and/or safe operation of the system. The list is generated from the FMEA.

Indenture Level The levels of system hierarchy that identify or describe the relative complexity of a system. The levels progress from the more complex (system) to the simpler (part/component) divisions level (MIL-STD-1629A on FMEAs). The hierarchy is the organizational structure defining dominant and subordinate relationships between subsystems down to the lowest component/piece part.

Risk Priority Number (RPN) A risk ranking index for reliability, where RPN = (probability of occurrence) × (severity ranking) × (detection ranking).

16.5   FMEA THEORY

FMEA is a qualitative and quantitative analysis method used for the evaluation of potential failure modes. The FMEA is a technique that answers a series of questions:

- What can fail?
- How does it fail?
- How frequently will it fail?
- What are the effects of the failure?
- What is the reliability/safety consequence of the failure?

To conduct an FMEA, it is necessary to know and understand certain system characteristics:

- Mission
- System design
- Operational constraints
- Success and failure boundaries
- Credible failure modes and a measure of their probability of occurrence

Figure 16.1 depicts the FMEA concept. The subsystem being analyzed is divided into its relevant indenture levels, such as unit 1, unit 2, unit 3, and so on. Each unit is then further subdivided into its basic items. Each item is listed down in the left-hand column of the FMEA worksheet and individually analyzed. The concept is to breakdown the "entity" being analyzed

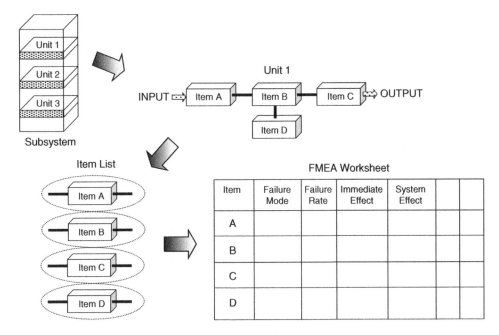

Figure 16.1 The FMEA concept.

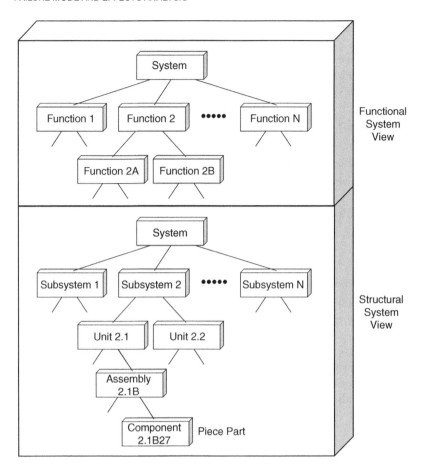

Figure 16.2   *Functional versus structural levels.*

into individual items. In effect, the subsystem is analyzed "top-down" when it is divided into indenture levels, and then it is analyzed "bottom-up" when each item is individually evaluated. An item can be a hardware part or component, or it can be a function. Each item is then singularly isolated and all potential failure modes for this item are listed in the first column of the FMEA. Each item is then evaluated in detail.

The primary building blocks of a system that an FMEA analyzes are the system hardware and the system functions, referred to as the system structural aspect and the system functional aspect. Figure 16.2 depicts the functional versus structural concept of a system, which is relevant for FMEA. The functional aspect defines how the system must operate and the functional tasks that must be performed. The structural aspect defines how the functions will be implemented via the hardware that actually carries out the system operations. System design and implementation progresses from the system functions down to the hardware piece parts.

Conceptually there are three approaches to performing an FMEA:

1. **Functional Approach**   The functional FMEA is performed on functions. The functions can be at any functional indenture level for the analysis: system, subsystem, unit, or assembly. This approach focuses on ways in which functional objectives of a system go unsatisfied or are erroneous. The functional approach is also applicable to the evaluation

of software through the evaluation of required software functions. The functional approach tends to be more of a system-level analysis.

2. **Structural Approach**   The structural FMEA is performed on hardware and focuses on potential hardware failure modes. The hardware can be at any hardware indenture level for the analysis: subsystem, unit, assembly, or part (component). The structural approach tends to be a detailed analysis at the component level.

3. **Hybrid Approach**   The hybrid FMEA is a combination of the structural and the functional approaches. The hybrid approach begins with the functional analysis of the system and then transitions to a focus on hardware, especially hardware that directly contributes to functional failures identified as safety-critical.

The functional approach is performed when the system is being defined by the functions that are to be accomplished. The structural hardware approach is performed when hardware items can be uniquely identified from schematics, drawings, and other engineering and design data. The hybrid approach combines both aspects, beginning with identification of important system functional failures and then identifying the specific equipment failure modes that produce those system functional failures.

16.5.1   FMEA Structural and Functional Models

The purpose of an FMEA is to evaluate potential design failure modes early in the development program to cost-effectively implement safety design corrections. To attain this objective, the FMEA must closely track the design as it progresses from conceptual to detailed.

Design depth and detail correlate to structural and functional decomposition of the system. A structural model of the system captures the static structure of the system comprising the hardware components. A functional model of the system captures the functions that must be performed in order for the system to achieve its goals and objectives. These two system views contrast *what* must be done (function) with *how* it is to be done (structure).

Figure 16.3 provides a brief example of a structural model and a functional model for a radio system and the failure modes that might be considered. These models also depict indenture levels for each type of model.

16.5.2   FMEA Product and Process FMEA

The FMEA is classified into a product FMEA or a process FMEA, depending upon the application. The product FMEA analyzes the design of a product or system by examining the way that the item's failure modes affect the operation of the product or system. The process FMEA analyzes the processes involved in the manufacture, use, and maintenance of a product. It examines the way that process methods affect the operation of the product or system. Both types of FMEA focus on design – design of the product or design of the process. The FMEA classification types, along with general failure mode areas, are presented in Figure 16.4.

16.5.3   FMEA Functional Failure Modes

Functional types of FMEAs evaluate system, subsystem, and unit functions. Functional failure modes are a little more abstract than hardware failure modes. The key is to consider each adverse state that is possible for each function.

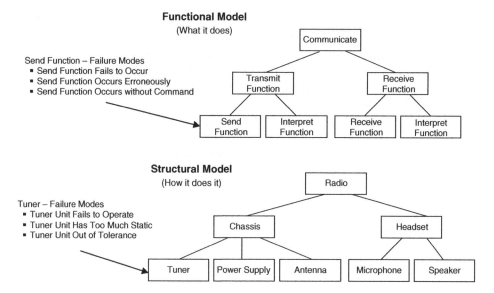

Figure 16.3 *Functional and structural models.*

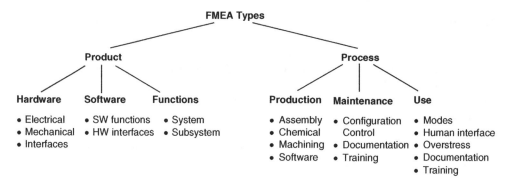

Figure 16.4 *FMEA types – product and process.*

Example functional failure modes may include, but are not limited to, the following:

1. Function fails to perform
2. Function performs incorrectly
3. Function performs prematurely
4. Function provides incorrect or misleading information
5. Function does not fail safe

16.5.4 FMEA Hardware Failure Modes

Hardware types of FMEAs consider both component catastrophic and component out-of-tolerance modes of failure. Catastrophic failure means complete component functional failure in the required mode of operation. For example, a resistor failing to open or shorted means that it no longer functions as intended. Out-of-tolerance failure refers to a failure mode where the component is functional, but not within specified operating boundaries. Example modes for a resistor might include too low resistance or too high resistance, but it still provides some level of resistance. An intermittent failure is a failure that is not continuous; the failure occurs in a cyclic on/off fashion.

The basic failure categories for hardware items include the following:

1. Complete failure

Partial failure (e.g., out of tolerance)

2. Intermittent failure

In a typical FMEA, these basic failure modes may be expressed as the following examples:

1. Open circuit
2. Short circuit
3. Out of tolerance
4. Leak
5. Hot surface
6. Bent
7. Oversize/undersize
8. Cracked
9. Brittle
10. Misaligned
11. Binding
12. Corroded
13. Failure to operate
14. Intermittent operation
15. Degraded operation
16. Loss of output

16.5.5   FMEA Software Failure Modes

Performing an FMEA on a mechanical or electrical system is generally more straightforward than performing an FMEA on software. Failure modes of components such as relays and resistors are generally well understood. Mechanical and electrical components fail due to aging, wear, or stress. For software, the situation is different, because software modules do not fail *per se*, they only display incorrect behavior. A software-oriented FMEA can address only incorrect behavior of software (i.e., the software fails to perform as intended).

A software FMEA (SFMEA) normally involves performing an analysis of the software functions. An SFMEA would follow the same basic steps as a hardware FMEA: set up a starting point, understand the design, make a list of typical failure modes, and then perform the analysis. Software failure modes would be seen as types of erroneous behavior, and not typos in the code. Distinguishing characteristics between the hardware and software FMEA are shown in Table 16.1.

TABLE 16.1   Hardware/Software FMEA Characteristics

Hardware	Software
Is performed at a part (component) level where failure rates can be obtained	Is only practical at the functional level.
System is considered free of failures at the start of operation	System is assumed to contain software faults at the start of operation.
Postulates failure modes due to aging, wear, or stress	Postulates failure modes according to functional failure
Analyzes failure consequence at the item level and the system level.	Analyzes failure consequence at the system level
States the criticality in measures of consequence severity and probability	States the criticality in measures of consequence severity, but probability cannot be determined
States hardware measures taken to prevent or mitigate failure consequence	States software measures taken to prevent or mitigate failure consequence
Software can cause hardware to fail	Hardware can cause software to fail

Example software functional failure modes may include, but are not limited to, the following:

1. Software function fails
2. Function provides incorrect results
3. Function occurs prematurely
4. Unsent messages
5. Messages sent too early or too late
6. Faulty message
7. Software stops or crashes
8. Software hangs
9. Software exceeds internal capacity
10. Software startup failure
11. Software function has slow response

16.5.6   Quantitative Data Sources

When performing a quantitative FMEA/FMECA, component failure rates are required. Although many models are available for performing reliability prediction analyses, each of these models was originally created with a particular application in mind. Table 16.2 describes the most widely used reliability prediction models in terms of their intended applications, noting both their advantages and their disadvantages. Note that there is no available failure rate data for software as there are no defined failure modes.

16.6   METHODOLOGY

Figure 16.5 shows an overview of the basic FMEA process and summarizes the important relationships involved. Based on the reliability theory, all components have inherent failure modes. The FMEA process evaluates the overall impact of each and every component

TABLE 16.2   Comparison of Reliability Prediction Models

Reliability Prediction Model	Application and Originating Country	Advantages	Disadvantages
MIL-HDBK-217 The Military Handbook for the Reliability Prediction of Electronic Equipment	Military and commercial, United States	Provides for both parts stress and parts count analysis of electronic parts. Can easily move from preliminary design stage to complete design stage by progressing from parts count to parts stress Includes models for a broad range of part types Provides many choices for environment types Well known and widely accepted	Does not consider other factors that can contribute to failure rate such as burn-in data, lab testing data, field test data, designer experience, wear, and so on. Considers only electronic parts
Telcordia (Bellcore) Reliability Prediction Procedure for Electronic Equipment (Technical Reference # TR-332 or Telcordia Technologies Special Report SR-332), AT&T Bell Labs	Commercial, United States	Offers analysis ranging from parts count to full parts stress through the use of calculation methods Considers burn-in data, lab testing data, and field test data Well known and accepted	Considers only electronic parts. Supports only a limited number of ground environments Fewer part models compared to MIL-HDBK-217 Does not account for other factors such as designer experience, wear, and so on.
The Handbook of Reliability Prediction Procedures for Mechanical Equipment (NSWC-98/LE1), Navy	Military and commercial, United States	Provides for analyzing a broad range of mechanical parts (seals, springs, solenoids, bearings, gears, etc.)	Limited to mechanical parts
HRD5 The Handbook for Reliability Data for Electronic Components Used in Telecommunication Systems	Telecommunications, United Kingdom	Similar to Telcordia Fairly broad range of part types modeled	Considers only electronic parts Not widely used
PRISM System Reliability Assessment Methodology developed by the Reliability Analysis Center (RAC)	Military and commercial, United States	Incorporates NPRD/EPRD database of failure rates Enables the use of process grading factors, predecessor data, and test or field data. Small, limited set of part types modeled	Newer standard, still gaining acceptance Considers only electronic parts Cannot model hybrids No reference standard available
NPRD/EPRD Nonelectronics Parts Reliability (NPRD) and Electronic Parts Reliability (EPRD) databases by RAC	Military and commercial, United States	Broad array of electronic and nonelectronic parts Based completely on field data	Consists entirely of databases of failure rates, not mathematical models

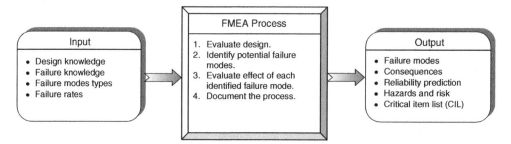

Figure 16.5 *FMEA overview.*

failure mode. The primary FMEA goal is to determine the effect on system reliability from component failures, but the technique can be extended to determine the effect on safety.

Input data for the FMEA include detailed hardware/function design information. Design data may be in the form of the design concept, the operational concept, and major components planned for use in the system and major system functions. Sources for this information include design specifications, sketches, drawings, schematics, function lists, functional block diagrams (FBDs), and/or reliability block diagrams (RBDs). Input data also include known failure modes for components, and failure rates for the failure modes. FMEA output information includes identification of failure modes in the system under analysis, evaluation of the failure effects, identification of hazards, and identification of system critical items in the form of a CIL.

Table 16.3 lists the basic steps in the FMEA process, which involves performing a detailed analysis of all item failure modes. A worksheet is utilized to document the FMEA as identified in the next section.

TABLE 16.3 FMEA Process

Step	Task	Description
1	Define system	Define scope and boundaries of the system. Define the mission, mission phases, and mission environments. Understand the system design and operation. Note that all steps are applicable for an SFMEA
2	Plan FMEA	Establish FMEA goals, definitions, worksheets, schedule, and process. Start with functional FMEA and then move to FMEA of hardware that is safety-critical (identified from functional FMEA) Divide the system under analysis into the smallest segments desired for the analysis. Identify items to be analyzed and establish indenture levels for items/functions to be analyzed
3	Select team	Select all team members to participate in FMEA and establish responsibilities. Utilize team member expertise from several different disciplines (e.g., design, test, manufacturing, etc.)
4	Acquire data	Acquire all of the necessary design and process data needed (e.g., functional diagrams, schematics, and drawings) for the system, subsystems, and functions for FMEA. Refine the item indenture levels for analysis. Identify realistic failure modes of interest for the analysis and obtain component failure rates
5	Conduct FMEA	1. Identify and list the items to be evaluated. 2. Obtain concurrence on the list and level of detail. 3. Transfer the list to the FMEA worksheet. 4. Analyze each item on the list by completing the FMEA worksheet questions. 5. Have the FMEA worksheets validated by a system designer for correctness.

(continued)

TABLE 16.3   (Continued)

Step	Task	Description
6	Recommend corrective action	Recommend corrective action for failure modes with unacceptable risk. Assign responsibility and schedule for implementing corrective action.
7	Monitor corrective action	Review test results to ensure that safety recommendations and System Safety Requirements are effective in mitigating hazards as anticipated.
8	Track Hazards	Transfer identified hazards into the Hazard Tracking System (HTS).
9	Document FMEA	Document the entire FMEA process on the worksheets. Update for new information and closure of assigned corrective actions.

16.7   FMEA WORKSHEET

The FMEA is a detailed analysis of potential failure modes. It is desirable to perform the FMEA using a form or worksheet to provide analysis structure, consistency, and documentation. The specific format of the analysis worksheet is not critical. Typically, matrix columnar or text type forms are utilized to help maintain focus and structure in the analysis. An FMEA that supports system safety and hazard analysis should contain the following information, as a minimum:

1. Failure mode
2. System effect of failure mode
3. System-level hazards resulting from failure
4. Mishap effect of hazards
5. Failure mode and/or hazard causal factors
6. How the failure mode can be detected
7. Recommendations (such as safety requirements/guidelines that can be applied)
8. The risk presented by the identified hazard

Many different FMEA worksheet formats have been proposed by different programs, projects, and disciplines over the years. Some different examples are shown below. Each form provides a different amount and type of information to be derived from the analysis. The specific form to be used may be determined by the customer, the System Safety Working Group, the safety manager, the reliability group, or the reliability/safety analyst performing the analysis. Typically, a program stays with the same FMEA worksheet over the life of the program. Therefore, it is important to ensure that relevant safety-related information is included in the FMEA worksheet.

Figure 16.6 uses a very basic FMEA worksheet format, primarily for use by the reliability organization.

Figure 16.7 illustrates a more complex FMEA worksheet format, which is also primarily for use by the reliability organization but provides needed system safety information.

Figure 16.8 is the preferred worksheet format for systems safety as it includes workspace for relevant safety related information as well as reliability information. Other worksheet formats may exist because different organizations often tailor their FMEA analysis worksheet to fit their particular needs.

The FMEA steps for this recommended worksheet are as follows:

1. *System*   This column identifies the system under analysis.
2. *Subsystem*   This column identifies the subsystem under analysis.

Failure Mode and Effects Analysis						
Component	Failure Mode	Failure Rate	Causal Factors	Immediate Effect	System Effect	RPN

RPN = Risk Priority Number (Reliability)

Figure 16.6 *Example FMEA worksheet 1: Reliability.*

Failure Mode and Effects Analysis									
Item	Failure Mode	Failure Rate	Causal Factors	Immediate Effect	System Effect	RPN	Method Of Detection	Current Controls	Recomm Action

Figure 16.7 *Example FMEA worksheet 2: Reliability.*

Failure Mode and Effects Analysis											
System: ① Subsystem: ② Mode/Phase: ③											
Item	Failure Mode	Failure Rate	Causal Factors	Immediate Effect	System Effect	Method Of Detection	Current Controls	Hazard	Risk	Recomm Action	
④	⑤	⑥	⑦	⑧	⑨	⑩	⑪	⑫	⑬	⑭	

Figure 16.8 *Recommended FMEA worksheet 3: Safety/Reliability.*

3. *Mode/Phase*   This column identifies the system mode or lifecycle phase under analysis.

4. *Item*   This column identifies the component, item, or function being analyzed. For hardware components, the component part number and descriptive title should be identified when possible. A description of the item's purpose or function is also useful and should be included.

5. *Failure Mode*   This column identifies all credible failure modes that are possible for the identified component, item, or function. This information can be obtained from various sources, such as historical data, manufacturer's data, experience, or testing. Since some components may have more than one failure mode, each mode must be listed and analyzed for its effect on the assembly and then on the subsystem.

6. *Failure Rate*   This column provides the failure rate or failure probability for the identified mode of failure. Some quantitative data sources are noted in Table 16.2. The source of the failure rate should also be provided, for future reference. These can be best judgments that are revised as the design process goes on. Care must be taken to make sure that the probability represents that of the particular failure mode being evaluated.

7. *Causal Factors*   This column identifies all of the possible factors that can cause the specific failure mode. Causal factors may include many different sources, such as physical failure, wear out, temperature stress, vibration stress, and so on. All conditions that affect a component or assembly should be listed to indicate whether there are special periods of operation, stress, personnel action, or combinations of events that would increase the probabilities of failure or damage.

8. *Immediate Effect*   This column identifies the most immediate and direct effect of the indicated failure mode. This is the low-level effect that occurs on the next item in the design.

9. *System Effect*   This column identifies the ultimate effect of the specific failure mode on the system. This is the high-level effect.

10. *Method of Detection*   This column identifies how the specific failure mode might be detected after it has occurred, and before resulting in any serious consequence. If a method of detection is possible, it may be used in the mitigating design.

11. *Current Controls*   This column identifies how the specific failure mode is prevented from happening, or how it is safely mitigated should it occur.

12. *Hazard*   This column identifies the specific hazard that is created as a result of the indicated failure mode. (Remember: document all hazard considerations, even if they are later proven to be nonhazardous).

13. *Risk*   This column provides a qualitative measure of mishap risk for the potential effect of the identified hazard, in terms of severity and probability. Note that reliability organizations use a risk priority number (RPN); however, an RPN is not useful for safety risk assessment. For system safety, the generally followed mishap risk index from MIL-STD-882 is used, which is shown below.

Severity	Probability
1 – Catastrophic	A – Frequent
2 – Critical	B – Probable
3 – Marginal	C – Occasional
4 – Negligible	D – Remote
	E – Improbable

14. *Recommended Action*   This column identifies methods for eliminating or mitigating the effects of the potential failure mode.

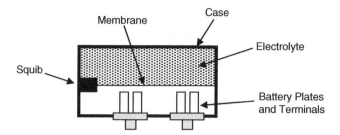

Figure 16.9   *FMEA example 1: Battery.*

16.8   FMEA EXAMPLE 1: HARDWARE PRODUCT FMEA

This is an example of a hardware-type FMEA used to evaluate the system design during the design development process.

Figure 16.9 depicts a missile battery that is inactive and inert until activated by a pyrotechnic squib. In this design, the electrolyte is separated from the battery plates by a frangible membrane. When battery power is desired, the squib is fired to break the membrane, and the released electrolyte energizes the battery.

The battery subsystem comprises the following components:

1. Case
2. Electrolyte
3. Battery plates and terminals
4. Membrane (separates the electrolyte from battery plates)
5. Squib (breaks open the membrane)

The FMEA worksheets for this battery design are shown in Tables 16.4 and 16.5.

16.9   FMEA EXAMPLE 3: FUNCTIONAL FMEA

This is an example of a functional-type FMEA analysis that concentrates on system and software functions, by evaluating the various functional failure modes.

Figure 16.10 depicts a generic aircraft landing gear system.

The GDnB button is pressed to lower the landing gear and the GupB button is pressed to raise the gear. When the gear is up, switch S1 sends a true signal to the computer; otherwise it sends a false signal. When the gear is down, switch S2 sends a true signal to the computer; otherwise, it sends a false signal. The purpose of the switches is for the system to know what position the landing gear is actually in, and prevent conflicting commands. S_{WOW} is the weight-on-wheels switch.

The major functions of the landing gear for both hardware and software functions include the following:

1. Gear up
2. Gear down

TABLE 16.4 Hardware FMEA of Battery; Worksheet 1

Failure Mode and Effects Analysis

System: Missile									Mode/Phase: Operation		
				Subsystem: Missile Battery							
Component	Failure Mode	Failure Rate	Causal Factors	Immediate Effect	System Effect	Method of Detection	Current Controls	Hazard	Risk	Recommended Action	
Case	Cracks	3.5×10^{-5} manuf. data	Manufacturing defect	Electrolyte leakage	No power output from battery	Inspection	QA	Fire source	2D	Add system sensor	
	Pinholes	1.1×10^{-9} manuf. data	Material defect	Electrolyte leakage	No power output from battery	Inspection	QA	Fire source	2E	Add system sensor	
Electrolyte	Leaks out of case	4.1×10^{-6} manuf. data	Case defect, pinholes	Electrolyte leakage	No power output from battery	Inspection	QA	Fire source	2D	Add system sensor	
	Wrong electrolyte used	1.0×10^{-5} manuf. data	Human error	Does not react with battery plates	No power output from battery	Inspection	QA	Unsafe battery reaction	2D		
Battery plates and terminals	Cracks	2.2×10^{-6} manuf. data	Material defect	Inadequate battery reaction	Insufficient power output from battery	None	None	None	4D		
	Breaks	1.0×10^{-9} manuf. data	Material defect	Inadequate battery reaction	No power output from battery	None	None	None	4E		

Analyst: Date: Page: 1/2

293

TABLE 16.5 Hardware FMEA of Battery; Worksheet 2

Failure Mode and Effects Analysis

System: Missile					Subsystem: Missile Battery			Mode/Phase: Operation		
Component	Failure Mode	Failure Rate	Causal Factors	Immediate Effect	System Effect	Method of Detection	Current Controls	Hazard	Risk	Recommended Action
Membrane	Cracks	3.5×10^{-5} manuf. data	Material defect	Electrolyte leakage	No power output from battery	Inspection	QA	Fire source	2D	Add system sensor
	Pinholes	1.1×10^{-9} manuf. data	Material defect	Electrolyte leakage	No power output from battery	Inspection	QA	Fire source	2E	Add system sensor
	Fails to rupture	4.1×10^{-6} manuf. data	Materiel defect	No electrolyte to battery plates	No power output from battery	Inspection	QA	None	2D	Add system sensor
Squib	Fails to fire	4.1×10^{-9} manuf. data	Material defect	No electrolyte to battery plates	No power output From battery	Inspection	None	None	2E	
	Fires prematurely	4.1×10^{-9} manuf. data	Material; RF energy; dropped	Premature electrolyte to battery plates	Premature battery activation	System sensor for power	System sensor for power	Premature missile power	2E	Add system sensor
Analyst:					Date:			Page: 2/2		

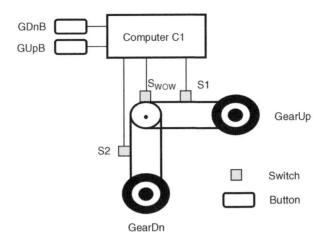

Figure 16.10   FMEA example 2: Aircraft landing gear system.

3. Perform self-test
4. Report system malfunction
5. Record self-test results

Tables 16.6 and 16.7 contain the FMEA worksheets evaluating the functions involved with the aircraft landing gear. Note that in the functional FMEA the failure rate column is blank, since rates are not available for functions.

16.10   FMEA LEVEL OF DETAIL

The FMEA level of detail applies to the function/hardware indenture level at which failures are postulated. Failures can be considered at any level, from the top system functions down to individual components. During the conceptual phase of system development, a high-level functional approach is particularly appropriate for eliminating design inadequacies. In latter system development phases, more detailed hardware or functional approaches are more appropriate for adequately implementing the design concept. Thus, the FMEA level of detail is affected by the phase of system development in which it is performed.

A less detailed analysis, completed at a time when it can contribute measurably to the adequacy of the system design, may be much more valuable than a more detailed analysis delivered at such a late date that implementation costs make changes unfeasible.

The functional approach is the system level approach for conducting an FMEA, is normally used when system hardware definition has not reached the point of identifying specific hardware items, and requires a less detailed analysis. This method of analysis is more adaptable to considering multiple failures and external influences such as software functions and human error. The hardware approach is the more rigorous and detailed method of conducting an FMEA and is normally used whenever the hardware items can be identified from engineering drawings. While the hardware approach is normally utilized from the part level up, it can be initiated at almost any indenture level.

TABLE 16.6 Functional FMEA of Landing Gear; Worksheet 1

Failure Mode and Effects Analysis

System: Aircraft | Subsystem: Landing Gear | Mode/Phase: Flight

Component	Failure Mode	Failure Rate	Causal Factors	Immediate Effect	System Effect	Method of Detection	Current Controls	Hazard	Risk	Recommended Action
Gear up	Fails to issue	N/A	Computer, software, wiring	landing gear	Aircraft flight with gear down	Sensor		Damage from drag	4C	
	Issues prematurely	N/A	Computer, software	If down, landing gear raises prematurely	Gear raises during taxi or takeoff	Sensor		Aircraft damage during taxi	2C	
Gear down	Fails to issue	N/A	Computer, software, wiring	Unable to lower landing gear	Aircraft must land with landing gear up	Sensor		Damage or injury during landing	2C	
	Issues prematurely	N/A	Computer, software	If up, landing gear lowers prematurely	Aircraft flight with gear down	Sensor		Damage from drag	2C	
Gear self-test (BIT – built-in Test)	Fails to occur	N/A	Computer, software, electronic faults	Landing gear fault not detected	Unable to raise or lower gear when desired, no warning	None		Possible unsafe state	4C	
	Occurs erroneously	N/A	Computer, software, electronic faults	Landing gear fault self-test data incorrect	No warning or incorrect warning regarding landing gear status	None		Possible unsafe state	4C	

Analyst: | Date: | Page: 1/2

TABLE 16.7 Functional FMEA of Landing Gear; Worksheet 2

Failure Mode and Effects Analysis

System: Aircraft							Subsystem: Landing Gear		Mode/Phase: Flight			
Component	Failure Mode	Failure Rate	Causal Factors	Immediate Effect	System Effect	Method of Detection	Current Controls	Hazard	Risk	Recommended Action
Report system malfunction	Fails to occur	N/A	Computer, software, electronic faults	BIT data not reported	No warning of fault state	Pilot report		None	4D	
	Occurs erroneously	N/A	Computer, software, electronic faults	BIT data reported incorrectly	No warning of fault state, or incorrect warning	Pilot report		None	4D	
Record self-test results	Fails to occur	N/A	Computer, software, electronic faults	BIT data not recorded	No recording of fault state	Data analysis		None	4D	
	Occurs erroneously	N/A	Computer, software, electronic faults	BIT data recorded incorrectly	No recording of fault state, or incorrect recording	Data analysis		None	4D	
Analyst:				Date:						Page: 2/2

16.11   FMEA ADVANTAGES AND DISADVANTAGES

The following are the advantages of the FMEA technique:

1. Is easily understood and performed.
2. Is relatively inexpensive to perform, yet provides meaningful results.
3. Provides rigor for focusing the analysis.
4. Provides a reliability prediction of the item being analyzed.
5. Has commercial software available to assist in the FMEA process.

The following are disadvantages of the FMEA technique:

1. Focuses on single-failure modes rather than failure mode combinations.
2. Is not designed to identify hazards unrelated to failure modes.
3. Provides limited examination of human error.
4. Provides limited examination of external influences and interfaces.
5. Requires expertise on the product or process under analysis.

16.12   COMMON FMEA MISTAKES TO AVOID

When first learning how to perform an FMEA, it is commonplace to commit some traditional errors. The following is a list of typical errors made when conducting an FMEA.

1. Not utilizing a structured approach with a standardized worksheet.
2. Not having a design team participate in the analysis in order to obtain all possible viewpoints.
3. Not fully investigating the complete effect of a failure mode.
4. Using the FMEA as a primary hazard analysis rather than a secondary hazard analysis.

16.13   FMEA SUMMARY

This chapter discussed the FMEA technique. The following are the basic principles that help summarize the concepts presented in this chapter:

1. The primary purpose of an FMEA is to identify potential failure modes and to evaluate the effect of these failures should they occur. The FMEA is primarily for reliability, but it can also be used for safety evaluations with some modification.
2. An FMEA is a qualitative and/or quantitative analysis tool. It can be used quantitatively to predict the failure rate of an assembly, unit, or subsystem.
3. The FMEA generally requires detailed design information.
4. FMEAs can be used to evaluate the design of hardware, software, functions, and processes.
5. The FMEA should not be the sole analysis for the identification of hazards but should be used in conjunction with other hazard analyses. FMEA should be a supplement to the DD-HAT and SD-HAT analyses.

6. FMEAs are most effectively performed through a team effort, involving all the disciplines involved with the product or process.

7. A safety-oriented FMEA provides the following information:
   - Failure modes
   - The immediate and system effects of the failure modes
   - Failure rates
   - Hazards resulting from the failure modes
   - Mishap risk assessment

FURTHER READINGS

Carlson, C. S., *Effective FMEAs*, John Wiley & Sons, 2012.

FMEA-3 Potential Failure Mode and Effects Analysis, Automotive Industry Action Group (AIAG), 3rd edition, July 2002 (equivalent of SAE J-1739).

IEC 60812, Analysis Techniques for System Reliability: Procedure for Failure Mode and Effects Analysis (FMEA), 2nd ed., 2001.

McDermott, R., Mikulak, R., and Beauregard, M., The Basics of FMEA, Productivity, Inc., 1996.

MIL-STD-1629A, Procedures for Performing a Failure Mode, Effects and Criticality Analysis, 1980.

OLB-71, Failure Mode and Effects Analysis, Engineering Industry Training Board (EITB), 1986, 41 Clarendon Road, Watford WD1 1HS, England.

SAE ARP5580, Recommended Failure Modes and Effects Analysis (FMEA) Practices for Non-Automobile Applications, July 2001.

SAE ARP4761, Guidelines and Methods for Conducting the Safety Assessment Process on Civil Airborne Systems and Equipment, Appendix G – Failure Mode and Effects Analysis, 1996.

Stamatis, D. H., *Failure Mode and Effect Analysis: FMEA from Theory to Execution*, Quality Press, American Society for Quality, 1995.

SAE Standard J-1739, Potential Failure Mode and Effects Analysis in Design (Design FMEA) and Potential Failure Mode and Effects Analysis in Manufacturing and Assembly Processes (Process FMEA) and Effects Analysis for Machinery (Machinery FMEA), August 2002.

STUK-YTO-TR 190, Failure Mode and Effects Analysis of Software-Based Automation Systems, August 2002, Finish Radiation and Nuclear Safety Authority.

Chapter 17

Hazard and Operability (HAZOP) Analysis

17.1 INTRODUCTION

Hazard and operability (HAZOP) analysis is a technique for identifying and analyzing hazards and operational concerns of a system. It is a very organized, structured, and methodical process for carrying out a hazard identification analysis of a system, from the concept phase through decommissioning. Although HAZOP is a relatively simple process in theory, the steps involved must be carefully observed in order to maintain the rigor of the methodology.

HAZOP analysis utilizes key guide words and system diagrams (design representations) to identify system hazards. Adjectives (guide words) such as *more*, *no*, *less*, and so on are combined with process/system conditions such as *speed, flow, pressure*, and so on in the hazard identification process. HAZOP analysis looks for hazards resulting from identified potential deviations in design operational intent. A HAZOP analysis is performed by a team of multidisciplinary experts in a brainstorming session under the leadership of a HAZOP team leader.

Some of the key components to a HAZOP analysis include the following:

- A structured, systematic, and logical process
- A multidisciplinary team with experts in many areas
- An experienced team leader
- The controlled use of system design representations
- The use of carefully selected system entities, attributes, and guide words to identify hazards

Hazard Analysis Techniques for System Safety, Second Edition. Clifton A Ericson, II.
© 2016 John Wiley & Sons, Inc. Published 2016 by John Wiley & Sons, Inc.

17.2 HAZOP ANALYSIS BACKGROUND

This analysis technique falls under the preliminary design hazard analysis type (PD-HAT) and the detailed design hazard analysis type (DD-HAT). Refer to Chapter 4 for a discussion on the analysis types. HAZOP analysis is also sometimes referred to as hazard and operability study (HAZOPS).

The purpose of HAZOP analysis is to identify the potential for system deviations from intended operational intent through the unique use of key guide words. The potential system deviations lead to possible system hazards.

HAZOP analysis is applicable to all types of systems and equipment, with analysis coverage given to subsystems, assemblies, components, software, procedures, environment, and human error. HAZOP analysis can be conducted at different abstraction levels, such as conceptual design, top-level design, and detailed component design. HAZOP analysis has been successfully applied to a wide range of systems, such as chemical plants, nuclear power plants, oil platforms, and rail systems. The technique can be applied to a system very early in design development and thereby help identify safety issues early in the design process. Early application helps system developers to design-in safety of a system during early development rather than having to take corrective action after a test failure or a mishap.

The HAZOP analysis technique, when applied to a given system by experienced personnel, should provide a thorough and comprehensive identification of hazards that exist in a given system or process. A basic understanding of the hazard analysis theory and system safety concepts is essential. Experience with the particular type of system is helpful in generating a complete list of potential hazards, so is experience in the HAZOP analysis process. The technique is uncomplicated and easily learned. An easily followed HAZOP analysis worksheet and instructions are provided in this chapter.

HAZOP analysis was initially developed for the chemical process industry and its methodology was oriented around process design and operations. The methodology can be extended to systems and functions with some practice and experience. The HAZOP analysis technique provides for an effective hazard analysis. In essence, the HAZOP analysis is not much different from preliminary hazard analysis (PHA) or subsystem hazard analysis (SSHA), except for the guide words used. HAZOP analysis could be utilized for the PHA and/or SSHA techniques.

17.3 HAZOP HISTORY

HAZOP analysis was formalized as an analysis technique by the Institute of Chemical Industry (ICI) in the United Kingdom in the early 1970s to assess safety risk in chemical process plants. HAZOP analysis has been subsequently developed and improved upon and commercial software is available to assist in the HAZOP analysis process.

The HAZOP analysis technique was initially developed by ICI, but it became more widely used within the chemical process industry after the Flixborough disaster in which a chemical plant explosion killed 28 people, many of them residents living nearby. Through the general exchange of ideas and personnel, the methodology was then adopted by the petroleum industry, which has a similar potential for major disasters. This was then followed by the food and water industries, where the hazard potential is as great, but of a different nature, the concerns being more to do with contamination rather than explosions or chemical releases.

17.4 HAZOP THEORY

HAZOP analysis entails the investigation of deviations from design intent for a process or system by a team of individuals with expertise in different areas, such as engineering, chemistry, safety, operations, and maintenance. The approach is to review the process/system in a series of meetings during which the multidisciplinary team "brainstorms" the system design methodically by following a sequence based on prescribed guide words and the team leader's experience. The guide words are used to ensure that the design is explored in every conceivable manner. The HAZOP analysis is based on the principle that several experts with different backgrounds can interact and better identify problems when working together than when working separately and combining their results.

Fault trees may be used to complement the HAZOP analysis process. However, the use of fault trees in this context does not imply that the trees should be quantified probabilistically since the purpose is only the identification of mishap scenarios.

In many ways, HAZOP analysis is similar to the PHA and SSHA in that it identifies hazards by evaluating the design by comparing system parameters to a list of key guide words that suggest modes of hazardous operation. The PHA and SSHA use hazard checklists in a manner similar to guide words.

The HAZOP analysis procedure involves taking a full description of a process/system and systematically questioning every part of it to establish how deviations from the design intent can arise. Once identified, an assessment is made as to whether such deviations and their consequences can have a negative effect upon the safe and efficient operation of the plant/system.

The HAZOP analysis is conducted through a series of team meetings led by the team leader. The key to a successful HAZOP is selection of the right team leader and the selection of the appropriate team members. This HAZOP analysis is applied in a structured way by the team, and it relies upon their imagination in an effort to discover credible causes of deviations from design intent. In practice, many of the deviations will be fairly obvious, such as pump failure causing a loss of circulation in a cooling water facility; however, a great advantage of the technique is that it encourages the team to consider other less obvious ways in which a deviation may occur. In this way, the analysis becomes much more than a mechanistic checklist type of review. The result is that there is a good chance that potential failures and problems will be identified that had not previously been experienced in the type of plant/system being studied.

Figure 17.1 shows an overview of the basic HAZOP process and summarizes the important relationships involved in the HAZOP process.

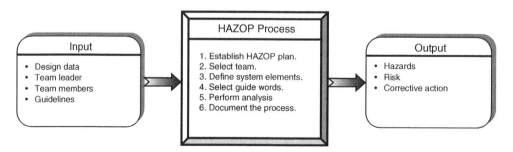

Figure 17.1 *The HAZOP process.*

17.5 HAZOP METHODOLOGY

Table 17.1 lists and describes the basic steps of the HAZOP process.

Some of the key components in a HAZOP analysis include the following:

- A planned process that is structured, systematic, and logical
- The correct composition of team members
- The correct team leader (this is a critical element)
- Teamwork
- HAZOP analysis training is vital
- The controlled use of design representations
- The planned use of entities, attributes, and guide words to identify hazards

HAZOP analysis is a time-consuming process, especially when many people are involved in the brainstorming sessions. Recommendations for avoidance or mitigation of the observed

TABLE 17.1 HAZOP Process

Step	Task	Description
1	Define system	Define the system scope and boundaries. Define the mission, mission phases, and mission environments. Understand the system design and operation. Note that all steps are applicable for a HAZOP on software
2	Plan HAZOP	Establish HAZOP analysis goals, definitions, worksheets, schedule, and process. Divide the system under analysis into the smallest segments desired for the analysis. Identify items to be analyzed and establish indenture levels for items/functions to be analyzed
3	Select team	Select a team leader and all team members to participate in HAZOP analysis and establish responsibilities. Utilize team member expertise from several different disciplines (e.g., design, test, manufacturing, etc.)
4	Acquire data	Acquire all of the necessary design and process data needed (e.g., functional diagrams, code, schematics, and drawings) for the system, subsystems, and functions. Refine the system information and design representation for HAZOP analysis
5	Conduct HAZOP	a. Identify and list the items to be evaluated b. Establish and define the appropriate parameter list c. Establish and define the appropriate guide word list d. Establish the HAZOP analysis worksheet e. Conduct the HAZOP analysis meetings f. Record the HAZOP analysis results on the HAZOP worksheets g. Have the HAZOP analysis worksheets validated by a system engineer for correctness
6	Recommend corrective action	Recommend corrective action for hazards with unacceptable risk. Assign responsibility and schedule for implementing corrective action
7	Monitor corrective action	Review the HAZOP at scheduled intervals to ensure that corrective action is being implemented
8	Track hazards	Transfer identified hazards into the Hazard Tracking System (HTS)
9	Document HAZOP	Document the entire HAZOP process on the worksheets. Update for new information and closure of assigned corrective actions

hazards cannot always be closed within the team meetings, thus action items often result. Since the basic purpose of a HAZOP study is to identify potentially hazardous scenarios, the team should not spend any significant time trying to engineer a solution if a potential problem is uncovered. If a solution to a problem is obvious, the team should document the recommended solution in both the HAZOP and the resulting HAR.

The HAZOP analysis is performed by comparing a list of system parameters against a list of guide words. This process stimulates the mental identification of possible system deviations from design intent and resulting hazards. Establishing and defining the system parameters and the guide words is a key step in the HAZOP analysis. The deviations from the intended design are generated by coupling the guide word with a variable parameter or characteristic of the plant, process, or system, such as reactants, reaction sequence, temperature, pressure, flow, phase, and so on. In other words:

$$Guideword + Parameter = Deviation$$

For example, when considering a reaction vessel in which an exothermic reaction is to occur and one of the reactants is to be added stepwise, the guide word "more" would be coupled with the parameter "reactant" and the deviation generated would be "thermal runaway." Systematic examinations are made of each part of a facility or system.

It should be noted that not all combinations of primary/secondary words are appropriate. For example, temperature/no (absolute zero or −273 °C) or pressure/reverse could be considered as meaningless.

The amount of preparation required for a HAZOP analysis depends upon the size and complexity of the facility or system. Typically, the data required consist of various drawings in the form of line diagrams, flow sheets, facility layouts, isometrics and fabrication drawings, operating instructions, instrument sequence control charts, logic diagrams, and computer code. Occasionally, there are facility manuals and equipment manufacturers' manuals. The data must be accurate and sufficiently comprehensive. In particular, for existing facilities, line diagrams must be checked to ensure that they are up to date and that modifications have not been made since the facility was constructed.

HAZOP analyses are normally carried out by a multidisciplinary team with members being chosen for their individual knowledge and experience in design, operation, maintenance or health, and safety. A typical team would have between four and seven members, each with a detailed knowledge of the way in which the facility or system is intended to operate. The technique allows experts in the process to bring their knowledge and expertise to bear systematically so that problems are less likely to be missed. It is essential that the team leader is an expert in the HAZOP technique. The team leader's role is to ensure that the team follows the procedure. The leader must be skilled in leading a team of people who may not want to focus on meticulous attention and detail. It is recommended that the team leader should be an independent person; the team leader should not be associated with program management. The team leader must have sufficient technical knowledge to guide the study properly but should not necessarily be expected to make a technical contribution. It is beneficial if team members have had some training in the HAZOP technique.

Many HAZOP studies can be completed in five to ten meetings, although for a small modification only one or two meetings may be necessary. However, for a large project it may take several months even with two or three teams working in parallel on different sections of the system. HAZOPs require major resources, which should not be underestimated. If HAZOP analyses are to be introduced into any organization for the first time, it may be appropriate to

TABLE 17.2   HAZOP Roles

	Postulate	Explore	Explain	Conclude	Record
Leader	Yes	Possibly	Possibly	Yes	
Expert		Yes	Yes		
Designer		Possibly	Yes		
User		Possibly	Yes		
Recorder		Possibly			Yes

apply the technique to one or two problems to find out whether it is useful and can be applied successfully. If the technique can be successfully applied, it can grow naturally and be applied to larger projects.

It is common practice to record each step of a HAZOP analysis. Analysis recording includes a data file, which is a copy of the data (flow diagrams, original and final process and instrument diagrams, running instructions, bar sheets, models, and so on) used by the team during the examinations sessions and marked by the study leader to show that they have been examined.

Key HAZOP activities or tasks and who should perform these tasks are delineated in Table 17.2.[1] It is important that these roles are performed by the team members.

17.5.1   Design Representations

A design representation models the system design and portrays the intention of the system designers through the features of the design. The design representation can take many forms and may be more or less detailed, depending on the stage of system development. The design representation can be either physical or logical. A physical model shows the physical real world layout of the system, such as through a drawing, schematic, or reliability block diagram. A logical design representation portrays the logical relationships between system elements in the form of how components should logically work, and can be represented by functional flow diagrams, data flow diagrams, and so on. An extensive HAZOP analysis will likely be utilized for both physical and logical design representations.

The study leader can use the design representation as a form of analysis control. The representation acts as an agenda for the study team meetings as the team sequentially evaluates each item in the design representation.

The use of design representation aids, such as functional block diagrams, reliability block diagrams, context diagrams, data flow diagrams, timing diagrams, and so on, greatly aids and simplifies the HAZOP analysis process. Each person on the team must understand the design representations utilized for the analysis.

17.5.2   System Parameters

A system comprises a set of components, and on a design representation a path between two components indicates an interaction or design feature that exists. An interaction can consist of a flow or transfer from one component to another. A flow may be tangible (such as a fluid) or intangible (such as an item of data). In either case, the flow is designed with certain properties,

---

[1]From "System Safety: HAZOP and Software HAZOP", F. Redmill, M. Chudleigh and J. Catmur, John Wiley and Sons, 1999.

TABLE 17.3   Example System Parameters

• Flow (gas, liquid, electrical current)	• Temperature
• Pressure	• Level
• Separate (settle, filter, centrifuge)	• Composition
• Reaction	• Mix
• Reduce (grind, crush, etc.)	• Absorb
• Corrode	• Erode
• Isolate	• Drain
• Vent	• Purge
• Inspection, surveillance	• Maintain
• Viscosity	• Shutdown
• Instruments	• Start-up
• Corrosion	• Erosion
• Vibration	• Shock
• Software data flow	• Density

which can be referred to as attributes or parameters, which affect how the system operates. These parameters are the key to identifying design deviations in a HAZOP analysis.

The correct operation of a system is determined by the parameters of the interactions and components maintaining their design values (i.e., design intent). Hazards can be identified by studying what happens when the parameters deviate from the design intent. This is the principle behind the HAZOP analysis.

Table 17.3 contains a list of example system parameters. The list below is purely illustrative, as the words employed in an actual HAZOP review will depend upon the plant or system being studied.

Note that some parameter words may not appear to be related to any reasonable interpretation of the design intent of a process. For example, one may question the use of the word corrode, on the assumption that no one would intend that corrosion should occur. Bear in mind, however, that most systems are designed with a certain life span, and implicit in the design intent is that corrosion should not occur, or if it is expected, it should not exceed a certain rate. An increased corrosion rate in such circumstances would be a deviation from the design intent.

17.5.3   Guide Words

Guide words help both to direct and to stimulate the creative process of identifying potential design deviations. Guide words may be interpreted differently in different industries and at different stages of the system's life cycle. The interpretation of a guide word must be in these contexts of these factors. The purpose of the interpretations is to enable the exploration of plausible deviations from the design intent.

HAZOP analysis guide words are short words used to stimulate the imagination of a deviation of the design intent. For example, for the parameter "data flow" in a computer system, the guide word "more" can be interpreted as more data is passed than intended, or data is passed at a higher rate than intended. For the parameter "wire" in a system, the guide word "more" can be interpreted as higher voltage or current than intended.

Table 17.4 contains an example list of HAZOP guide words.

Table 17.5 demonstrates how an abbreviated guide word table can be used to brainstorm the system design as part of the overall process. This table contains an example list of HAZOP guide words showing how they might be used prior to filling out a more detailed HAZOP

TABLE 17.4   Example HAZOP Guide Words

Guide Word	Meaning
No	The design intent does not occur (e.g., Flow/No), or the operational aspect is not achievable (Isolate/No)
Less	A quantitative decrease in the design intent occurs (e.g., Pressure/Less)
More	A quantitative increase in the design intent occurs (e.g., Temperature/More)
Reverse	The opposite of the design intent occurs (e.g., Flow/Reverse)
Also	The design intent is completely fulfilled, but in addition some other related activity occurs (e.g., Flow/Also indicating contamination in a product stream, or Level/Also meaning material in a tank or vessel which should not be there)
Other	The activity occurs, but not in the way intended (e.g., Flow/Other could indicate a leak or product flowing where it should not, or Composition/Other might suggest unexpected proportions in a feedstock)
Other than	Complete substitution
Fluctuation	The design intention is achieved only for a part of the time (e.g., an air-lock in a pipeline might result in Flow/Fluctuation)
Early	The timing is different from the intention. Usually used when studying sequential operations, this would indicate that a step is started at the wrong time or done out of sequence
Late	Timing concern similar to Early
As well as	Qualitative modification/increase
More than	Exceeds a specified value
Part of	Only some of the design intention is achieved
Reverse	Logical opposite of the design intention occurs
Where else	Applicable for flows, transfers, sources, and destinations
Before/after	The step (or some part of it) is effected out of sequence
Faster/slower	The step is done/not done with the right timing
Fails	Fails to operate or perform its intended purpose
Inadvertent	Function occurs inadvertently or prematurely (i.e., unintentionally)

worksheet. Each of the table cells under guide word should stimulate ideas for hazards and also where more detailed information may be required.

17.5.4   Deviation from Design Intent

Since the HAZOP analysis is based on searching for *deviations from design intent*, it is important to understand this concept. All systems are designed with an overall purpose in mind. For an industrial plant system, it may be to produce a certain tonnage per year of a particular chemical, to manufacture a specified number of cars, to process and dispose of a certain volume of effluent per annum, and so on. For a weapon system, the purpose is to intentionally hit the intended target. These are the primary design intents for these systems; however, a secondary intent would be to operate the system in the safest and most efficient manner possible.

In order to achieve its goals, each subsystem of the system must consistently function in a particular manner. It is this manner of performance that could be classified as the design intent for that particular item.

To illustrate, imagine that as part of the overall production requirement for a system is the need for a cooling water facility. Meeting this requirement would involve circulating water in a pipe system that is driven by a pump. A simplified statement as to the design intent of this small section of the plant would be "to continuously circulate cooling water at an initial temperature of $x°$ C and at a rate of n gallons per hour." It is usually at this low level of design intent that a

TABLE 17.5 Example of HAZOP Guide Word versus System Parameters

Parameter	Guide Word						
	More	Less	None	Reverse	As Well As	Part Of	Other Than
Flow	High flow	Low flow	No flow	Reverse flow	Deviating concentration	Contamination	Deviating material
Pressure	High pressure	Low pressure	Vacuum		Delta-p		Explosion
Temperature	High temperature	Low temperature					
Level	High level	Low level	No level		Different level		
Time	Too long/too late	Too short/too soon	Sequence step skipped	Backwards	Missing actions	Extra actions	Wrong time
Agitation	Fast mixing	Slow mixing	No mixing				
Reaction	Fast reaction/runaway	Slow reaction	No reaction				Unwanted reaction
Start-up/Shut-down	Too fast	Too slow			Actions missed		Wrong recipe
Draining/Venting	Too long	Too short	None		Deviating pressure	Wrong timing	
Vibrations	Too low	Too high	None				Wrong frequency

HAZOP analysis is directed. The use of the word "deviation" now becomes easier to understand. A deviation or departure from the design intent in the case of the cooling facility would be a failure of circulation, or the water being at too high an initial temperature. Note the difference between a deviation and its cause. In this case, failure of the pump would be a cause, not a deviation.

17.6 HAZOP WORKSHEET

The HAZOP analysis technique is a detailed hazard analysis utilizing structure and rigor. It is desirable to perform the HAZOP analysis using a specialized worksheet. Although the format of the analysis worksheet is not critical, typically, matrix or columnar type worksheets are used to help maintain focus and structure in the analysis. The HAZOP analysis sessions are primarily reported in the HAZOP worksheets, in which the different items and proceedings are recorded. As a minimum, the following basic information is required from the HAZOP analysis worksheet:

1. Item under analysis
2. Guide words
3. System effect if guide word occurs
4. Resulting hazard or deviation (if any)
5. Risk assessment
6. Safety requirements for eliminating or mitigating the hazards

The recommended HAZOP analysis worksheet is shown in Figure 17.2. This particular HAZOP analysis worksheet utilizes a columnar-type format. Other worksheet formats may exist because different organizations often tailor their HAZOP analysis worksheet to fit their particular needs. The specific worksheet to be used may be determined by the system safety program (SSP), system safety working group, or the HAZOP analysis team performing the analysis.

HAZOP Analysis										
No.	Item	Function/ Purpose	Parameter	Guide Word	Consequence	Cause	Hazard	Risk	Recommendation	Comments
1	2	3	4	5	6	7	8	9	10	11

Figure 17.2 *The recommended HAZOP worksheet.*

The following instructions describe the information required under each column entry of the HAZOP worksheet:

1. *No*   This column identifies each HAZOP line item and is used for purposes of reference to any part of the analysis.
2. *Item*   This column identifies the process, component, item, or function being analyzed.
3. *Function/Purpose*   This column describes the item's purpose or function in the system so that the operational intent is understood.
4. *Parameter*   This column identifies the system parameter that will be evaluated against the guide words.
5. *Guide Word*   This column identifies the guide words selected for the analysis.
6. *Consequence*   This column identifies the most immediate and direct effect of the guide word occurring, usually in terms of a system deviation from design intent.
7. *Cause*   This column identifies all of the possible factors that can cause the specific deviation. Causal factors may include many different sources, such as physical failure, wear out, temperature stress, vibration stress, and so on. All conditions that affect a component or assembly should be listed to indicate whether there are special periods of operation, stress, personnel action, or combinations of events that would increase the probabilities of failure or damage.
8. *Hazard*   This column identifies the specific hazard that is created as a result of the specific consequence or deviation. (Remember: document all hazard considerations, even if they are later proven to be nonhazardous.)
9. *Risk*   This column provides a qualitative measure of mishap risk for the potential effect of the identified hazard. Risk measures are a combination of mishap severity and probability, and the recommended qualitative values from MIL-STD-882 are shown below.

Severity	Probability
1 – Catastrophic	A – Frequent
2 – Critical	B – Probable
3 – Marginal	C – Occasional
4 – Negligible	D – Remote
	E – Improbable

10. *Recommendation*   This column provides for any recommendations for hazard mitigation which are evident from the HAZOP analysis, such as design or procedural safety requirements.
11. *Comments*   This column provides for any pertinent comments to the analysis that need to be remembered for possible future use.

17.7   HAZOP EXAMPLE 1

Figure 17.3 contains an example water-pumping system for this HAZOP analysis example. In this system, water is supplied from a common source to three steam generators. For successful system operation, two of the three generators must be operational. Some redundancy has been

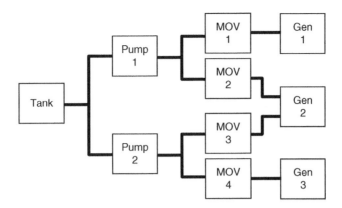

Figure 17.3 *Example 1 system diagram.*

designed into the system to help achieve this operational requirement. The pumps are electricity driven that pump water from the tank to the motor-operated valves (MOVs). The MOVs are opened and closed by electrical power. The pumps, MOVs, and generators are all monitored and controlled by a single common computer, and they are powered from a common electricity source.

After carefully reviewing all of the design representations for this system, the following design parameters have been selected for HAZOP analysis: fluid, pressure, temperature, electricity, and steam.

The HAZOP analysis worksheet for this example is shown in Table 17.6. Note that the analysis worksheet was not completed for the entire set of parameters and guide words, but only for a few in order to demonstrate the HAZOP analysis technique.

17.8 HAZOP EXAMPLE 2

HAZOP example 2 is a software-oriented HAZOP analysis.

Figure 17.4 depicts the software design for a hypothetical missile system. The software segment shown is for the missile fire control system, which handles missile status data and missile command data.

The HAZOP analysis worksheet for this example is shown in Table 17.7.

17.9 HAZOP ADVANTAGES AND DISADVANTAGES

The HAZOP technique affords the following advantages:

1. HAZOP analysis is easily learned and performed.
2. HAZOP analysis does not require considerable technical expertise for technique application.
3. HAZOP analysis provides rigor for focusing on system elements and hazards.
4. HAZOP analysis is a team effort with many viewpoints.
5. Commercial software is available to assist in HAZOP analysis.

TABLE 17.6 HAZOP Worksheet for Example 1

HAZOP Analysis

No.	Item	Function / Purpose	Parameter	Guide Word	Consequence	Cause	Hazard	Risk	Recommendation	Comments
1	Pipes	To carry water through system	Fluid	No	Loss of fluid, system failure; equipment damage	Pipe leak; pipe rupture	Equipment damage	2D		
2				More	Pressure becomes too high resulting in pipe rupture	No pressure relief valves in system	Equipment damage	2C	Add pressure relief valves to system	
3				Less	Insufficient water for operation of generators	Pipe leak; pipe rupture	Equipment damage	2D		
4				Reverse	Not applicable			--		
5	Electrical power	To provide electricity to operate Pumps, MOVs and Generators	Electricity	No	Loss of power to operate system components	Power grid loss; circuit breakers trip	Loss of system operation	2D	Provide source of emergency backup power	
6				More	Trips circuit breakers	Power surge	Loss of system operation	2C	Provide for fault detection and isolation	
7				Less	Insufficient power to adequately operate system components	Power grid fault	Equipment damage	2D	Provide source of emergency backup power	
8				Reverse	Not applicable			--		

Analyst: Date: Page: 1 of 1

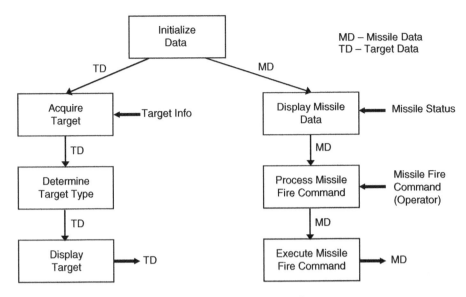

Figure 17.4 *Example 2 system diagram.*

The following are disadvantages of the HAZOP technique:

1. HAZOP analysis focuses on single events rather than combinations of possible events.
2. The HAZOP analysis's focus on guide words allows it to overlook some hazards not related to a guide word.
3. HAZOP analysis training is essential for optimum results, especially for the facilitator.
4. The HAZOP analysis can be time consuming and thus expensive.

17.10 COMMON HAZOP ANALYSIS MISTAKES TO AVOID

When first learning how to perform a HAZOP analysis, it is commonplace to commit some typical errors. The following is a list common of errors made during the conduct of an HAZOP analysis:

1. Not selecting an experienced and trained team leader
2. Not selecting the appropriate team
3. Not adequately planning, scheduling, or funding the HAZOP analysis

17.11 HAZOP SUMMARY

This chapter discussed the HAZOP analysis technique. The following are the basic principles that help summarize the discussion in this chapter:

1. The primary purpose of HAZOP is to identify deviations from design intent that can lead to the occurrence of an undesired event or hazard.

TABLE 17.7 HAZOP Worksheet for Example 2

HAZOP Analysis

No.	Item	Function / Purpose	Parameter	Guide Word	Consequence	Cause	Hazard	Risk	Recommendation	Comments
1	Missile Fire Control	Performs missile status and control	Missile Data	No (None)	Loss of missile status to operator	Hardware fault; software error	Unsafe missile	2D		
2				More/Less (wrong)	Missile status to operator is incorrect	Hardware fault; software error	Equipment damage	2D		
3				Early/Late (Timing)	Missile status to operator is incorrect	Hardware fault; software error	Equipment damage	2D		
4			Missile Command	No (None)	Loss of missile control	Hardware fault; software error	Unable to safe missile	2D		
5				More/Less (wrong)	Operator command to missile is incorrect	Hardware fault; software error	Inadvertent launch command	1D	Add command status checks to design	
6				Early/Late (Timing)	Operator command to missile is incorrect	Hardware fault; software error	Unable to safe missile	2D		
Analyst:					Date:				Page: 1 of 1	

2. HAZOP analysis requires an experienced team leader in conjunction with an appropriately selected team.
3. The use of design representation aids, such as functional block diagrams, reliability block diagrams, context diagrams, and so on, greatly aids and simplifies the HAZOP analysis process.

FURTHER READINGS

The following are references on HAZOP:

Chemical Industries Association, *A Guide to Hazard and Operability Studies*, Chemical Industries Association 1977.

IEC 61882 Ed. 1.0 b:2001, Hazard and operability studies (HAZOP studies) - Application guide Paperback – August 19, 2007.

Kletz, Trevor A., *HAZOP and Hazan: Identifying and Assessing Process Industry Hazards*, 4th edition, CRC Press, Taylor & Francis, Inc., 1999.

Kletz, Trevor, Hazop and Hazan, Taylor & Francis, ISBN-10: 1560328584, The Institution of Chemical Engineers; 4th edition (July 28, 2006).

Nolan, Dennis P., *Application of HAZOP and What-If Safety Reviews to the Petroleum, Petrochemical and Chemical Industries*, Noyes Publications, 1994.

Redmill, F., Chudleigh, M. and Catmur, J., *System Safety: HAZOP and Software HAZOP*, John Wiley and Sons, Inc., 1999.

Swann, C. D. and Preston, M. L., Twenty Five Years of HAZOPS, *J. Loss Prev.*, **8** (6):349–353 (1995).

Chapter 18

Event Tree Analysis (ETA)

18.1 ETA INTRODUCTION

Event tree analysis (ETA) is an analysis technique for identifying and evaluating the sequence of events in a potential accident scenario following the occurrence of an initiating event. ETA utilizes a visual logic tree structure known as an event tree (ET). The objective of ETA is to determine whether the initiating event will develop into a serious mishap or if the event is sufficiently controlled by the safety systems and procedures implemented in the system design. An ETA can result in many different possible outcomes from a single initiating event, and it provides the capability to obtain a probability for each outcome.

18.2 ETA BACKGROUND

The ETA technique falls under the system design hazard analysis type (SD-HAT) and should be used as a supplement to the system hazard analysis. Refer to Chapter 4 for a description of the analysis types. The ETA is a very powerful tool for identifying and evaluating all of the system consequence paths that are possible after an initiating event occurs. The ETA model shows the probability of the system design resulting in a safe operation path, a degraded operation path, and an unsafe operation path.

The purpose of ETA is to evaluate all of the possible outcomes that can result from an initiating event. Generally, there are many different outcomes possible from an initiating event, depending upon whether design safety systems work properly or malfunction when needed. ETA provides a probabilistic risk assessment (PRA) of the risk associated with each potential outcome.

Hazard Analysis Techniques for System Safety, Second Edition. Clifton A Ericson, II.
© 2016 John Wiley & Sons, Inc. Published 2016 by John Wiley & Sons, Inc.

ETA can be used to model an entire system, with analysis coverage given to subsystems, assemblies, components, software, procedures, environment, and human error. ETA can be conducted at different abstraction levels, such as conceptual design, top-level design, and detailed component design. ETA has been successfully applied to a wide range of systems, such as nuclear power plants, spacecraft, and chemical plants. The technique can be applied to a system very early in design development and thereby helps identify safety issues early in the design process. Early application helps system developers design-in the safety of a system during early development rather than having to take corrective action after a test failure or a mishap.

The ETA technique, when applied to a given system by an experienced analyst, helps to thoroughly identify and evaluate all of the possible outcomes resulting from an initiating event (IE). A basic understanding of ETA and the FTA theory is essential to develop an ETA model. In addition, it is crucial for the analyst to have a detailed understanding of the system. Overall, ETA is very easy to learn and understand. Proper application depends on the complexity of the system and the skill of the analyst. Applying the ETA technique to the evaluation of a system design is not a difficult process; however, it does require an understanding of FTA and the probability theory.

A cause–consequence analysis (CCA) is very similar to ETA and is a possible alternative technique. Additionally, multiple FTAs could be performed to obtain the same results as an ETA. The ETA produces many different potential outcomes from a single event, whereas the FTA only evaluates the many causes of a single outcome.

The use of an ETA is recommended for a PRA of the possible outcomes resulting from an initiating event. The resulting risk profiles provide management and design guidance on areas requiring additional safety countermeasures design methods.

18.3 ETA HISTORY

ETA is a binary form of a decision tree for evaluating the various multiple decision paths in a given problem. ETA appears to have been developed during the WASH-1400 (see Ref. [1]) nuclear power plant safety study (circa 1974). The WASH-1400 team realized that a nuclear power plant PRA could be achieved by FTA, however, the resulting fault trees (FTs) would be very large and cumbersome and they, therefore, established ETA to condense the analysis into a more manageable picture, while still utilizing FTA.

18.4 ETA DEFINITIONS

ETA is based on the following definitions:

Accident Scenario An accident scenario is a series of events that ultimately result in an accident. The sequence of events begins with an initiating event and is (usually) followed by one or more pivotal events that lead to the undesired end state.

Initiating Event (IE) An initiating event is a failure or undesired event that initiates the start of an accident sequence. The IE may result in a mishap, depending upon successful operation of the hazard countermeasure methods designed into the system. Refer to Chapter 3 on the hazard theory for information on the components of a hazard.

Pivotal Events   Pivotal events are key intermediary events between the IE and the final mishap. These are the failure/success events of the design safety methods established to prevent the IE from resulting in a mishap. If a pivotal event works successfully, it stops the accident from occurring and is referred to as a mitigating event. If a pivotal event fails to work, then the accident is allowed to progress and is referred to as an aggravating event.

Probabilistic Risk Assessment (PRA)   PRA is a comprehensive, structured, and logical analysis method for identifying and evaluating risk in a complex technological system. The detailed identification and assessment of accident scenarios, with a quantitative analysis is the PRA goal.

Event Tree (ET)   An ET is a graphical model of an accident scenario that yields multiple outcomes and outcome probabilities. ETs are one of the most used tools in a PRA.

A common definition of risk in the PRA discipline is that risk is based upon a set of triplets[1]:

1. Accident scenarios – what can go wrong?
2. Scenarios frequencies – how likely is it?
3. Scenario consequences – what are the consequences?

18.5   ETA THEORY

When performing a PRA, identifying and developing accident scenarios are fundamental to the concept of risk evaluation. The process begins with a set of initiating events (IEs) that perturb the system (i.e., cause it to change its operating state or configuration). For each IE, the analysis proceeds by determining the additional failure modes necessary to lead to the undesirable consequences. The consequences and frequencies of each scenario are computed for the individual IEs and the collection of probabilities forms a risk profile for the system.

ETs are used to model accident scenarios. An ET starts with the IE and progresses through the scenario via a series of pivotal events (PEs) until an end state is reached. The PEs are failures or events that mitigate or aggravate the scenario. The frequency (i.e., probability) of the PE can be obtained from an FTA of the event.

The PRA theory relates very closely to the standard system safety terminology. An accident scenario is equivalent to a hazard, the scenario frequency is equivalent to hazard probability, and the scenario outcome is equivalent to hazard severity.

Risk management involves the identification and prevention or reduction of adverse accident scenarios and the promotion of favorable scenarios. Risk management requires understanding the elements of adverse scenarios so that their components can be prevented or reduced, and an understanding of favorable scenarios in order that their components can be enhanced or promoted.

An accident scenario typically contains an IE and (usually) one or more pivotal events leading to an end state as shown in Figure 18.1.

As modeled in most PRAs, an IE is a perturbation that requires some kind of response from operators and/or one or more systems to prevent an undesired consequence. The pivotal events include successes or failures of these responses, or possibly the occurrence or nonoccurrence of external conditions or key phenomena. The end states are formulated according to the decisions

[1]S. Kapan and B. J. Garrick, *On the Quantitative Definition of Risk*, *Risk Analysis*, **1**: 11–37 (1981).

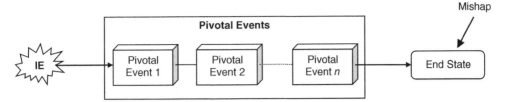

Figure 18.1 *Accident scenario concept.*

being supported by the analysis. Scenarios are classified into end states according to the kind and severity of consequences, ranging from completely successful outcomes to losses of various kinds, such as follows:

- Loss of life or injury/illness to personnel
- Damage to, or loss of, equipment or property (including software)
- Unexpected or collateral damage as a result of tests
- Failure of mission
- Loss of system availability
- Damage to the environment

An ET distills the pivotal event scenario definitions and presents this information in a tree structure that is used to help classify scenarios according to their consequences. The headings of the ET are the IE, the pivotal events, and the end states. The tree structure below these headings shows the possible scenarios ensuing from the IE, in terms of the occurrence or nonoccurrence of the pivotal events. Each distinct path through the tree is a distinct scenario. According to a widespread but informal convention, where pivotal events are used to specify system success or failure, the "down" branch is considered to be "failure."

The ET concept is shown in Figure 18.2.

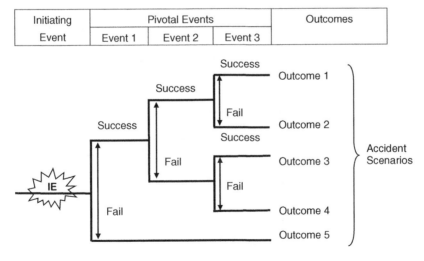

Figure 18.2 *Event tree concept.*

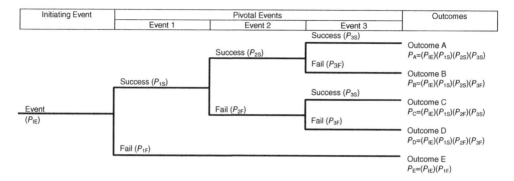

Initiating Event	Pivotal Events			Outcomes
	Event 1	Event 2	Event 3	

Figure 18.3 *ETA concept.*

In most ETs, the pivotal event splits are binary: a phenomenon either does or does not occur, a system either does or does not fail. This binary character is not strictly necessary; some ETs split into more than two branches. What is necessary is that distinct paths be mutually exclusive and quantified as such (at least to the desired level of accuracy).

An example of ET structure with quantitative calculations is displayed in Figure 18.3. The ET model logically combines all of the system-design safety countermeasure methods intended to prevent the IE from resulting in a mishap. A side effect of the analysis is that many different outcomes can be discovered and evaluated. Note how the ET closely models the scenario concept shown in Figure 18.1.

18.6   ETA METHODOLOGY

Figure 18.4 shows an overview of the basic ETA process and summarizes the important relationships involved in the ETA process.

The ETA process involves utilizing detailed design information to develop ETDs for specific IEs. In order to develop the ETD, the analyst must have first established the accident scenarios, IEs, and pivotal events of interest. Once the ETD is constructed, failure frequency data can be applied to the failure events in the diagram. Usually, this information is derived from FTA of the failure event. Since $1 = P_S + P_F$, the probability of success can be derived from the probability of failure calculation. The probability for a particular outcome is computed by multiplying the event probabilities in the path.

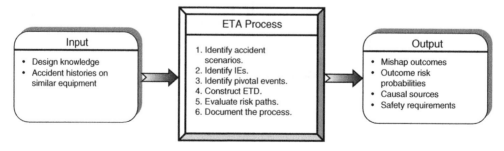

Figure 18.4 *ETA overview.*

TABLE 18.1 ETA Process

Step	Task	Description
1	Define the system	Examine the system and define the system boundaries, subsystems, and interfaces
2	Identify the accident scenarios	Perform a system assessment or hazard analysis to identify the system hazards and accident scenarios existing within the system design
3	Identify the initiating events	Refine the hazard analysis to identify the significant IEs in the accident scenarios. IEs include events such as fire, collision, explosion, pipe break, toxic release, and so on
4	Identify the pivotal events	Identify the safety barriers or countermeasures involved with the particular scenario that are intended to preclude a mishap
5	Build the event tree diagram	Construct the logical ETD, starting with the IE, then the PEs and completing with the outcomes of each path
6	Obtain the failure event probabilities	Obtain or compute the failure probabilities for the PEs on the ETD. It may be necessary to use FTs to determine how a PE can fail and to obtain the probability
7	Identify the outcome risk	Compute the outcome risk for each path in the ETD
8	Evaluate the outcome risk	Evaluate the outcome risk of each path and determine if the risk is acceptable
9	Recommend corrective action	If the outcome risk of a path is not acceptable, develop design strategies to change the risk
10	Document ETA	Document the entire ETA process on the ETDs. Update for new information as necessary

Table 18.1 lists and describes the basic steps of the ETA process, which involves performing a detailed analysis of all the design safety features involved in a chain of events that can result from the initiating event to the final outcome.

Complex systems tend to have a large number of interdependent components, redundancy, standby systems, and safety systems. Sometimes, it is too difficult or cumbersome to model a system with just an FT; therefore, PRA studies have combined the use of FTs and ETDs. The ETD models accident/mishap cause–consequence scenarios and FTs model complex subsystems to obtain the probability of these subsystems failing. An accident scenario can have many different outcomes, depending on which PEs fail and which function correctly. The ET/FT combination models this complexity very well.

The goal of ETA is to determine the probability of all the possible outcomes resulting from the occurrence of an IE. By analyzing all possible outcomes, it is possible to determine the percentage of outcomes that lead to the desired result, and the percentage of outcomes that lead to the undesired result.

ETs can be used to analyze systems in which all components are continuously operating, or analyze systems in which some or all of the components are in the standby mode – those that involve sequential operational logic and switching. The starting point (referred to as the initiating event) disrupts normal system operation. The event tree displays the sequences of events involving success and/or failure of the system components.

In the case of standby systems, and in particular, safety and mission-oriented systems, the ET is used to identify the various possible outcomes of the system following a given IE that is generally an unsatisfactory operating event or situation. In the case of continuously operated

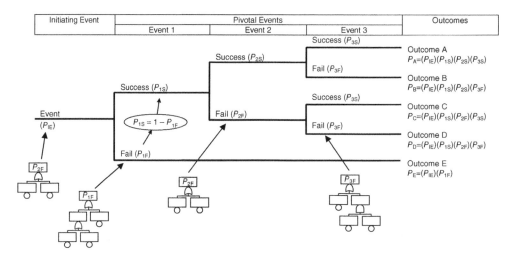

Figure 18.5   *ETD development.*

systems, these events can occur (i.e., components can fail) in any arbitrary order. In the event tree analysis, the components can be considered in any order since they do not operate chronologically with respect to each other.

ETA is based on binary logic, in which an event either has or has not happened or a component has or has not failed. It is valuable in analyzing the consequences arising from a failure or undesired event. An ET begins with an IE, such as a component failure, increase in temperature/pressure, or a release of a hazardous substance that can lead to an accident. The consequences of the event are followed through a series of possible paths. Each path is assigned a probability of occurrence and the probability of the various possible outcomes can be calculated.

The ETD is a diagram modeling all possible events that follow an originating failure or undesired event. The originating event can be a technical failure or an operational human error. The objective is to identify the chain of events following one or more specified basic events, in order to evaluate the consequences and determine if the event will develop into a serious accident or if the consequences are sufficiently controlled by the safety systems and procedures implemented. The results can, therefore, be recommendations to increase the redundancy or to conduct modifications on the safety systems.

The ETA begins with the identified IE listed at the left side of the diagram in Figure 18.5. All safety design methods or countermeasures are then listed at the top of the diagram as contributing events. Each safety design method is evaluated for the contributing event under the following criteria: (a) operates successfully and (b) fails to operate. The resulting diagram combines all of the various success/failure event combinations and fans out to the right in a sideways tree structure. Each success/failure event can be assigned a probability of occurrence, and the final outcome probability is the product of the event probabilities along a particular path. Note that the final outcomes can range from safe to catastrophic, depending upon the chain of events involved.

18.7 ETA WORKSHEET

The primary worksheet for an ETA is the event tree diagram (ETD), which provides the following information:

1. Initiating event
2. System pivotal events
3. Outcomes
4. Event and outcome probabilities

Figure 18.5 demonstrates the typical ETD. Each event is divided into two paths, success and failure. The success path always is the top path and the failure path is the lower path. The ETD has only one IE, which is identified at the far left of the diagram. As many contributing events as necessary to fully describe the system are listed at the top of the diagram. The more the contributing events involved, the larger the resulting ETD and the more tree branches required.

18.8 ETA EXAMPLE 1

Figure 18.6 contains an example ETA for a fire detection and suppression system in an office building. This ETA analyzes all the possible outcomes of a system fire. The IE for the ET is "fire starts." Note the range of outcomes resulting from the success or failure of the safety subsystems (pivotal events).

Note from this example that when computing the success/failure probability for each contributing PE that the PE states must always sum to 1.0, based on the reliability formula $P_{SUCCESS} + P_{FAILURE} = 1$. Also, note that in this case there are three contributing PEs that generate five possible different outcomes, each with a different probability.

18.9 ETA EXAMPLE 2

Figure 18.7 contains an example ETA for an automobile system, where the car battery has failed. The dead battery is the IE that begins the scenario analysis.

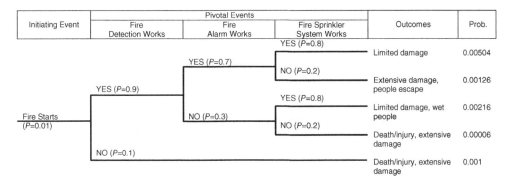

Figure 18.6 *ETA for example 1.*

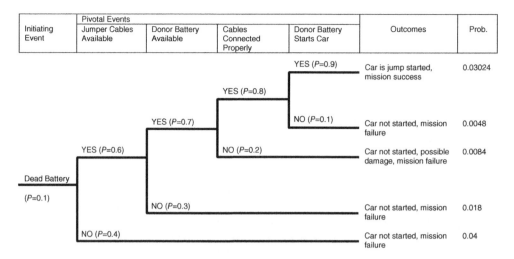

Figure 18.7   *ETA for example 2.*

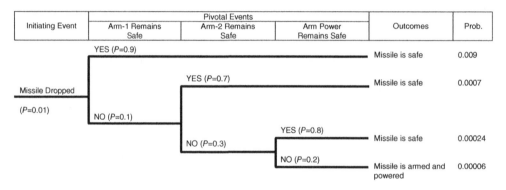

Figure 18.8   *ETA for example 3.*

18.10   ETA EXAMPLE 3

Figure 18.8 contains an example ETA for a missile system. The IE is the missile being dropped during handling or transportation.

18.11   ETA EXAMPLE 4

Figure 18.9 contains an example ETA for a nuclear power plant system. The IE is a pipe break in the cooling subsystem.

18.12   ETA ADVANTAGES AND DISADVANTAGES

There are significant advantages of the ETA technique listed as follows:

1. Structured, rigorous, and methodical approach.
2. A large portion of the work can be computerized.

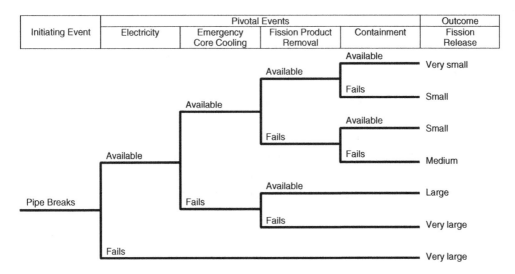

Initiating Event	Pivotal Events				Outcome
	Electricity	Emergency Core Cooling	Fission Product Removal	Containment	Fission Release

Figure 18.9 *ETA for example 4.*

3. Can be effectively performed on varying levels of design detail.
4. Visual model displaying cause/effect relationships.
5. Relatively easy to learn, do, and follow.
6. Models complex system relationships in an understandable manner.
7. Follows fault paths across system boundaries.
8. Combines hardware, software, environment, and human interaction.
9. Permits probability assessment.
10. Commercial software is available.

The following are significant disadvantages of the ETA technique:

1. An ETA can have only one initiating event; therefore, multiple ETAs will be required to evaluate the consequences of multiple initiating events.
2. ETA can overlook subtle system dependencies when modeling the events.
3. Partial successes/failures are not distinguishable.
4. Requires an analyst with some training and practical experience.

18.13 COMMON ETA MISTAKES TO AVOID

When first learning how to perform an ETA, it is commonplace to commit some typical errors. The following is a list of typical errors made during the conduct of an ETA:

1. Not identifying the proper IE
2. Not identifying all of the contributing pivotal events

18.14   SUMMARY

This chapter discussed the ETA technique. The following are the basic principles that help summarize the discussion in this chapter:

1. ETA is used to model accident scenarios and to evaluate the various outcome risk profiles resulting from an initiating event.
2. ETA is used to perform a PRA of a system.
3. The ETA diagram provides structure and rigor to the ETA process.
4. ETA can be a supplement to the SD-HAT.
5. FTs are often used to determine the causal factors and probability for failure events in the ETA.

REFERENCE

1. N. C. Rasmussen, *Reactor Safety Study: An assessment of accident risks in US Commercial Nuclear Power Plants*, WASH-1400, Nuclear Regulatory commission, 1975.

FURTHER READINGS

Andrews, J. D. and Dunnett, S. J., Event Tree Analysis Using Binary Decision Diagrams, *IEEE Trans. Reliab.*, **49** (2):230–238 (2000).

Henley, E. J. and Kumamoto, H., *Probabilistic Risk Assessment and Management for Engineers and Scientists*, 2nd edition, IEEE Press, 1996.

Kapan, S. and Garrick, B. J., On the Quantitative Definition of Risk, *Risk Analysis*, **1**:11–37 (1981).

NASA, *Fault Tree Handbook with Aerospace Applications*, NASA, version 1 2002.

Papazoglou, I. A., Functional Block Diagrams and Automated Construction of Event Trees, *Reliab. Eng. Syst. Safety*, **61** (3):185–214 (1998).

Chapter *19*

Cause–Consequence Analysis

19.1 INTRODUCTION

Cause–consequence analysis (CCA) is an analysis methodology for identifying and evaluating the sequence of events resulting from the occurrence of an initiating event. CCA utilizes a visual logic tree structure known as a cause–consequence diagram (CCD). The objective of CCA is to determine whether the initiating event will develop into a serious mishap or if the event is sufficiently controlled by the safety systems and procedures implemented in the system design. A CCA can result in many different possible outcomes from a single initiating event, and it provides the capability to obtain a probability for each outcome.

CCA is a method of risk assessment that provides a means of graphically displaying interrelationships between consequences and their causes. Safety design features that are intended to arrest accident sequences are accounted for in the CCA.

19.2 CCA BACKGROUND

This analysis technique falls under the system design hazard analysis type (SD-HAT). Refer to Chapter 4 for a discussion on hazard analysis types.

The purpose of CCA is to identify and evaluate all possible outcomes that can result from an initiating event (IE). An IE is an event that starts an accident sequence that may result in an undesirable consequence. Generally, there are many different outcomes possible from an IE, depending upon whether design safety systems work properly or malfunction when needed. CCA provides a probabilistic risk assessment (PRA) of the risk associated with each potential outcome.

Hazard Analysis Techniques for System Safety, Second Edition. Clifton A Ericson, II.
© 2016 John Wiley & Sons, Inc. Published 2016 by John Wiley & Sons, Inc.

CCA can be used to model an entire system, with analysis coverage given to subsystems, assemblies, components, software, procedures, environment, and human error. CCA can be conducted at different abstraction levels, such as conceptual design, top-level design, and detailed component design. CCA has been successfully applied to a wide range of systems, such as nuclear power plants, spacecraft, and chemical plants. The technique can be applied to a system very early in design development and thereby helps identify safety issues early in the design process. Early application helps system developers to design-in safety of a system during early development rather than having to take corrective action after a test failure or a mishap. CCA could be a supplement to the SD-HAT and the detailed design hazard type (DD-HAT).

The CCA technique, when applied to a given system by an experienced analyst, is thorough at identifying and evaluating all possible outcomes resulting from an IE, and combining them together in a visual diagram. A basic understanding of CCA and fault tree analysis (FTA) theory is essential to developing a CCA model. In addition, it is crucial for the analyst to have a detailed understanding of the system. As system complexity increases, increased knowledge and experience in CCA and FTA is also required. Overall, CCA is very easy to learn and understand. Proper application depends on the complexity of the system and the skill of the analyst.

CCA is a very powerful tool for identifying and evaluating all of the system consequence paths that are possible after an IE occurs. The CCA model will show the probability of the system design resulting in a safe operation path, a degraded operation path, and an unsafe operation path.

The use of a CCA is recommended for a PRA of the possible outcomes resulting from an initiating event. The resulting risk profiles provide management and design guidance on areas requiring additional safety countermeasures design methods. The CCD provides a means for the analyst to organize the system design into a manner showing the failure behavior of the system. CCA emphasizes the fact that an IE has many possible causes and many possible consequences, and the CCD displays these relationships.

19.3   CCA HISTORY

The CCA methodology was developed at RISO National Laboratories, Denmark, in the 1970s, specifically to aid in the reliability and risk analysis of nuclear power plants in Scandinavian countries. The method was developed to assist in the cause–consequence accident analysis of key system components. Some analysts feel that the technique is superior to an event tree analysis (ETA), which is also capable of identifying all possible consequences of a given critical event.

19.4   CCA DEFINITIONS

CCA is based on the following definitions:

Accident Scenario   An accident scenario is a series of events that ultimately result in an accident, mishap, or undesired outcome. The sequence of events begins with an initiating event and is (usually) followed by one or more intermediate events that lead to the undesired end state or outcome.

Initiating Event A failure or undesired event that initiates the start of an accident sequence. The IE may result in a mishap, depending upon the successful operation of the hazard countermeasure methods designed into the system. Refer to Chapter 3 on the hazard theory for information on the components of a hazard.

Consequence (Outcome) The outcome resulting from the occurrence of a series of incremental system successes and failures. System safety analysts are generally concerned with outcomes that result in mishaps, while reliability analysts generally are concerned with system unavailability outcomes.

Intermediate Event An intermediate event, as the name implies, takes place in the cause–consequence sequence of events. These are the failure/success events of the design safety methods established to prevent the IE from resulting in a mishap. If an intermediate event works successfully, it stops the accident scenario and is referred to as a mitigating event. If an intermediate event fails to work, the accident scenario is then allowed to progress and is referred to as an aggravating outcome event. The intermediate event is similar to the pivotal event in ETA (refer to Chapter 18).

Probabilistic Risk Assessment PRA is a compre*hensive, structured,* and logical analysis method for identifying and evaluating risk in a complex technological system. The detailed identification and assessment of accident scenarios, with a quantitative analysis is the PRA goal.

19.5   CCA THEORY

When performing a CCA, identifying and developing accident scenarios is fundamental to the concept. CCA is very similar to ETA (see Chapter 18). The theory is to first identify all of the IEs that have significant safety impact or concern and then perform a CCA on each IE. Hazard analysis is a methodology for identifying IEs of concern. A CCD is constructed to model the sequence of events that can result from an IE, taking into account the success or failure of design safety features intended to prevent undesirable consequences. The CCD combines the causes and consequences of each IE, resulting in a model with many different possible outcomes from a single IE.

An accident scenario contains an IE and (usually) one or more intermediate events leading to an end state, or outcome. A scenario contains an IE and (usually) one or more pivotal events leading to an end state as shown in Figure 19.1.

Figure 19.2 shows the CCA concept. Through the use of FT logic and CCD logic, probability values can be obtained from the CCA. Note that FTA is used to determine the event causes of failure and their failure probability. To determine the probability of an outcome, the event probabilities in the outcome path are multiplied together.

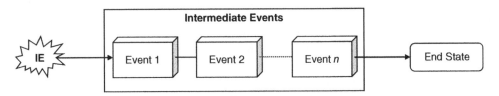

Figure 19.1 *Accident scenario concept.*

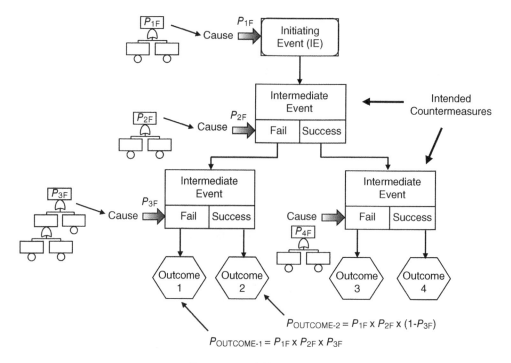

Figure 19.2 CCA overview.

19.6 CCA METHODOLOGY

Table 19.1 describes the basic steps of the CCA process. This process involves identifying and evaluating the sequence events that are possible after the occurrence of a given IE.

Complex systems tend to have a large number of interdependent components, redundancy, standby systems, and safety systems. Sometimes, it is too difficult or cumbersome to model a system with just an FT, so PRA studies have combined the use of FTs and CCDs. The CCD models accident/mishap cause–consequence scenarios, and FTs model complex subsystems to obtain the probability of these subsystems failing. An accident scenario can have many different outcomes, depending on which components fail and which function correctly. The CCD/FT combination models this complexity very well.

The goal of CCA is to determine the probability of all the possible outcomes resulting from the occurrence of an IE. By analyzing all possible outcomes, it is possible to determine the percentage of outcomes that lead to the desired result, and the percentage of outcomes that lead to the undesired result.

The CCD is a diagram modeling all of the possible events that follow an IE. The IE can be a design failure or an operational human error. The objective is to identify the chain of events following one or more specified intermediate events, in order to evaluate the consequences and determine whether the event will develop into a mishap or not are sufficiently controlled by the safety systems implemented. The results can, therefore, be recommendations to increase the redundancy or to modifications to the safety systems.

The CCA begins with the identified IE listed at the top of the diagram. All safety design methods or countermeasures are then listed sequentially below, in the form of decision boxes. The decision box provides two possible paths: (a) operates successfully and (b) fails to operate.

TABLE 19.1   CCA Process

Step	Task	Description
1	Define the system	Examine the system and define the system boundaries, subsystems, and interfaces
2	Identify the accident scenarios	Perform a system assessment or hazard analysis to identify the system hazards and accident scenarios existing within the system design
3	Identify the initiating events	Refine the hazard analysis to identify the significant IEs in the accident scenarios. IEs include events such as fire, collision, explosion, pipe break, toxic release, and so on
4	Identify the intermediate events	Identify the safety barriers or countermeasures involved with the particular scenario that are intended to preclude a mishap. These become the intermediate events
5	Build the CCA diagram	Construct the logical CCD, starting with the IE, then the intermediate events, and completing with the outcomes of each path
6	Obtain the failure event probabilities	Obtain or compute the failure probabilities for the intermediate failure events on the CCD. It may be necessary to use FTs to determine how an event can fail and to obtain the failure probability
7	Identify the outcome risk	Compute the risk for each outcome in the CCD. To determine the probability of an outcome, the event probabilities in the outcome path are multiplied together
8	Evaluate the outcome risk	Evaluate the outcome risk of each path and determine if the risk is acceptable
9	Recommend corrective action	If the outcome risk of a path is not acceptable, develop design strategies to change the risk
10	Hazard tracking	Enter identified hazards, or supporting data, into the hazard tracking system (HTS)
11	Document CCA	Document the entire CCA process on the CCDs. Update for new information as necessary

The resulting diagram combines all of the various success/failure event combinations and fans out in a downward tree structure. Each success/failure event can be assigned a probability of occurrence, and the final outcome probability is the product of the event probabilities along a particular path. Note that the final outcome severity can range from safe to catastrophic, depending upon the chain of events.

The CCA identifies all of the causes of an undesired event and all of the possible outcomes resulting from this event. The CCA documents the failure logic involved in mishaps resulting from an IE. CCA also provides a mechanism for time delays and sequencing between events.

A failure dependency arises when the same fault event exists in more than one FT structure on the same path in the CCD. Repeated failures often influence more than one decision box in the CCA. As in FTA, repeated failure events must be properly accounted for in the mathematical evaluation; otherwise, the final probability calculation will be incorrect. In order to resolve this problem, the failure event is extracted from the various FTs and placed on the CCD (see example 2 below).

An IE is an event that starts a certain operational sequence or an event that activates safety systems. The IE must be carefully selected for the CCA. Potential IEs are identified from hazard analyses and known system problems.

19.7 CCD SYMBOLS

Figure 19.3 shows the CCD symbols along with their definitions.

Symbol	Name	Purpose
Initiating Event	Initiating Event Box	An independent event that can initiate a sequence of events leading to an accident or mishap.
Function / Yes / No	Intermediate Event (Decision Box)	An event that represents the functionality of a component or subsystem; generally, a safety feature. The function is either successful or it fails.
⟨Outcome⟩	Consequence Box	Represents the outcome of a series of events.
FT-n ⇨	Fault Tree Pointer	Identifies the fault trees of the IE and intermediate events. These FTs model the cause of event failure and provide a probability calculation.
T = xx	Time Delay Box	Identifies a time delay that must take place.
 (OR gate symbol)	OR Gate	Combines IE and/or Decision Box logic when necessary. At least one input is required to produce an output.
 (AND gate symbol)	AND Gate	Combines IE and/or Decision Box logic when necessary. All inputs are required to produce an output.

Figure 19.3   *CCA symbols.*

19.8   CCA WORKSHEET

The primary worksheet for a CCA is the CCD, which provides the following information:

1. Initiating event
2. Intermediate events (i.e., pivotal events)
3. Outcomes
4. Event and outcome probabilities
5. Timing

Figure 19.2 shows a generic CCD structure. Each intermediate event is divided into two paths, success and failure. The CCA has only one IE, which is identified, at the top of the diagram, and as many intermediate events as necessary to fully describe the system as traced through by the CCD. The more intermediate events involved the larger the resulting CCA and the more branches required.

19.9   CCA EXAMPLE 1: THREE-COMPONENT PARALLEL SYSTEM

Figure 19.4 contains a CCA for an example system comprised of three components in parallel. Successful system operation requires successful operation of any one, or more, of the three components. The IE for this example is "power applied to the system."

System   CCD

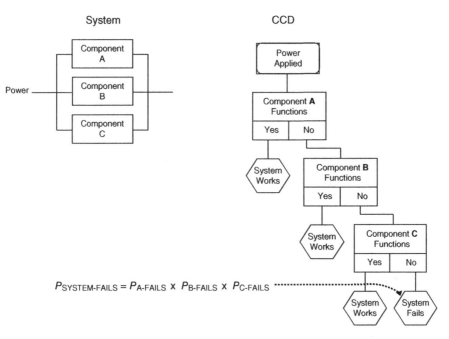

$P_{\text{SYSTEM-FAILS}} = P_{\text{A-FAILS}} \times P_{\text{B-FAILS}} \times P_{\text{C-FAILS}}$

Figure 19.4 *Example 1: three-component parallel system and CCD.*

19.10 CCA EXAMPLE 2: GAS PIPELINE SYSTEM

Example 2 involves an example gas pipeline system as shown in Figure 19.5. It has been determined from hazard analyses that an undesired IE is a high-pressure surge in the pipeline. The CCD shows the various outcomes that are possible from this IE.

Table 19.2 lists the components in this example system and describes their function. This portion of the system is intended to prevent a pipeline rupture if a gas pressure surge occurs in the line. Sensor S1 detects high pressure and sends a warning signal to the master computer C1. When C1 receives the pressure warning, it sends out signals to master valves MV1, MV2, and MV3 to immediately close. Should this primary pressure detection and control system fail, there is a backup pressure detection and control system that functions in a similar manner.

Figure 19.6 contains the CCD for this example gas pipeline system.

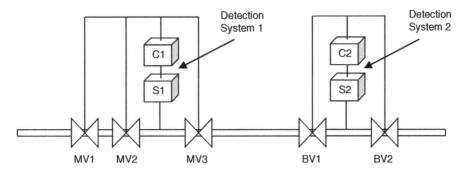

Figure 19.5 *Example 2: gas pipeline system diagram.*

TABLE 19.2 System Components for Gas Pipeline

Label	Name	Function
S1	Sensor 1	Senses high pressure and sends warning to C1
C1	Computer 1	Sends command to the three master valves to close
MV1	Master valve 1	When open allows gas flow; when closed stops gas flow
MV2	Master valve 2	When open allows gas flow; when closed stops gas flow
MV3	Master valve 3	When open allows gas flow; when closed stops gas flow
S2	Sensor 2	Senses high pressure and sends warning to C2
C2	Computer 2	Sends command to the two backup valves to close
BV1	Backup valve 1	When open allows gas flow; when closed stops gas flow
BV2	Backup valve 2	When open allows gas flow; when closed stops gas flow

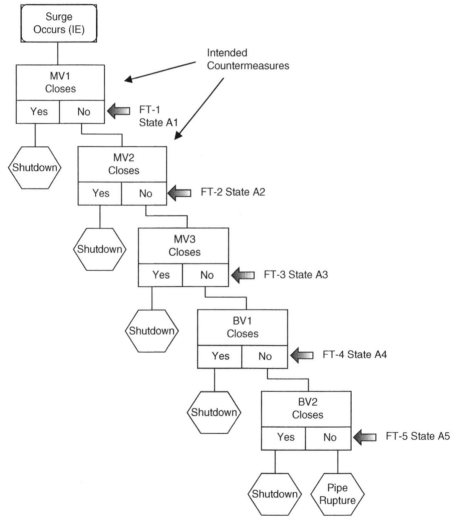

Figure 19.6   Example 2: CCD (version 1).

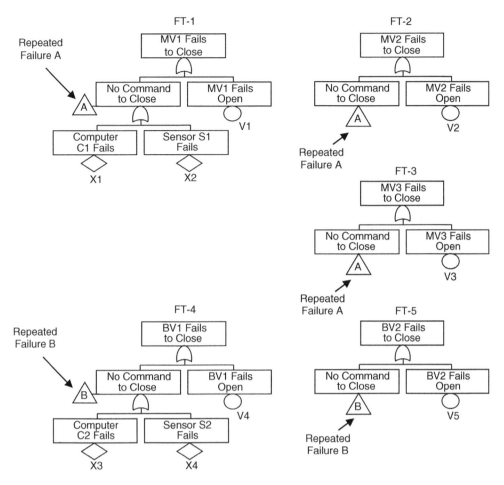

Figure 19.7 *Example 2: CCD fault trees (version 1).*

Figure 19.7 contains the FTs that support the CCD shown in Figure 19.6.

19.10.1 Reducing Repeated Events

The CCA version 1 fault trees shown in Figure 19.7 reveal that there are some repeated failures in the CCD events. FT transfer symbol A indicates that this FT branch in FT-1 also occurs in FT-2 and FT-3 and transfer symbol B in FT-4 also occurs in FT-5. This means that if the CCD events were to be multiplied together, there would be an error in the calculation because of the repeated events being in each of the event calculations.

In order to derive a correct calculation, the repeated events must be mathematically reduced. Chapter 15 on FTA describes how to do this for FTs. In CCA, the reduction can be achieved through Boolean reduction or by revising the CCD accordingly, as shown in Figures 19.8 and 19.9.

Figure 19.8 breaks the original CCD into two separate branches via the OR gate at the top of the diagram. This revised structure eliminates the repeated events in all paths.

Figure 19.9 contains the FTs that support the version 2 CCD shown in Figure 19.8.

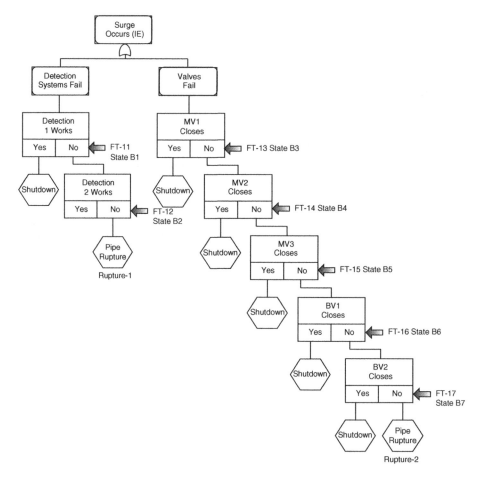

Figure 19.8 *Example 2: CCD revised (version 2).*

The mathematical equations for computing the probability of pipe rupture for the two CCD versions are as follows:

Version 1

$$P_{\text{RUPTURE}} = P_{\text{IE}} \times P_{\text{A1}} \times P_{\text{A2}} \times P_{\text{A3}} \times P_{\text{A4}} \times P_{\text{A5}}$$
$$= (P_{\text{IE}})(P_{\text{X1}} + P_{\text{X2}} + P_{\text{V1}})(P_{\text{X1}} + P_{\text{X2}} + P_{\text{V2}})$$
$$(P_{\text{X1}} + P_{\text{X2}} + P_{\text{V3}})(P_{\text{X3}} + P_{\text{X4}} + P_{\text{V4}})(P_{\text{X3}} + P_{\text{X4}} + P_{\text{V5}})$$

Version 2

$$P_{\text{RUPTURE}} = P_{\text{RUPTURE-1}} + P_{\text{RUPTURE-2}}$$
$$= (P_{\text{IE}} \times P_{\text{B1}} \times P_{\text{B2}}) + (P_{\text{IE}} \times P_{\text{B3}} \times P_{\text{B4}} \times P_{\text{B5}} \times P_{\text{B6}} \times P_{\text{B7}})$$
$$= (P_{\text{IE}})[(P_{\text{B1}} \times P_{\text{B2}}) + (P_{\text{B3}} \times P_{\text{B4}} \times P_{\text{B5}} \times P_{\text{B6}} \times P_{\text{B7}})]$$
$$= (P_{\text{IE}})[(P_{\text{X1}} + P_{\text{X2}})(P_{\text{X3}} + P_{\text{X4}}) + (P_{\text{V1}} \times P_{\text{V2}} \times P_{\text{V3}} \times P_{\text{V4}} \times P_{\text{V5}})]$$

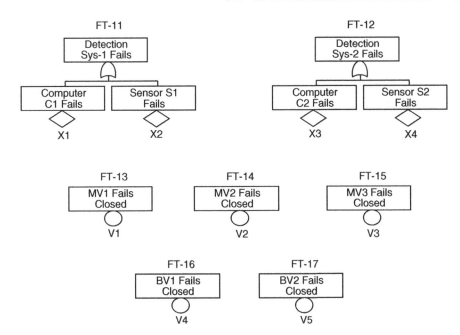

Figure 19.9 *Example 2: CCD revised fault trees (version 2).*

Note that version 1 contains the repeated events X1, X2, X3, and X4 that will have to be reduced via Boolean algebra in order to obtain a correct solution. However, in version 2 the repeated events have been reduced via the modified CCD structure.

19.11 CCA ADVANTAGES AND DISADVANTAGES

The CCA technique has the following advantages:

1. Structured, rigorous, and methodical approach.
2. A large portion of the work can be computerized.
3. Visual model displaying cause/effect relationships.
4. Relatively easy to learn, do, and follow.
5. Models complex system relationships in an understandable manner.
6. Combines hardware, software, environment, and human interaction.
7. Permits probability assessment.
8. Multiple outcomes are analyzed.
9. Time sequences of events are treated.

The following are disadvantages of the CCA technique:

1. A CCA can have only one initiating event; therefore, multiple CCA will be required to evaluate the consequence of multiple initiating events.
2. Requires an analyst with some training and practical experience.

19.12 COMMON CCA MISTAKES TO AVOID

When first learning how to perform a CCA, it is commonplace to commit some typical errors. The following are some typical errors made during the conduct of a CCA:

1. Not identifying the proper IE.
2. Not identifying all of the contributing intermediate or pivotal events.
3. Not developing the correct CCA model of the system.

19.13 SUMMARY

This chapter discussed the CCA technique. The following basic principles help summarize the discussion in this chapter:

1. CCA is used to model accident scenarios and to evaluate the various outcome risk profiles resulting from an initiating event.
2. A quantitative CCA is used to perform a PRA of a system. A qualitative CCA is used to help recognize design weaknesses.
3. The CCD provides structure and rigor to the CCA process.
4. A CCA would make a useful supplement to the SD-HAT and the DD-HAT analyses.

FURTHER READINGS

A Cause–Consequence IEEE Society of U.S. Chart of a Redundant Protection System, *IEEE Trans. Reliab.*, **24** (1): (1975).

Andrews, J. D. and Ridley, L. M., Application of the Cause–Consequence Diagram Method to Static Systems, *Reliab. Eng. Syst. Safety*, **75**:47–58 (2002).

Andrews, J. D. and Ridley, L. M., Reliability of Sequential Systems Using the Cause–Consequence Diagram Method, *Proceedings of the Institution of Mechanical Engineers*, **215**: Part E, 207–220 (2001).

Danish Atomic Energy Commission, *The Cause–Consequence Diagram Method as a Basis for Quantitative Accident Analysis*, Danish Atomic Energy Commission RISO-M-1374, 1971.

Danish Atomic Energy Commission, *Interlock Design Using Fault Tree Analysis and Cause–Consequence Analysis*, Danish Atomic Energy Commission, RISO-M-1890, 1977.

Kaufman, L. M., Fussell, J. B., Wagner, D. P., Arendt, J. S., Rooney, J. J., Crowley, W. K., and Campbell, D. J., Improving System Safety through Risk Assessment, Proceedings 1979 Annual Reliability and Maintainability Symposium, 160–164 1979.

Chapter 20

Common Cause Failure Analysis

20.1 INTRODUCTION

Common cause failure analysis (CCFA) is an analysis methodology for identifying common causes of multiple failure events. A common cause failure (CCF) is a single-point failure (SPF) that destroys independent redundant designs. The objective of CCFA is to discover common cause vulnerabilities in the system design that can result in the common failure of redundant subsystems and to develop design strategies to mitigate these types of hazards.

An example of a CCF would be the compelled failure of two independent, and redundant, flight control computers due to the failure of a common circuit breaker in the system design providing electrical power.

CCFs create the subtlest type of hazards because they are not always obvious, making them difficult to identify. The potential for this type of event exists in any system architecture that relies on redundancy or uses identical components or software in multiple subsystems. CCF vulnerability results from system failure dependencies inadvertently designed into the system.

If a CCF is overlooked, total system risk is understated because the probability of this hazard is not included in the total risk calculation. If the common cause dependency is in a critical subsystem, CCFs could contribute to a significant impact in overall system risk.

CCFs can be caused from a variety of sources, such as the following:

a. Common weakness in design redundancy
b. The use of identical components in multiple subsystems
c. Common software design
d. Common manufacturing errors

Hazard Analysis Techniques for System Safety, Second Edition. Clifton A Ericson, II.
© 2016 John Wiley & Sons, Inc. Published 2016 by John Wiley & Sons, Inc.

    e. Common requirements errors

    f. Common production process errors

    g. Common maintenance errors

    h. Common installation errors

    i. Common environmental factor vulnerabilities

20.2 CCFA BACKGROUND

This analysis technique falls under the system design hazard analysis type (SD-HAT). Refer to Chapter 4 for a discussion on hazard analysis types.

The purpose of CCFA is to identify CCF vulnerabilities in the system design that eliminate or bypass design redundancy, where such redundancy is necessary for safe and reliable operation. Once CCFs are identified and evaluated for risk, defense strategies mitigating critical CCFs can be established and implemented. CCFA also provides a methodology for determining the quantitative risk presented by CCFs. An alternative name for this analysis technique is common mode failure (CMF) analysis.

CCFA can be applied to any type of system, but it is particularly useful for safety-critical systems using design redundancy. The CCFA technique, when applied to a given system by an experienced analyst, is thorough at identifying and evaluating all of the possible CCFs in a system.

A basic understanding of CCFA and FTA theory is essential to developing a CCFA model. In addition, it is crucial for the analyst to have a detailed understanding of the system. As system complexity increases, increased knowledge and experience in CCFA and FTA is required. Proper application depends on the complexity of the system and the skill of the analyst.

Applying CCFA to the analysis of a system design is not a trivial process. It is more difficult than an analysis technique such as a PHA, primarily because it requires an understanding of FTA along with extensive data collection and analysis of CCFA components.

CCFA is a very powerful tool for identifying and evaluating potential CCFs in a system design, and it is the only tool to date that provides any rigor to the identification of CCF events. If CCFs are not included in system risk analysis, total system risk is understated because the probability of this type hazard is not included in the total risk calculation. If common cause dependencies exist in a critical subsystem, CCFs could contribute to a significant impact on overall system risk.

The use of CCFA is recommended for system safety programs (SSPs) to support the goal of identifying and mitigating all CCF modes. CCFA is recommended as part of a probabilistic risk assessment (PRA), particularly in order to obtain a truer view of system risk. CCFA is especially applicable to the evaluation of redundant designs in safety-critical applications. CCFA is an analysis technique specified in the SAE ARP-4754 that is commonly imposed by the FAA.

20.3 CCFA HISTORY

Since the inception of system safety, there has always been a concern with regard to CCFs and how to identify them. Many analysts attempted to identify CCFs with hit-or-miss brute force analyses without utilizing any sort of coherent methodology. It was probably not until 1988 when reference [1] was published and 1998 when reference [2] was published by the US Nuclear Regulatory Commission that CCFA became more of a formalized analysis technique with a coherent and comprehensive framework.

20.4 CCFA DEFINITIONS

In order to understand CCFA, it is necessary to define some common terms, which will help to provide a better grasp of the complications involved in CCF theory. The relevant terms are explained in this section.

20.4.1 Independent Event

Events are independent when the outcome of one event does not influence the outcome of the second event (probability theory). To find the probability of two independent events both occurring, multiply the probability of the first event by the probability of the second event; for example, $P(A \text{ and } B) = P(A) \cdot P(B)$. For example, find the probability of tossing two number cubes (dice) and getting a 3 on each one. These events are independent: $P(3) \cdot P(3) = (1/6) \cdot (1/6) = 1/36$. The probability is $1/36$.

20.4.2 Dependent Event

Events are dependent when the outcome of one event directly affects or influences the outcome of the second event (probability theory). To find the probability of two dependent events both occurring, multiply the probability of A and the probability of B after A occurs: $P(A \text{ and } B) = P(A) \cdot P(B \text{ given } A)$ or $P(A \text{ and } B) = P(A) \cdot P(B|A)$. This is known as conditional probability. For example, a box contains a nickel, a penny, and a dime. Find the probability of choosing first a dime and then, without replacing the dime, choosing a penny. These events are dependent. The first probability of choosing a dime is $P(A) = 1/3$. The probability of choosing a penny is $P(B|A) = 1/2$ since there are now only two coins left. The probability of both is $1/3 \cdot 1/2 = 1/6$. Keywords such as "not put back" and "not replace" suggest that events are dependent.

20.4.3 Independence (in Design)

Independence is a design concept that ensures the failure of one item does not cause the failure of another item [3]. This concept is very important in many safety and reliability analysis techniques due to the impact on logic and mathematics. Many models, such as FTA, assume event independence.

20.4.4 Dependence (in Design)

Dependence involves a design whereby the failure of one item directly causes, or leads to, the failure of another item. This refers to when the functional status of one component is affected by the functional status of another component. CCF dependencies normally stem from the way the system is designed to perform its intended function. Dependent failures are those failures that defeat redundancy or diversity, which are intentionally employed to improve reliability and/or safety.

In some system designs, dependency relationships can be very subtle, such as in the following cases [4]:

1. *Standby Redundancy* When an operating component fails, a standby component is put into operation, and the system continues to function. Failure of an operating component causes a standby component to be more susceptible to failure because it is now under load.

2. *Common Loads* When failure of one component increases the load carried by other components. Since the other components are now more likely to fail, we cannot assume statistical independence.

3. *Mutually Exclusive Events* When the occurrence of one event precludes the occurrence of another event.

Two failure events A and B are said to be dependent if $P(A \text{ and } B) \neq P(A)P(B)$. In the presence of dependencies, often, but not always, $P(A \text{ and } B) > P(A)P(B)$. This increased probability of two (or more) events is why CCFs are of concern.

20.4.5 Common Cause Failure

A CCF is the failure (or unavailable state) of more than one component due to a shared cause during the system operation. Viewed in this fashion, CCFs are inseparable from the class of dependent failures [3]. It is an event or failure that bypasses or invalidates redundancy or independence (ARP-4761).

A CCF is the simultaneous failure of multiple components due to a common or shared cause, for example, when two electrical motors become inoperable simultaneously due to a common circuit breaker failure that provides power to both motors. CCFs include CMFs but CCF is much larger in scope and coverage. Components that fail due to a shared cause normally fail in the same functional mode. CCFs deal with causes other than just design dependencies, such as environmental factors, human error, and so on. Ignoring the effects of dependency and CCFs can result in overestimation of the level of reliability and/or safety.

For system safety, a CCF event consists of item/component failures that meet the following criteria:

1. Two or more individual components fail or are degraded such that they cannot be used when needed, or used safely if still operational.

2. The component failures result from a single *shared cause* and *coupling mechanism.*

20.4.6 Common Mode Failure

A common mode failure (CMF) is the failure of multiple components in the same mode [3]. This is an event, which simultaneously affects a number of elements otherwise, considered to be independent [2]. For example, a set of identical resistors from the same manufacturer may all fail in the same mode (and exposure time) due to a common manufacturing flaw.

The term CMF, which was used in the early literature and is still used by some practitioners, is more indicative of the most common symptom of the CCF, but it is not a precise term for describing all of the different dependency situations that can result in a CCF event. A CMF is a special case of a CCF, or a subset of a CCF.

20.4.7 Cascading Failure

A cascading failure is a failure event for which the probability of occurrence is substantially increased by the existence of a previous failure [5]. Cascading failures are dependent events, where the failure of one component causes the failure of the next component in line, similar to the falling domino effect.

20.4.8 Mutually Exclusive Events

Two events are mutually exclusive if the occurrence of one event precludes the occurrence of the other. For example, if the event "switch A fails closed" occurs, then the event "switch A fails open" cannot possibly occur.

20.4.9 CCF Root Cause

The root cause is the most basic reason for the component failure, which if corrected, would prevent recurrence. Example CCF root causes include events such as heat, vibration, moisture, and so on. The identification of a root cause enables the analyst to implement design defenses against CCFs.

20.4.10 CCF Coupling Factor

A coupling factor is a qualitative characteristic of a group of components or piece parts that identifies them as susceptible to the same causal mechanisms of failure. Such factors include similarity in design, location, environment, mission and operational, maintenance, and test procedures. The coupling factor(s) is part of the root cause for a CCF. The identification of coupling factors enables the analyst to implement defenses against common root cause failure vulnerabilities.

20.4.11 Common Cause Component Group

A group of components or piece parts with a common susceptibility to CCF is called a common cause component group (CCCG) and the following guidelines help identify CCCGs:

1. When identical, functionally nondiverse, and active components are used to provide redundancy, these components should always be assigned to a CCCG, one for each group of identical redundant components (e.g., resistor groups, pump groups, thermostat groups, etc.).
2. The initial assumption of independent failure modes among diverse components is a good one if it is supported by operating experience. However, when diverse redundant components have piece parts that are identically redundant, the components should not be assumed fully independent. One approach in this case is to break down the component boundaries and identify the common piece parts as a CCCG (e.g., pumps can be identical except for their power supply).
3. In system reliability analysis, it is frequently assumed that certain passive components can be omitted, based on the argument that active components dominate. In applying these screening criteria to common cause analysis, it is important to include events such as debris blockage of redundant or even diverse pump strainers.

20.5   CCFA THEORY

Many systems utilize subsystem design redundancy to ensure that a specific function occurs upon demand. The idea is that two separate and independent subsystems are much less likely to fail from independent failures than a single independent subsystem. System designs have become so complex, however, that occasionally a dependency is inadvertently built into the redundancy design. One form of dependency is the CCF event that can cause failure of both redundant subsystems. A CCF is effectively an SPF that nullifies independent redundant subsystem designs.

For example, a DC-10 crash occurred when an engine exploded and a fan blade from the engine cut two independent and separated hydraulic lines. Aircraft control depended upon system hydraulics and the design, therefore, intentionally had two independent and redundant hydraulic systems. Though the redundant hydraulic lines were physically separated by a large distance, the exploding of the engine was the common cause SPF resulting in the loss of both critical hydraulic subsystems.

Figure 20.1 demonstrates the CCF concept where the root cause is an SPF and the coupling factor is the design vulnerability to an SPF. In this simplified example, two computers are used in parallel to ensure that safety-critical output is provided when necessary. Only one computer is necessary for system success, but should one computer fail (independently), the second computer takes over the operation.

Note that in this example system, a common electrical power source is utilized for both computers. The electrical power is a CCF source. Both computers are dependent on the same electrical power source. If power fails, then both computers will instantly fail with a probability of $P = 1.0$. The conditional probability of the dependent event, *Computer Fails Given Failure of Power*, is $P = 1.0$.

Figure 20.2 demonstrates a slightly different CCF concept. In this example, the two computers are supplied power by different and independent sources, to eliminate the power dependency. However, there is the possibility of both computers being exposed to a strong RF energy field, which can cause computer upset or failure. Since both computers are identical and manufactured under the same specifications, they are both susceptible to failing in the same mode from the common energy source.

In this example, the root cause is the presence of RF energy and the coupling factor is the design vulnerability of the safety-critical components to RF energy.

Figure 20.3 demonstrates another CCF concept. In this example, two redundant subsystems perform the same system function. One component in each subsystem uses a different (diverse) operational method (e.g., a mechanical fuze and an electrical fuze), to ensure that they do not fail simultaneously in the same mode due to a CCF. However, even with diversity in design,

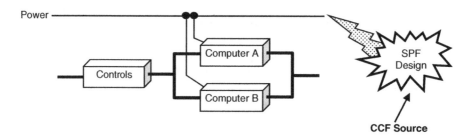

Figure 20.1   *Example redundant system.*

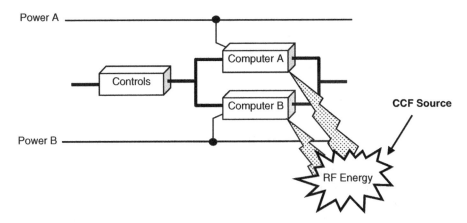

Figure 20.2 *Improved redundant system.*

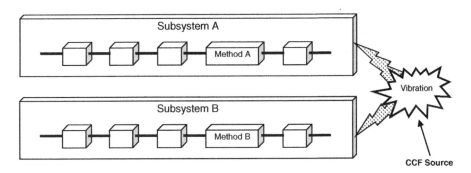

Figure 20.3 *Diverse redundant system.*

both items are vulnerable to an external CCF source, such as vibration, which can cause both items to fail.

In this example, the root cause is the presence of external vibration and the coupling factor is the design vulnerability of the safety-critical components to external vibration.

Figure 20.4 demonstrates the cascading CCF concept. In this example, several items are connected in series, with an interdependency between items. If item A should partially fail, it places a heavier load on item B, possibly exceeding item B's design load limits. As a result, item B either fails or passes a heavier load onto item C, and so on. For example, a steel beam plate may have seven rivets holding two beams together. If two rivets fail, the load weakens the plate–beam connection. Or, several electrical components may be connected in series. If one

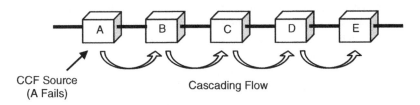

Figure 20.4 *Cascading CCF.*

component partially fails, it may pass a higher than designed-for current to the next component, resulting in a cascading effect on the circuit.

In this example, the root cause is the failure of component A and the coupling factor is the design vulnerability of the safety-critical components to a stress level higher than the design limit.

As demonstrated in the above figures, the definition of CCF is closely related to the general definition of a dependent failure. Two events, A and B, are said to be dependent if

$$P(A \text{ and } B) \neq P(A)P(B)$$

In the presence of dependencies, often, but not always, $P(AB) > P(A)P(B)$. Therefore, if A and B represent failure of a safety function, the actual probability of failure of both may be higher than the expected probability calculated based on the assumption of independence. In cases where a system provides multiple layers of defense against total system or functional failure, the presence of dependence may translate into a reduced level of safety and reliability if the dependence is ignored.

Dependencies can be classified in many different ways. Dependencies are categorized based on whether they stem from intended intrinsic functional physical characteristics of the system or are due to external factors and unintended characteristics. Therefore, CCF dependence is either intrinsic or extrinsic to the system.

Intrinsic dependency refers to dependencies where the functional status of one component is affected by the functional status of another. These dependencies normally stem from the way the system is designed to perform its intended function. This type of dependency is based on the type of influence that components have on each other.

Extrinsic dependency refers to dependencies where the couplings are not inherent and not intended in the designed functional characteristics of the system. Such dependencies are often physically external to the system, such as vibration, heat, RF, environment, mission changes beyond original design limits, and so on.

20.6 CCFA METHODOLOGY

CCFs result from the presence of two key factors in a system design:

1. A root cause of component failure (i.e., the particular reason(s) for failure of each component that failed in the CCF event).
2. A coupling factor (or factors) that create the conditions for multiple components to be involved in the CCF event and be affected by the same root cause.

For example, using two switches in a system that is both susceptible to failure from vibration creates a common cause failure situation. Failure from vibration is the root cause. The vibration and both components designed for exposure to the same vibration are the coupling factors. However, if a switch can be used that is not susceptible to vibration or the vibration is eliminated, the CCF situation is avoided.

As another example, consider the failure of two identical redundant electronic devices due to exposure to excessively high temperatures. This CCF event is the result of susceptibility of each of the devices to heat (considered to be the root cause in this case), and, a result of both units being identical and also being exposed to the same harsh environment (coupling factors).

TABLE 20.1 CCFA Process Methodology

Step	Task	Description
1	Define the system	Examine the system and define the system boundaries, subsystems, and interfaces. Identify analysis boundaries
2	Develop initial system logic model	Development of an initial component-level system logic model (e.g., fault tree) that identifies major contributing components
3	Screening analysis	Screen system design and data for identification of CCF vulnerabilities and CCF events
4	Detailed CCF analysis	Place CCF components in FT and perform a qualitative and quantitative analysis to assess CCF risk
5	Evaluate the outcome risk	Evaluate the outcome risk of each CCF event and determine if the risk is acceptable
6	Recommend corrective action	If the outcome risk is not acceptable, develop design strategies to countermeasure CCF effect and change system risk
7	Track hazards	Transfer identified hazards into the hazard tracking system (HTS)
8	Document CCFA	Document the entire CCFA process, including the system level FTs. Update for new information as necessary

Since the use of identical components in redundancy formation is a common strategy to improve system safety and reliability, coupling factors stemming from similarities of the redundant components are often present in system designs, leading to vulnerability to CCF events. CCF events of identical redundant components, therefore, merit special attention in risk and reliability analysis of such systems.

The characterization of CCF events in terms of susceptibilities and coupling factors provides an effective means of assessing the CCF phenomenon and evaluating the need for and effectiveness of defenses against them.

There are several different CCFA models that can be applied to the evaluation of CCFs; however, the fault tree analysis (FTA) model seems to be the best and most used methodology. The FTA model methodology is presented herein.

Table 20.1 lists and describes the basic steps of the CCFA process using the FTA approach.

Step 1 is input to the process and steps 5, 6, 7, and 8 are outputs of the CCFA process.

Steps 2, 3, and 4 comprise the analysis portion of the CCFA process. These steps are described next.

20.6.1 CCFA Process Step 2: Initial System Fault Tree Model

The development of a system-level FTA is a key step in CCFA process. The initial system fault tree (FT) logic model is developed as a basic system model that identifies the primary contributing components to fault events leading to the undesired top-level event. The initial FT is developed around basic independent failure events, which provides a first approximation of cut sets and probability.

Many component failure dependencies among the components are not accounted for explicitly in the first approximation FT model, resulting in an underestimation of the risk of the FTs top-level event. As CCF events are identified in step 3, the analyst expands the FT model in step 4 to include identified CCFs.

The FT model is much more complete and accurate with CCF events included. The FT is recomputed to reevaluate the criticality, sensitivity, and probability of the CCF within the FT.

The revised FT probability risk estimate that includes CCFs provides a correct risk estimate over the first approximation FT.

Refer to Chapter 15 for a full discussion on FTA and guidance on developing an FT model and performing quantitative calculations.

20.6.2 CCFA Process Step 3: Common Cause Screening

The purpose of screening is to identify CCF vulnerabilities in the system design, and to identify the specific CCF events and components that are required in the FT model. An analysis is performed to identify a list of potential susceptibilities of the system and the CCF components involved. During the screening analysis, it is important not to discount any potential CCF susceptibilities. An effective CCF screening analysis should involve the following activities:

- Review of system design and operating practices
- Review of operating historical experience (if available)
- Review of other similar systems
- Evaluation of root cause-defense and coupling factor-defense methods

The most efficient approach to identifying CCF system susceptibilities is to focus on identifying coupling factors, regardless of defenses that might be in place. The resulting list will be a conservative assessment of the system susceptibilities to CCFs. A coupling mechanism is what distinguishes CCFs from multiple independent failures. Coupling mechanisms are suspected to exist when two or more component failures exhibit similar characteristics, both in the cause and in the actual failure mechanism. The analyst, therefore, should focus on identifying those components of the system that share common characteristics.

When identifying CCF coupling factors, remember that a

1. CCF root cause is the most basic reason or reasons for the component failure, which if corrected, would prevent recurrence.
2. CCF coupling factor is a characteristic of a group of components or piece parts that identifies them as susceptible to the same causal mechanisms of failure. Such factors include similarity in design, location, environment, mission and operational, maintenance, and test procedures.

A list of common factors is provided in Table 20.2. This list is a tool to help identify the presence of identical components in the system and most commonly observed coupling factors. Any group of components that share similarities in one or more of these characteristics is a potential point of vulnerability to CCF.

Coupling factors can be divided into four major classes:

- Hardware based
- Operation based
- Environment based
- Software based

Hardware-based coupling factors are factors that propagate a failure mechanism among several components due to identical physical characteristics. An example of hardware-based

TABLE 20.2 Key Common Cause Attributes

Characteristic	Description
Same design	The use of the same design in multiple subsystems can be the source of a CCF coupling factor vulnerabilities. This is particularly true of software design
Same hardware	The use of identical components in multiple subsystems resulting in a vulnerability of multiple subsystems
Same function	When the same function is used in multiple places, it may require identical or similar hardware that provides CCF vulnerabilities
Same staff	Items are vulnerable to the same installation, maintenance, test or, operations staff that can make common errors
Same procedures	Items are vulnerable to the same installation, maintenance, test, or operations procedures, which may have common errors
Redundancy	When redundant items are identical, they are vulnerable to the same failure modes, failure rates, and CCF coupling factors
Same location	Items are located in the same physical location, making them vulnerable to the same undesired conditions (fire, water, shock, etc.)
Same environment	Items are vulnerable to the same undesired environmental conditions (fire, water, shock, electromagnetic radiation, dust, salt, etc.)
Same manufacturer	Components have the same manufacturer, making all components vulnerable to the same failure modes and failure rates
Common requirements	Common requirements for items or functions may contain common errors that generate CCF vulnerabilities
Common energy sources	Items with common energy sources (e.g., electrical, chemical, hydraulic, etc.) generate CCF vulnerabilities
Common data sources	Items with common data sources generate CCF vulnerabilities, particularly in software design
Common boundaries	Items that share a common boundary (physical, functional, logical, etc.) may have CCF vulnerabilities

coupling factors is failure of several residual heat removal (RHR) pumps because of the failure of identical pump air deflectors. There are two subcategories of hardware-based coupling factors: (1) hardware design and (2) hardware quality (manufacturing and installation).

Hardware design coupling factors result from common characteristics among components determined at the design level. There are two groups of design-related hardware couplings: system level and component level. System-level coupling factors include features of the system or groups of components external to the components that can cause propagation of failures to multiple components. Features within the boundaries of each component cause component-level coupling factors.

Table 20.3 lists some example coupling factors in the hardware design category.

The operation-based coupling factors are those factors that propagate a failure mechanism due to identical operational characteristics among several components. Table 20.4 lists some example coupling factors in the operational category.

The environment-based coupling factors are those factors that propagate a failure mechanism via identical external or internal environmental characteristics. Table 20.5 lists some example coupling factors in the environmental category.

Software-based coupling factors are factors that propagate a failure mechanism among several components due to a common software module. For example, three separate aircraft flight control displays may be controlled by a common software module, with common data inputs. Table 20.6 lists some example coupling factors in the software category.

TABLE 20.3   Hardware-Based Component-Coupling Factors

Characteristic	Description
Same physical appearance	This refers to cases where several components have the same identifiers (e.g., same color, distinguishing number, letter coding, and/or same size/shape). These conditions could lead to misidentification by the operating or maintenance staff
System layout/configuration	This refers to the arrangement of components to form a system. Component arrangement may result in independent systems being dependent upon a common source
Same component internal parts	This refers to cases where several components could fail because they each use similar or identical internal subcomponents. A manufacturing flaw in a subcomponent could affect the entire lot
Same maintenance, test, and/or calibration characteristics	This refers to cases where several components are maintained, tested, and/or calibrated by the same set of procedures. A flaw in the procedures could affect an entire lot of components
Manufacturing attributes	This refers to the same manufacturing staff, quality control procedure, manufacturing method, and material being used for an entire lot of components. The same flaw applies to all components equally, thus they could all be expected to fail equally
Construction and installation attributes	This refers to the same construction/installation staff, procedures, and testing being applied to an entire lot of components. The same flaw applies to all components equally, thus they could all be expected to fail equally

TABLE 20.4   Operation-Based Component-Coupling Factors

Characteristic	Description
Same operating staff	This refers to cases where the same operator or operators are assigned to operate all trains of a system, increasing the probability that operator errors will affect multiple components simultaneously
Same operating procedure	This refers to cases where the same operating procedures govern the operation of all physically or functionally identical components. Any deficiency in the procedures could affect all the components
Same maintenance, test, and/or calibration schedule	This refers to cases where the same maintenance/test/calibration schedule is applied at the same time for identical components. Any deficiency could affect all the components
Same maintenance, test, and/or calibration staff	This refers to cases where the same maintenance/test/calibration staff being responsible for identical components. Any deficiency could affect all the components
Same maintenance, test, and/or calibration procedures	This refers to cases where the same maintenance/test/calibration procedure is applied on identical components. Any deficiency could affect all the components

Additional methods and/or tools that can be used to identify CCFs include the following:

1. Network tree diagrams used in sneak circuit analysis (see Chapter 25).
2. Zonal analysis analyzes the major zones of a system (see Refs. [2] and [5]).
3. Connector bent pin analysis and wire shorts (see Chapter 29).

It should be noted, however, that these techniques are oriented for particular types of CCF types and, therefore, are not all inclusive of all CCF types.

TABLE 20.5   Environment-Based Component-Coupling Factors

Characteristic	Description
Same system location	This refers to all redundant systems/components being exposed to the same environmental stresses because of the same system (e.g., flood, fire, high humidity, earthquake, etc.)
Same component location	This refers to all redundant systems/components being exposed to the same environmental stresses because of the component location within the system (e.g., vibration, heat, human error, etc.)
Internal environment/ working medium	This refers to the common exposure of components in terms of the medium of their operation, such as internal fluids, oils, gases, and so on

TABLE 20.6   Software-Based Coupling Factors

Characteristic	Description
Common algorithms	This refers to multiple hardware items that are driven by a single or common software module containing the control algorithm
Common data	This refers to multiple hardware items that are provided data by a single or common software module
Common requirements	This refers to multiple software modules that are developed using common software design requirements

20.6.3   CCFA Process Step 4: Detailed CCF Analysis

This step is a qualitative and/or quantitative evaluation of the revised FT model that has had all credible CCF causal factor events incorporated into the model. This step utilizes the results of steps 2 and 3. The intent of this step is to perform a system risk evaluation of the identified CCFs after they have been placed in the FT model. This involves identifying the appropriate CCF events, placing them in the FT and obtaining the failure frequency or probability for these events.

A common cause basic event (CCBE) is an event involving failure of a specific set of CCF components due to a common cause. For instance, in a system of three redundant components A, B, and C, the CCBEs are C_{AB}, C_{AC}, C_{BC}, and C_{ABC}, and are defined as follows:

A = single independent failure of component A (a basic event).

B = single independent failure of component B (a basic event).

C = single independent failure of component C (a basic event).

C_{AB} = failure of components A and B (and not C) from common causes (a CCBE).

C_{AC} = failure of components A and C (and not B) from common causes (a CCBE).

C_{BC} = failure of components B and C (and not A) from common causes (a CCBE).

C_{ABC} = failure of components A, B, and C from common causes (a CCBE).

Figure 20.5 shows a system design where three components operate in parallel redundancy, and only successful operation of one component is necessary for system success. This figure also shows the initial system FT model of the independent failure events.

CCBEs for this three-component system are C_{AB}, C_{AC}, C_{BC}, and C_{ABC}. However, since failure of all three components is necessary for system failure, C_{AB}, C_{AC}, and C_{BC} have no

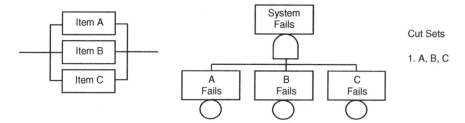

Figure 20.5 *Redundant system and initial FT model.*

system impact, but C_{ABC} does have an impact as this is the CCBE that causes failure of all three components. Figure 20.6 shows the revised FT model that incorporates the CCBE "C_{ABC}."

Note that in Figure 20.5 the FT produces only one cut set, which is a three-order cut set, indicating that the probability of system failure should be small. In Figure 20.6, the FT produces two cut sets, the original three-order cut set and an SPF cut set. Depending upon the probability of the SPF CCF event, the probability could be much higher than the initial FT probability.

Figure 20.7 shows a system design where three components operate in parallel redundancy, and two of the three components must operate for system success. This figure also shows the initial system FT model of the independent failure events.

The FT model in Figure 20.7 uses a k-of-n gate to show the required failure combinations. Figure 20.8 shows a modified FT where all of the combination events are modeled explicitly.

Figure 20.9 shows the revised FT model that incorporates the CCBEs for the particular system design parameters.

Table 20.7 provides the results for the FT model of the 2/3 redundant system design. Note from these results that when the CCBEs are added to the FT the number of cut sets increases, which in turn increase the probability of occurrence.

Note that the CCBEs are identified and used only according to the impact they have on the specific sets of components within the CCCGs and system function. In the first example (Figure 20.4), the CCF events C_{AB}, C_{AC}, and C_{BC} were not used because they would not cause system failure. However, they were used in the second example (Figure 20.6) because they could cause system failure. The analysts must watch out for subtleties such as this when conducting CCFA.

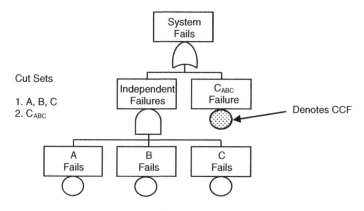

Figure 20.6 *CCBEs added to FT model.*

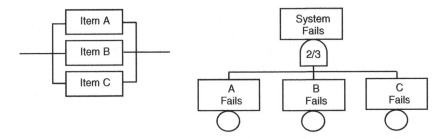

Figure 20.7   *Redundant 2 of 3 system and initial FT model.*

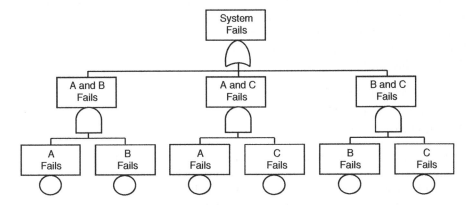

Figure 20.8   *Modified FT model.*

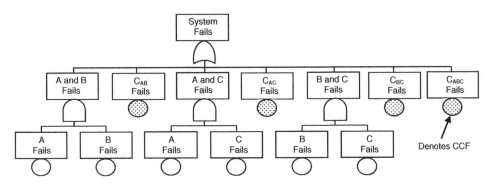

Figure 20.9   *CCBEs added to FT model.*

TABLE 20.7   FT Results for 2/3 Redundant System

Cut Sets and First-Order Probability Approximation Equation	
Initial FT Model	FT Model with CCBEs Added
{A, B}; {A, C}; {B, C}	{A, B}; {A, C}; {B, C}
	{C_{AB}}; {C_{AC}}; {C_{BC}}
	{C_{ABC}}
$P = P(A)P(B) + P(A)P(C) + P(B)P(C)$	$P = P(A)P(B) + P(A)P(C) + P(B)P(C)$
	$+ P(C_{AC}) + P(C_{AC}) + P(C_{BC}) + P(C_{ABC})$

It can be seen that the CCF expansion of an FT results in a proliferation of the number of cut sets, which may create practical difficulties when dealing with complex systems. However, in most cases standard fault tree computer codes for cut set determination and probabilistic quantification can be applied without concern about dependencies or size due to CCFs. If, after careful screening, the number of CS's is still unmanageable, a practical solution is to prune the FT of low-probability events.

20.7 CCF DEFENSE MECHANISMS

To understand a defense strategy against a CCF event, it is necessary to understand that defending against a CCF event is no different than defending against an independent failure that has a single root cause of failure (i.e., SPF). In the case of a CCF event, more than one failure occurs, and the failures are related through a coupling mechanism.

There are three methods of defense against a CCF:

1. Defend against the CCF root cause.
2. Defend against the CCF coupling factor.
3. Defend against both items 1 and 2.

A defense strategy against root causes is usually somewhat difficult because components generally have an inherent set of failure modes that cannot be eliminated. However, the failure modes can be protected via redundancy, diversity, and barriers.

A defense strategy for coupling factors typically includes diversity (functional, staff, and equipment), barriers, personnel training, and staggered testing and maintenance.

20.8 CCFA EXAMPLE

Figure 20.10 contains an example water pumping system for this CCFA example. In this system, water is supplied from a common source to three steam generators. For a successful system operation, two of the three generators must be operational. Some redundancy has been designed into the system to help achieve this operational requirement. The pumps are electricity driven that pump water from the tank to the motor-operated valves (MOVs). The MOVs are opened and closed by electrical power. The pumps, MOVs, and generators are all monitored and controlled by a single common computer, and they are powered from a common electrical source.

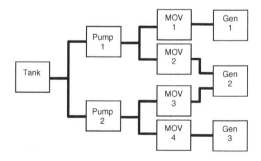

Figure 20.10 Example system diagram.

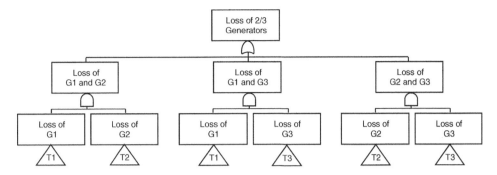

Figure 20.11 *Example system – top-level FT.*

Figure 20.11 is the top-level FT for the undesired event *loss of two out of three generators*. This top-level FT applies to both the preliminary (without CCFs) and the final (with CCFs) subtrees.

Figure 20.12*a–c* contains three preliminary version subtrees for system analysis *without* consideration given to CCFs.

Figures 20.13*a–c* contains three preliminary version subtrees for system analysis *with* consideration given to CCFs.

This example problem has been simplified in order to demonstrate the CCF concept. The FTs in Figure 20.13 contain CCF events for the pumps, MOVs, and generators. These CCFs are denoted by the double diamond symbol in the FTs.

Through screening of the system design and the preliminary FT, it has been determined that the factors shown in Table 20.8 contribute to CCF events. In a real-world situation, each of these factors (power, computer, software, etc.) would create an individual CCF event. However, for the purpose of simplification, each of these factors has been combined into a single CCF event for the pumps, MOVs, and generators.

Table 20.9 contains the basic failure rate data for the basic fault events in the three preliminary FT branches shown in Figure 20.12. The component basic failure rate and exposure (operating) times are provided, along with the computed probability of failure for the events. A 1 h operating time was used to keep the calculations simple.

Table 20.10 contains the qualitative and quantitative results of the preliminary version 1 FT. This FT yielded 22 cut sets (CS's), with the lowest CS probability being 1.00E-10 for failure of generator 1 AND generator 2 (also, G1 AND G3, G2, AND G3).

Table 20.11 contains the basic failure rate data for the basic fault events in the three preliminary FT branches shown in Figure 20.13. The values are identical to Table 20.9, except additional events and rates have been added for the three identified CCFBEs.

Table 20.12 contains the qualitative and quantitative results of the final version 2 FT containing the CCF events. This FT yielded 25 CS's, with the lowest CS probability for the CCF events. Note that all the CSs in this table are the same as in Table 20.10, except the CCF CSs are now included. Note that even though the CCF events failure rates were much smaller than the basic event failure rates, the CCF events are more likely to occur because they are SPFs (i.e., all of the other CS are two and three order, except for tank failure).

The system failure probability calculations can be summarized as follows:

- P(failure without CCFs) $= 7.83 \times 10^{-10}$
- P(failure with CCFs) $= 6.78 \times 10^{-9}$

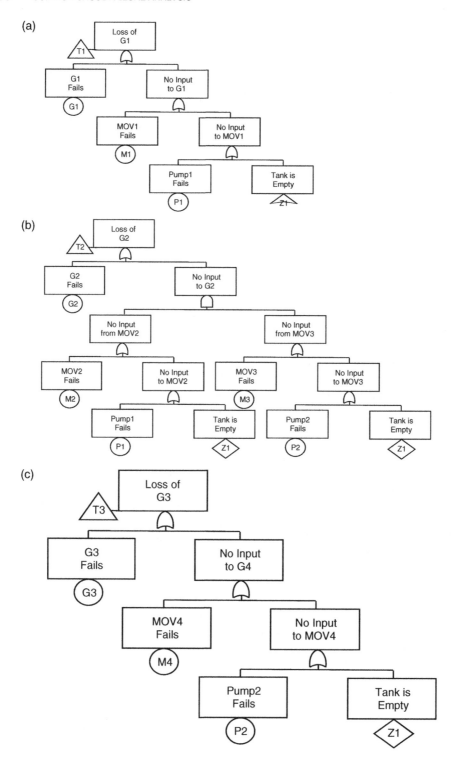

Figure 20.12 *(a) Subtree T1 without CCFs (version 1). (b) Subtree T2 without CCFs (version 1). (c) Subtree T3 without CCFs (version 1).*

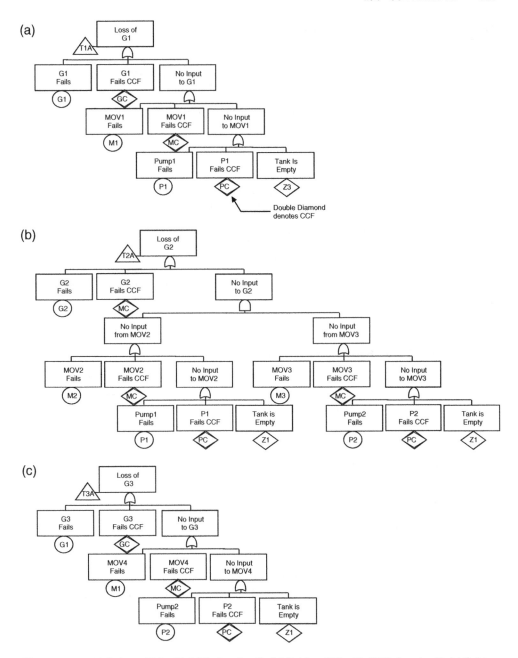

Figure 20.13 *(a) Subtree T1A with CCFs (version 2). (b) Subtree T2A with CCFs (version 2). (c) Subtree T3A with CCFs (version 2).*

TABLE 20.8   Component-Coupling Factors.

Coupling Factor	Effect	Impact On
Electrical power	Common power to pumps, common power to MOVs, and common power to generators. Pumps, MOVs, and generator power is separate	Pumps MOVs Generators
Computer control	Common computer sensing and control to pumps, MOVs, and generators	Pumps MOVs Generators
Software	Common computer software for pumps, MOVs, and generators	Pumps MOVs Generators
Manufacturing	Common manufacturing of pumps, common manufacturing of MOVs, and manufacturing of generators	Pumps MOVs Generators
Maintenance	Common maintenance procedures for pumps, common maintenance procedures for MOVs, and common maintenance procedures for generators	Pumps MOVs Generators
Installation	Common installation procedures for pumps, common installation procedures for MOVs, common installation procedures for generators	Pumps MOVs Generators

TABLE 20.9   Basic Event Data for Version 1 FT

Event	Failure Rate	Exposure Time (Hours)	Probability
P1	1.00E-06	1	1.00E-06
P2	1.00E-06	1	1.00E-06
M1	4.00E-06	1	4.00E-06
M2	4.00E-06	1	4.00E-06
M3	4.00E-06	1	4.00E-06
M4	4.00E-06	1	4.00E-06
G1	1.00E-05	1	1.00E-05
G2	1.00E-05	1	1.00E-05
G3	1.00E-05	1	1.00E-05
Z1	2.50E-10	1	2.50E-10

These probability numbers indicate how CCFs can have a significant impact on the total system failure frequency. When CCFs are ignored in an analysis, the total failure frequency can be understated, making a risk assessment incorrect.

20.9   CCFA MODELS

Several different models have developed that can be used for CCF evaluation. These models include the following:

1. Beta factor (BF) model [7].
2. Basic parameter (BP) Model [8].
3. Multiple Greek letter (MGL) Model [9].

TABLE 20.10   Results of Version 1 FT

CS No.	CS			Probability
1	G1	G2		1.00E-10
2	G1	G3		1.00E-10
3	G2	G3		1.00E-10
4	Z1			2.50E-10
5	G1	P2		1.00E-11
6	G2	P1		1.00E-11
7	G2	P2		1.00E-11
8	G3	P1		1.00E-11
9	M1	M4		1.60E-11
10	G1	M4		4.00E-11
11	G2	M1		4.00E-11
12	G2	M4		4.00E-11
13	G3	M1		4.00E-11
14	P1	P2		1.00E-12
15	M1	P2		4.00E-12
16	M2	P2		4.00E-12
17	M3	P1		4.00E-12
18	M4	P1		4.00E-12
19	G1	M2	M3	1.60E-16
20	G3	M2	M3	1.60E-16
21	M1	M2	M3	6.40E-17
22	M2	M3	M4	6.40E-17

4. Binomial failure rate (BFR) model [10].
5. System fault tree model (see Refs. [3] and [5]) approach, which is presented in this chapter.

20.10   CCFA ADVANTAGES AND DISADVANTAGES

The CCFA technique has the following advantages:

1. Structured, rigorous, and methodical approach.

TABLE 20.11   Basic Event Data for Version 2 FT

Event	Failure Rate	Exposure Time (Hours)	Probability
P1	1.00E-06	1	1.00E-06
P2	1.00E-06	1	1.00E-06
M1	4.00E-06	1	4.00E-06
M2	4.00E-06	1	4.00E-06
M3	4.00E-06	1	4.00E-06
M4	4.00E-06	1	4.00E-06
G1	1.00E-05	1	1.00E-05
G2	1.00E-05	1	1.00E-05
G3	1.00E-05	1	1.00E-05
Z1	2.50E-10	1	2.50E-10
PC	3.00E-09	1	3.00E-09
MC	2.00E-09	1	2.00E-09
GC	1.00E-09	1	1.00E-09

TABLE 20.12   Results of Version 2 FT

CS No.	CS			Probability
1	GC			1.00E-09
2	MC			2.00E-09
3	PC			3.00E-09
4	G1	G2		1.00E-10
5	G1	G3		1.00E-10
6	G2	G3		1.00E-10
7	Z1			2.50E-10
8	G1	P2		1.00E-11
9	G2	P1		1.00E-11
10	G2	P2		1.00E-11
11	G3	P1		1.00E-11
12	M1	M4		1.60E-11
13	G1	M4		4.00E-11
14	G2	M1		4.00E-11
15	G2	M4		4.00E-11
16	G3	M1		4.00E-11
17	P1	P2		1.00E-12
18	M1	P2		4.00E-12
19	M2	P2		4.00E-12
20	M3	P1		4.00E-12
21	M4	P1		4.00E-12
22	G1	M2	M3	1.60E-16
23	G3	M2	M3	1.60E-16
24	M1	M2	M3	6.40E-17
25	M2	M3	M4	6.40E-17

2. Identifies fault events that can bypass safety-critical redundant designs.
3. Permits probability assessment of CCF.
4. Probability assessment of CCF provides a truer view of system risk.

The CCFA technique has the following disadvantages:

1. Requires an analyst with some training and practical experience.
2. Sometimes avoided because of the complexity and cost.
3. Does not identify all system hazards, only those associated with CCFs.

20.11   COMMON CCFA MISTAKES TO AVOID

When first learning how to perform a CCFA, it is commonplace to commit some traditional errors. The following is a list of typical errors made during the conduct of a CCFA:

1. Not conducting a thorough investigation of CCF factors, events, or groups.
2. Not evaluating all redundant subsystems for CCF vulnerability,
3. Not using an FTA for visualization of CCF events,

20.12 SUMMARY

This chapter discussed the CCFA technique. The following basic principles help summarize the discussion in this chapter:

1. The primary purpose of the CCFA is to identify single-failure events that defeat design redundancy, where the redundancy is primarily intended to assure operation in safety-critical applications.
2. If a CCF is overlooked, total system risk is understated because the probability of the CCF hazard is not included in the total risk calculation.
3. CCFA should be a supplement to the SD-HAT.
4. Some CCFA identified hazards may require more detailed analysis by other techniques (e.g., FTA) to ensure that all causal factors are identified.

REFERENCES

1. A. Mosleh et al., Procedures for Treating Common Cause Failures in Safety and Reliability Studies: Procedural Framework and Examples, Volume 1, NUREG/CR-4780, January 1988, US NRC, Washington DC.
2. A. Mosleh, D. M. Rasmuson, and F.M. Marshall, Guidelines on Modeling Common-Cause Failures in Probabilistic Risk Assessment, NUREG/CR-5485, INEEL/EXT-97-01327, 1998, U.S. NRC, Washington DC.
3. Probabilistic Risk Assessment Procedures Guide for NASA Managers and Practitioners, NASA, August, 2002.
4. H. Kumanoto and E. J. Henley, *Probabilistic Risk Assessment and Management for Engineers and Scientists*, 2nd edition, IEEE Press, 1996.
5. ARP-4754, SAE Aerospace Recommended Practice, Certification Considerations for Highly-Integrated or Complex Aircraft Systems, 1996.
6. ARP-4761, SAE Aerospace Recommended Practice, Guidelines and Methods for Conducting the Safety Assessment Process on Civil Airborne Systems and Equipment, 1996.
7. K. N. Fleming et al., Classification and Analysis of Reactor Operating Experience Involving Dependent Events, Pickard, Lowe and Garrick, Inc., PLG-0400, prepared for EPRI, Feb., 1985.
8. K. N. Fleming, A Reliability Model for Common Mode Failure in Redundant Safety Systems, Proceedings of the 6th Annual Pittsburgh Conference on Modeling and Simulation, General Atomic Report GA-A13284, April, 1975, pp. 20–25.
9. Pickard, Lowe and Garrick, Inc., Seabrook Station Probabilistic Safety Assessment, prepared for Public Service Company of New Hampshire and Yankee Atomic Electric Company, PLG-0300, Dec., 1983.
10. PRA Procedures Guide: A Guide to the Performance of Probabilistic Assessments for Nuclear Power Plants, USNRC, NUREG/CR-2300, 1983, Appendix B.

FURTHER READINGS

Groen, F. J., Mosleh, A., and Smidts, C., Automated Modeling and Analysis of Common Cause Failure with QRAS, Proceedings of the 20th International System Safety Conference, pp. 38–44, 2002.

Kaufman, L. M., Bhide, S., and Johnson, B. W., Modeling of Common-Mode Failures in Digital Embedded Systems, Proceedings 2000 Annual Reliability and Maintainability Symposium, pp. 350–357, 2000.

Procedures for Treating Common Cause Failures in Safety And Reliability Studies: Analytical Background and Techniques, Volume 2, NUREG/CR-4780, January 1989, US NRC, Washington DC.

Procedures for Analysis of Common Cause Failures in Probabilistic Safety Analysis, NUREG/CR-5801, SAND91-7087, April 1993, US NRC, Washington DC.

Rankin, J. P., A Practical Common Cause Failure Analysis, Proceedings of the 6th International System Safety Conference, pp. 8-4.1 through 8.4–13, 1983.

Rankin, J. P., Common Cause Failure Analysis: Why Interlocked Redundant Systems Fail, Proceedings of the Turbine Powered Executive Aircraft Meeting, April 9–11, 1980, SAE, paper 800631.

Chapter *21*

Software Hazard Analysis

21.1 SwHA INTRODUCTION

Software hazard analysis (SwHA) is a safety analysis methodology for identifying and assessing the potential hazards software may present to a system. All projects involving software should have a software safety program, typically referred to as a software system safety program (SwSSP), which is comprised of several different elements, of which SwHA is one of the main elements.

The purpose of the SwHA is to identify system hazards involving software. Software can do damage only when it is combined with hardware; the hardware causes damage and the software is a contributing causal factor. Therefore, SwHA requires a true systems approach that simultaneously evaluates software, hardware, and functions.

In addition to hazard identification, the SwHA provides enough information to size and scope the amount of software safety effort that will be required by the SwSSP. It should be noted that SwHA is both a hazard analysis and an assessment of the potential safety criticality of software in the system design. The level of software criticality provides an indication of how many additional test and analysis tasks must be performed on the software by the software development process in order to attain a level of assurance that the software should be safe.

The SwHA involves analyzing different aspects of the software, which include the following:

1. Software requirements
2. Software functions
3. Software code modules
4. Software risk criticality

Hazard Analysis Techniques for System Safety, Second Edition. Clifton A Ericson, II.
© 2016 John Wiley & Sons, Inc. Published 2016 by John Wiley & Sons, Inc.

21.2 SwHA BACKGROUND

This analysis technique falls under the system design hazard analysis type (the basic analysis types are described in Chapter 4). There are no alternative names for this technique. It should, however, be conducted as early as possible in the program and it highly depends on other analyses, such as the PHA, SSHA, and SHA.

SwHA can be performed on any type or size of system involving software. It provides a list of hazards, the software modules contributing to the hazards, and the safety criticality level of the software modules. The SwHA technique, when applied to a given system by experienced safety personnel, should provide a thorough and comprehensive identification software-related hazards.

The technique is uncomplicated and easily learned. Standard, easily followed SwHA worksheets and instructions are provided in this chapter. A SwHA performed early in the system development cycle will help ensure that software safety is adequately addressed by the project.

It is recommended that a SwHA be performed on every new or modified system in order to determine if the system contains software, the size of the software, and the safety criticality of the software. Sometimes, system developers overlook the fact that a system may contain firmware, which is considered software, and therefore requires a SwSSP.

Software has many unique characteristics that make it quite different from hardware. Some of these characteristics affect SwHA and the overall software safety process. When performing a SwHA, software must be treated slightly differently from hardware due to the complexity, special attributes, and unique nature of software.

The following are some software safety principles that have been established from the unique characteristics of software. These principles are useful when applying the software safety process and when identifying software-related hazards:

- Software does not have hard discrete failures such as hardware, instead it has *functional* failures.
- Software by itself is not hazardous; it is only hazardous in a system when performing system functions involving hardware.
- Hardware causes the damage in a mishap (i.e., explosives, radiation equipment, flight controls, chemicals, etc.); however, software can be an initiating or controlling factor.
- When identifying software-related hazards, the key is to look for safety-related hardware–software relationships, since software can contribute to hazards only via hardware.
- SwHA requires multiple perspectives: system, hardware interfaces, human interfaces, and functions.
- Hardware faults can induce software functional failures (modifies or fails software intent).
- Software can have errors and still function (safely or unsafely).
- Not all software errors are safety related.
- Software hazard risk is difficult (if not impossible) to quantify because software errors or functional failures cannot be quantified.
- Software always works exactly as coded, but complexity makes comprehension difficult.
- Sometimes, software does more than intended or expected; unintended functions are a safety concern.

The unique nature of software causes the existence of two enigmas associated with software safety and SwHA, which in turn cause the need for a more diverse approach in order to ensure adequate safety of software. The two stumbling blocks are as follows:

1. Software functional failures (and hazards) can be postulated, but they cannot usually be definitively proven by specific identifiable causal factors in the software design or code. Typically, testing is required.
2. Failure rates cannot be determined for software functional failures; therefore, hazard risk cannot be calculated for software-related risk assessments when hazards are identified.

Because of these two software safety dilemmas, software presents potential mishap risk value that is unknown and cannot be precisely determined. Therefore, the most pragmatic way to ensure that software is safe is by applying a bilateral safety approach consisting of (1) software functional coverage and (2) software developmental coverage.

The software functional coverage scheme focuses on the functional design to provide hazard identification and mitigation assurance and safety-critical function (SCF) identification and assurance. The software development coverage scheme utilizes the software development process to assist in the forced focus on specific development tasks that ensure higher quality software that is presumably safer. The theory is that if the software is developed to a specified set of rigorous requirements, analyses, tests, and development procedures, then the resulting product will present acceptable safety risk.

When all the appropriate hazard mitigation tasks and software development tasks are successfully performed, the overall software mishap risk is "judged" to be acceptable. This bilateral approach is a strategy intended to broadly cover all aspects of software that can impact safety. This software safety scheme provides a "presumed level of assurance" that the software has received complete safety coverage and the risk presented by the software is deemed acceptable. Software safety assurance requires visibility of both the product and the process. It should be noted that these software safety tasks do not provide a quantitative estimate of the potential mishap risk associated with the software. What this approach does provide is a level of confidence that the software can be considered as being safe. Software-related hazards can be accepted for risk based on the conclusions drawn from the safety case, where the safety case is built upon the results of both the functional and the developmental completion evidence.

21.3 SwHA HISTORY

Software safety and SwHA were first conceived in the early 1970s when it was recognized that software could contribute to mishaps and that the use of safety-critical software was increasing. Since then, many technical papers and books have been written on software safety. In addition, several industry standards and guidelines have been published. Refer to the reference section for a list of relevant material.

21.4 SwHA THEORY

Figure 21.1 shows an overview of the basic SwHA concept and summarizes the important relationships involved. The SwHA process consists of identifying software modules in the system design and identifying the hazards in these modules, along with their safety criticality.

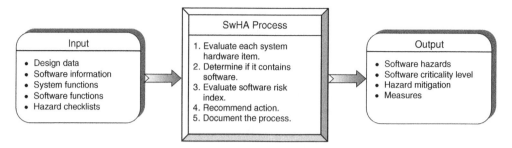

Figure 21.1   *SwHA concept.*

SwHA is based on the following definitions:

Software   Software is a combination of associated computer instructions and computer data that enable a computer to perform computational or control functions. Embedded software is software developed to control a hardware device. It may be executed on a processor contained within a device or an external computer connected to the hardware to be controlled. Software includes computer programs, procedures, rules, and any associated documentation pertaining to the operation of a computer system EIA SEB-6A, System Safety Engineering in Software Development, Apr 1990.

Firmware   Firmware is software that resides in a nonvolatile medium that is read-only in nature, which cannot be dynamically modified by the computer during processing (write protected during operation) (EIA SEB6-A).

It should be noted that software safety provisions that apply to software also apply to firmware. Firmware code cannot be ignored from a safety standpoint.

21.5   SwHA METHODOLOGY

Figure 21.2 contains a functional diagram of the SwHA methodology. This process begins by acquiring design information in the form of the design concept, the operational concept, major components planned for use in the system, and major system functions. Sources for this

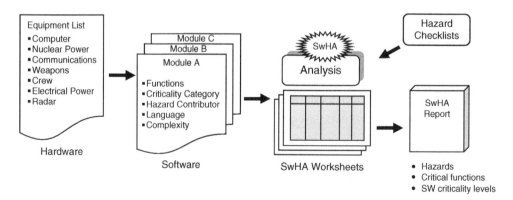

Figure 21.2   *SwHA methodology.*

TABLE 21.1   SwHA Process

Step	Task	Description
1	Define the system	Examine the system and define the system boundaries, subsystems, functions and interfaces
2	Identify the hardware/ functions	Identify and list all of the major system elements or components. Understand their functions and purpose
3	Identify software or firmware	Identify if the hardware components will contain embedded or resident software
4	Identify the software modules	Identify the software modules that will be used by the hardware components to achieve the system functions. This may require some analysis and predictions if not already done by the project
5	Identify software module functions	Identify the functions being performed in each software module
7	Identify software-related hazards	Identify the hazards resulting from incorrect or nonperformance of the software functions. The hazard description should include causes and effects of the hazard
8	Establish software criticality	While evaluating software hazards, establish the software criticality of the software modules. This should support the software criticality index and level of rigor tasks
9	Recommend corrective action	Provide recommendations for design measures that should be incorporated in order to mitigate the hazards
10	Document SwHA	Document the entire SwHA process on the worksheets. Update for new information as necessary

information could include statement of work (SOW), statement of objectives (SOO), design specifications, sketches, drawings, schematics, concept of operations, software design (modules and functions), and so on.

The SwHA is generally performed as early as possible in the design life cycle. The analysis starts by looking at the hardware, and then gradually moving into software associated with the hardware. The SwHA process considers all software functions for the effect of undesired conditions, such as function fails, function malfunctions, inadvertent function, partial function, and so on.

Table 21.1 lists and describes the basic steps of the SWSA process.

21.6   SwHA WORKSHEET

The SwHA technique utilizes a worksheet to provide structure and rigor. Although the exact format of the assessment worksheet is not critical, typically, matrix- or columnar-type worksheets are used to help maintain focus and composition in the analysis. As a minimum, the following basic information is required from the SwHA worksheet:

1. Software module under assessment
2. Purpose or function of the module
3. Hazards that the software modules present
4. Safety criticality of the software function with rationale
5. Recommended design features to mitigate the hazard

The recommended SwHA worksheet is shown in Figure 21.3. This particular SwHA worksheet utilizes a columnar-type format. Other worksheet formats may exist because

Software Hazard Analysis					
Software Module	Module Functions	Software-Related Hazards	Function Criticality/Rationale	Mitigation Recommendation	Comments
①	②	③	④	⑤	⑥

Figure 21.3   *Recommended SwHA worksheet.*

different organizations often tailor their SwHA worksheet to fit their particular needs. The specific worksheet to be used may be determined by the system safety working group, the safety integrated product team (IPT), or the safety analyst performing the SwHA.

The columns in this SwHA worksheet are described as follows:

1. *Software Module*   This column identifies the software module under investigation.
2. *Module Functions*   This column lists all of the functions in the software module and describes the purpose of the function.
3. *Software-Related Hazards*   This column identifies the hazards that the identified software module function may be a contributing factor. If there is no hazard visibility, then this column may be left blank.
4. *Function Criticality/Rationale*   This column provides a measure of the relative safety significance expected for the software function and the level of safety rigor that will be required of the software. Refs. 1 and 2 provide an index referred to as the software hazard criticality index (SwCI) that classifies the relative risk level of software modules.
5. *Mitigation Recommendation*   This column provides for any recommendations that are immediately evident that can be used to mitigate the identified hazard. Recommendations may also include existing safety requirements.
6. *Comments*   This column provides for any pertinent comments to the analysis that need to be documented for possible future use. For example, this column can be used to record that a software module is a commercial-off-the-shelf (COTS) item.

21.7   SOFTWARE CRITICALITY LEVEL

The aerospace industry approach to software safety is based primarily on three standards:

1. DoD Joint Software Systems Safety Handbook (JSSSH), original December 1999, version 1.0 August 2010 (see Ref. 1).

2. MIL-STD-882E, Standard Practice for System Safety, 2012 (see Ref. 2).

3. ANSI/GEIA-STD-0010, Standard Best Practices for System Safety Program Development and Execution (see Further Readings).

A successful software safety program is based on hazard mitigation assurance and design development assurance. The hazard analysis process identifies and mitigates software-related hazards. The software design development assurance process increases the confidence that the software will perform as specified to software system safety and performance requirements while reducing the number of contributors to hazards that may exist in the system. Both processes are essential in reducing the likelihood of software initiating a propagation pathway to a hazardous condition or mishap.

SwHA fulfills the hazard mitigation assurance aspect of system safety. The SwHA process is described in this chapter. Emphasis is placed on the context of the *system* and how software contributes to, or mitigates, potential failures, hazards, and mishaps. From the perspective of the system safety engineer and the SwHA process, software is considered a subsystem. In most instances, the system safety engineers will perform the SwHA process in conjunction with the software development team.

The design development assurance aspect is fulfilled by the software development team when they perform a prescribed set of LOR tasks. When the LOR tasks are successfully performed, the software is considered to be acceptably safe. LOR tasks are based upon the safety criticality of a software module; the more critical the module the more LOR tasks that must be performed. Software criticality is established from a set of tables contained in the standards listed above. A software criticality index (SwCI) is determined based on hazard severity and a software control category (SCC). The SwCI determines which LOR tasks must be performed.

The SCC, hazard severity, SwCI, and LOR tasks are provided by the above-mentioned standards and also shown below. These tables are provided here because it is important that the SwHA help provide the information necessary to determine the SCC, hazard severity, and SCI for each software module. Determining the SwCI index involves three steps.

Step 1   Identify the control category for the software module using the criteria in Table 21.2.

Step 2   Identify the severity of the mishap that the software would pertain to, using the criteria in Table 21.3.

Step 3   Using the classifications from steps 1 and 2, determine the SHRI for the software using the criteria in Table 21.4.

Step 4   Based on the SwCI, the risk level for the software is determined using the criteria in Table 21.5.

Unlike the hardware-related hazard risk index (HRI), a low software SwCI index number does not mean that a design is unacceptable. Rather, it indicates that greater resources must be applied to the analysis and testing of the software in order to provide safety assurance.

21.8   SwHA EXAMPLE

In order to demonstrate the SWSA methodology, the hypothetical Ace Missile System, from Chapter 8, will be used here. This system is comprised of the missile and the weapon control system (WCS) as shown in Figure 21.4.

TABLE 21.2   Software Control Code (CC) Categories

Level	Name	Description
1	Autonomous (AT)	• Software functionality that exercises autonomous control authority over potentially safety-significant hardware systems, subsystems, or components without the possibility of predetermined safe detection and intervention by a control entity to preclude the occurrence of a mishap or hazard (*This definition includes complex system/software functionality with multiple subsystems, interacting parallel processors, multiple interfaces, and safety-critical functions that are time critical*)
2	Semiautonomous (SAT)	• Software functionality that exercises control authority over potentially safety-significant hardware systems, subsystems, or components, allowing time for predetermined safe detection and intervention by independent safety mechanisms to mitigate or control the mishap or hazard (*This definition includes the control of moderately complex system/software functionality, no parallel processing, or few interfaces, but other safety systems/mechanisms can partially mitigate. System and software fault detection and annunciation notifies the control entity of the need for required safety actions*) • Software item that displays safety-significant information requiring immediate operator entity to execute a predetermined action for mitigation or control over a mishap or hazard. Software exception, failure, fault, or delay will allow, or fail to prevent, mishap occurrence (*This definition assumes that the safety-critical display information may be time-critical, but the time available does not exceed the time required for adequate control entity response and hazard control*)
3	Redundant fault tolerant (RFT)	• Software functionality that issues commands over safety-significant hardware systems, subsystems, or components requiring a control entity to complete the command function. The system detection and functional reaction includes redundant, independent fault tolerant mechanisms for each defined hazardous condition (*This definition assumes that there is adequate fault detection, annunciation, tolerance, and system recovery to prevent the hazard occurrence if software fails, malfunctions, or degrades. There are redundant sources of safety-significant information, and mitigating functionality can respond within any time-critical period*) • Software that generates information of a safety-critical nature used to make critical decisions. The system includes several redundant, independent fault tolerant mechanisms for each hazardous condition, detection, and display
4	Influential	• Software generates information of a safety-related nature used to make decisions by the operator, but does not require operator action to avoid a mishap
5	No safety impact (NSI)	• Software functionality that does not possess command or control authority over safety-significant hardware systems, subsystems, or components and does not provide safety-significant information. Software does not provide safety-significant or time-sensitive data or information that requires control entity interaction. Software does not transport or resolve communication of safety-significant or time-sensitive data

TABLE 21.3 Hazard Severity Categories

Description	Cat	Mishap Definition
Catastrophic	I	Could result in death, permanent total disability, loss exceeding $1 M, or irreversible severe environmental damage that violates law or regulation
Critical	II	Could result in permanent partial disability, injuries, or occupational illness that may result in hospitalization of at least three personnel, loss exceeding $200 K but less than $1 M, or reversible environmental damage causing a violation of law or regulation
Marginal	III	Could lead to injury or occupational illness resulting in one or more lost work days(s), loss exceeding $10 K but less than $200 K, or mitigable environmental damage without violation of law or regulation where restoration activities can be accomplished
Negligible	IV	Could result in injury or illness not resulting in a lost work day, loss exceeding $2 K but less than $10 K, or minimal environmental damage not violating law or regulation

TABLE 21.4 Software Criticality Index (SwCI)

Software Safety Criticality Matrix

Software Control Category	Severity Category			
	Catastrophic (1)	Critical (2)	Marginal (3)	Negligible (4)
1	SwCI 1	SwCI 1	SwCI 3	SwCI 4
2	SwCI 1	SwCI 2	SwCI 3	SwCI 4
3	SwCI 2	SwCI 3	SwCI 4	SwCI 4
4	SwCI 3	SwCI 4	SwCI 4	SwCI 4
5	SwCI 5	SwCI 5	SwCI 5	SwCI 5

TABLE 21.5 Software Risk Levels

SwCI	Risk Level	Software LOR Tasks and Risk Assessment/Acceptance
SwCI 1	High	If SwCI 1 LOR tasks are unspecified or incomplete, the contributions to system risk will be documented as HIGH and provided to the PM for decision. The PM shall document the decision of whether to expend the resources required to implement SwCI 1 LOR tasks or prepare a formal risk assessment for acceptance of a HIGH risk
SwCI 2	Serious	If SwCI 2 LOR tasks are unspecified or incomplete, the contributions to system risk will be documented as SERIOUS and provided to the PM for decision. The PM shall document the decision of whether to expend the resources required to implement SwCI 2 LOR tasks or prepare a formal risk assessment for acceptance of a SERIOUS risk
SwCI 3	Medium	If SwCI 3 LOR tasks are unspecified or incomplete, the contributions to system risk will be documented as MEDIUM and provided to the PM for decision. The PM shall document the decision of whether to expend the resources required to implement SwCI 3 LOR tasks or prepare a formal risk assessment for acceptance of a MEDIUM risk
SwCI 4	Low	If SwCI 4 LOR tasks are unspecified or incomplete, the contributions to system risk will be documented as LOW and provided to the PM for decision. The PM shall document the decision of whether to expend the resources required to implement SwCI 4 LOR tasks or prepare a formal risk assessment for acceptance of a LOW risk
SwCI 5	Not Safety	No safety-specific analyses or testing is required

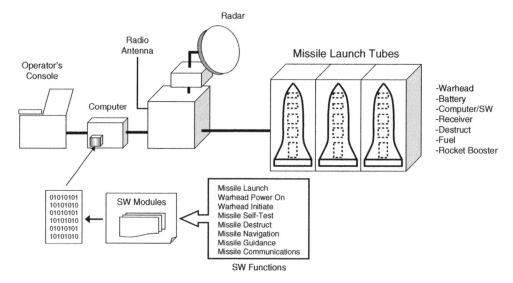

Figure 21.4 *Ace Missile System with software modules.*

The SwHA will utilize the following list of TLMs obtained from the PHA in Chapter 8, plus some additional TLMs obtained from other hazard analyses:

- Inadvertent W/H explosives initiation
- Inadvertent launch
- Incorrect target
- Inadvertent missile destruct
- Missile destruct fails
- Missile fire
- Personnel injury
- Unknown missile state
- Inadvertent explosives detonation
- Inadvertent ignition/detonation
- Incorrect targeting
- Inadvertent power to W/H

The SwHA will be performed on the following list of software modules:

- Missile launch
- Warhead power-on
- Warhead initiate
- Missile self-test
- Missile destruct
- Missile navigation
- Missile guidance
- Missile communications

Tables 21.6–21.8 contain SwHA worksheets for the Ace Missile System.

TABLE 21.6 SwHA Worksheet for the Ace Missile System (Page 1)

Software Hazard Analysis

Software Module	Functions	Software-Related Hazards	Function Criticality/ Rationale	Mitigation Recommendation	Comments
Missile launch	Missile launch function	—			
	When the SW module receives all of the appropriate input data, and data verification, it generates a launch command	SW generates an inadvertent missile launch command. Invalid command; unexpected launch results in death, injury, and damage, plus system loss	SC function Sev = 1 SCC = 3 SwCI = 2	No COTS software allowed for this function	
Warhead power-on	Warhead power-on function	—			
	After missile launch, the SW generates a warhead power-on signal that provides electrical power to the warhead	SW generates an inadvertent warhead power-on command during ground operations The warhead is one step closer to inadvertent detonation. Invalid command	SC function Sev = 1 SCC = 3 SwCI = 2		
		SW generates a premature warhead Power-On command after launch. The warhead is one step closer to inadvertent detonation. Invalid command	SC function Sev = 1 SCC = 3 SwCI = 2		
Warhead initiate	Warhead initiate function	—			
	After missile launch the SW generates a warhead initiate signal. Specific checks must pass in order to issue a valid command	SW generates an inadvertent warhead initiate command during ground operations. The warhead is one step closer to inadvertent detonation. Invalid command	SC function. Sev = 1 SCC = 3 SwCI = 2		Firmware in warhead.
		SW generates a premature warhead initiate command after launch. Assuming warhead power-on has been issued the warhead detonate Invalid command	SC function Sev = 1 SCC = 3 SwCI = 2		

Analyst:

Page number: 1 of 3

TABLE 21.7   SwHA Worksheet for the Ace Missile System (Page 2)

Software Hazard Analysis

Software Module	Functions	Software-Related Hazards	Function Criticality/ Rationale	Mitigation Recommendation	Comments
Missile self-test	Missile self-test function	—			
	A test of critical missile components prior to launch to determine if missile is safe	SW issues a safe state when it is not true Missile is not safe and blows up during launch, resulting in death/injury	SC function Sev = 1 SCC = 3 SwCl = 2		
Missile destruct	Missile destruct function	—			
	A destruct command is sent to missile in order to abort the missiles flight. The order to destruct is received from the operator	SW issues an inadvertent destruct command. Personnel in vicinity resulting in death/injury	SC function Sev = 1 SCC = 3 SwCl=		
		SW fails to issue destruct command as directed by operator. Missile strikes objects when not intended, resulting in death/injury	SC function Sev = 1 SCC = 3 SwCl = 2		
Missile navigation	Compute nav. path function	—			
	Navigation commands are computed based on target data and sent to the flight control Guidance SW.	Nav. errors in SW resulting in target error, causing collateral damage of death/injury	SC function Sev = 1 SCC = 3 SwCl = 2		Kahlman filter is COTS SW
	Display nav. path function	—			
	The computed navigation path is displayed to the operator for validity	Errors in display SW cause operator to see incorrect target path and choosing to erroneously destroy missile. Possible death/ injury worst case	SC function Sev = 1 SCC = 3 SwCl = 2		

Analyst:

TABLE 21.8   SwHA Worksheet for the Ace Missile System (Page 3)

Software Hazard Analysis

Software Module	Functions	Software-Related Hazards	Function Criticality/ Rationale	Mitigation Recommendation	Comments
Missile guidance	Compute guidance function	—			
	Flight control commands are computed based on nav. data and sent to the flight control guidance HW	Errors in SW cause flight control error, resulting incorrect target, causing collateral damage of death/injury	SC function Sev = 1 SCC = 3 SwCI = 2		
Missile communications	Missile comm. function	—			
	Missile sends missile data and status info to operator	SW error causes data error resulting in operating seeing incorrect data. Confused operator may destruct missile. Possible death/injury worst case	SC function Sev = 1 SCC = 3 SwCI = 2	Commands require strict data structure. Bit error checks present	
	Missile receives commands from operator	SW error causes data error resulting in missile receiving or interpreting incorrect data. Possible inadvertent destruct interpretation resulting in death/injury	SC function Sev = 1 SCC = 3 SwCI = 2	Commands require strict data structure. Bit error checks present	
Analyst:				Page number: 3 of 3	

Results and conclusions from the SwHA for the Ace Missile System are summarized as follows:

1. The system contains both software and firmware.
2. The system software involves safety-critical software.
3. Some software modules have been identified that are potential casual factors for the identified top-level mishaps.
4. The high-risk categories for some software modules indicate that system safety requirements (SSRs) are needed for safe software design.
5. More detailed hazard analysis will be required to ensure safe system software.
6. Some of the software modules will have a high level of complexity due to the functions being performed, such as Kahlman filtering for navigation equations, missile launch functions, missile warhead initiation functions, and so on.
7. Some COTS software has been identified in the system, meaning that special safety analyses will have to be performed to ensure that the COTS software does not degrade safety-critical and safety-related software.

21.9 SOFTWARE FAULT TREE ANALYSIS

Software fault tree analysis (SFTA) is the application of fault tree analysis (FTA) to software. FTA is a well-recognized root cause analysis tool for identifying the detailed causal factors to a problem of interest, such as a hazard, mishap, or undesired event. SFTA follows all of the standard rules for FTA; see Chapter 15 for detailed information on the FTA process.

The intent of SFTA is to apply FTA to software and thereby identify all potential hazard causal factors in the code. However, it should be noted that SFTA is not a single golden bullet solution for the development of safe software; it is instead a tool in the arsenal of the safety engineer to be used in the SwS process when needed and when appropriate. Software fault tree analysis is a topic in SwS that is often misunderstood and misused. The use of FTA methodology in the analysis of hardware is a viable application because hardware is straightforward and has discrete and identifiable failure modes that easily translate to a fault tree. Software, on the other hand, is not quite so straightforward as hardware. As shown in Chapter 3, software is quite different from hardware. For this reason, SFTA must be applied more judiciously.

Since software code does not have failure modes, the resulting software fault tree is strained for reality. Also, if every line of code were to be analyzed by SFTA, the resulting fault tree is essentially nothing more than another flow diagram of the code. There are many static code evaluators that can do a better job in trying to identify software code bugs.

Since failure probability numbers are not available for software functional failures, a quantitative SFTA is not possible. This typically means that generating fault tree cut sets is typically incorrect or not useful. People have tried using values of zero or one for functional failures, but this typically also results in misunderstood results. Applying an assumed failure rate, between zero and one, has also been tried, but since the data is not valid this can lead to misunderstood results.

The most reasonable approach for SFTA is to apply it as a systems FTA that is inclusive of hardware, software, humans, and environmental factors. A SFTA is best used to analyze a system and find the software "weak points" (i.e., functional failures) that are of safety concerns.

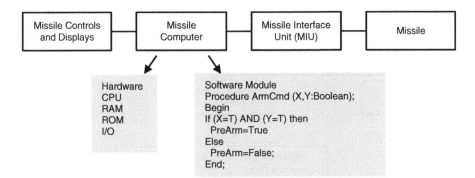

Figure 21.5 *System functional diagram.*

This will identify the software code modules, and perhaps code points, where further more detailed analysis using another tool is more productive. This system SFTA approach can be used to evaluate software architecture and software requirements.

The system FTA approach models the entire system starting from the hardware and going down into the system to the possible software causal areas. This approach is actually a standard FTA of a system, except it is carried down into the system to the point where the applicable software modules are identified. Those software modules are then handled by other means, such as the SwHA and the LOR tasks. As is typical in FTA, the analysis begins with the top undesired event, which is usually a potential hazard, mishap under study, or potential event that requires detailed investigation.

Figure 21.5 contains a simple functional diagram of the example missile system under investigation. The analysis follows the system architecture for all items in the design that can contribute to the undesired event. The concept is to let the analysis path take its natural course through the system hardware, software, and human interfaces.

The SFTA begins with the undesired event and follows standard FTA rules and methods. Figure 21.6 contains an FTA, or SFTA, of the missile system for the top undesired event of "Missile receives inadvertent arm command."

The advantages of the system-level SFTA approach include the following:

- Provides a system view covering both hardware and software.
- FTA provides a deductive and methodical approach for analyzing critical paths, and all parts of that path.
- Provides an evaluation of software/hardware interfaces.
- Indicates what the safety-critical software modules are and where they interface with the hardware.

21.10 SwHA ADVANTAGES AND DISADVANTAGES

There are the following advantages of the SwHA technique:

1. Structured, rigorous, and methodical approach.
2. Can be effectively performed on varying levels of design detail.

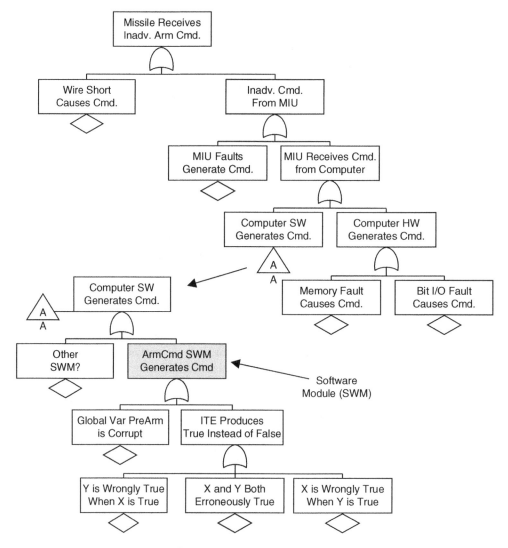

Figure 21.6 *Missile system SFTA.*

3. Relatively easy to learn, perform, and follow.
4. Provides a cursory software risk assessment.

The SwHA technique has only one disadvantage:

1. Requires an analyst with some knowledge of software and knowledge and experience in software safety.

21.11 SwHA MISTAKES TO AVOID

When first learning how to perform an SWSA, it is commonplace to commit some typical errors. The following is a list of typical errors made during the conduct of a SwHA:

1. Not including subsystems and software provided by vendors and subcontractors.
2. Not adequately addressing all of the system software, particularly firmware.
3. Not including COTS software.
4. Not understanding or underestimating the safety-critical nature of the software modules.
5. Failing to establish a reasonable and credible list of LOR tasks early in the program, since these tasks will affect cost and schedule.

21.12 SwHA SUMMARY

This chapter discussed the SwHA technique. The following basic principles help summarize the discussion in this chapter:

1. SwHA is used to identify software-related hazards, to determine the safety significance of the software, and to determine the potential size and extent of a SwSSP.
2. The SwHA worksheet provides structure and rigor to the SwHA process.
3. SwHA should be performed early in the program life cycle to assist in planning and funding of the SwSSP.
4. Whereas hardware safety is risk based, software safety is assurance based (also sometimes referred to as integrity based).
5. Hardware safety focuses primarily on mitigating hazard risk to an acceptable level; software safety involves mitigating identified hazards and developing software to an appropriate level of rigor.
6. In the case of software, actual hazard risk cannot be calculated, thus acceptable risk is nebulous and is based on a diverse level of rigor assurance process.
7. The software safety assurance process focuses on functional hazard assurance and software development assurance. In essence, the development of safe software involves a mix of hazard analysis, design safety requirements scrutiny, safety-critical function (SCF) scrutiny, significant testing, and the use of rigorous software development methods and tools.
8. In addition to a SwHA, a system FTA combining both hardware and software interfaces with the hardware help in the safety process; the FTA both helps identify safety-critical software modules and helps in determining where hazard mitigations are most effective.

REFERENCES

The following are some important SwHA references for further software safety and SwHA related information:

1. Department of Defense *Joint Software System Safety Handbook*, revised 2010.

2. MIL-STD-882, Standard Practice for System Safety, original version 15 July 1969, version B Notice 1: 30 March 1984 contains the first software safety material Task 301, version C: 19 January 1993 contains update, version D: 10 February 2000 dropped detailed information, version E (2012) adds software safety detail back, plus additional safety tasks.

3. EIA SEB-6A, System Safety Engineering in Software Development, Apr 1990.

FURTHER READINGS

AMCOM, AMCOM Regulation 385-17: AMCOM Software System Safety Policy, 15 March 2008 (Army).

ANSI/GEIA, ANSI/GEIA-STD-0010, Standard Best Practices for System Safety Program Development and Execution 2010.

Electronic Industries Association, Electronic Industries Association (EIA) SEB-6A: System Safety Engineering in Software Development, Apr 1990.

Ericson, C. A., Software and System Safety: Proceeding of the 5th International System Safety Conference, 1981.

Ericson, C. A., *Software Safety Primer*, CreateSpace, Charleston, NC 2013.

Hardy, T., *Software and System Safety*, AuthorHouse, 2012.

Herrmann, D. S., *Software Safety and Reliability: Techniques, Approaches and Standards of Key industrial Sectors*, IEEE Computer Society, 1999.

Hilderman, V. and Baghai, T., *Avionics Certification: A Complete Guide to DO-178 and DO-254*, Avionics Communications Inc., 2007.

IEEE Std-1228, IEEE Standard for Software Safety Plans, 17 March 1994.

Leveson, N. G., *Safeware: System Safety and Computers*, Addison-Wesley Publishing Company, 1995.

NASA, NASA-GB-8719.13: NASA Software Safety Guidebook, March 31, 2004. Replaces NASA-GB-1740.13-96, NASA Guidebook for Safety-Critical Software – Analysis and Development, 1996.

NASA, NASA-STD-8719.13B: Software Safety Standard, NASA Technical Standard, July 8, 2004 (Ver. A Sept 1997).

Rierson, L., *Developing Safety-Critical Software: A Practical Guide for Aviation Software and DO-178C Compliance*, Leanna, CRC Press, 2013.

RTCA, RTCA/DO-178, Software Considerations in Airborne Systems and Equipment Certification, 1992; version C in 2011.

Chapter *22*

Process Hazard Analysis

22.1 PHA INTRODUCTION

Process hazard analysis (PHA) is an analysis technique for evaluating industrial process type systems, such as a chemical plant or an oil processing plant. A PHA provides information intended to assist designers, managers, and employees in making decisions for improving plant safety. A PHA is directed toward analyzing hazards leading to such undesired events as fires, explosions, releases of toxic or flammable chemicals, and major spills of hazardous chemicals. The PHA focuses on equipment, instrumentation, utilities, human actions, and external factors that might adversely impact the process.

A processing plant is a system that specializes in processing operations where certain products are manufactured (processed) from other basic commodities. For example, crude oil is processed into gasoline, diesel, and other oil-based products. Or, tuna fish are processed into canned tuna. Quite often processing plants involve highly hazardous chemicals and chemical combinations that must occur at a critical rate and quantity, thereby making them safety-critical.

Note that the acronym *PHA* for process hazard analysis is also the same acronym for preliminary hazard analysis, thus the acronym must be used carefully to avoid confusion.

22.2 PHA BACKGROUND

This analysis technique falls under the detailed design hazard analysis type (DD-HAT). The basic analysis types are described in Chapter 4. The PHA is performed over a period of time, continually being updated and enhanced as more design information becomes available.

Hazard Analysis Techniques for System Safety, Second Edition. Clifton A Ericson, II.
© 2016 John Wiley & Sons, Inc. Published 2016 by John Wiley & Sons, Inc.

The PHA serves the following purposes:

a. Provide a design safety focus specifically on a processing plant and its operations.
b. Identify and mitigate hazards in the plant design and processing operations.

The PHA is applicable to analysis of all types of systems, equipment, and facilities. The PHA evaluates safety during production, operation, maintenance, and disposal. The PHA technique, when applied to a given system by experienced safety personnel, should provide a thorough and comprehensive identification of hazards that exist in a given system. A basic understanding of hazard analysis theory is essential, so is system safety concept. Experience with the particular type of system is helpful in generating a complete list of potential hazards. The technique is uncomplicated and easily learned. Standard, easily followed PHA worksheets and instructions are provided in this chapter.

22.3 PHA HISTORY

The PHA technique was established by the chemical process industry due to the safety-critical nature of many chemical plants. The chemical industry decided to establish their own analysis process rather than utilize existing system safety analysis techniques. The hazard and operability (HAZOP) technique was developed and promoted as the primary tool for meeting the requirements of a PHA for processing plants.

22.4 PROCESSING MISHAPS

Processing plant accidents are a serious safety concern. Processing plants are specialized systems that often involve safety-critical components and/or operation. The following are some well-known process plant accidents that have occurred over the years, which demonstrate the safety concern involved:

a. Year 1974, Flixborough, England
   - Cyclohexane release and explosion; pipe rupture
   - Number of workers killed 28; injured 36
   - Offsite injuries 53
b. Year 1984, Bhopal gas tragedy, India
   - Toxic material released (highly toxic methyl isocyanate).
   - Plant was in the middle of a highly populated area.
   - About 2,800 immediate fatalities.
   - Many other offsite injuries; 20,000+ total surrounding the facility.
c. Year 1984, Mexico City, Mexico
   - Explosion of flammable LPG in a tank
   - About 300 fatalities (mostly offsite)
   - Damages $20 million
d. Year 1988, Norco, LA
   - Explosion of flammable hydrocarbon vapors

- Seven onsite fatalities, forty-two injured
- Damages $400 million+

e. Year 1989, Pasadena, TX, Phillips 66 Houston Chemical Complex
- Explosion and fire, flammable ethylene/isobutene vapors in a 10" line
- Number of fatalities 23; injured 130
- Damage $800 million+

f. Year 2005, BP Products, Texas City, TX
- Refinery fire and explosion
- Workers killed 15
- Injured 180
- Major property damage
- $50.6 million in fines

g. Year 1998, Anacortes, WA
- Equilon Oil Refinery explosion and fire
- Six fatalities

22.5 PROCESS SAFETY MANAGEMENT

Due to the increase in safety-critical industrial process accidents, the process safety management (PSM) discipline was established and made an integral part of OSHA Occupational Safety and Health Standards in 1992. It is formally known as Process Safety Management of Highly Hazardous Chemicals (29 CFR 1910.119). Typically, PSM applies to most industrial processes containing 10,000+ pounds of hazardous material. PSM is the proactive and systematic identification, evaluation, and mitigation or prevention of chemical releases that could occur as a result of failures in process, procedures, or equipment.

The PSM process is built upon the following 14 elements:

1. Process safety information
2. Employee involvement
3. Process hazard analysis
4. Operating procedures
5. Training
6. Contractors
7. Prestart-up safety review
8. Mechanical integrity
9. Hot work
10. Management of change
11. Incident investigation
12. Emergency planning and response
13. Compliance audits
14. Trade secrets

Note that item 3 is process hazard analysis. PSM requirements for PHA include the following:

- The use of one or more established methodologies appropriate to the complexity of the process.
- Performed by a team with expertise in engineering and process operations.
- Includes personnel with experience and knowledge specific to the process being evaluated and the hazard analysis methodology being used.

The suggested PHA techniques provided by PSM guidelines are as follows:

- Checklist
- What-if
- What-if/checklist
- HAZOP
- FMEA
- FTA

Note that PSM guidelines do not call out a specific PHA methodology or worksheet, the guideline recommend use of one or more of the above techniques. As will be discussed below, some of these techniques are not really adequate and perhaps additional techniques should be added to the list.

22.6 PHA THEORY

A processing plant is a system, one that specializes in processing operations where certain products are manufactured (processed) from other basic commodities. Therefore, most standard hazard analysis methods for a system are applicable to processing plants. The PHA focuses on the identification of potential hazards resulting from the design and operational aspects of a process plant.

Figure 22.1 shows an overview of the basic PHA process and summarizes the important relationships involved in the PHA process. This process consists of utilizing both design information and known hazard information to identify hazards. Known hazardous elements

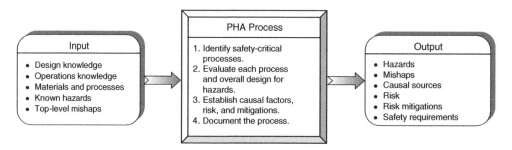

Figure 22.1 PHA overview.

and mishap lessons learned are compared with the system design to determine if the design concept contains any of these potential hazard elements.

The PHA process involves the following:

1. Identifying hazards in the system design
2. Identifying hazards in the operational design
3. Determining the safety-critical processes
4. Determining the critical quantities or exposure levels involved, based on the use, quantity, and type of substance/agent used
5. Establishing hazard risk
6. Establishing design mitigation methods to eliminate or reduce exposures to acceptable levels
7. Including software and human error in the PHA analyses

22.7 PHA METHODOLOGY

Table 22.1 lists and describes the basic steps of the PHA process, which involves performing a detailed analysis of the processing plant design and operations.

TABLE 22.1 PHA Process

Step	Task	Description
1	Acquire design information	Acquire all of the design, operational, and manufacturing data for the system
2	Establish the environments involved	Determine all of the environments the system will be involved with
3	Acquire process plant hazard checklists	Acquire checklists of known hazard sources, such as chemicals, materials, processes, and so on. Also, acquire checklists of known human limitations in the operation of systems, such as noise, vibration, heat, and so on
4	Acquire regulatory information	Acquire all regulatory data and information that are applicable to process hazards
5	Identify hazard sources	Examine the system and identify all potentially hazardous sources and processes within the system. Include quantity and location when possible. Utilize the checklists
6	Identify hazards	Identify and list potential hazards created in the system design resulting from the hazard sources
7	Identify hazard barriers	Identify design mitigation methods or barriers in the path of the hazard source. Also, identify existing design features to eliminate or mitigate the hazard
8	Evaluate system risk	Identify the level of mishap risk presented, both with and without design controls in the system design
9	Recommend corrective action	Determine if the design controls present are adequate, and if not, recommend controls that should be added to reduce the mishap risk
10	Track hazards	Transfer newly identified hazards into the HTS. Update the HTS as hazards; hazard causal factors and risk are identified in the PHA
11	Document PHA	Document the entire PHA process on the worksheets. Update for new information as necessary

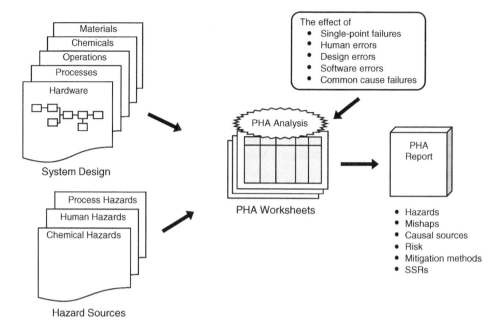

Figure 22.2   *PHA Methodology.*

The thought process behind the PHA methodology is shown in Figure 22.2. The idea supporting this process is that different kinds of design information are used to facilitate hazard identification. Known hazards and top-level mishaps (TLMs) are the starting point for the PHA. The next step is to employ the use of hazard checklists and undesired mishap checklists. Of particular interest are safety-critical components and operations.

The PHA strongly depends upon the use of hazard checklists. Hazard checklists are generic lists of known hazardous items and potentially hazardous designs or situations and should not be considered complete or all-inclusive. Checklists are intended as a starting point to help trigger the analyst's recognition of potential hazard sources from past lessons learned.

22.8   PHA WORKSHEET

Since a processing plant is nothing more than a system that specializes in processing operations, a new specialized hazard analysis worksheet is not necessarily required. Several exiting hazard analysis techniques could be effectively utilized. In fact, the suggested PHA techniques recommended by PSM guidelines are as follows:

- Checklist analysis
- What-if analysis
- What-if/checklist analysis
- HAZOP analysis
- Failure mode and effects analysis (FMEA)
- Fault tree analysis (FTA)

The problem with these recommended methods is that they are not all thorough and complete hazard analysis methodologies. The checklist, what-if, what-if/checklist, FMEA, and FTA are not complete hazard analyses; they are partial or supporting hazard analysis techniques. They are useful for identifying some hazards, but they are not designed to be thorough or complete in recognizing all hazards.

Some exiting methods that are recommended, but which have not been identified by the PSM guidelines, include the following, each of which are described elsewhere in this book:

1. Preliminary hazard analysis (PHA)
2. System hazard analysis (SHA)
3. Operations and support hazard analysis (O&SHA)
4. Software hazard analysis

There is no single unique hazard analysis worksheet for a process hazard analysis, which is more of an objective than a technique. This objective can be achieved using three specific methodologies and worksheets that already exist. System design coverage can be achieved using the preliminary hazard analysis or the SHA. Human operational hazards can be identified using the O&SHA, and system software hazards can be identified using the software hazard analysis. All three are really necessary for a complete safety assessment; all three of these techniques should be applied together in order to conduct a thorough process hazard analysis.

As a minimum, the following basic information is required from the PHA:

1. Hazards existing in the system's design
2. Hazards existing in the system's human interface
3. Hazards existing in the system's software design
4. Dispersion area of toxic materials if inadvertently released
5. Risk to operational personnel and individuals external to the plant
6. Plant hazards resulting from design errors, human errors, and software errors
7. Hazard causal factors (materials, processes, excessive exposures, etc.)
8. Risk assessment (before and after design safety features are implemented)
9. Safety features for eliminating or mitigating the hazards
10. Derived safety requirements for documenting and implementing the safety features

22.9 SUPPORTING NOTES

When conducting a PHA, information pertaining to the technology of the process should include at least the following:

- A block flow diagram or simplified process flow diagram
- Process chemistry and its properties
- Maximum intended inventory
- Safety upper and lower limits for such items as temperatures, pressures, flows, or compositions

An evaluation of the consequences of deviations, including those effecting the safety and health of the employees

Information pertaining to hazards of the highly hazardous chemicals in the process should consist of at least the following:

- Toxicity information
- Permissible exposure limit
- Physical data
- Reactivity data
- Corrosivity data
- Thermal and chemical stability data
- Hazardous effects of inadvertent mixing of different materials that could foreseeably occur

Information pertaining to the equipment in the process should include the following:

- Materials of construction
- Piping and instrument diagram (P&ID)
- Electrical classification
- Relief system design and design basis
- Ventilation system design
- Design codes and standards employed
- Energy balances
- Emergency procedures
- Safety system (for example interlocks, detection, or suppression systems)

When performing the PHA some initial process safety TLMs to utilize include the following:

- Fires
- Explosions
- Unintended release of chemicals, energy, or hazardous materials
- Undesired chemical reactions
- Personnel injuries
- Damage from leaks, spills, corrosion, and so on
- Unintended steam release
- Electrocution

22.10 PHA ADVANTAGES AND DISADVANTAGES

The PHA technique has the following advantages:

1. The PHA is easily and quickly performed.
2. The PHA does not require considerable expertise for technique application.
3. The PHA is relatively inexpensive, yet provides meaningful results.
4. The PHA provides rigor for focusing on system hazards.

There are no notable disadvantages of the PHA technique.

22.11   COMMON PHA MISTAKES TO AVOID

When first learning how to perform a PHA, it is commonplace to commit some traditional errors. The following is a list of typical errors made during the conduct of a PHA.

1. All aspects of the process are not considered.
2. The hazard identified is incorrectly described.
3. The hazard description is not detailed enough.
4. Design mitigation factors not stated or provided.
5. Design mitigation factors that do not address the actual causal factor(s).
6. Overuse of project specific terms and abbreviations.

22.12   SUMMARY

This chapter discussed the PHA technique. The following basic principles help summarize the discussion in this chapter:

1. The primary purpose of PHA is to identify hazards.
2. The use of a hazard checklist greatly aids and simplifies the PHA process.
3. The use of the recommended PHA worksheet aids the analysis process and provides documentation of the analysis.

FURTHER READINGS

There are no significant references for the PHA technique that describe it in detail. Many textbooks on system safety discuss the methodology in general, but do not provide detailed explanation with examples

Center for Chemical Process Safety (CCPS) of the American Institute of Chemical Engineers, Guidelines for Hazard Evaluation Procedures, 3rd edition, 2008.

Crowl, D. A. and Louver, J. F., *Chemical Process Safety: Fundamentals with Applications*, Prentice Hall, 2012.

Lees' Loss Prevention in the Process Industries: Hazard Identification, Assessment and Control (3 Volumes), Sam Mannan, 4th edition, August 17, 2012

Chapter *23*

Test Hazard Analysis

23.1 THA INTRODUCTION

The "requirements/design/build/test/deploy" paradigm is considered a traditional way to build systems. Testing is an action by which the operability, supportability, and/or performance capability of an item is verified when subjected to controlled conditions that are real or simulated. These verifications often use special test equipment or instrumentation to obtain very accurate quantitative data for analysis. Testing is an important aspect of system development and is generally considered part of the systems engineering process.

Test hazard analysis (THA) is a safety analysis for evaluating the safety of testing and test operations. Test operations can run the gamut from large-scale development test programs to small special tests for problem resolution and subsystem verification.

THA does not require a special analysis technique; there are several different existing hazard analysis techniques that can be utilized, such as system hazard analysis (SHA), operating and support hazard analysis (O&SHA), and HAZOP. When performing a THA, the key is to focus on the specific test operations, test procedures, and test equipment when using the SHA, O&SHA, or HAZOP worksheets.

Note that the acronym *THA* is the same for both test hazard analysis and threat hazard analysis; use the acronym carefully to avoid confusion.

23.2 THA BACKGROUND

This analysis technique falls under the operations design hazard analysis type (OD-HAT) because it evaluates test procedures and tasks performed by humans. It also includes the system

Hazard Analysis Techniques for System Safety, Second Edition. Clifton A Ericson, II.
© 2016 John Wiley & Sons, Inc. Published 2016 by John Wiley & Sons, Inc.

design involved in the test tasks. The basic analysis types are described in Chapter 4. There are no alternative names for this type of analysis.

The THA is applicable to the analysis of all types of test operations, procedures, tasks, and functions. It can be performed on draft procedural instructions or detailed instruction manuals. A basic understanding of hazard analysis theory is essential so is the knowledge of system safety concepts. Experience with or a good working knowledge of the particular type of system and subsystem is necessary in order to identify and analyze hazards that may exist within test procedures and instructions. The methodology is uncomplicated and easily learned.

The THA technique serves the following purpose:

a. Prevent mishaps during test activities.

b. Provide a safe set of test plans and procedures.

c. Identify and mitigate hazards existing in test plans and procedures.

d. Ensure test equipment is safe for use during test activities.

23.3 THA HISTORY

There is no formal history for this analysis. It is logical to evaluate test activities for safety; however, not all programs call it out as a program requirement. Experience has shown that mishap can and have occurred during testing; therefore, THA is highly recommended in order to ensure safety in all test activities.

23.4 THA THEORY

Testing is part of the acquisition development process that is performed on a system during development to verify and validate that the system performs all of its operational, functional, and performance requirements. Testing confirms system performance against documented capability needs and requirements and verifies that the system is safe for intended use. Other purpose-specific types of testing are also performed, such as environment qualification testing (temperature, humidity, explosive atmosphere, etc.), insensitive munition (IM) testing, explosives safety testing, reliability testing, and so on.

On some projects, testing is referred to as the test and evaluation (T&E) program. The T&E program is performed on a system to verify that the system meets all of its operational functional and performance requirements. T&E confirms performance against documented requirements and verifies that the system is safe and will perform its intended function. T&E is a formal demonstration that all requirements are met and satisfied.

US document DoD 5000.1 states that "T&E is the principal tool with which progress in system development is measured." T&E is structured to support the defense acquisition process and the user by providing essential information to decision-makers, assessing attainment of technical performance parameters, and determining whether systems are operationally effective, suitable, and survivable for intended use.

T&E involves verifying and validating mishap risk reduction and mitigation through appropriate analysis, testing, or inspection. It is the responsibility of the program manager (PM) to ensure that the selected mishap risk mitigation approaches will result in the expected residual mishap risk. To provide this assurance, the system test effort should verify the performance of the mitigation actions. In addition, the system test effort may identify new hazards during testing.

Tests and demonstrations validate SSRS and safety features of the system. Integration of safety testing into the appropriate system test and demonstration plans, to the maximum extent possible, is desirable. Where costs for safety testing would be prohibitive, engineering analyses, analogy, laboratory test, functional mock-ups, or model simulation may verify safety characteristics or procedures.

A T&E test program typically involves the following stages:

1. *Test Planning* This involves identifying the test requirements, test schedule, test conditions, tests resources, and required test documentation.
2. *Test Preparation* This involves the preparation necessary for performing the testing and includes the following:
   a. Selection of test items
   b. Development of detailed test and evaluation procedures
   c. Test site selection
   d. Selection of test personnel and training
   e. Obtaining the necessary test facilities and resources
   f. Obtaining the necessary test support equipment
3. *Test Performance* This involves conducting the tests per the detailed test procedures.
4. *Test Evaluation and Reporting* This involves evaluating and documenting the test results, including documentation of the test data.

A T&E test program is typically for a large-scale development program for complex and/or safety-critical systems. There are other test programs that are smaller and less formal, such as for reliability accelerated life testing (ALT), highly accelerated life testing (HALT), or reliability qualification testing (RQT).

The test and evaluation master plan (TEMP) is the basic planning document for all T&E related to a particular major system acquisition. It provides the basis and authority for all other detailed T&E planning documents. The TEMP identifies all critical technical parameters and operational issues and describes the objectives, responsibilities, resources, and schedules for all completed and planned T&E. The TEMP is prepared during conceptual design and usually includes coverage of test objectives, technical and operating characteristics of the system, critical issues and interfaces, developmental test and evaluation requirements, operational test and evaluation requirements, test resource requirements, test procedures, and test reporting. The TEMP documents the overall structure and objectives of the T&E program and is continually updated as necessary. It provides the framework within which to generate detailed T&E plans. It is the responsibility of the system safety program (SSP) to ensure that system safety test plans are incorporated into the TEMP.

Figure 23.1 shows an overview of the basic THA process and summarizes the important relationships involved in the THA process. This process consists of utilizing both design information and known hazard information to identify hazards. Known hazardous elements and mishap lessons learned are compared with the system design to determine if the design concept contains any of these potential hazard elements.

The THA process involves the following:

1. Providing a safety focus from a test activity viewpoint.
2. Identifying test-oriented hazards caused by design, hardware failures, software errors, human error, timing, and the like.

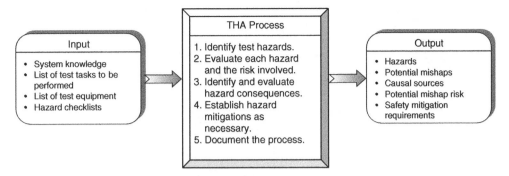

Figure 23.1   *THA overview.*

3. Assessing the potential operations mishap risk.

4. Identifying design system safety requirements (SSRs) to mitigate operational task hazards.

5. Ensuring all test operational procedures and tasks are safe.

23.5   THA METHODOLOGY

Table 23.1 lists and describes the basic steps of the THA process, which involves performing a detailed analysis of all potentially hazardous test activities.

TABLE 23.1   THA Process

Step	Task	Description
1	Acquire test information	Acquire information on the planned test, along with a complete list of the specific tasks to be performed, and the test equipment involved
2	Acquire design information	Acquire all of the design and operational data for the system, particularly for the test being performed. Also, acquire design information on all test equipment that is planned for use
3	Acquire hazard checklists	Acquire checklists of known hazard sources, such as chemicals, materials, processes, and so on. Also, acquire checklists of known human limitations in the operation of systems, such as noise, vibration, heat, and so on
4	Acquire regulatory information	Acquire all regulatory data and information that are applicable to the particular job under review
5	Identify hazard sources	Examine the system and the tasks to be performed and identify all potentially hazardous sources and processes within the system
6	Identify hazards	Identify and list potential hazards created by the tasks to be performed using hazard knowledge, hazard checklists, potential human errors, and equipment failure modes
7	Evaluate system risk	Identify the level of mishap risk presented to personnel, both with and without design controls in the system design, in order to understand risk changes
8	Recommend corrective action	Determine if the hazard controls present are adequate, and if not, recommend controls that should be added to reduce the mishap risk
9	Track hazards	Transfer newly identified hazards to the HTS. Update the HTS as hazards; hazard causal factors and risk are identified in the THA
10	Document THA	Document the entire THA process on the worksheets. Update for new information is necessary

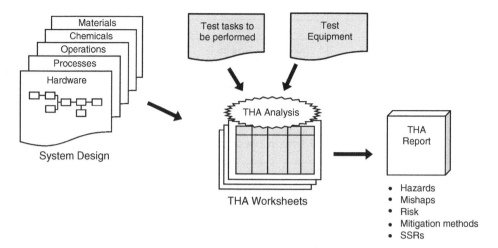

Figure 23.2 *THA methodology.*

The thought process behind the THA methodology is shown in Figure 23.2.

The idea supporting the THA process is that mishaps can happen during test operations if the associated hazards are not identified and mitigated. It is imperative that all test procedures and test equipment be evaluated for safety.

When performing the THA, the primary objective of testing is to ensure that the system operates successfully and safely. However, the THA must consider more than just system operation. The focus of a THA should be on the following considerations:

- Safe system operation – no hazards in the basic system design
- Safe test operations – no hazards created by the special test setups and test steps
- Safe test equipment – no hazards created by the use, or failure, of test equipment

23.6 THA WORKSHEET

There is no unique HA worksheet for a THA analysis; a THA is more of an objective than a technique. A THA can be performed using any one of a number of HA worksheets that already exist. For example, the following hazard analysis techniques can be utilized to achieve a THA: preliminary hazard analysis (PHA), subsystem hazard analysis, SHA, O&SHA, and/or HAZOP. Other HA techniques can be used as found necessary and applicable.

As a minimum, the following basic information should be obtained from the THA:

1. Test-related hazards
2. Hazard causal factors
3. Hazard risk
4. Information to determine if further causal factor analysis is necessary for a particular hazard
5. Recommended risk mitigation methods

The O&SHA methodology appears to be the most closely aligned technique for evaluated hazards associated with testing and test procedures. The O&SHA is primarily for operational and maintenance tasks and procedures, which are very similar in nature to test activities, procedures, and tasks.

23.7 THA CONSIDERATIONS

Testing is a method for system verification and validation. There are numerous definitions for the terms *verification* and *validation*. Most dictionaries identify these terms as being synonymous, yet they are used in different contexts in the system development process when referring to system verification and validation (V&V). Typical V&V definitions are as follows:

23.7.1 Verification

1. The act of reviewing, inspecting, testing, checking, auditing, or otherwise establishing and documenting whether items, processes, services, or documents conform to specified requirements.
2. The process of determining whether the products of a given phase of the software development life cycle fulfill the requirements established during the previous phase.
3. Am I building the right system?

23.7.2 Validation

1. The evaluation of software at the end of the software development process to ensure compliance with user requirements.
2. Validation is, therefore, end-to-end verification.
3. Am I building the system right?

Verification determines whether the requirements for a system are complete and correct and that the outputs of an activity fulfill the requirements or conditions imposed on them in the previous activities. Verification includes coverage of process, requirements, design, code, integration, and documentation. Validation determines whether the final, as-built system fulfills its specific intended use.

Verification involves using reviews, stress and performance testing, observations, and demonstrations as applicable to ensure that the requirements are being addressed at each phase of the development life cycle. Validation places a stronger emphasis on ensuring that the system will perform as intended in the operational environment. The Defense Acquisition Glossary states that validation is the precursor to approval.

In the system safety arena these terms should be interpreted as follows:

Verification   Verify incorporation of SSRs, safety devices, safety features, and so on into the design (i.e., Am I building the right system, by correctly implementing SSRs?).

Validation   Validate that the SSRs, safety devices, safety features, and so on work effectively (i.e., Am I building the system right, by ensuring the mishap mitigation methods are effective?).

The verification aspect of system safety verification and validation (SSVV) involves ensuring that the SSRs are correctly implemented in the system design, and the validation aspect of SSVV involves ensuring that the SSRs successfully and adequately mitigate mishap risk. In system safety, validation of SSRs, safety devices, safety features, and so on is required before a hazard action record can be closed.

23.8   TESTING IN THE SYSTEM DEVELOPMENT LIFE CYCLE

During system development, many different types of testing are conducted. Figure 23.3 illustrates the various types of tests performed during system development and the associated safety involvement.

Testing can be subdivided into the following categories or levels[1]:

Type 1   This level of testing is performed during the early phases of detailed design to verify certain performance and physical design characteristics of components and subsystem. It may be performed on breadboards, bench test models, engineering models, engineering software, service test models, and so on. Although these tests are not formal demonstrations in an operational environment, results from these tests help verify design concepts or necessitate design changes.

Type 2   This level of testing is performed during the latter part of detailed design when preproduction prototype equipment is available. Prototype equipment is similar to production equipment but is not necessarily fully qualified at this point in time.

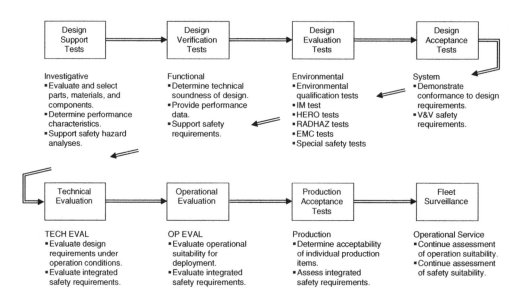

Figure 23.3   *Safety as part of integrated testing.*

[1]System Engineering and Analysis, Chapter 6 – System Test and Evaluation, B. S. Blanchard and W. J. Fabrycky, Prentice Hall, 3rd edition, 1998.

Type 3   This level of testing is performed after initial system qualification and prior to the completion of production. This is the first time that all elements of the system are operated and evaluated on an integrated basis with support equipment and formal operator and maintenance procedures. These tests are formal demonstrations in an operational environment, or an environment that is close to the operational environment. System performance and operational readiness characteristics (availability, dependability, effectiveness, etc.) are determined from this level of testing.

Type 4   This level of testing is performed during the product use phase to gain further insight into a specific area. It may be performed to test new operational parameters, technology insertion, technology upgrade, and so on.

23.9   TYPES OF TESTING

Testing involves different categories and processes. There are four basic test categories. They are as follows:

- *Development Test*   Conducted on new items to demonstrate proof of concept or feasibility.
- *Qualification Test*   Conducted to prove that the system design meets its requirements with a predetermined margin above expected operating conditions, for instance by using elevated environmental conditions for hardware.
- *Acceptance Test*   Conducted prior to transition such that the customer can decide that the system is ready to change ownership from supplier to acquirer.
- *Operational Test*   Conducted to verify that the item meets its specification requirements when subjected to the actual operational environment.

There are also many other test categories that are more specific in nature, such as follows:

- *Quality Test*   Conducted to verify and validate the quality of a product. They include tests such as statistical control, manufacturing control, and reliability.
- *Human Test*   Conducted to verify that the human operator can effectively operate the system as designed. These tests also determine if the operator is likely to easily commit errors and the prediction of human reliability.
- *Safety Test*   Conducted to specifically validate specific safety designs that must meet specific safety requirements for particular types of equipment.

23.9.1   Standard Development Test Types

Many different types of test are performed during system development. Table 23.2 lists the more common types of tests.

23.9.2   Performance Tests

Table 23.3 lists the more common types of tests used for performance testing of hardware.

23.9.3   Software Performance Tests

Table 23.4 lists the more common types of tests used for software performance testing.

TABLE 23.2   Standard System Development Test Types

Test	Purpose
Performance tests	These tests are performed to verify individual system performance characteristics. For example, these tests will verify that an electric motor will provide the necessary output. These tests also verify form, fit, and interchangeability
Environmental qualification tests	These tests verify that the equipment will operate properly in various environments to which it may be exposed, such as temperature cycling, shock, vibration, humidity, wind, salt spray, sand, dust, fungus, acoustic noise, explosive atmospheres, and electromagnetic interference
Structural tests	These tests are performed to determine material characteristics relative to stress, strain, fatigue, bending, torsion, and general decomposition
Reliability qualification tests	These tests are performed to determine the mean time between failure (MTBF) and the mean time between maintenance (MTBM), modes of failure, and so on
Maintainability demonstration tests	These tests are performed to evaluate various elements of logistics support, such as maintenance tasks, task times, maintenance procedures, maintenance skill levels, and so on
Support equipment compatibility tests	These tests are performed to evaluate test equipment, support equipment, and transportation and handling equipment
Personnel test and evaluation	These tests are performed to evaluate human system integration (HSI), personnel skill needs, and training requirements
Technical data verification	These tests are performed to evaluate operations and maintenance procedures
Software verification tests	These tests are performed to evaluate operations and maintenance software. This includes software performance, hardware–software compatibility, software reliability and maintainability, and packaging and documentation

23.9.4   Special Safety-Related Testing

In addition to design requirements testing, safety requirements testing, and special case safety testing, there is a category of testing known as safety-related tests. These are tests that must be performed to ensure compliance with safety criteria and standards for specific design areas. Table 23.5 provides a list of safety-related tests.

23.10   THA SAFETY GOALS

SSVV is an important element in the overarching SSP. As such, the testing and evaluation processes each require distinct system safety responsibilities. These responsibilities include the following:

1. The *testing* effort consists of designing tests and test procedures, performing specific tests, and collecting data on the system under test. The important safety functions that are performed for the testing effort include the following:
   a. Ensure that mishaps/incidents do not occur during the test through review and analysis of the test design, procedures, equipment, facilities, and instrumentation to identify test hazards.
   b. Provide safety input to test procedures (e.g., hazards, cautions, warnings, etc.)

TABLE 23.3 Hardware Performance Tests

Test	Purpose
Component testing	Component testing involves the testing of a part or component to demonstrate hazard control at the lowest level. For example, testing a diode to verify that it survives expected temperature requirements
Subassembly testing	Subassembly testing involves testing parts that are used together in a subassembly. This level of testing ensures that the parts work safely together and that the subassembly works safely and does not generate any previously unforeseen hazards
Assembly testing	Assembly testing involves testing a complete assembly. This level of testing ensures the safe operation of an assembly, and verifies that assembly design control measures work effectively. Assembly testing verifies and validates assembly level SSRs
Subsystem testing	Subsystem testing involves testing a single complete subsystem. This level of testing ensures the safe operation of an entire subsystem comprised of one or more assemblies, and verifies that subsystem design control measures work effectively. Subsystem testing verifies and validates subsystem-level SSRs, and encompasses both hardware and software SSRs. Subsystem testing is primarily performed in a laboratory to ensure the subsystem is ready for higher level testing
Integration testing	Integration testing involves testing several subsystems integrated together. This level of testing ensures the safe operation of one or more subsystems integrated together, and verifies that design control measures work effectively. Integration testing verifies and validates subsystem- and system-level SSRs, and encompasses both hardware and software SSRs. Integration testing is primarily performed in a laboratory; however, it is sometimes also accomplished in the field
System testing	This level of testing ensures the safe operation of one or more subsystems integrated together, and verifies that any design control measures work effectively. System testing verifies and validates subsystem- and system-level SSRs, and encompasses both hardware and software SSRs. System testing may start out in a laboratory, but eventually it must involve testing of the system in its real environment
Regression testing	Regression testing involves performing a specific set of tests on an item to ensure it provides a known set of results. Regression testing is performed when an item (hardware or software) has been modified after already having completed testing. This type of testing ensures that the modifications have not degraded performance or safety, without repeating the entire suite of test previously performed on the item
Fault injection testing	Fault injection testing involves intentionally inserting or injecting hardware failure modes and/or software errors into a system during test operation to determine the overall effect. It is a method that is used to determine if design safety control measures work effectively, and also to identify previously unforeseen hazards
Free play testing	Free play testing involves allowing system operators to randomly do anything they want during a test, such as pressing buttons randomly, performing tasks out of sequence, not performing intended tasks, and so on. It is a method that is used to determine if unidentified hazards still exist in the system design and if design safety control measures work effectively

c. Identify hardware safety functions and devices that require testing.

d. Ensure that software SSRs are appropriately tested by the test program.

e. Ensure that special safety-related tests are performed when necessary.

f. Establish special precautions for tests requiring that safety controls, barriers, and safety devices be disabled in order to perform the tests.

TABLE 23.4 Software Performance Tests

Test	Purpose
Unit (module) testing	Unit testing is the testing of individual units (or modules) that comprise a computer software configuration item (CSCI). For each software unit, test cases and test procedures (including inputs, expected results, and evaluation criteria) are developed, inspected, and executed. Test cases cover all aspects of the unit's design and ensure that the algorithms and logic employed by each unit are correct and the unit satisfied its specified requirements
Unit integration testing	Unit integration testing is the formal testing of combined units that form a CSCI. For unit integration testing, test cases and test procedures (including inputs, expected results, and evaluation criteria) are developed, inspected, and executed. The test cases should cover all aspects of the architectural design
CSCI integration testing	CSCI integration testing is the formal testing of combined CSCIs
CSCI/HWCI integration testing	CSCI/HWCI integration testing is the formal testing of CSCIs combined with hardware configuration items (HWCIs) to determine whether they work together as intended, and continuing this process until all CSCIs and HWCIs in the system are integrated and tested
System testing	System testing is performed to demonstrate to the acquirer that system requirements have been met. This level of testing ensures the safe operation of one or more subsystems integrated together, and verifies that any design control measures work effectively. System testing verifies and validates subsystem- and system-level SSRs, and encompasses both hardware and software. System testing may start out in a laboratory, but eventually it must involve testing of the system in its real environment
Formal qualification testing (FQT)	FQT is performed to demonstrate that CSCI requirements have been met. FQT includes strict identification and configuration control of software and hardware, clear test objective definition and pass/fail criteria, test result reporting with pass/fail results, and anomalies. The testable entity does not include the use of any software not to be delivered to operational environments, (e.g., instrumented code). The person(s) responsible for qualification testing of a given CSCI shall not be the persons who performed detailed design or implementation of that CSCI
Stress testing	Stress testing attempts to overload the system to determine the effect of the software's performance. For example, the system may be subjected to data input variables that exceed the expected conditions or design limitations. Also, the software may be operated continuously for many more hours that expected in a normal mission
Free play testing	Free play testing consists of the execution of nonstructured (nonplanned) test scenarios. The test operator performs operations that might be out of sequence, randomly presses buttons, and so on to determine the effect of a simulated operator error. Free play testing is performed with CM controlled software. Additional safety testing is performed as part of free play when requested by the system safety group
Regression testing	When existing (tested) software is changed, additional testing is required to verify and validate the changes. Rather than reperforming the entire original testing, a subset of selected tests is performed. The specific set of tests selected should be based on the new or modified requirements. The objective is to test the software for expected results, and to also test for new unexpected results that have been inadvertently created

TABLE 23.4.   (*Continued*)

Test	Purpose
Fault injection testing	Fault injection testing involves intentionally inserting or injecting hardware failure modes and/or software errors into a system during test operation to determine the overall effect. It is a method that is used to determine if design safety control measures work effectively, and also to identify previously unforeseen hazards
Decision coverage testing	Testing that verifies that every point of entry and exit in a program has been invoked at least once and every decision in the program has taken on all possible outcomes at least once (refer to DO-178B). The purpose is to determine which code structure was not exercised by the requirements based testing
Modified condition/decision coverage testing	Testing that verifies that every point of entry and exit in a program has been invoked at least once and every condition in a decision in the program has taken all possible outcomes at least once, every decision in the program has taken on all possible outcomes at least once, and each condition in a decision has been shown to independently affect that decision's outcome (refer to DO-178B). A condition is shown to independently affect a decision's outcome by simply varying that condition while holding fixed all other possible conditions. The purpose is to determine which code structure was not exercised by the requirement-based testing

TABLE 23.5   Safety-Related Tests

Test	Purpose
Electrostatic discharge (ESD) test	To evaluate the safety response of unit to the repeated application of high-potential electrostatic energy discharges at selected points. For explosives and ESD-sensitive electronics
Vibration test	To evaluate the safety response of a unit to the expected vibration environment it will encounter when in system operational usage. Refer to MIL-STD-810 (Series) and/or MIL-STD-310 (Series).
Temperature and humidity test	To evaluate the safety response of a unit to the expected temperature and humidity environment it will encounter when in system operational usage. Refer to MIL-STD-810 (Series) and/or MIL-STD-310 (Series).
Explosive atmosphere test	To evaluate the safety response of unit to normal operation in an explosive atmosphere environment it will encounter when in system operational usage. Refer to MIL-STD-810 (Series) and/or MIL-STD-310 (Series)
Forty foot drop test	To evaluate the safety response of a unit to the stress load associated with a free-fall impact from various attitudes, onto a striking plate from a vertical distance of 40 feet. For systems containing explosives. Refer to MIL-STD-2105 (Series)
Fast cook off (FCO) test	To evaluate the safety response of a unit to a rapid increase in temperature. For systems containing explosives. Refer to MIL-STD-2105 (Series)
Slow cook off (SCO) test	To evaluate the safety response of a unit to a slow increase in temperature. For systems containing explosives. Refer to MIL-STD-2105 (Series)
Fragment impact tests	To evaluate system IM detonation due to impact and penetration by high-velocity fragments. Refer to MIL-STD-2105 (Series)

(*continued*)

TABLE 23.5.   (*Continued*)

Test	Purpose
Bullet impact (BI) Tests	To evaluate system IM detonation due to the impact and penetration by bullets. For systems containing explosives. Refer to MIL-STD-2105 (Series)
Spall impact tests	To evaluate system IM detonation due to the impact of hot spall fragments. Refer to MIL-STD-2105 (Series)
Shape charge jet impact tests	To evaluate system IM detonation due to a shaped charge jet impact. Refer to MIL-STD-2105 (Series)
Sympathetic detonation tests	To evaluate sympathetic detonation between identical weapons in their storage or transport configuration
Catapult and arrested landing test	To evaluate the safety response of a unit to stress loads associated with acceleration and deceleration forces inherent in launch and recovery of carrier-based aircraft
Separation from aircraft on landing test	To evaluate the safety response of a unit to severe mechanical and frictional forces encountered when released from an aircraft during landing
RADHAZ (radiation hazard to personnel) test	To measure and plot the electromagnetic radiation from a system emitter to determine the unsafe hazardous levels to personnel zone relative to the emitting device. Refer to MIL-R-9673 (Series)
HERO (hazards of electromagnetic radiation to ordnance) test	To measure and evaluate the susceptibility of a unit's electro-explosive devices to initiation or dudding by specified electromagnetic environment levels that are expected during the system's life cycle
Noise-level test	To measure and evaluate the undesired noise levels emitting from a unit to determine the hazardous effect on personnel. Refer to MIL- STD -1472 (Series)
Fail safe design test	To confirm that probable modes of system and/or equipment failure will fail to a safe state precluding an undesirable hazardous condition
Toxicity test	To determine the presence of suspected toxic materials hazardous to personnel
EMC (electromagnetic compatibility) test	To measure and evaluate the levels of electromagnetic radiation from a system emitter and determine if there is an undesirable effect on the proper operation of system equipment
EMI (electromagnetic interference) test	To evaluate subsystems and equipment and determine if there is an undesirable electronic disruption from internal and/or external EMR energy
High-intensity radio frequency (HIRF) test	To ensure safe operation of aircraft safety-critical and flight electronics in an HIRF environment, and determine safe/unsafe HIRF levels

2. The *evaluation* effort consists of compiling the data from the test effort and making determinations on system performance and the adequacy of design requirements. The important safety functions that must be performed during the evaluation effort include the following:

   a. Determine if all software SSRs were tested and implemented in the software build and performed satisfactorily.

   b. Determine if the safety features of the system functioned as expected and controlled identified hazards to an acceptable level of risk.

c. Identify previously unidentified hazards that might be uncovered through testing.

d. Assist in the correction of inadequate safety features or resolution of new hazards when identified.

Some of the special safety tests that may be required of a program will not be known until specific hazards and hazard mitigation methods have been identified.

In order to effectively support the T&E program from a system safety standpoint, the SSP is responsible for the following activities:

1. Coordinate with test planning to determine testing milestones in order to ensure that safety activities are completed in time to support testing.
2. Schedule safety analysis, evaluation, and approval of test plans and other test documents to ensure that safety is covered during all testing.
3. Prepare safety inputs to operating and test procedures, and to the TEMP.
4. Ensure all software SSRs will be tested.
5. Ensure that other safety-related testing is performed, such as environmental testing, hazardous atmosphere testing, and so on. MIL-STD-810 and MIL-STD-310 provide test method details for environmental tests.
6. Analyze test facilities, equipment, installation of test equipment, and instrumentation prior to the start of testing for the identification of hazards.
7. Identify any hazards unique to the test environment.
8. Identify hazard control measures for identified test hazards.
9. Identify test data that will be of use to safety to close identified hazards.
10. Review test documentation to ensure incorporation of safety requirements, warnings, and cautions.
11. Review test results to determine if safety goals have been met or if any new hazards have been introduced by the test conditions.
12. Collect data on the effectiveness of operating procedures and any safety components or controls of the system.
13. Compile and evaluate safety-related test data.
14. Make a determination about the safety of the system by determining if the safety features have controlled identified hazards as expected, and to an acceptable level of risk.
15. Evaluate the safety compatibility of planned tests with existing systems, equipment, and COTS items.
16. Identify deficiencies and needs for modifications.
17. Evaluate lessons learned from previous tests of new or modified systems or tests of comparable systems to identify possible hazards or restrictions on test conditions.
18. Evaluate tests where safety features must be inhibited in order to perform the test, and ensure that suitable safety features or procedures are in place to mitigate mishap risk associated with the special test.
19. Coordinate with the range safety officer when testing at a range is required.

20. Prepare a safety assessment report (SAR) of the system design prior to a test milestone that demonstrates that all system hazards have been mitigated to an acceptable level of mishap risk.

23.11 THA ADVANTAGES AND DISADVANTAGES

The following are advantages of the THA technique:

1. The THA is easily and quickly performed.
2. The THA does not require considerable expertise for technique application.
3. The THA is relatively inexpensive, yet provides meaningful results.
4. The THA provides rigor for focusing on system tests.
5. The THA quickly provides an indication of where major test hazards will exist.
6. Existing HA worksheets and methods can be utilized or modified, such as the O&SHA.

There are no notable disadvantages of the THA technique.

23.12 COMMON THA MISTAKES TO AVOID

When first learning how to perform a THA, it is commonplace to commit some traditional errors. The following is a list of typical errors made during the conduct of a THA.

1. The hazard analysis fails to include test equipment.
2. The hazard analysis fails to look for errors and incongruities in the test procedures that affect safety.
3. The hazard description is not detailed enough.
4. Design mitigation factors not stated or provided.
5. Design mitigation factors do not address the actual causal factor(s).
6. Overuse of project-specific terms and abbreviations.

23.13 SUMMARY

This chapter discussed the THA technique. The following basic principles that help summarize the discussion in this chapter:

1. The primary purpose of THA is to identify and mitigate test activity hazards prior to conducting any testing.
2. Testing is the form of V&V.
3. Existing HA worksheets can be applied, primarily the O&SHA.
4. Large programs apply a TEMP for defining and developing test activities.
5. Testing provides an opportunity to identify hazards that may have been overlooked by design hazard analyses.
6. It is important that test equipment also be evaluated by the THA.

FURTHER READINGS

Department of Defense, MIL-STD-1472E: Design Criteria Standard: Human Engineering, March 1998.

Department of Defense, MIL-STD-2105C: Notice 1, Hazard Assessment Tests for Non-Nuclear Munitions, 23 July 2003.

Department of Defense, MIL-STD-810E: Test Method Standard for Environmental Engineering Considerations and Laboratory Tests, July 1989.

MIL-HDBK-310: Global Climatic Data for Developing Military Products, June 1997. (Replaces MIL-STD-210, Climatic Information to Determine Design and Test Requirements for Military Systems and Equipment.)

Navy Insensitive Munitions Office, NAVSEAINST 8020.5C: Qualification and Final (Type) Qualification Procedures for Navy Explosives (High Explosives, Propellants, Pyrotechnics and Blasting Agents).

Chapter *24*

Fault Hazard Analysis

24.1   FHA INTRODUCTION

Fault hazard analysis (FHA) is an analysis technique for identifying those hazards arising from component failure modes. It is accomplished by examining the potential failure modes of subsystems, assemblies, or components and determining which failure modes can form undesired states that could result in a mishap.

Note that fault hazard analysis shares the same acronym, *FHA*, with functional hazard analysis. Typically, this does not cause much confusion because fault hazard analysis is not used extensively any more.

24.2   FHA BACKGROUND

The FHA technique falls under the detailed design hazard analysis type (DD-HAT) analysis. The basic hazard analyses types are described in Chapter 4. The purpose of FHA is to identify hazards through the analysis of potential failure modes in the hardware that comprises a subsystem.

The FHA is applicable to analysis of all types of systems and equipment. The technique can be implemented to analyze a subsystem, a system, or an integrated set of systems. The FHA can be performed at any level from the component level to the system level. It is hardware oriented and not suited for software analysis.

The FHA is a thorough technique for evaluating potential failure modes. However, it has the same limitations as the FMEA. It looks at single failures and not combinations of failures.

Hazard Analysis Techniques for System Safety, Second Edition. Clifton A Ericson, II.
© 2016 John Wiley & Sons, Inc. Published 2016 by John Wiley & Sons, Inc.

FHAs generally overlook hazards that do not result entirely from failure modes, such as poor design, timing errors, and so on.

The conduct of an FHA requires a basic understanding of hazard analysis theory, failure modes, and a detailed understanding of the system under analysis. The methodology is similar to failure mode and effects analysis (FMEA). Although the FHA is a valuable hazard analysis technique, the subsystem hazard analysis (SSHA) has replaced the FHA. The SSHA methodology includes considering failure modes for safety implications, and thus it accomplishes the same objective as the FHA.

The FHA technique is not recommended for general usage. Other safety analysis techniques are more cost-effective for the identification of hazards and root causes, such as the SSHA. The FHA should be used only when a rigorous analysis of all component failure modes is required.

The FHA technique is uncomplicated and easily mastered using the worksheets and instructions provided in this chapter.

24.3 FHA HISTORY

The Boeing Company developed the FHA in 1965 for the Minuteman program as a variation of the FMEA technique. It was developed to allow the analyst to stop the analysis at a point where it becomes clear that a failure mode did not contribute to a hazard, whereas the FMEA requires complete evaluation of all failure modes.

24.4 FHA THEORY

The FHA is a qualitative and/or quantitative analysis method, and can be used exclusively as a qualitative analysis or, if desired, expanded to a quantitative one for individual component failure modes. The FHA requires a detailed investigation of the subsystems to determine which components can fail leading to a hazard and resultant effects to the subsystem and its operation.

The FHA answers a series of questions:

- What can fail?
- How it can fail?
- How frequently will it fail?
- What are the effects of the failure?
- What hazards result as a consequence of failure?

The FHA considers total functional and out-of-tolerance modes of failure. For example, a 5%, 5 K ohm (plus or minus 250 ohm) resistor can have such functional failure modes as "failing open" or "failing short," while the out-of-tolerance modes might include "too low resistance" or "too high a resistance."

To conduct an FHA, it is necessary to know and understand the following system characteristics:

- Equipment mission
- Operational constraints
- Success and failure boundaries
- Realistic failure modes and their probability of occurrence

The general FHA approach involves the following:

- Analyzing each component.
- Analyzing all component failure modes.
- Determining if failure mode directly causes hazard.
- Determining if the failure mode has an effect on subsystem and system.
- Determining if the component failure can be induced by another component.

The FHA approach utilizing a columnar form with specially selected entries provides optimum results. This approach establishes a means for systematically analyzing a system or subsystem design for the identification of hazards. In addition to identifying hazards, data in the FHA form provides useful information for other safety analyses, such as the fault tree analysis.

The purpose of the FHA is to identify hazards existing within a subsystem due to potential hardware component failure. This is accomplished by examining the causes and effects of subsystem component failures.

24.5   FHA METHODOLOGY

Figure 24.1 shows an overview of the basic FHA process and summarizes the important relationships involved. The FHA process evaluates the overall impact of each and every component failure mode in a subsystem. The primary FHA goal is to identify hazards from component failures.

Input data for the FHA includes detailed hardware/function design information. Design data may be in the form of the design concept, the operational concept, and major components planned for use in the system, and major system functions. Sources for this information include design specifications, sketches, drawings, schematics, function lists, functional block diagrams (FBDs), and/or reliability block diagrams (RBDs). Input data also includes know failure modes for components, and failure rates for the failure modes. The FHA output information includes the identification of failure modes in the system under analysis, the evaluation of the failure effects, and the identification of hazards.

Table 24.1 lists the basic steps in the FHA process.

In performing an FHA, the idea is to evaluate the components within each subsystem. The system is first subdivided into subsystems; the subsystems are then subdivided into major components or black boxes, whose failure modes can then be evaluated. This FHA approach is shown in Figure 24.2, which contains a hypothetical system, consisting of two subsystems, shown in FBD format.

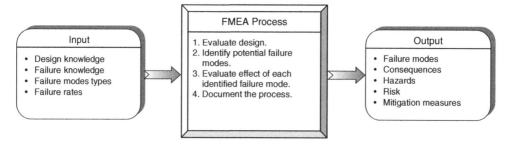

Figure 24.1 FHA overview.

TABLE 24.1 FHA Process

Step	Task	Description
1	Define system	Define scope and boundaries of the system. Establish indenture levels for items to be analyzed
2	Plan FHA	Establish FHA goals, definitions, worksheets, schedule, and process. Define credible failures of interest for the analysis
3	Acquire data	Acquire all of the necessary design and process data needed for the FHA. Refine the item indenture levels for analysis. Data can include functional diagrams, schematics, and drawings for the system, subsystems, and functions. Sources for this information could include design specifications, functional block diagrams, sketches, drawings, and schematics
4	Partition system	Divide the system under analysis into smaller logical and manageable segments, such as subsystems, units, or functional boxes
5	Conduct FHA	For analyses performed down to the component level, a complete component list with the specific function of each component is prepared for each module as it is to be analyzed. Perform the FHA on each item in the identified list of components. This step is further expanded in the next section. Analysis identifies: • Failure mode • Immediate failure effect • System-level failure effect • Potential hazard and associated risk
6	Recommend corrective action	Recommend corrective action for failure modes with unacceptable risk or criticality to program manager for action
7	Monitor corrective action	Review the FHA at scheduled intervals to ensure that corrective action is being implemented
8	Document FHA	Documentation of the entire FHA process, including the worksheets. Update for new information and closure of assigned corrective actions

The next step is to identify and evaluate all credible failure modes for each component within the black box or subsystem. For instance, in subsystem 1, component B may fail "open," The effects of this failure mode on components A and C are determined, so are the effects on the subsystem interface with subsystem 2.

Secondary factors that could cause component B to fail open are identified. For instance, excessive heat radiated from component C may cause component B to fail open.

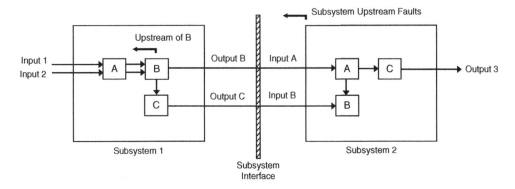

Figure 24.2 *Example system interface.*

Events "upstream" of component B that could directly command component B to fail open are identified. These types of events are usually a part of the normal sequence of planned events, except they occur at the wrong time and may not be controllable once they occur on their own. For example, a short circuit in component A may output from component A the signal that commands component B to respond in the open mode.

When the FHA is completed, the effects of failures in subsystem 1 will terminate at the interface, and the upstream events commanding failures in subsystem 2 will begin from the interface. Hence, it is possible to determine interface hazards by comparing the "effects" of subsystem 1 with the "upstream events" of subsystem 2. This is an indirect result of the FHA.

24.6 FHA WORKSHEET

The FHA is a formal and detailed hazard analysis utilizing structure and rigor. It is desirable to perform the FHA using a worksheet. Although the format of the analysis worksheet is not critical, a recommended FHA format is shown in Figure 24.3. This is the form that was successfully used on the Minuteman missile weapon system program.

The intended content for each column is described as follows:

1. *Component* This column identifies the major functional or physical hardware components within the subsystem being analyzed. The component should be identified by part number and descriptive title.
2. *Failure Mode* This column identifies all credible failure modes that are possible for the identified component. This information can be obtained from the FMEA, manufacturer's data, or testing. (*Note*: This column matches the "primary" cause question in an FTA.)
3. *Failure Rate* This column provides the failure rate or failure probability for the identified mode of failure. The source of the failure rate should also be provided for future reference.
4. *Operational Mode* This column identifies the system phase or mode of operation during the indicated failure mode.
5. *Effect on Subsystem* This column identifies the direct effect on the subsystem and components within the subsystem for the identified failure mode.

Fault Hazard Analysis									
Subsystem_____			Assembly/Unit_____				Analyst_____		
Component	Failure Mode	Failure Rate	System Mode	Effect on Subsystem	Secondary Causes	Upstream Command Causes	MRI	Effect on System	Remarks
①	②	③	④	⑤	⑥	⑦	⑧	⑨	⑩

Figure 24.3 *Recommended FHA worksheet.*

6. *Secondary Causes*   This column identifies secondary factors that may cause the component to fail. Abnormal and out-of-tolerance conditions may cause the component failure. Component tolerance levels should be provided. Also, environmental factors or common cause events may be a secondary cause for failure. (*Note*: This column matches the "secondary" cause question in an FTA).

7. *Upstream Command Causes*   This column identifies those functions, events, or failures that directly force the component into the indicated failure mode. (*Note*: This column matches the "command" cause question in an FTA).

8. *Mishap Risk Index (MRI)*   This column provides a qualitative measure of mishap risk for the potential effect of the identified hazard, given that no mitigation techniques are applied to the hazard. Risk measures are a combination of mishap severity and probability, and the recommended values from MIL-STD-882 are shown in the following table.

Severity	Probability
I – Catastrophic	A – Frequent
II – Critical	B – Probable
III – Marginal	C – Occasional
IV – Negligible	D – Remote
	E – Improbable

9. *Effect on System*   This column identifies the direct effect on the system of the indicated component failure mode.

10. *Remarks*   This column provides for any additional information that may be pertinent to the analysis.

24.7   FHA EXAMPLE

In order to demonstrate the FHA technique, the same hypothetical small missile system from Chapter 7 on preliminary hazard list (PHL) analysis will be used in this example. A system diagram is shown in Figure 24.4 and a conceptual list of system components, functions, energy sources, and phases is shown in Figure 24.5. The missile battery will be selected as the subsystem for this FHA.

Typically, an FHA would be performed on each of the component subsystem designs. For this FHA example, the battery subsystem has been selected for evaluation using the FHA technique. The battery design is shown in Figure 24.6.

In this design, the electrolyte is contained separately from the battery plates by a frangible membrane. When battery power is desired, the squib is fired, thereby breaking the electrolyte housing and releasing electrolyte into the battery, thus energizing the battery.

The battery subsystem is comprised of the following components:

1. Case
2. Electrolyte
3. Battery plates
4. Frangible container separating electrolyte from battery plates
5. Squib that breaks open the electrolyte container

The battery FHA is shown in Table 24.2.

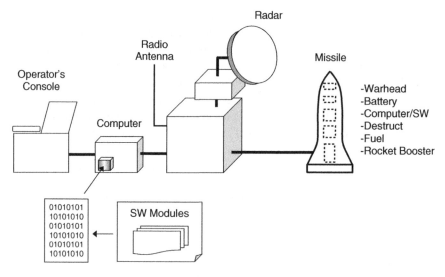

Figure 24.4   *Ace Missile System.*

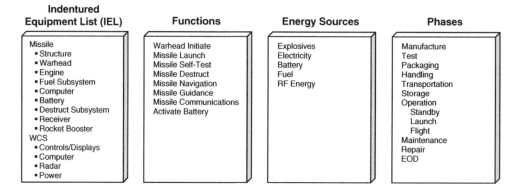

Figure 24.5   *Missile system component list and function list.*

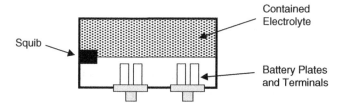

Figure 24.6   *Example missile battery design.*

TABLE 24.2   FHA Worksheet for Battery

| | Fault Hazard Analysis | | | | | | | | |
| | Subsystem: Missile | | Assembly/Unit: Battery | | | | Analyst: | | Date: |
Component	Failure Mode	Failure Rate	System Mode	Effect on Subsystem	Secondary Causes	Upstream Command Causes	MRI	Effect On System	Remarks
Battery squib	Squib fails to ignite	3.5×10^{-5} Manuf. data	Flight	No power output from battery	Excessive shock	No ignition command	4C	Dud missile	Safe
Battery electrolyte	Squib ignites inadvertently	1.1×10^{-9} Manuf. data	Ground operations	Battery power is inadvertently applied	Heat; shock	Inadvertent ignition command	2C	Unsafe system state	Further analysis required
Battery Power	Electrolyte leakage	4.1×10^{-6} Manuf. data	Ground operations	Corrosion; gases; fire	Excessive shock; puncture	Manufacturing defect	2C	Unsafe system state	Further analysis require
Battery Case	Premature power output	1.0×10^{-10} Manuf. data	Ground operations	Power is inadvertently applied to missile electronics	None	Electrolyte leakage into battery cells	2C	Unsafe system state	Further analysis required
	No power output	2.2×10^{-6} Manuf. data	Flight	No power output to missile electronics	Battery damage	Broken cables	4C	Dud missile	Safe
	Case leaks	1.0×10^{-12} Manuf. data	Flight	No power output	Excessive Shock		4C	Dud missile	Safe
			Ground operations	Corrosion; gases: fire	Excessive shock		2C	Unsafe state	Further analysis required

413

The following conclusions can be derived from the FHA worksheet contained in Table 24.2:

1. The failures with a risk level of 2C indicate that the failure mode leaves the system in an unsafe state, which will require further analysis to evaluate the unsafe state and design mitigation to reduce the risk.
2. The failures with a risk level of 4C indicate that the failure mode leaves the missile in a state without power, resulting in a dud missile (not a safety problem).

24.8 FHA ADVANTAGES AND DISADVANTAGES

The FHA technique has the following advantages:

1. FHAs are more easily and quickly performed than other techniques (e.g. FTA).
2. FHAs can be performed with minimal training.
3. FHAs are inexpensive.
4. FHAs force the analyst to focus on system elements and hazards.

The following are disadvantages of the FHA technique:

1. The FHAs focus on single-failure modes and not combinations of failure modes.
2. The FHA focuses on failure modes, overlooking other types of hazards (e.g. human errors).
3. FHAs are not applicable to software since software has no failure modes.

24.9 COMMON FHA MISTAKES TO AVOID

When first learning how to perform a FHA, it is commonplace to commit some traditional errors. The following is a list of typical errors made during the conduct of an FHA:

1. Not fully understanding the FHA technique.
2. Using the FHA technique when another technique might be more appropriate.

24.10 SUMMARY

This chapter discussed the FHA technique. The following basic principles help summarize the discussion in this chapter:

1. The primary purpose of the FHA is to identify hazards by focusing on potential hardware failure modes. Every credible single-failure mode for each component is analyzed to determine if it can lead to a hazard.
2. The FHA is a qualitative and/or quantitative analysis tool.
3. The use of a functional block diagram greatly aids and simplifies the FHA process.

FURTHER READINGS

Ericson, C. A., Boeing Document D2-113072-2. System Safety Analytical Technology: Fault Hazard Analysis, 1972.

Harris, R. W., Fault Hazard Analysis, USAF-Industry System Safety Conference, Las Vegas, Feb. 1969.

Chapter 25

Sneak Circuit Analysis

25.1 SCA INTRODUCTION

Sneak circuit analysis (SCA) is an analysis technique for identifying a special class of hazards known as *sneak circuits or sneak electrical paths*. SCA is accomplished by examining electrical circuits (or command/control functions) and searching out unintended electrical paths (or control sequences) that, without component failure, can result in the following:

1. Undesired operations
2. Desired operations but at inappropriate times
3. Inhibited desired operations

A sneak circuit is a latent path or condition in an electrical system that inhibits a desired condition or initiates an unintended or unwanted action. This condition is not caused by component failures, but has been inadvertently designed into the electrical system to occur as normal operation. Sneak circuits often exist because subsystem designers lack the overall system visibility required to electrically interface all subsystems properly. When design modifications are implemented, sneak circuits frequently occur because changes are rarely submitted to the rigorous testing that the original design undergoes. Some sneak circuits are evidenced as "glitches" or spurious operational modes and can be manifested in mature, thoroughly tested systems after long use. Sometimes, sneaks are the real cause of problems thought to be the result of electromagnetic interference or grounding "bugs." SCA can be applied to both hardware and software designs.

Hazard Analysis Techniques for System Safety, Second Edition. Clifton A Ericson, II.
© 2016 John Wiley & Sons, Inc. Published 2016 by John Wiley & Sons, Inc.

25.2 SCA BACKGROUND

This analysis technique falls under the detailed design hazard analysis type (DD-HAT) and/or the system design hazard analysis type (SD-HAT). Refer to Chapter 4 for further information on analysis types.

The purpose of the SCA is to identify latent paths that can cause the occurrence of unwanted functions or inhibit desired functions, assuming all components are functioning properly. Sneak paths are of concern because they may result in unintended paths or control sequences in electrical/electronic systems that might result in hazards, undesired events, or inappropriately timed events.

SCA is applicable to control and energy delivery circuits of all kinds (e.g., electrical, hydraulic, pneumatic, etc.), although electronic/electrical are the most common. Systems benefiting from SCA include solid-state electronic devices, relay logic systems, and digital systems. It has been used extensively for the evaluation of rocket propulsion, spacecraft, missiles, aircraft, and computers.

SCA can be implemented on a small subsystem, a complete functional system, or an integrated set of systems. Analysis is based on "as built" documentation, in the form of final schematics and drawings. The preferred start time to begin SCA is during engineering development and prior to critical design review (CDR). However, SCA can be performed during any phase of the program where sufficiently detailed design drawings are available. If performed too early, the results may be meaningless, and if performed too late, design changes may be too costly.

The technique, when applied to a given system, is thorough at identifying all previously known types of sneak circuits. The sneak types form the sneak clue list, and as new sneak types are identified, they are added to the clue list. The SCA will not be able to identify paths that lie outside the known clue list; however, a skilled analyst may be able to do so.

SCA is a somewhat a difficult technique to learn, understand, and master, primarily due to the lack of public domain information on the topic. The technique must be mastered, the material understood, and there must be detailed requisite knowledge of the process being modeled. Detailed knowledge of the SCA process, along with all of the clues for identifying sneak paths, is required. SCA is not an analysis for the uninitiated. There is commercially available software that evaluates computer-aided design (CAD) electrical schematics for sneak circuits, thereby making it a slightly simpler process. These software packages typically contain proprietary sneak clues not available to the general public.

The SCA technique enjoys a favorable reputation among control system designers who recognize it as a disciplined approach to the discovery of inadvertent design flaws. A disadvantage is that the technique lends itself to application after significant design and developmental engineering effort has been expended. This makes any sizable design change a relatively expensive consideration. Even with its good reputation, its usage seems to be infrequent and limited to a few programs, primarily due to the costs involved. The use of SCA is not widespread in the system safety discipline because of the proprietary sneak clues, the large amount of effort required, and its focus on sneak hazards. SCA must generally be subcontracted to companies that have the capability, or it must be performed using commercial software packages that are available.

The actual methodology is very straightforward and simple in concept. However, the "clues" used for identifying sneak paths are considered proprietary by each company that has developed SCA capability. Even the companies selling commercial SCA applications will not reveal the clues used; they are built into the program for automatic usage by the program.

A company creating their own SCA technology will not know if their capability is 100% successful without significant research and experience.

As an alternative to an SCA, a hazard analysis that identifies safety-critical circuits, followed by a detailed analysis of those circuits. However, it may not be possible to identify all sneaks without a clue list.

Although a very powerful analysis tool, the benefits of an SCA are not as cost-effective to the system safety analyst as other hazard analysis tools. Other safety analysis techniques, such as SSHA and fault tree analysis, are more cost-effective for the identification of hazards and root causes. SCA is highly specialized, and assists only in a certain niche of potential safety concerns dealing with timing and sneak paths. The technique is not recommended for every day safety analysis usage, and should be used when required for special design or safety-critical concerns. There are the following specific reasons for performing an SCA:

1. The system is *safety-critical* or *high consequence and requires significant analysis coverage to provide safety assurance* (e.g., safe and arm devices, fuzes, guidance systems, launch commands, fire control system, etc.).
2. When an independent design analysis is desired.
3. The cause of unresolved problems (e.g., accidents, test anomalies, etc.) cannot be found via other analysis techniques.

It is important to perform an SCA as early as possible in order to cost-effectively influence system design. Yet, there is a trade-off involved since SCA can be somewhat costly, it should be done only once on a mature design.

25.3   SCA HISTORY

The original SCA technique using topographs and sneak clues was invented and developed by the Boeing Company in Houston, Texas. Beginning in 1967, Boeing developed the methodology for use on the Apollo and Skylab systems for NASA. As the technique proved successful, it was applied to many other types of systems and software. It was originally applied to electrical/electronic systems. It was found to be very successful and was later applied to software, with some limited success.

25.4   SCA DEFINITIONS

In order to facilitate a better understanding of SCA, some definitions for specific terms are in order. The following are the basic SCA terms:

Sneak   A latent path or condition in an electrical system that inhibits a desired condition or initiates an unintended or unwanted action through normal system operation, without failures involved.

Sneak Clues   A checklist of items or clues that helps the analyst recognize and identify a sneak. The analyst compares the clues to the network tree and topograph to recognize sneaks. The clue list has been developed from past experience and research.

Node   A node is an electrical circuit component, such as a resistor, transistor, switch, and so on.

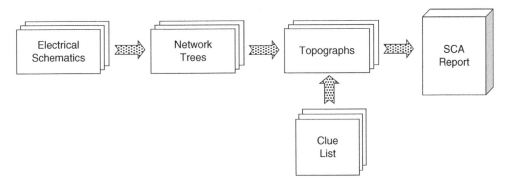

Figure 25.1 *SCA concept.*

Nodal Sets   Nodal sets are the set of interconnected nodes that make up each circuit.

Paths   Paths are the route that results when all of the nodes in a nodal set are connected together.

Network Trees   A network tree is a diagram that represents a simplified version of the system circuitry, created by selective deletion of extraneous circuitry detail to reduce system complexity, while retaining all circuit elements.

Topograph   A topograph is the topological pattern that appears in each network tree.

25.5   SCA THEORY

The purpose of SCA is to identify sneak paths in electrical circuits, software, and so on that result in unintended operation or inhibited operation of a system function. There are several ways by which this can be achieved, such as systematic inspection of detailed circuit diagrams, manually drawing simplified diagrams for manual examination, or by using the automated topgraph-clue method developed by Boeing. This chapter focuses on the topograph-clue method because it is more structured and rigorous, and it was the genesis for the SCA concept.

The theory behind the automated topgraph-clue method of SCA is conceptually very simple, and is portrayed in Figure 25.1. Electrical circuit diagrams are transformed into network trees through the use of special computer programs. The network trees are then reduced down into topographs. The topographs are evaluated in conjunction with clue lists to identify sneak circuits. Although the concept appears to be very simple, there is much more complexity actually involved in the process.

25.6   SCA METHODOLOGY

Figure 25.2 shows an overview of the basic SCA process and summarizes the important relationships involved. The input data primarily consists of detailed electrical schematics and wire lists. The output from the SCA consists of a sneak report identifying all of the identified sneak-type problems and concerns.

The basic definition for a sneak circuit is that it is a designed-in signal or current path that causes unwanted functions or modes of operation to occur, or which inhibits a desired function

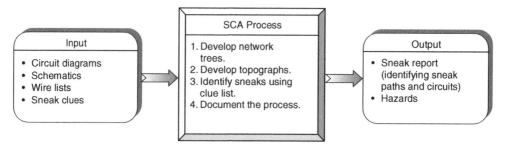

Figure 25.2   *SCA overview.*

from occurring. Sneak circuit conditions are latent in nature, that is, they are always present but not always active; they do not depend upon component failures.

The analysis method is accomplished by examining circuits (or command/control functions), searching out unintended paths (or control sequences) that, without component failure, can result in undesired operations, or in desired operations at inappropriate times, or which can inhibit desired operations.

Experienced SCA analysts indicate the sneak circuits are primarily caused by the following:

a. *Design Oversight*   Large and complex systems make complete overview extremely difficult. As a result, latent sneak paths are accidentally built into the design and are not immediately recognized.

b. *Changes*   Revision to the original design that corrects one design problem but inadvertently creates a new sneak path in the design.

c. *Incompatible Design*   Designs prepared by independent designers or design organizations may be incompatible with respect to the inadvertent creation of a sneak circuit in the integrated system.

d. *Fixes*   Malfunctions observed during testing are occasionally corrected by field fixes that not only fix the immediate problem but also generate a sneak condition that is not immediately recognized.

e. *Human Error*   Human error can contribute to a sneak condition when specified tasks are performed improperly, out of sequence, or when unanticipated modes of operation are performed.

In general, the SCA process involves seven basic steps, as shown in Table 25.1. These are the major tasks that must be performed in order to conduct an SCA. If a commercial software package is used, it will perform some of the steps for the analyst.

Each of these seven steps is described in greater detail in the following sections.

25.6.1   Step 1: Acquire Data

The first step is to acquire data that represents the "as built" circuitry of the system, as closely as possible. Functional schematics, integrated schematics, and system-level schematics do not always accurately represent the constructed hardware. Detailed manufacturing and installation schematics must be used because these drawings specify exactly what is built, contingent upon quality control checks, tests, and inspections. The final analysis results are only as good as the

TABLE 25.1 Sneak Circuit Analysis Process

Step	Task	Description
1	Acquire Data	Understand system; acquire current design data
2	Code data	Format data for computer
3	Process data	Computer processing of input data
4	Produce network trees	Generate network tree diagrams of system
5	Identify topographs	Identify various topographs within network trees
6	Perform analysis	Apply clues and identify sneak problems
7	Generate SCA report	Report problems and recommended solutions, and document the entire SCA process

input data that is used. The data requirements necessitate performance of an SCA after the detail level circuitry design is available, typically after a program's preliminary design review. If the analysis is performed earlier in the program, subsequent design changes can invalidate much of the analysis results. On the other hand, while SCA is applicable to mature operational systems, the analysis ideally should be applied at very early stages of the system development for it to minimize integration problems and to permit design changes early in the development process before manufactured hardware requires modifications. The most cost-effective period in a project's life cycle for performing an SCA is after the detailed circuit design has been completed but before all hardware assembly and testing have been accomplished.

25.6.2 Step 2: Code Data

Computer automation has played a major role in SCA from its inception. Computer programs have been developed to allow for encoding of schematics and wire lists into simple "to-from" connections. The analyst must first study the system intensively and partition it at key points, such as power and ground busses, to avoid large and unnecessarily cumbersome network trees. Once partitioned, rigorous encoding rules for both schematic and wire list data are used to ensure that an accurate representation of the circuitry is maintained. The same rules are applied on all electrical continuity data regardless of source, format, or encoding analyst. The single resultant format for all continuity data ensures that the computer can tie all connected nodes together accurately to produce nodal sets. Note that the coding rules are either proprietary or internal parts of a commercial computer program.

25.6.3 Step 3; Process Data

When the system schematics are encoded and input into the computer, the computer program performs all the necessary checks, calculations, and SCA processing to generate the needed data. At this stage, the computer may discover errors in the coded data and require the analyst to make corrections in the input data.

The SCA program recognizes each reference designator/item/pin as a single point or node. This node is tied to other nodes as specified by the input data. The computer connects associated nodes into paths and collects associated paths into nodal sets. The nodal sets represent the interconnected nodes that make up each circuit. Each node is categorized by the computer as "special" or "nonspecial." Special nodes include all active circuit elements such as switches, loads, relays, and transistors; nonspecial nodes are interconnecting circuit elements such as connectors, terminal boards, and tie points. Before a nodal set can be output for analysis, the set is simplified by the elimination of the nonspecial nodes. This simplification removes all nonactive circuit elements from the nodal sets while leaving the circuit functionally intact. This simplified nodal set is then output from the computer.

First, the SCA program generates plots of each nodal set. Then, the program generates other output reports including (a) path reports wherein every element of each path is listed in case the analyst needs to trace a given path node by node; (b) an output data index that lists every item and pin code, and provides the nodal set and path numbers in which each appears; and (c) matrix reports that list each node of a nodal set, lists all labels associated with active circuit elements of that nodal set, and provides cross-references to other related circuitry (e.g., a relay coil in one nodal set will be cross-referenced to its contacts that may appear in another nodal set). Once these subsidiary reports are generated, the reports and the nodal set plots are turned over to sneak circuit analysts for the next stage of the analysis, network tree production.

25.6.4 Step 4: Produce Network Trees

Network trees are the end result of all data manipulation activities undertaken in preparation for the actual analysis. The SCA program derives network trees from the circuit under analysis. The network trees represent a simplified version of the system circuitry by employing selective deletion of extraneous circuitry detail to reduce system complexity, while retaining all circuit elements pertinent to an understanding of all system operational modes. All power sources are drawn at the top of each network tree with grounds appearing at the bottom. The circuit is oriented on the page such that current flow would be directed from top to bottom down the page.

To produce completed network trees, the analyst begins with the computer nodal set plots. Occasionally, these must be redrawn to ensure "down-the-page" current flow, then thoroughly labeled and cross-referenced with the aid of matrix reports and the other computer output reports. If these simple guidelines are followed in the production of network trees, identification of the basic topographs (step 5) is greatly simplified.

25.6.5 Step 5: Identify Topographs

The analyst must next identify the basic topological patterns (topographs) that appear in each network tree. There are five basic topograph patterns: (a) single line (no node), (b) ground dome, (c) power dome, (d) combination dome, and (e) "H" pattern, as illustrated in Figure 25.3.

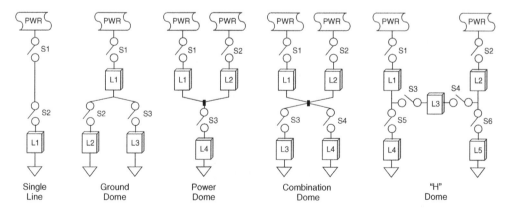

Figure 25.3 General SCA topograph patterns.

One of these patterns, or several together, will characterize the circuitry of any given network tree. Although at first glance a given circuit may appear more complex than these basic patterns, closer inspection reveals that the circuit is composed of these basic patterns in combination. While examining each intersect node in the network tree, the SCA analyst must identify the pattern or patterns containing that node and apply the basic clues that have been found to typify sneak circuits involving that particular pattern. When every intersect node in the topograph has been examined, all sneak circuit conditions within the network tree will have been uncovered.

25.6.6 Step 6: Perform Analysis

Associated with each topograph pattern is a list of clues to help the analyst identify sneak circuit conditions. These lists were first generated by Boeing during the original study of historical sneak circuits and were updated and revised during the first several years of applied SCA. Now, the lists provide a guide to all possible design flaws that can occur in a circuit containing one or more of the five basic topograph patterns, subject to the addition of new clues associated with new technological developments. The clue list consists of a series of questions, either imbedded in the SCA program or known to the analyst, used to identify sneak paths. For example, the single-line topograph shown Figure 25.4 would have clues such as follows:

a. Is switch Sl open when load Ll is desired?
b. Is switch Sl closed when load L1 is not desired?
c. Does the label Sl reflect the true function of Ll?

Sneak circuits are rarely encountered in single-line topographs because of their simplicity. This is an elementary example given primarily as the default case that covers circuitry not included by the other topographs. With each successive topograph, the clue list becomes longer and more complicated. The clue list for the "H" pattern includes over 100 clues. This pattern,

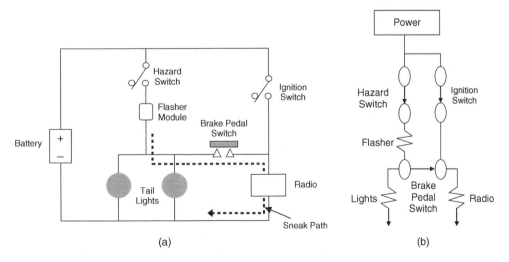

Figure 25.4 *Sneak path example (automotive circuit). (a) Electrical circuit and (b) network tree.*

because of its complexity, is associated with more sneak circuits than any other pattern. Almost half the critical sneak circuits identified to date can be attributed to the "H" pattern. Such a design configuration should be avoided whenever possible. The possibility of current reversal through the "H" crossbar is the most commonly used clue associated with "H" pattern sneak circuits.

25.6.7   Step 7: Generate SCA Report

Sneak circuit analysis of a system produces the following four general categories of outputs: (a) drawing error reports, (b) design concern reports, (c) sneak circuit reports, and (d) network trees and supplementary computer output reports.

Drawing error reports disclose document discrepancies identified primarily during the data-encoding phase of the SCA effort. Design concern reports describe circuit conditions that are unnecessary or undesirable but are not actual sneak circuits. These would include single failure points, unsuppressed inductive loads, unnecessary components, and inadequate redundancy provisions. A number of such conditions usually are identified whenever an analyst examines a circuit at the level of detail required for a formal SCA.

Sneak circuit reports delineate the sneak conditions identified during the analysis. These reports fall into the following broad categories:

1. *Sneak Paths*   Sneak paths are latent paths in designed circuitry that even without component failures permit unwanted functions to occur, or inhibit desired functions from occurring. A sneak path is one that permits current or energy to flow along an unsuspected path or in an unintended direction.

2. *Sneak Timing*   Sneak timing involves a latent path that modifies the intended timing or sequencing of a signal, creating an inappropriate system response. Functions are inhibited or occur at an unexpected or undesired time. For example, faulty ignition timing can ruin the performance of an automobile engine.

3. *Sneak Labels*   Sneak labels refers to the lack of precise nomenclature or instructions on controls or operating consoles that can lead to operator error. A label on a switch or control device could cause incorrect actions to be taken by operators.

4. *Sneak Indicators*   Sneak indicators are an indication that causes ambiguous or incorrect operator displays; a false or ambiguous system status resulting from an improper connection or control of display devices. For example, a warning light might glow while the item it monitors is not failed, or vice versa.

5. *Sneak Procedures*   Sneak procedures refer to ambiguous wording, incomplete instructions, lack of caution notes, or similar deficiencies that might result in improper operator action under contingency operations.

25.7   EXAMPLE 1: SNEAK PATH

Figure 25.4 illustrates an automotive circuit with a sneak path. Note the "H" topograph pattern in this example. Design intent was for the car radio to be powered only through the ignition switch, and for the taillights to be powered from the ignition through the brake switch or via the hazard light switch. A reverse current condition occurs when the hazard switch is closed and the brake switch closes. The radio blinks on and off even though the ignition switch is off.

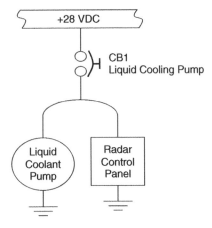

Figure 25.5 *Sneak label example.*

Although not a hazardous situation, this demonstrated how sneak circuits could exist through design errors.

25.8 EXAMPLE 2: SNEAK LABEL

Figure 25.5 illustrates a sneak label example that was discovered on an aircraft radar system. In this example, the circuit breaker provides power to two disparate systems, but the circuit breaker label reflects only one of the systems. An operator attempting to remove power from the liquid coolant pump would inadvertently remove power from the entire radar system.

25.9 EXAMPLE 3: SNEAK INDICATOR

Figure 25.6 illustrates a sneak indicator example that was discovered on a sonar power supply system. In this example, the indicator lamps for motor ON and OFF do not truly monitor and reflect the actual status of the motor. Switch S3 could be in the position shown, providing motor ON indication even though switches S1 or S2, or relay contacts K1 or K2, could be open and inhibiting motor operation.

25.10 EXAMPLE SNEAK CLUES

Clues are applied to the software network trees and topographs to identify design problems and sneak circuits. Table 25.2 shows a few example clues that are used in the analysis. Companies that are experienced in SCA generally have a much longer list of clues, which they usually consider proprietary. The list of clues is one of the strengths behind SCA.

25.11 SOFTWARE SNEAK CIRCUIT ANALYSIS

Software sneak circuit analysis (SSCA) is the extension of the hardware SCA technique to analyze software, because sneak paths can also exist within computer software code. A software

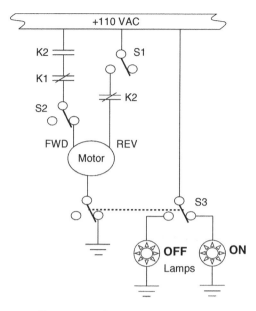

Figure 25.6 *Sneak indicator example.*

sneak path is a latent path or condition in software code that inhibits a desired condition or initiates an unintended or unwanted action.

The purpose of SSCA is to discover software code logic that could cause undesired program outputs, incorrect program operation, or incorrect sequencing/timing. When software controls a safety-critical function, an SSCA can help detect sneak paths that would result in a mishap.

TABLE 25.2   Example Electrical SCA Clues

No.	Clue Questions
1	Can a switch be open when a load is required
2	Can a switch be closed when a load is not required
3	Can the clocks be unsynchronized when they are required to be synchronized
4	Does the label indicate the true condition
5	Is anything else expected to happen that is not indicated by the label
6	Can current flow in the wrong direction? Current reversal concern
7	Can a relay open prematurely under certain conditions
8	Can power be cutoff at a wrong time because of an energized condition in the interfacing circuit
9	Can a circuit lose its ground because of interfacing circuits
10	Is relay race possible
11	Is feedback possible
12	Conflicting commands possible
13	Ambiguous labels
14	False indicators
15	Are intermittent signals possible
16	Incorrect signal polarities
17	Load exceeds drive capabilities
18	Incorrect wiring interconnections
19	Improper voltage levels
20	Is counter initialized (inaccurate count concern)
21	Is latch initialized (initial system state undefined concern)

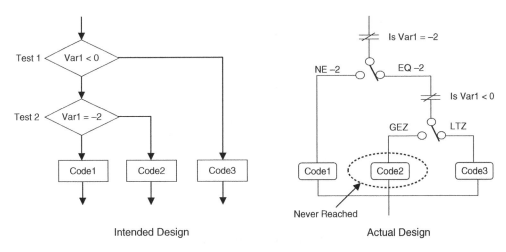

Figure 25.7 *Software SCA example.*

The technique was invented and developed by the Boeing. Following the successful application of SCA on hardware, Boeing research in 1975 showed that the same methodology could also be applied to software.

The following are significant advantages of the SSCA technique:

1. SSCA is a rigorous approach based on the actual software code.
2. SSCA permits large portions of the work to be computerized.
3. SSCA can locate sneak paths that are difficult to find via other techniques.
4. SSCA works equally well on different programming languages.
5. SSCAA does not require the execution of the software.

The overall methodology is identical to that of hardware SCA. The same basic seven-step SCA process applies, except the system and data consist of software code instead of electrical circuits. The program source code is converted into network trees and topographs, which the analyst evaluates using software clues. In this case, the network trees are based on electrical pseudo-circuit models.

Figure 25.7 shows example software sneak that is often cited in various SCA literature. The intended design is shown on the left as a flow diagram. The topographs developed from a network tree is shown on the right. This SSCA discovered that the actual code was not implemented correctly as designed. Test 1 and test 2 were inadvertently interchanged in the code and thus the circled branch would never be executed.

Although SSCA did discover this code error without actually exercising the code, it might also have been found through peer code review or through the analysis of a flow diagram generated from the code. Module testing might have also discovered the error.

Clues are applied to the software network trees and topographs just as done in hardware SCA. Table 25.3 shows a few example software clues that are used in the analysis. Companies that are experienced in SSCA generally have a much longer list of clues, which they consider proprietary.

TABLE 25.3   Example Software SCA Clues

No.	Clue
1	Unused paths
2	Inaccessible paths
3	Improper initialization
4	Lack of data storage/usage synchronization
5	Bypass of desired paths
6	Improper branch sequencing
7	Potential undesirable loops
8	Infinite looping
9	Incorrect sequencing of data processing
10	Unnecessary (redundant) instructions

25.12   SCA ADVANTAGES AND DISADVANTAGES

The following are advantages of the SCA technique:

1. Rigorous approach based on detailed schematics.
2. A large portion of the analysis can be computerized.
3. Can locate sneak paths that are difficult to find via manual techniques.
4. Commercial CAD/SCA software packages are available for performing SCA.

Although a strong and powerful technique, SCA has the following disadvantages:

1. SCA is somewhat of a proprietary technique. Only corporations that have researched and developed the tool have the clues that are necessary for identifying the sneak paths. The clues are not available in any public domain form. Therefore, the cost of SCA can be considerable.
2. Entering the data for the SCA is a time-consuming process. Therefore, it is usually done only once during the program development, and this is usually with detailed design information. Consequently, any identified design changes can be more costly than if identified earlier in the program development life cycle.
3. SCA does not identify all system hazards, but only those dealing with sneak paths.
4. SCA requires an experienced analyst to actually recognize the sneak paths from the clues and topological diagrams.
5. SCA considers only normal component operation; it does not consider component failures.

25.13   COMMON SCA MISTAKES TO AVOID

When first learning how to perform an SCA, it is commonplace to commit some traditional errors. The following is a list of typical errors made during the conduct of an SCA:

1. Not obtaining the necessary training.
2. Using the complex SCA technique when a simpler technique might be more appropriate.
3. Trying to apply the SCA technique without all of the established clues.

25.14 SUMMARY

This chapter discussed the SCA technique. The following basic principles help summarize the discussion in this chapter:

1. SCA is a specialized tool for identifying specific classes of hazards, such as those dealing with hardware timing issues and sneak electrical paths in the hardware.
2. SCA simplifies a complex circuit into network tree diagrams and topographs that can be easily analyzed using a set of clues (usually proprietary) available to the SCA analyst.
3. SCA and SSCA require special computer programs to generate the networks and topographs for the analysis.
4. The SCA analyst must have a complete set of sneak clues in order to ensure a complete analysis.
5. SSCA simplifies complex computer programs into network tree diagrams and topographs that can be analyzed using a set of clues (generally proprietary) available to the SSCA analyst.

FURTHER READINGS

Although there are many papers and references on SCA, the following are probably the most relevant:

Apollo Spacecraft Sneak Circuit Analysis Plan, NASA; SB08-P-108; NASW-1650, 1968.

Browne, J., Jr., Benefits of Sneak Circuit Analysis for Defense Programs, Proceedings of the Third International System Safety Conference, pp. 303–320, Oct., 1977.

Buratti, D. L., Pinkston, W. E., and Simkins, R. O., Sneak Software Analysis, RADC-TR-82-179, June, 1982, pp. 1–6.

Carter, A. H., Budnik, K. T., and Douglass, S. R., Computer Produced Drawings for Circuit Analysis, *Proc. Ann. RM Symp.*, 1985, 224–229.

Clardy, R. C., Sneak Circuit Analysis: An Integrated Approach, Proceedings of the Third International System Safety Conference, pp. 377–387, Oct., 1977.

Clardy, R. C., Sneak Circuit Analysis Development and Application, 1976 Region V IEEE Conference Digest, April 1976.

Clardy, R. C., Sneak Circuit Analysis, *Reliability and Maintainability of Electronic Circuits*, Computer Science Press, pp. 223–241, 1980.

Godoy, S. G. and Engels, G. J., Sneak Circuit and Software Sneak Analysis, *J. Aircraft*, **15**(8), 509–513, (1978).

Hill, E. J., Sneak Circuit Analysis of Military Systems, Proceedings of the Second International System Safety Conference, pp. 351–372, July, 1975.

Peyton, B. H. and Hess, D. C., Software Sneak Analysis, Seventh Annual Conference of the IEEE/ Engineering in Medicine and Biology Society, pp. 193–196, 1985.

Price, C. J., Snooke, N., and Landry J., Automated Sneak Identification, *Eng. Appl. Artif. Intel.*, **9**(4), 423–427 (1995).

Rankin, J. P., Origins, Application and Extensions of Sneak Circuit Analysis on Space Projects *Hazard Prevent.*, **33**(2), 2nd Q, 24–30 (1997).

Rankin, J. P., Sneak Circuit Analysis, *Nucl. Safety*, **15**(5), 461–468 (1973).

Rankin, J. P., Sneak Circuit Analysis, Proceedings of the 1st International System Safety Conference pp. 462–482, 1973.

Sneak Circuit Analysis Guideline for Electro-Mechanical Systems, NASA Practice No. PD-AP-1314, Oct. 1995.

Chapter 26

Markov Analysis

26.1 MA INTRODUCTION

Markov analysis (MA) is an analysis technique for modeling system state transitions and calculating the probability of reaching various system states from the model. MA is a tool for modeling complex system designs involving timing, sequencing, repair, redundancy, and fault tolerance. MA is accomplished by drawing system state transition diagrams and examining these diagrams for understanding how certain undesired states are reached, and their relative probability. MA can be used to model system performance, dependability, availability, reliability, and safety. MA describes failed states and degraded states of operation where the system is either partially failed or in a degraded mode, where some functions are performed while others are not.

Markov chains are random processes in which changes occur only at fixed times. However, many of the physical phenomena observed in everyday life are based on changes that occur continuously over time. Examples of these continuous processes are equipment breakdowns, arrival of telephone calls, and radioactive decay. *Markov processes* are random processes in which changes occur continuously over time, where the future depends only on the present state and is independent of history. This property provides the basic framework for investigations of system reliability, dependability, and safety. There are several different types of Markov processes. In a semi-Markov process, time between transitions is a random variable that depends on the transition. The discrete and continuous time Markov processes are special cases of the semi-Markov process (as will be further explained).

26.2 MA BACKGROUND

This analysis technique falls under the system design hazard analysis type (SD-HAT) and should be used as a supplement to the SD-HAT analysis. Refer to Chapter 4 for a description

Hazard Analysis Techniques for System Safety, Second Edition. Clifton A Ericson, II.
© 2016 John Wiley & Sons, Inc. Published 2016 by John Wiley & Sons, Inc.

of the analysis types. The purpose of MA is to provide a technique to graphically model and evaluate systems components, in order to resolve system reliability, safety, and dependency issues. The graphical model can be translated into a mathematical model for probability calculations. The strength of MA is its ability to precisely model and numerically evaluate complex system designs, particularly those involving repair and dependencies.

MA can be used to model the operation, or failure, of complex system designs. MA models can be constructed on detailed component designs or at a more abstract subsystem design level. MA provides a very detailed mathematical model of system failure states, state transitions, and timing. The MA model quickly becomes large and unwieldy as system size increases and is, therefore, usually used only on small system applications, or systems abstracted to a smaller more manageable model.

MA can be applied to a system early in development and thereby identify design issues early in the design process. Early application will help system developers to design-in safety and reliability of a system during early development rather than having to take corrective action after a test failure, or worse yet, a mishap.

MA is a somewhat difficult technique to learn, understand, and master. A high-level understanding of mathematics is needed to apply the methodology. The technique must be mastered, the material understood, and there must be detailed requisite knowledge of the process being modeled. MA generally requires an analyst very experienced with the technique and the mathematics involved.

Although a very powerful analysis tool, MA does not appear to provide as strong of a benefit to the system safety analyst as do other analysis tools that are available. It is more often used in reliability for availability modeling and analysis. MA does not identify hazards; its main purpose is to model state transitions for better understanding of system operation and calculating failure state probabilities. MA models can quickly become excessively large and complex, thereby forcing simplified models of the system. MA is recommended primarily only when extremely precise probability calculations are required.

Fault tree analysis (FTA) is recommended for most analysis applications because the fault tree combinatorial model is easier to generate from the system design, and the resulting probability calculations are equal or very close to results from MA models. FTA can be used to model extremely large complex systems, which would be impossible by MA.

26.3 MA HISTORY

The Markov chain theory derives its name from the Russian mathematician Andrei A. Markov (1856–1922), who pioneered a systematic investigation of mathematically describing random processes. The semi-Markov process was introduced in 1954 by Paul Levy to provide a more general model for probabilistic systems.

26.4 MA DEFINITIONS

In order to facilitate a better understanding of MA, some definitions for specific terms are in order. The following are the basic MA terms:

State The condition of a component or system at a particular point in time (i.e., operational state, failed state, degraded state, etc.).

Connecting Edge   A line or arrow that depicts a component changing from one system state to a different state, such as transitioning from an operational state to a failed state.

State Transition Diagram   A state transition diagram is a directed graph representation of system states, transitions between states, and transition rates. These diagrams contain sufficient information for developing the state equations, which are used for probability calculations. The state transition diagram is the backbone of the technique.

Combinatorial Model   A graphical representation of a system that logically combines system components together according to the rules of the particular model. Various types of combinatorial models that are available include reliability block diagrams (RBDs), fault tree s (FTs) and success trees. In an MA, the state transition diagram is the combinatorial model.

Deterministic Process   A deterministic process or model predicts a single outcome from a given set of circumstances. A deterministic process results in a sure or certain outcome and is repeatable with the same data. A deterministic model is sure or certain, and is the antonym of random.

Stochastic Process   A stochastic process or model predicts a set of possible outcomes weighted by their likelihoods or probabilities. A stochastic process is a random or chance outcome.

Markov Chain   These are sequences of random variables in which the future variable is determined by the present variable, but is independent of the way in which the present state arose from its predecessors (the future is independent of the past given the present). The Markov chain assumes discrete states and a discrete time parameter, such as a global clock.

Markov Process   The Markov process assumes states are continuous and evaluates the probability of jumping from one known state into the next logical state until the system has reached the final state. For example, the first state is everything in the system working, the next state is the first item failed, and this continues until the final system failed state is reached. The behavior of this process is that every state is memory-less, meaning that the future state of the system depends only on its present state. In a stationary system, the probabilities that govern the transitions from state to state remain constant, regardless of the point in time when the transition occurs.

Semi-Markov Process   The semi-Markov process is similar to that of a pure Markov model, except that the transition times and probabilities depend upon the time at which the system reached the present state. The semi-Markov model is useful in analyzing complex dynamical systems, and is frequently used in reliability calculations.

26.5   MA THEORY

The changes of state of the system are called transitions, and the probabilities associated with various state changes are called transition probabilities. The process is characterized by a state space, a transition matrix describing the probabilities of particular transitions, and an initial state (or initial distribution) across the state space. By convention, we assume all possible states and transitions have been included in the definition of the process, so there is always a next state, and the process does not terminate.

A discrete time random process involves a system that is in a certain state at each step, with the state changing randomly between steps. The steps are often thought of as moments in time, but they can equally well refer to physical distance or any other discrete measurement. Formally, the steps are the integers, and the random process is a mapping of these to states. The Markov property states that the conditional probability distribution for the system at the next step (and in fact at all future steps) depends only on the current state of the system, and not on the state of the system at previous steps.

Since the system changes randomly, it is generally impossible to predict with certainty the state of a Markov chain at a given point in the future. However, the statistical properties of the system's future can be predicted. In many applications, it is these statistical properties that are important.

A Markov chain example is the dietary habits of a person who eats only grapes, cheese, or lettuce, and whose dietary habits conform to the following rules:

- He eats exactly once a day.
- If he ate cheese today, tomorrow he will eat lettuce or grapes with equal probability.
- If he ate grapes today, tomorrow he will eat grapes with probability 1/10, cheese with probability 4/10, and lettuce with probability 5/10.
- If he ate lettuce today, he will not eat lettuce again tomorrow but will eat grapes with probability 4/10 or cheese with probability 6/10.

This person's eating habits can be modeled with a Markov chain since its choice tomorrow depends solely on what he ate today, not what he ate yesterday or even farther in the past. One statistical property that could be calculated is the expected percentage, over a long period, of the days on which the person will eat grapes.

In modelling system failures, MA utilizes a state transition diagram or directed graph that portrays in a single diagram the operational and failure states of the system. The state diagram is flexible in that it can serve equally well for a single component or an entire system. The diagram provides for representation of system states, transitions between states, and transition rates. These diagrams contain sufficient information for developing the state equations, which when resolved provide the probability calculations for each state.

Figure 26.1 illustrates the overall MA process.

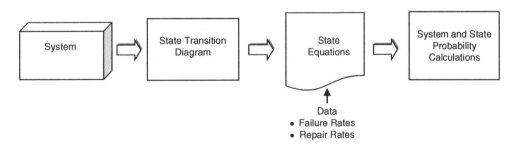

Figure 26.1 *MA process.*

TABLE 26.1   MA Steps

Step	Task	Description
1	Define the system	Examine the system and define the system boundaries, subsystems, and interfaces
2	Identify the system states	Establish the goals of the MA and determine the system and component states of interest
3	Construct state diagram	Construct the state diagram for all of the identified system states. Show the transitions between states and transition rates
4	Develop mathematical equations	Develop the mathematical equations from the state diagram
5	Solve mathematical equations	Solve the mathematical equations through manual or computer techniques
6	Evaluate the outcome	Evaluate the outcome of the MA analysis
7	Recommend corrective action	Make recommended design changes as found necessary from the MA analysis
8	Hazard tracking	Enter identified hazards, or hazard data, into the hazard tracking system
9	Document MA	Document the entire MA process, including state diagrams, equations, transition rates, and mathematical solution

26.6   MA METHODOLOGY

Table 26.1 lists and describes the basic steps of the MA process.

26.6.1   State Transition Diagram Construction

Although the basic rules guiding the construction of a state diagram are simple, a good understanding of the system being analyzed is necessary. Construction of a state diagram begins with an examination of the system and a determination of the possible states in which it may exist.

Figure 26.2 shows the symbols utilized in MA modeling.

Specifically, the state diagram is constructed as follows:

1. Begin at the left of the diagram with a state (circle) identified as S1. All equipment is initially good (operational) in this state.

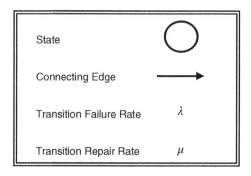

Figure 26.2   MA symbols.

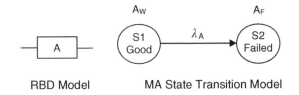

RBD Model   MA State Transition Model

Figure 26.3 *MA model: one-component system with no repair.*

2. Study the consequences of each failing element (any component, circuit, or channel defined as a single failure) in each of its failure modes. Group as a common consequence any that result in removing the same or equivalent circuitry from operation.
3. Assign new states (circles) and identify as S2, S3, S4, and so on, for the unique consequences of step 2.
4. Connect arrows from S1, to each of the new states, and note on each arrow the failure rate or rates of the element or elements whose failure determined transition to the new state.
5. Repeat steps 2, 3, and 4 for each of the new states failing only the elements still operational in that state. Continuously observe for cases where the failures may cause transition to one of the states formerly defined.
6. Continue the process until the initial equipment is totally nonoperational.

To limit the state diagram to a reasonable size, without a major sacrifice in accuracy, longer paths between the initial operational state and the system failure state may be truncated. For example, if one path to a system failure consists of three transitions and another is five transitions, then the longer path may be truncated. The effect of this approximation must be examined in the final model to ensure minimal impact.

Figure 26.3 shows an example MA state transition model for a one-component system with no repair. The RBD shows the system design complexity. In this MA model, only two states are possible, the operational state and the failed state. The starting state is S1, in which the system is operational (good). In state S2, the system is failed. The transition from state S1 to state S2 is based on the component failure rate λ_A. Note that A_W indicates component A *working* and A_F indicates component A *failed*. The connecting edge with the notation λ_A indicates the transitional failure of component A.

Figure 26.4 shows an example MA model for a one-component system with repair. Note how component A can return from the failed state to the operational state at the repair transition rate μ_A.

RBD Model   MA State Transition Model

Figure 26.4 *MA model: one-component system with repair.*

$\dot{\underline{P}} = [A]\,\underline{P}$ where P and P are $n \times 1$ column vectors and [A] is an $n \times n$ matrix.

$\underline{P} = \exp[A]t \cdot \underline{P}(0)$ where $\exp[A]t$ is an $n \times n$ matrix and $P(0)$ is the initial probability vector describing the initial state of the system.

Figure 26.5   *Markov state equations.*

26.6.2   State Equation Construction

A stochastic process is a random process controlled by the laws of probability that involve the "dynamic" part of probability theory, which includes a collection of random variables, their interdependence, their change in time, and limiting behavior. The most important variables in analyzing a dynamic process are those of rate and state. MA models are representations of a stochastic process.

A Markov process is completely characterized by its transition probability matrix, which is developed from the transition diagram. In safety and reliability work, events involve failure and repair of components. The transitional probabilities between states are a function of the failure rates of the various system components. A set of first-order differential equations is developed by describing the probability of being in each state in terms of the transitional probabilities from and to each state. The number of first-order differential equations will equal the number of system states. The mathematical formula is shown in Figure 26.5 and the solution then becomes one of solving the differential equations.

Figure 26.6 shows a Markov transition diagram for a two-component system comprised of components A and B. A_W indicates component A working and A_F indicates component A failed. States S1, S2, and S3 are noted a "good" indicating the system is operational. State S4 is noted as "failed' meaning the system is now in the failed state.

The Markov differential equations are developed by describing the probability of being in each system state at time $t + \Delta t$ as a function of the state of the system at time t. The probability

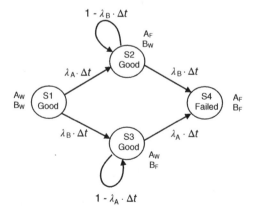

Figure 26.6   *Markov transition diagram: two-component system.*

of being in state S1 at some time $t + \Delta t$ is equal to the probability of being in state S1 at time t and not transitioning out during Δt. This equation can be written as follows:

$$P1(t + \Delta t) = P1(t) \cdot [1 - (\lambda_A + \lambda_B) \cdot \Delta t]$$

The probability of being in state S2 at time $t + \Delta t$ is equal to the probability of being in state S1 at time t and transitioning to state S2 in Δt plus the probability of being in state S2 at time t and not transitioning out during Δt. This equation can be written as

$$P2(t + \Delta t) = P1(t) \cdot \lambda_A \cdot \Delta t + P2(t)(1 - \lambda_B \cdot \Delta t)$$

All of the state equations are generated in a similar manner, resulting in the following equations:

$$P1(t + \Delta t) = P1(t) \cdot [1 - (\lambda_A + \lambda_B) \cdot \Delta t]$$

$$P2(t + \Delta t) = P1(t) \cdot \lambda_A \cdot \Delta t + P2(t)(1 - \lambda_B \cdot \Delta t)$$

$$P3(t + \Delta t) = P1(t) \cdot \lambda_B \cdot \Delta t + P3(t)(1 - \lambda_A \cdot \Delta t)$$

$$P4(t + \Delta t) = P2(t) \cdot \lambda_B \cdot \Delta t + P3(t) \cdot \lambda_A \cdot \Delta t + P4(t)$$

Rearranging the equations results in

$$[P1(t + \Delta t) - P1(t)]/\Delta t = -(\lambda_A + \lambda_B) \cdot P1(t)$$

$$[P2(t + \Delta t) - P2(t)]/\Delta t = \lambda_A \cdot P1(t) - \lambda_B \cdot P2(t)$$

$$[P3(t + \Delta t) - P3(t)]/\Delta t = \lambda_B \cdot P1(t) - \lambda_A \cdot P3(t)$$

$$[P4(t + \Delta t) - P4(t)]/\Delta t = \lambda_B \cdot P2(t) + \lambda_A \cdot P3(t)$$

Taking the limit as $\Delta t \rightarrow 0$ results in

$$dP1(t)/\Delta t = -(\lambda_A + \lambda_B) \cdot P1(t)$$

$$dP2(t)/\Delta t = \lambda_A \cdot P1(t) - \lambda_B \cdot P2(t)$$

$$dP3(t)/\Delta t = \lambda_B \cdot P1(t) - \lambda_A \cdot P3(t)$$

$$dP4(t)/\Delta t = \lambda_B \cdot P2(t) + \lambda_A \cdot P3(t)$$

In matrix form this becomes

$$
\begin{vmatrix} dP1(t)/\Delta t \\ dP2(t)/\Delta t \\ dP3(t)/\Delta t \\ dP4(t)/\Delta t \end{vmatrix} = \begin{vmatrix} -(\lambda_A + \lambda_B) & 0 & 0 & 0 \\ \lambda_A & -\lambda_B & 0 & 0 \\ \lambda_B & 0 & -\lambda_A & 0 \\ 0 & \lambda_B & \lambda_A & 0 \end{vmatrix} \bullet \begin{vmatrix} P1(t) \\ P2(t) \\ P3(t) \\ P3(t) \end{vmatrix}
$$

Solution of these equations provides the probability of being in each state.

26.7   MA EXAMPLES

26.7.1   Markov Chain

A Markov model can look at a long sequence of rainy and sunny days, and analyze the likelihood that one kind of weather is followed by another kind. Let us say it was found that 25% of the time, a rainy day was followed by a sunny day, and 75% of the time a rainy day was followed by more rain. Additionally, sunny days were followed 50% of the time by rain and 50% by sun. Given this data, a new sequence of statistically similar weather can be generated from the following steps:

1. Start with today's weather.
2. Given today's weather, choose a random number to pick tomorrow's weather.
3. Make tomorrow's weather "today's weather" and go back to step 2.

A sequence of days would result, which might look like as follows:

Sunny–Sunny–Rainy–Rainy–Rainy–Rainy–Sunny–Rainy–
Rainy–Sunny–Sunny . . .

The "output chain" would statistically reflect the transition probabilities derived from the observed weather. This stream of events is called a Markov chain.

26.7.2   Markov Model of Two-Component Series System with No Repair

Figure 26.7 shows an example MA model for a two-component series system with no repair. The RBD indicates that successful system operation requires successful operation of both components A and B. If either component fails, the system fails.

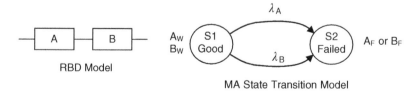

Figure 26.7   *MA model: two-component series system with no repair.*

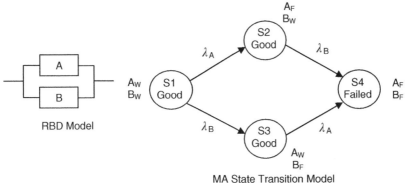

Figure 26.8 *MA model: two-component parallel system with no repair.*

In this MA model, two states are possible. The starting state is S1, whereby the system is good (operational) when both A and B are working. Transition to state S2 occurs when either A fails or B fails. In state S2, either A and B are failed and the system is failed.

26.7.3 Markov Model of Two-Component Parallel System with No Repair

Figure 26.8 shows an example MA model for a two-component parallel system with no repair. The RBD indicates that successful system operation requires only successful operation of either component A or B. Both components must fail to result in system failure.

In this MA model, four states are possible. The starting state is S1, whereby the system is good (operational) when both A and B are working. Based on A failure rate λ_A, it transitions to the failed state S2. In state S2, A is failed, while B is still good. In state S3, B is failed, while A is still good. In state S4, both A and B are failed. In states S1, S2, and S3, the system is good, while in state S4 the system is failed.

26.7.4 Markov Model of Two-Component Parallel System with Component Repair

Figure 26.9 shows the MA model for a two-component parallel system with component repair, but no system repair. As in Figure 26.8, this MA model has four possible states.

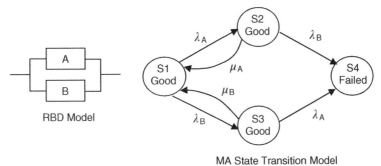

MA State Transition Model

Figure 26.9 *MA model: two-component parallel system with component repair.*

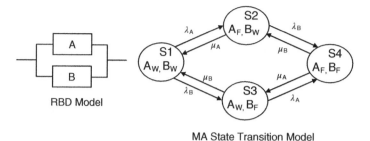

Figure 26.10 *MA model: two-component parallel system with system/component repair.*

In this system design, and in a MA model, if the system transitions to state S2, but A is repaired before B fails, then the system is returned to state S1. Conversely, if the system is in state S3, the system returns to state S1 if component B is repaired before component A fails. The connecting edge with the notation μ_A indicates repair of component A and μ_B indicates repair of component B.

26.7.5 Markov Model of Two-Component Parallel System with Component/System Repair

Figure 26.10 shows the MA model for a two-component parallel system with component repair and/or system repair. In this system design, even after system failure occurs, one or both components can be repaired, thereby making the system operational again.

As in the previous figure, this MA model has four possible states. In this system design, and corresponding MA model, if the system transitions to state S2, but A is repaired before B fails, then the system is returned to state S1. Conversely, if the system is in state S3, the system returns to state S1 if component B is repaired before component A fails. If state S4 is reached, the system can be repaired through repair of A and/or B.

26.7.6 Markov Model of Two-Component Parallel System with Sequencing

Figure 26.11 shows the MA model for a two-component parallel system where system failure occurs only when the components fail in a specific sequence. In this system design, A monitors B (i.e., checks on the status) such that if B fails, the fault is detected by A and is immediately

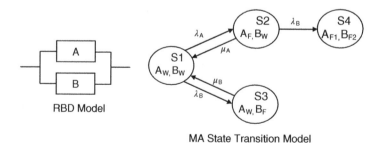

Figure 26.11 *MA model: two-component parallel system with sequencing.*

repaired. If A fails before B, then it cannot detect failure of B and initiate the repair of B and system failure occurs. Note, this model assumes that B is always repaired before A can fail, thereby maintaining an operational system.

In this MA model, four states are possible. The starting state is S1, whereby the system is good (operational) when both A and B are working. In state S3, component B has failed while A is still working. In this state, repair is the only option, thereby taking the system back to state S1. In state S2, component A is failed while B is working. If A is repaired before B fails, the system can be restored to state S1, otherwise the system will continue to operate until component B fails, thereby taking the system to state S4, which is system failure.

26.8 MA and FTA COMPARISONS

MA and FTA are often competing techniques. Each technique has its advantages and disadvantages. This section demonstrates both MA and FTA models and probability calculations for the same system design complexities. Comparing the two methods side by side helps to illustrate some of the strengths and weaknesses of each methodology.

Figure 26.12 compares MA and FTA for a two-component series system. The conclusion from this comparison is that both methods provide the same results (i.e., the equations are identical). For most analysts, the FTA model is easier to understand and the FTA mathematics are easier to solve.

Figure 26.13 compares MA and FTA for a two-component parallel system. The conclusion from this comparison is that both methods provide the same results (i.e., the equations are identical). For most analysts, the FTA model is easier to understand and the FTA mathematics are easier to solve.

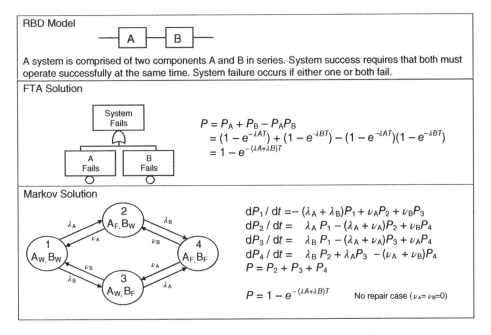

Figure 26.12 *MA and FTA comparison for a two-component series system.*

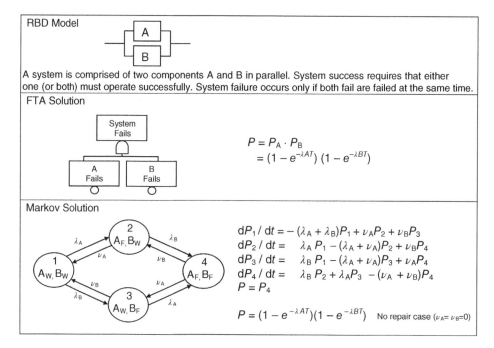

RBD Model

A system is comprised of two components A and B in parallel. System success requires that either one (or both) must operate successfully. System failure occurs only if both fail are failed at the same time.

FTA Solution

$$P = P_A \cdot P_B$$
$$= (1 - e^{-\lambda AT})(1 - e^{-\lambda BT})$$

Markov Solution

$$dP_1 / dt = -(\lambda_A + \lambda_B)P_1 + \nu_A P_2 + \nu_B P_3$$
$$dP_2 / dt = \lambda_A P_1 - (\lambda_A + \nu_A)P_2 + \nu_B P_4$$
$$dP_3 / dt = \lambda_B P_1 - (\lambda_A + \nu_A)P_3 + \nu_A P_4$$
$$dP_4 / dt = \lambda_B P_2 + \lambda_A P_3 - (\nu_A + \nu_B)P_4$$
$$P = P_4$$

$$P = (1 - e^{-\lambda AT})(1 - e^{-\lambda BT}) \quad \text{No repair case } (\nu_A = \nu_B = 0)$$

Figure 26.13   *MA and FTA comparison for a two-component parallel system.*

Figure 26.14 compares MA and FTA for a two-component sequence parallel system.

The conclusion from this comparison is that the resulting equations for each model are different. The FT equation is an approximation. The numerical comparison table shows calculations for different time intervals using the same failure rates. The results contained in this table show that the two models produce very close results up to about 1 million hours of operation. This indicates that the FTA approximation produces very good results. For most analysts, the FTA model is easier to understand and the FTA mathematics are easier to solve.

Figure 26.15 compares MA and FTA for a partial monitor with coverage system. This is a coverage type problem, whereby the monitor does not provide complete coverage of the circuit being monitored.

The conclusion from this comparison is that the resulting equations for each model are different. The FT equation is an approximation. The numerical comparison table shows calculations for different time intervals using the same failure rates. The results contained in this table show that the two models produce very close results up to about 10 000 h of operation. This indicates that the FTA approximation produces very good results. For most analysts, the FTA model is easier to understand and the FTA mathematics are easier to solve.

26.9   MA ADVANTAGES AND DISADVANTAGES

The MA technique has the following advantages:

1. MA provides a precise model representation for special design complexities, such as timing, sequencing, repair, redundancy, and fault tolerance.

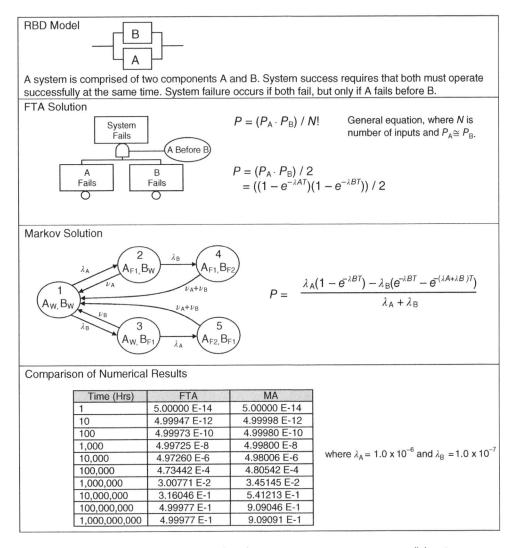

Figure 26.14 *MA and FTA comparison for a two-component sequence parallel system.*

2. MA is a good tool for modeling and understanding system operation, as well as potential system failure states and repair.

3. MA can be applied to a system very early in development and thereby identify safety issues early in the design process.

4. There are commercial software packages available to assist in MA modeling and probability calculations.

Although a strong and powerful technique, MA analysis has the following disadvantages:

1. MA is a secondary hazard analysis tool as it does not identify system hazards; it only evaluates identified hazards in more detail.

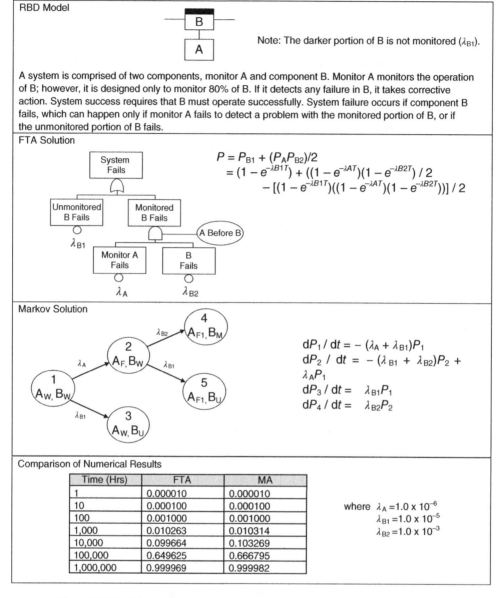

RBD Model

Note: The darker portion of B is not monitored (λ_{B1}).

A system is comprised of two components, monitor A and component B. Monitor A monitors the operation of B; however, it is designed only to monitor 80% of B. If it detects any failure in B, it takes corrective action. System success requires that B must operate successfully. System failure occurs if component B fails, which can happen only if monitor A fails to detect a problem with the monitored portion of B, or if the unmonitored portion of B fails.

FTA Solution

$$P = P_{B1} + (P_A P_{B2})/2$$
$$= (1 - e^{-\lambda B1 T}) + ((1 - e^{-\lambda AT})(1 - e^{-\lambda B2 T})/2$$
$$- [(1 - e^{-\lambda B1 T})((1 - e^{-\lambda AT})(1 - e^{-\lambda B2 T}))]/2$$

Markov Solution

$$dP_1/dt = -(\lambda_A + \lambda_{B1})P_1$$
$$dP_2/dt = -(\lambda_{B1} + \lambda_{B2})P_2 + \lambda_A P_1$$
$$dP_3/dt = \lambda_{B1}P_1$$
$$dP_4/dt = \lambda_{B2}P_2$$

Comparison of Numerical Results

Time (Hrs)	FTA	MA
1	0.000010	0.000010
10	0.000100	0.000100
100	0.001000	0.001000
1,000	0.010263	0.010314
10,000	0.099664	0.103269
100,000	0.649625	0.666795
1,000,000	0.999969	0.999982

where $\lambda_A = 1.0 \times 10^{-6}$
$\lambda_{B1} = 1.0 \times 10^{-5}$
$\lambda_{B2} = 1.0 \times 10^{-3}$

Figure 26.15 *MA and FTA comparison for a partial monitor with coverage system.*

2. MA is not a root cause analysis tool. It is a tool for evaluating the most effective methods for combining components together.

3. MA requires an experienced analyst to generate the graphical models and probability calculations.

4. The MA model quickly becomes large and complex; thus, it is more limited to small systems or a high-level system abstraction.

26.10 COMMON MA MISTAKES TO AVOID

When first learning how to perform a MA, it is commonplace to commit some traditional errors. The following is a list of typical errors made during the conduct of an MA:

1. Not obtaining the necessary training.
2. Using the complex MA technique when a simpler technique, such as FTA, might be more appropriate.
3. Failing to recognize that the transitions (probabilities) of changing from one state to another are assumed to remain constant. Thus, a Markov model is used only when a constant failure rate and repair rate assumption is justified.
4. Failing to recognize that the transition probabilities are determined only by the present state and not the system's history. This means future states of the system are assumed to be independent of all but the current state of the system. The Markov model allows only the representation of independent states.
5. Using an MA as a primary hazard analysis technique when it is only a secondary hazard analysis technique (refer to Chapter 4 for a discussion on primary technique versus secondary techniques).

26.11 SUMMARY

This chapter discussed the MA technique. The following basic principles help summarize the discussion in this chapter:

1. MA is a tool for modeling complex system designs involving timing, sequencing, repair, redundancy, and fault tolerance.
2. MA provides both graphical and mathematical (probabilistic) system models.
3. MA models can easily become too large in size for comprehension and mathematical calculations, unless the system model is simplified. Computer tools are available to aid in analyzing more complex systems.
4. MA is recommended only when very precise mathematical calculations are necessary.
5. MA is a secondary-type hazard analysis and should be used to supplement the SD-HAT analysis when more detailed information is required.

FURTHER READINGS

The following are major references for Markov analysis:

Bolch, G., Greiner, S., de Meer, H., and Trivedi, K. S., *Queueing Networks and Markov Chains*, Wiley–Blackwell, 2nd edition, 2006.

Ericson, C. A. and Andrews, J. D., Fault Tree and Markov Analysis Applied to Various Design Complexities, Proceedings of the 18th International System Safety Conference, 2000, pp. 324–335.

Faraci Jr., V., Calculating Probabilities of Hazardous Events (Markov vs. FTA), Proceedings of the 18th International System Safety Conference, 2000, pp. 305–323.

International Electrotechnical Commission, IEC 61165, Application of Markov Techniques, 1995.

Pukite, J. and Pukite, P., *Modeling for Reliability Analysis: Markov Modeling for Reliability, Maintainability, Safety and Supportability Analyses of Complex Computer Systems*, IEEE Press, 1998.

Chapter *27*

Petri Net Analysis

27.1 PNA INTRODUCTION

Petri net analysis (PNA) is an analysis technique for identifying hazards dealing with timing, state transitions, sequencing, and repair. PNA consists of drawing graphical Petri net (PN) diagrams and analyzing these diagrams to locate and understand design problems.

A Petri net consists of places, transitions, and arcs. Arcs run from a place to a transition or vice versa, never between places or between transitions. The places from which an arc runs to a transition are called the input places of the transition; the places to which arcs run from a transition are called the output places of the transition. Graphically, places in a Petri net may contain a discrete number of marks called tokens. Any distribution of tokens over the places will represent a configuration of the net called a marking. In an abstract sense relating to a Petri net diagram, a transition of a Petri net may fire if it is enabled, that is, there are sufficient tokens in all of its input places; when the transition fires, it consumes the required input tokens and creates tokens in its output places. A firing is atomic, that is, it is a single noninterruptible step.

Unless an execution policy is defined, the execution of Petri nets is nondeterministic: when multiple transitions are enabled at the same time, any one of them may fire. Since firing is nondeterministic, and multiple tokens may be present anywhere in the net (even in the same place), Petri nets are well suited for modeling the concurrent behavior of distributed systems.

Models of system performance, dependability, and reliability can be developed using PN models. PNA is very useful for analyzing properties such as reachability, recoverability, deadlock, and fault tolerance. The biggest advantage of Petri nets, however, is that they can link hardware, software, and human elements in the system.

Hazard Analysis Techniques for System Safety, Second Edition. Clifton A Ericson, II.
© 2016 John Wiley & Sons, Inc. Published 2016 by John Wiley & Sons, Inc.

PNA may be used to evaluate safety-critical behavior of control system software. In this situation, the system design and its control software is expressed as a timed PN. Subsets of the PN states are designated as possible unsafe states. The PN is augmented with the conditions under which those states are unsafe. A PN reachability graph (RG) will then determine if those states can be reached during the software execution.

27.2   PNA BACKGROUND

The PNA technique falls under the system design hazard analysis type (SD-HAT) and should be used as a supplement to the SD-HAT analysis. Refer to Chapter 4 for a description of the analysis types. The purpose of the PNA is to provide a technique to graphically model system components at a wide range of abstraction levels, in order to resolve system reliability, safety, and dependency issues. The graphical model can then be translated into a mathematical model for probability calculations.

PNs can be used to model a system, subsystem, or a group of components. PNA can be used to model hardware or software operation, or combinations thereof. To date, the application of PNA for system safety has been limited to the examination of software control systems. Its use has rarely been applied to large systems. PNA can be used to develop reliability models of system operation.

The significant advantage of the technique is that it can be applied to a system very early in development, thereby identify timing issues that may effect safety early in the design process. PNA application helps system developer's design-in safety and system quality during early development, eliminating the need to take corrective action after a test failure or mishap.

PNA is somewhat difficult to learn, understand, and master. A graduate-level understanding of mathematics and computer science is needed for the application of PNA. The analyst must master the technique and have a detailed knowledge of the process being modeled. The PNA technique is suited for use by theoretical mathematicians. The PN model quickly becomes large and unwieldy as system size increases, and is therefore usually used only on small system applications.

The use of PNA is not widespread in the system safety discipline because of its difficulty to use, its limitation to smaller problems and its limitation in scope to timing-type problems. PNA is highly specialized and assists only in a certain niche of potential safety concerns dealing with timing and state reachability. The technique is recommended only for special design safety concerns.

27.3   PNA HISTORY

The concept of Petri nets has its origin in Carl Adam Petri's doctoral dissertation *Kommunikation mit Automaten* submitted in 1962 to the Faculty of Mathematics and Physics at the Technische Universität Darmstadt, Germany. His thesis developed this graph-based tool for modeling the dynamics of systems incorporating switching. Subsequent research by many individuals has provided the means for using Petri's concepts as the basis for system modeling in many different applications, such as reliability, dependency, safety, and business models. Some sources state that Petri nets were actually invented in August 1939 by Petri, at the age of 13, for the purpose of describing chemical processes.

27.4 PNA DEFINITIONS

In order to facilitate a better understanding of PNA, some definitions for specific terms are in order. The following are basic PNA terms:

Transition Transitions represent a system event that must occur. Transitions contain a switching delay time. When all of the inputs to the transition have a token, the transition event is *enabled*, and *it occurs* or *switches* after *the gi*ven delay time. The delay time represents the actual system operational design. A delay time is placed in each transition node. Immediate transitions have a delay time equal to zero ($D = 0$). When the transition node *switches* or *fires*, all input nodes lose their token and all output nodes receive a token.

Place Places are used to represent the input and output nodes to a transition. Places contain the tokens.

Token Tokens represent timing in the system logic. As a PN model develops, the tokens build sequencing into the model. Tokens are analogous to currency, they are used for transactions.

Connecting Edge Connecting edges connect the places and transitions together, to build the logical model.

State The static condition (or state) of the PN model before and after the firing of a transition. A PN model will have a finite number of states, based on the model design.

Reachability A system can have many different possible states; reachability refers to the system's capability to reach any or all of those states during operation. As designed, the system may not be able to reach some states.

Repair Repair refers to the capability to physically repair a failed component and restore it to an operational state.

27.5 PNA THEORY

PNA utilizes a diagram or directed graph that portrays in a single diagram the operational states of the system. The state diagram is flexible in that it can serve equally well for a subsystem or an entire system. The diagram provides for representation of system states, transitions between states, and timing.

Figure 27.1 illustrates the overall PNA process.

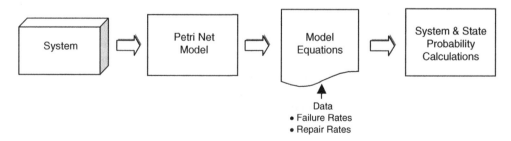

Figure 27.1 PNA process.

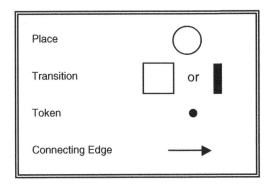

Figure 27.2 *PN model symbols.*

A Petri net is a graphical and mathematical modeling tool. It consists of places, transitions, and arcs that connect them. Input arcs connect places with transitions, while output arcs start at a transition and end at a place. There are other types of arcs, for example, inhibitor arcs. Places can contain tokens; the current state of the modeled system (the marking) is given by the number (and type if the tokens are distinguishable) of tokens in each place. Transitions are active components. They model activities that can occur (the transition fires), thus changing the state of the system (the marking of the Petri net). Transitions are allowed only to fire if they are enabled, which means that all the preconditions for the activity must be fulfilled (there are enough tokens available in the input places). When the transition fires, it removes tokens from its input places and adds some at all of its output places. The number of tokens removed/added depends on the cardinality of each arc. The interactive firing of transitions in subsequent markings is called *token game*.

Figure 27.2 shows the symbols used in making a PN model. All PN models can be constructed with just these components and using the firing rules listed below.

A PN is a bipartite directed graph (digraph). It consists of two types of nodes: places (drawn as circles), which can be marked with tokens (drawn as a dot), and transitions (drawn as squares or bars), which are marked by the time, D, it takes to delay the output of tokens. If $D = 0$, the transition time is immediate; otherwise, it is timed. PNs dealing with transition times are often referred to as timed PNs. Timed Petri nets are models that consider timing issues in the sequencing. The practice of integrating timed Petri nets with software fault tree analysis has recently become popular.

The movement of tokens is governed by the *firing rules* as follows:

1. A transition is enabled when all of the places with edges pointing to it are marked with a token.
2. After a delay $D \geqslant 0$, the transition switches or fires.
3. When the transition fires, it removes the token from each of its input places and adds the token to each of its output places.
4. A place can have multiple tokens.
5. The number of tokens in a PN is not necessarily constant.
6. Tokens move along the edges at infinite speed.
7. The transition time D can be either random or deterministic.

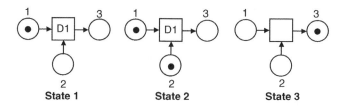

Figure 27.3   *Example PN model with three transition states.*

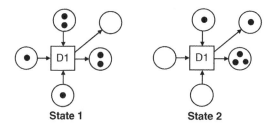

Figure 27.4   *Example PN model with multiple tokens.*

The PNA has a static part and a dynamic part. The static part consists of the places, transitions, and edges. The dynamic part involves the marking of places with the tokens when the transition firing occurs. In the PN model, places represent events that correspond to discrete system states. The transitions represent logic gates. The marking of a PN model at a given moment represents the state at that moment in time.

Figure 27.3 shows an example PN model with three transition states. In state 1, place 1 has a token but place 2 does not. Nothing can happen until place 2 receives a token. In state 2, place 2 receives a token. Now transition D1 has both inputs fulfilled, so after delay D1 it fires. State 3 shows the final transition, whereby D1 has fired, it has removed the two input tokens (places 1 and 2) and given an output token to place 3.

Figure 27.4 shows another example PN model with two transition states. This example shows how places can have multiple tokens, and how tokens are added and subtracted depending on the model. In this example, transition D1 has tokens for all input places in state 1; it, therefore, fires and transitions into state 2. State 2 shows that each input place loses a token and each output place receives a token. Multiple tokens can be used for modeling counting loops or redundant components in a system design. Transition D1 cannot fire again because all of its inputs do not have tokens now.

Figure 27.5 shows how AND gates and OR gates are modeled via PN models. Note that the transitions have a delay time of zero and are, therefore, immediate.

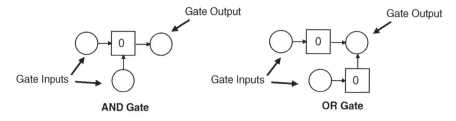

Figure 27.5   *PN model of AND gate and OR gate.*

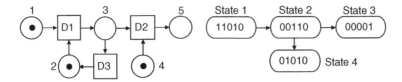

Figure 27.6 *PN with reachability graph.*

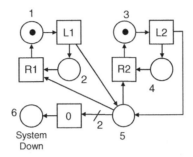

Figure 27.7 *PN of two-component system with repair.*

A powerful feature of PNs is their corresponding state graphs, otherwise referred to as reachability graphs. RGs show all of the possible system states that can be reached by a PN model. Knowing which states can be reached and which cannot is valuable information for reliability and safety evaluations. For example, it is important to know if a system can reach a suspected high-risk state.

Figure 27.6 shows an example PN model along with its corresponding reachability graph. In this RG, state 1 is the initial state, with places 1, 2, and 4 each having a token. State 2 shows the result of transition D1. State 3 shows the result when transition time D2 < D3, while state 4 shows the result when transition time D2 > D3.

With five places in this model, one might assume that with each place having a binary value of 0 or 1 that there would be $2^5 = 32$ possible states. But as shown by the reachability graph, only four states are actually reachable.

Figure 27.7 shows a PN model for a system design with two redundant components with repair. This system is comprised of two redundant components operating simultaneously. Successful operation requires that only one component remain functional. When a component fails, it undergoes repair. If the second component fails while the first is being repaired, then the system is failed. Note the special PN terminology in this PN model. L1 indicates life operation until random failure of component 1, while R1 indicates component 1 during repair duration. The/2 notation indicates that two tokens are required for this transition to occur.

Figure 27.8 shows the RG for the PN model in Figure 27.7. This type of RG is referred to as a cyclic RG with an absorbing state. Note that with two components in this system, four possible states are expected, but in this case the eventual marking of place 6 defines a 5th state. Note also that since place 5 requires two tokens for switching, these state numbers are nonbinary.

PNA may be used to evaluate safety-critical behavior of control system software. In this situation, the system design and its control software is expressed as a timed PN. A subset of the PN states are designated as possible unsafe states. The PN is augmented with the conditions

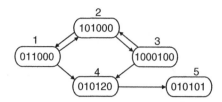

Figure 27.8 *RG for two-component system with repair.*

TABLE 27.1 PNA Steps

Step	Task	Description
1	Define the system	Examine the system and define the system boundaries, subsystems, and interfaces
2	Identify the system states	Establish the goals of the PNA and determine the system and component states of interest
3	Construct state diagram	Construct the Petri diagrams for all of the identified system states. Show the transitions between states and transition rates
4	Develop mathematical equations	Develop the mathematical equations from the state diagram
5	Solve mathematical equations	Solve the mathematical equations through manual or computer techniques
6	Evaluate the outcome	Evaluate the outcome of the PNA analysis
7	Recommend corrective action	Make recommended design changes as found necessary from the PNA analysis
8	Hazard tracking	Enter identified hazards, or hazard data, into the hazard tracking system
9	Document PNA	Document the entire PNA process, including state diagrams, equations, transition rates, and mathematical solution

under which those states are unsafe. A PN reachability graph will then determine if those states can be reached. If the unsafe cannot be reached, then the system design has been proven to not have that particular safety problem.

27.6   PNA METHODOLOGY

Table 27.1 lists and describes the basic steps of the PNA process.

27.7   PNA EXAMPLE

The hypothetical missile fire control system shown in Figure 27.9 will be used to demonstrate a PN model.

System operation is as follows:

1. The operator presses the ARM button to initiate arming and firing process.
2. When the computer receives ARM button signal, the computer initiates a system self-test to determine if the system is in a safe state and all arming and firing parameters are met.
3. If the self-test passes, the computer generates a fire signal when the FIRE button is pressed; if the self-test does not pass, then firing is prohibited by the computer.

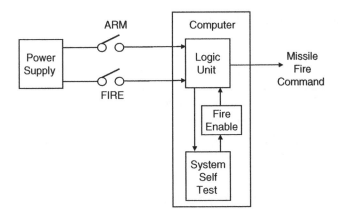

Figure 27.9 *Example missile fire sequence system.*

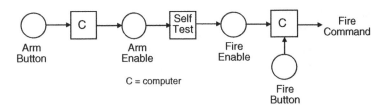

Figure 27.10 *PN for missile fire sequence system.*

Figure 27.10 shows how a PN model might be applied to analyze an actual weapon system design.

The PN model for the missile fire sequence reveals the following safety information:

1. A valid fire signal requires three interlocks: ARM command, FIRE command, and FIRE enable signal.
2. The computer is a single-point item that processes all three interlocks and, therefore, susceptible to single-point failures, both in hardware and in software.

27.8 PNA ADVANTAGES AND DISADVANTAGES

The PNA technique has the following advantages:

1. PNs can be used to model an entire system, subsystem, or system components at a wide range of abstraction levels, from conceptual to detailed design.
2. When a PN model has been developed for analysis of a particular abstraction, its mathematical representation can support automation of the major portions of the analysis.
3. PNA is a good tool for modeling and understanding system operation.

The following are disadvantages of the PNA technique:

1. PNA only identifies system hazards dealing with timing and state change issues.
2. PNA is limited for hazard analysis because it does not identify root causes and it does not identify all hazards; it is a secondary hazard analysis tool for a detailed analysis of an already identified hazard or a suspected hazard.
3. PNA requires an analyst experienced in PN graphical modeling.
4. PNA models quickly becomes large and complex, thus is more suitable to small systems or high-level system abstractions.

27.9 COMMON PNA MISTAKES TO AVOID

When first learning how to perform a PNA, it is commonplace to commit some traditional errors. The following is a list of typical errors made during the conduct of a PNA:

1. Not obtaining the necessary training in PNA.
2. Using the complex PNA technique when a simpler technique might be more appropriate.
3. Using a PNA as a primary hazard analysis technique when it is only a secondary hazard analysis technique (refer to Chapter 4 for a discussion on primary technique versus secondary techniques).

27.10 SUMMARY

This chapter discussed the PNA technique. The following basic principles help summarize the discussion in this chapter:

1. PNA models the timing and sequencing operation of the system.
2. PNA is a secondary-type hazard analysis tool for supporting a special class of hazards, such as those dealing with timing, state transitions, and repair.
3. PNA provides both a graphical and mathematical model.
4. PNA only requires places, tokens, transitions, and connecting edges to model a system.
5. PNA can easily become too large in size for understanding, unless the system model is simplified.
6. For system safety applications, PNA is not a general purpose hazard analysis tool and should be used only in situations to evaluate suspected timing, state transition, sequencing, and repair hazards.

FURTHER READINGS

The following are major references for Petri analysis:

Agerwala, T., Putting Petri Nets to Work, *IEEE Computer*, 85–94 (1979).

Carl Adam, P, and Wolfgang, R., Petrti Net, *Scholarpedia*, **3**(4):6477 (2008).

Malhotra, M, and Trevedi, K., Dependability Modeling Using Petri Nets, *IEEE Trans. Reliab.*, **44**, 428–440 (1995).

Murata, T., Petri Nets: Properties, Analysis and Applications, *Proc. IEEE*, **77**(4): (1989).

Petri Nets Properties, Analysis and Applications, *Proc. IEEE*, **77**, 541–580 (1989).

Schneeweiss, W. G., *Petri Nets for Reliability Modeling*, LiLoLe-Verlag, 1999.

Schneeweiss, W. G., Tutorial: Petri Nets as a Graphical Description Medium for Many Reliability Scenarios, *IEEE Trans. Reliab.*, **50**(2):159–164 (2001).

Chapter *28*

Barrier Analysis

28.1 BA INTRODUCTION

Barrier analysis (BA) is an analysis technique for identifying hazards specifically associated with hazardous energy sources. BA provides a tool to evaluate the unwanted flow of (hazardous) energy to target (personnel or equipment) through the evaluation of barriers preventing the hazardous energy flow.

BA is a powerful and efficient system safety analysis tool for the discovery of hazards associated with energy sources. The sequentially structured procedures of BA produce consistent, logically reasoned, and less subjective judgments about hazards and controls than many other analysis methods available. However, BA is not comprehensive enough to serve as the sole hazard analysis of a system, as it may miss critical human errors or hardware failures not directly associated with energy sources.

28.2 BA BACKGROUND

Because the BA technique is unique with a limited scope of coverage, it does not completely fulfill the requirements of any one of the seven basic hazard analyses types described in Chapter 4. However, BA is often used to support the system design hazard analysis type (SD-HAT), detailed design hazard analysis type (DD-HAT), or preliminary design hazard analysis type (PD-HAT) analyses. The BA technique is also known as the energy trace and barrier analysis (ETBA) or the energy trace analysis.

Many system designs cannot eliminate energy sources from the system since they are a necessary part of the system. The purpose of BA is to evaluate these energy sources and

Hazard Analysis Techniques for System Safety, Second Edition. Clifton A Ericson, II.
© 2016 John Wiley & Sons, Inc. Published 2016 by John Wiley & Sons, Inc.

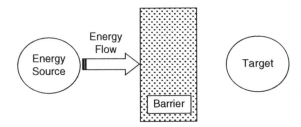

Figure 28.1 *Barrier between energy source and target.*

determine if potential hazards in the design have been adequately mitigated through the use of energy barriers.

Figure 28.1 illustrates the concept of a barrier providing separation between the energy source and the target. The simple concept of barrier analysis and its graphical portrayal of accident causation is a powerful analysis tool. It should be noted that unwanted energy source from single source may attack multiple targets. Also, in some situations, multiple barriers may be required for optimum safety.

BA is implemented by identifying energy flow paths that may be hazardous and then identifying or developing the barriers that must be in place to prevent the energy flow from damaging equipment or injuring personnel. There are many different types and methods of energy barriers that can be applied in a system design, that is, physical barrier (barricade), procedural barrier, or a time barrier. Barriers serve as countermeasures to control probability and/or severity of personnel injury or system damage.

BA is a generally applicable methodology for analysis of systems of all types. It is used to ensure disciplined, consistent, and efficient procedures for the discovery of energy hazards in a system. It can also be used during accident investigations to help develop and understand damage scenarios. BA lends itself to overviews of energies in systems and guides the search for specific hazards or risks that require more detailed analysis.

BA is capable of producing detailed analyses of hazards in new or existing systems. By meticulously and logically tracking energy flow paths sequentially, into, within, and out of a system, BA facilitates a thorough analysis for each specific energy type. A thorough understanding of energy sources in the system and their behaviors is necessary, so is a good understanding of the system design and operation. The BA technique is uncomplicated and easily learned. Standard easily followed BA forms and instructions are provided in this chapter.

28.3 BA HISTORY

BA is based on a useful set of concepts introduced by William Haddon, Jr. [1]. These concepts have been adopted and improved upon by others until the technique has evolved into a useful safety analysis tool.

28.4 BA DEFINITIONS

In order to facilitate a better understanding of BA, some definitions for specific terms are in order. The following are the basic BA-related terms:

28.4.1 Energy Source

Any material, mechanism, or process that contains potential energy that can be released is an energy source. The concern is that the released energy may cause harm to a potential target.

28.4.2 Energy Path

Energy path is the path of energy that flows from source to target.

28.4.3 Energy Barrier

Any design or administrative method that prevents a hazardous energy source from reaching a potential target in sufficient magnitude to cause damage or injury is an energy barrier. Barriers separate the target from the source by various means involving time or space. Barriers can take many forms, such as physical barriers, distance barriers, timing barriers, procedural barriers, and the like.

28.5 BA THEORY

BA is based on the theory that when hazardous energy sources exist within a system, they pose a hazardous threat to certain targets. Placing barriers between the energy source and the target can mitigate the threat to targets. This concept is illustrated in Figure 28.2, which also shows some example types of energy sources, barriers, and threats.

BA involves the meticulous tracing of energy flows through the system. BA is based on the premise that a mishap is produced by unwanted energy exchanges associated with energy flows through barriers into exposed targets. The BA process begins with the identification of energy sources within the system design. Diagrams are then generated tracing the energy flow from its source to its potential target. The diagram should show barriers that are in place to prevent damage or injury. If no barriers are in place, then safety design requirements must be generated to establish and implement effective barriers.

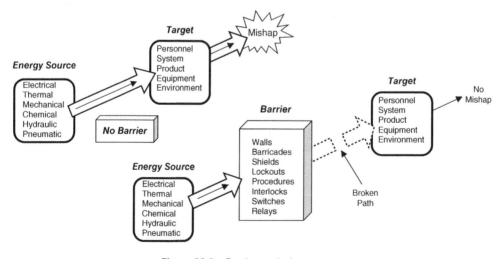

Figure 28.2 *Barrier analysis concept.*

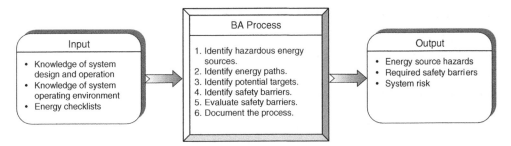

Figure 28.3 *Barrier analysis overview.*

28.6 BA METHODOLOGY

Figure 28.3 shows an overview of the basic BA process and summarizes the important relationships involved.

Table 28.1 lists and describes the basic steps of the BA process and summarizes the important relationships involved. Remember that a worksheet is utilized during this analysis process.

TABLE 28.1 Barrier Analysis Process

Step	Task	Description
1	Identify energy sources	Examine the system and identify all potentially hazardous energy sources. Include energy quantity and location when possible. Utilize energy source checklists. Examples include explosives, electromagnetic radiation, hazardous materials, electricity, and so on
2	Identify single energy paths	Any potentially harmful energy flow path to the target (e.g., people, equipment, facilities, and the environment) likely to result in a mishap
3	Identify multiple energy paths	Multiple energy paths to the target where more than one energy path is required to reach the target and cause a mishap (e.g., both the mechanical and the electrical arm functions of a fuze)
4	Identify targets	For each energy source, trace its travel through the system, from beginning to end. Identify all potential targets that can be injured or damaged by the hazardous energy sources. Utilize diagrams
5	Identify the target vulnerability	The vulnerability of the target to the unwanted energy flow. For example, an inadvertent application of $+28$ VDC will have little effect on a human but will destroy a microprocessor
6	Identify safety barriers	Identify all barriers in the path of the energy source, or identify barriers that should be present. Evaluate the impact of potential failure of barriers The lack of barriers and/or the effectiveness of existing barriers. For example, if the heat shields on the space shuttle fall off during re-entry, then the shuttle and crew could be lost
7	Evaluate system risk	Identify the level of mishap risk presented to the target by the energy source, both with barriers and without barriers in the system design
8	Recommend corrective action	Determine if the barrier controls present are adequate, and if not, recommend barriers that should be added to reduce the mishap risk. The need for more detailed analysis by other techniques (e.g., FTA) to ensure that all hazard causal factors are identified and mitigated
9	Track hazards	Transfer identified hazards into the hazard tracking system
10	Document BA	Document the entire BA process on the worksheets. Update for new information as necessary

28.6.1 Example Checklist of Energy Sources for BA

Table 28.2 contains an example of an energy checklist. If the system design contains any energy sources in this list, then a specific energy has been identified for BA.

TABLE 28.2 Energy Checklist (Sample)

Category	Energy Sources
Acoustical Radiation	Equipment noise
	Ultrasonic cleaners
	Alarm devices and signal horns
Atmospheric	Wind velocity, density, and direction
	Rain (warm/cold/freezing)
	Snow/hail/sleet
	Lightning/electrostatic
	Particulates/dusts/aerosols/powders
	Sunshine/solar
	Acid rain, vapor/gas clouds
	Air (warm/cold/freezing, inversion)
	Moisture/humidity
Chemical (acute and chronic sources)	Anesthetic/asphyxiant
	Corrosive/dissolving/solvent/lubricating
	Decomposable/degradable
	Deposited materials/residues
	Detonable
	Oxidizing/combustible/pyrophoric
	Polymerizable
	Toxic/carcinogenic/embryo toxic
	Waste/contaminating (air/land/water)
	Water reactive
Corrosive	Chemicals, acids, caustics
	Decon solutions
	"Natural" chemicals (soil, air, and water)
Electrical	Battery banks
	Diesel generators
	High lines
	Transformers
	Wiring
	Switch gear
	Buried wiring
	Cable runs
	Service outlets and fitting
	Pumps, motors, and heaters
	Power tools and small equipment
	Magnetic fields
	AC or DC current flows
	Stored electrical energy/discharges
	Electromagnetic emissions/RF pulses
	Induced voltages/currents
	Control voltages/currents
Etiologic Agents	Viral
	Parasitic
	Fungal
	Bacterial
	Biological toxins

(continued)

TABLE 28.2. *(Continued)*

Category	Energy Sources
EMR and particular radiations	Lasers, masers, and medical X-rays
	Radiography equipment and sources
	Welding equipment
	Electron beam
	Blacklight (e.g. Magniflux)
	Radioactive sources, contamination, waste, and scrap
	Storage areas, plug storage
	Activation products, neutrons
Explosive or pyrophoric	Caps, primer cord, explosives
	Electrical squibs
	Power metallurgy, dusts
	Hydrogen and other gases
	Nitrates, peroxides, perchlorates
	Carbides, superoxides
	Metal powders, plutonium, and uranium
	Zirconium
	Enclosed flammable gases
Flammables	Chemicals, oils, solvents, and grease
	Hydrogen (battery banks), gases
	Spray paint, solvent vats
	Coolants, rags, plastics, and foam
	Packing materials
Kinetic – linear	Cars, trucks, railroads, and carts
	Dollies, surfaces, and obstructions
	Crane loads in motion, shears
	Presses, Pv blowdown
	Power-assisted driving tools
	Projectiles, missiles/aircraft in flight
	Rams, belts, and moving parts
	Shears and presses
	Vehicle/equipment movement
	Springs, stressed members
Kinetic – rotational	Centrifuges, motors, and pumps
	Flywheels, gears, and fans
	Shop equipment (saws, grinders, drills, etc.)
	Cafeteria and laundry equipment
	Rotating machinery/gears/wheels
	Moving fan/propeller blades
Mass, gravity, and height	Human effort
	Stairs, lifts, and cranes
	Sling, hoists, elevators, and jacks
	Bucket and ladder
	Lift truck, pits, and excavations
	Vessels, canals, and elevator doors
	Crane cabs, scaffolds, and ladders
	Trips and falls
	Falling/dropped objects
	Suspended objects
Noise/vibration	Noise
	Vibration

(continued)

TABLE 28.2.   *(Continued)*

Category	Energy Sources
Nuclear	Vaults, temporary storage areas
	Casks, hot cells, and reactor areas
	Criticality potential in process
	Laboratories and pilot plants
	Waste tanks and piping, basins, and canals
Pressure – volume/K constant	Boilers and heated surge tanks
	Autoclaves
	Test loops and facilities
	Gas bottles and pressure vessels
	Coiled springs, stressed members
	Gas receivers
	Overpressure ruptures/explosions thermal cycling
	Vacuum growth cryogenic
	Liquid spill/flood/buoyancy
	Expanding fluids/fluid jets
	Uncoiling object
	Ventilating air movement
	Trenching/digging/earth moving
Terrestrial	Earthquake
	Floods/drowning
	Landslide/avalanche
	Subsidence
	Compaction
	Cave-ins
	Underground water flows
	Glacial
	Volcanic
Thermal (except radiant)	Convection, furnaces
	Heavy metal weld preheat
	Gas heaters, lead melting pots
	Electrical wiring and equipment
	Exposed steam pipes and valves
	Steam exhausts
Thermal radiation	Furnaces and boilers
	Steam lines
	Lab and pilot plant equipment
	Heaters
	Solar
	Radiant/burning/molten
	Conductive
	Convective/turbulent evaporative/expansive heat/cool
Toxic pathogenic	Toxic chemicals, check MSDS
	Exhaust gases
	Oxygen-deficient atmosphere
	Sand blasting and metal plating
	Decon and cleaning solutions
	Bacteria, molds, fungi, and viruses
	Pesticides, herbicides, and insecticides
	Chemical wastes and residues

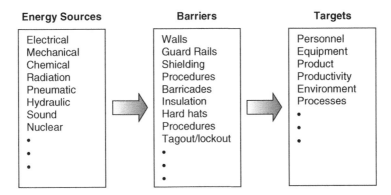

Figure 28.4 *Generic BA components.*

28.6.2 BA Considerations

Figure 28.4 summarizes generic components of BA: energy sources, barriers, and targets. These lists are starting points for a BA. Each component of BA must be well understood and evaluated in the system context.

After an energy source has been identified, there are a series of questions that can be answered that assist in identifying hazardous designs. Table 28.3 contains a list of some of the typical questions that must be answered by the BA.

BA verifies the adequacy of engineered or administrative barriers. In this context, engineered safety features are considered "hard" barriers, while administrative controls such as procedures, warning signs, and supervisory checks are "soft" barriers. Because hard barriers are more difficult to bypass than soft barriers, the former is preferred. However, soft barriers may be all that can be used in some situations; therefore, an array of complementary soft barriers are often used to better ensure energy containment.

Barriers may be categorized by their function, location, and/or type. Figure 28.5 provides some examples.

Haddon [1] postulated the concept that one or more barriers can control the harmful effects of energy transfer. Expanding on Haddon's work, analysts have identified the following barrier mechanisms in order of precedence:

1. Eliminate the hazardous energy from the system (e.g., replace with alternative).
2. Reduce the amount of energy (e.g., voltages, fuel storage, etc.).
3. Prevent the release of energy (e.g., strength of containment of the energy).

TABLE 28.3   BA Hazard Discovery Checklist

Energy Flow Changes	Changes in Barriers
1. Flow too much/too little/none at all	1. Barrier too strong/too weak
2. Flow too soon/too late/not at all	2. Barrier designed wrong
3. Flow too fast/too slowly	3. Barrier too soon/too late
4. Flow blocked/builtup/release	4. Barrier degraded/disturbed
5. Wrong form/wrong-type input or flow	5. Barrier impedes flow/enhances flow
6. Cascading effects of release	6. Wrong barrier-type selected
7. Flow conflicts with another energy flow	7. Barrier failed completely

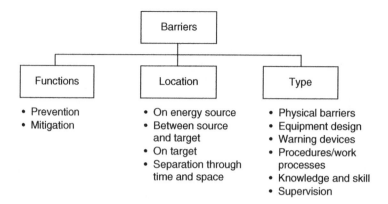

Figure 28.5 *Example barrier categorizations.*

4. Reduce the rate of release of energy (e.g., slow down burning rate, speed, etc.).

5. Prevent the buildup of released energy (e.g., pressure relief valve).

6. Control improper energy input (e.g., electrical energy through supercooled environment).

7. Separate in space or time the energy from the target (e.g., electric lines out of reach).

8. Interpose material barriers (e.g., insulation, guards, and safety glasses).

9. Modify shock concentration surfaces (e.g., round off and make soft).

10. Strengthen the target to withstand the energy (e.g., earthquake-proof structures).

11. Limit the damage of energy release (e.g., prompt signals and action, sprinklers, etc.).

12. Train personnel to prevent energy release (e.g., warnings, procedures, etc.).

These successive methods are called "energy barriers." The energy barriers may be a physical obstruction or they may be a written or verbal procedure that is put into place as a means of separating the energy from the persons or objects in time or space. Substituting a less harmful energy may be a way to "limit the energy" or "prevent the buildup."

These 12 barrier mechanisms are expanded in Table 28.4.

TABLE 28.4   Barrier Mechanisms

Barrier Mechanism Strategy for Managing Harmful Energy Flow	Implementation of Mechanism
Eliminate the energy source • Exclude (remove) energy concentration Reduce the amount of energy • Limit quantity and/or level of energy	• Eliminate from design • Replace with alternative design • Store heavy loads on ground floors • Lower dam height • Reduce system design voltage/operating pressure • Use small(er) electrical capacitors/pressure accumulators • Reduce/control vehicle speed • Monitor/limit radiation exposure • Substitute less energetic chemicals

(continued)

TABLE 28.4.   *(Continued)*

Barrier Mechanism Strategy for Managing Harmful Energy Flow	Implementation of Mechanism
Prevent release of energy	• Heavy-walled pipes/vessels • Interlocks • Tagout–lockout • Double-walled tankers • Wheel chocks
Reduce the rate of release of energy • Modify rate of release of energy	• Flow restrictors in discharge lines • Resistors in discharge circuits • Fuses/circuit breakers • Ground fault circuit interrupters
Prevent the buildup of released energy	• Use pressure relief valves • Control chemical reactions
Control improper energy input • Keep energy source within specifications • Prevent the combining of energy sources	• Separate hypergolic fuel sources
Separate energy from target in time and/or space	• Evacuate explosives test areas • Impose explosives safety quantity–distance rules • Install traffic signals • Use yellow no-passing lines on highways • Control hazardous operations remotely
Isolate by interposing a material barrier	• Concrete road barrier • Safety eye glasses
Modify shock concentration surfaces • Modify target contact surface or basic structure	• Rounded corners • Padding
Strengthen potential target to withstand the energy	• Earthquake-proof structure • Nuclear reaction containment facility
Limit the damage of energy release	• Building sprinkler systems • Aircraft fire suppression systems
Train personnel to prevent energy release	• Warning notes • Special procedures • Safety training

28.7 BA WORKSHEET

The BA is a detailed hazard analysis of energy sources and their potential effect on system personnel and/or equipment. It is desirable to perform the BA using a form or worksheet to provide analysis structure, consistency, and documentation. The specific format of the analysis worksheet is not critical. Typically, columnar-type worksheets are utilized to help maintain focus and structure in the analysis. As a minimum, the following basic information should be obtained from the analysis worksheet:

1. System energy sources that provide a threat.
2. Targets within the system that are susceptible to damage or injury from the energy sources.
3. Barriers in place that will control the energy hazard.
4. Barriers that are recommended to control the energy hazard.
5. System risk for the energy-barrier hazard.

Barrier Analysis						
Energy Source	Energy Hazard	Target	IHRI	Barrier	FHRI	Comments
①	②	③	④	⑤	⑥	⑦

Figure 28.6 *Recommended BA worksheet.*

The recommended BA worksheet is shown in Figure 28.6. This particular BA worksheet utilizes a columnar-type format. Other worksheet formats may exist because different organizations often tailor their analysis worksheet to fit their particular needs. The specific worksheet to be used may be determined by the system safety program (SSP), system safety working group, or the safety analysis customer.

The following instructions describe the information required under each column entry of the BA worksheet:

1. *Energy Source* This column identifies the hazardous energy source of concern.
2. *Energy Hazard* This column identifies the type of energy-related hazard (i.e., the energy path) involved with the identified energy source. The hazard should describe the hazard effect and mishap consequences and all of the relevant causal factors involved. All possibilities of hardware faults, software errors, and human error should be investigated.
3. *Target* This column identifies the target, or targets, that can be adversely affected by the energy source if barriers are not in place and a mishap occurs.
4. *Initial mishap risk index (IMRI)* This column provides a qualitative measure of mishap risk for the potential effect of the identified hazard, given that no mitigation techniques are applied to the hazard. Risk measures are a combination of mishap severity and probability, and the recommended values from MIL-STD-882 are shown below.

Severity	Probability
1 – Catastrophic	A – Frequent
2 – Critical	B – Probable
3 – Marginal	C – Occasional
4 – Negligible	D – Remote
	E – Improbable

5. *Barrier* This column establishes recommended preventive measures to eliminate or control identified hazards. Safety requirements in this situation generally involve the

addition of one or more barriers to keep the energy source away from the target. The preferred order of precedence for design safety requirements is as shown below.

Order of Precedence

1 – Eliminate the hazard through design measures or reduce the
    hazard mishap risk through design measures
2 – Reduce the hazard mishap risk through the use of safety devices
3 – Reduce the hazard mishap risk through the use of warning devices
4 – Reduce the hazard mishap risk through special safety training
    and/or safety procedures

6. *Final Mishap Risk Index (FMRI)*   This column identifies the "final" mishap risk given that the barriers or safety features are in place to mitigate the hazard. This risk assessment will show the risk improvement due to barriers in the system. The same risk matrix table as used to evaluate column 4 is also used here.

7. *Comments*   This column provides a place to record useful information regarding the hazard or the analysis process that are not noted elsewhere.

28.8   BA EXAMPLE

In order to demonstrate the BA methodology, the hypothetical water heating system shown in Figure 28.7 will be analyzed for energy-barrier hazards. Table 28.5 contains a list of system components and establishes if they are energy sources of safety concern.

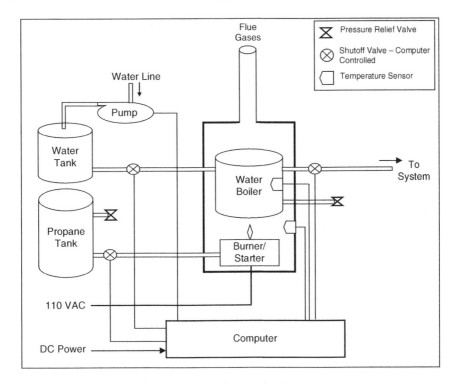

Figure 28.7   *Example water heating system.*

TABLE 28.5 List of Energy Sources for Water Heating System

System Component	Hazardous Energy Source	Hazard Potential	Barrier
Propane tank	Yes	Yes	Yes
Propane gas	Yes	Yes	Yes
Water tank	Yes	Yes	Yes
Water	Yes	Yes	Yes
Water boiler	Yes	Yes	Yes
Electricity	Yes	Yes	Yes
Gas burner	Yes	Yes	Yes
Computer	No	Yes	Yes

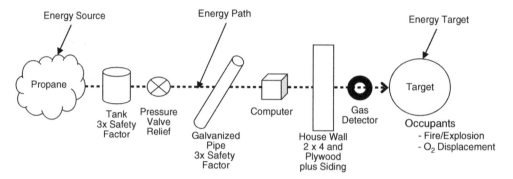

Figure 28.8 *Propane energy path with barriers.*

Figure 28.8 contains a diagram of the energy path for the propane energy source. This diagram shows all of the energy barriers in the system design.

Tables 28.6 and 28.7 contain the worksheets for a partial BA of this example system. Two of the system components, propane and water, were selected for demonstration of the BA technique.

TABLE 28.6 Example BA: Worksheet 1

Barrier Analysis							
Energy Source	Hazard	Target	IHRI	Barrier	FHRI	Comments	
Propane	Fire/explosion causing death, injury, and/or damage	Personnel/ facility	1C	• Isolate tank safe distance from facility • Use protected lines • Minimize ignition sources	1E		
	High-pressure release causing death, injury, and/or damage	Personnel/ facility	1C	• Isolate tank safe distance from facility • Use protected lines • Isolate lines from personnel • Use pressure relief valve	1E		
	Oxygen replacement causing death	Personnel	1C	• Use propane with smell detection (e.g., mercaptan) • Use gas detector	1E		

TABLE 28.7 Example BA: Worksheet 2

Barrier Analysis						
Energy Source	Hazard	Target	IHRI	Barrier	FHRI	Comments
Water	High-temperature causing tank explosion, which results in death, injury, and/or damage	Personnel/ facility	1C	• Isolate boiler tank • Use protected lines	1E	
	High-pressure causing tank explosion, which results in death, injury, and/or damage	Personnel/ facility	1C	• Isolate tank safe distance from facility • Use protected lines • Isolate lines from personnel • Use pressure relief valve	1E	
	Flood causing damage	Facility	2C	• Isolate tank safe distance from facility • Use water detector	2E	

28.9 BA ADVANTAGES AND DISADVANTAGES

The BA technique the following advantages:

1. BA is simple to grasp and use.
2. BA has a pictorial benefit that aids analysts in visualizing hazards.
3. BA is a relatively inexpensive analysis tool.
4. Most energy sources are easily recognized (e.g., explosives, electricity, springs, and compressed gas).

The following are disadvantages of the BA technique:

1. BA is limited by the ability of the analyst to identify all the hazardous energy sources.
2. BA does not identify all system hazards, only those associated with energy sources.
3. Not all sources of harm to targets are readily recognizable as energy sources (e.g., asphyxiate gases, pathogenic organisms, etc.).

28.10 COMMON BARRIER ANALYSIS MISTAKES TO AVOID

When first learning how to perform a BA, it is commonplace to commit one or more of the following errors:

1. Not identifying all of the energy sources within the system
2. Not evaluating the potential failure of energy barriers

3. Not evaluating the possible cascading effects of energy sources
4. Not identifying/understanding all of the energy paths
5. Not considering the complete system (i.e., taking too narrow a view of energy paths)
6. Using a BA as a primary hazard analysis technique when it is only a secondary hazard analysis technique as it does not provide complete hazard coverage (refer to Chapter 4 for a discussion on primary technique versus secondary technique).

28.11   SUMMARY

This chapter discussed the BA technique. The following basic principles help summarize the discussion in this chapter:

1. BA involves a detailed focus on potentially hazardous energy sources within the system design, and intentional barriers for mitigating the energy hazards.
2. BA should be a supplement to the PD-HAT, DD-HAT, and SD-HAT.
3. Some hazards identified through BA may require more detailed analyses by other techniques (e.g., FTA) to ensure that all hazard causal factors are identified and mitigated.
4. The use of worksheets provides structure and rigor to the BA process, and energy flow diagrams aid the analysis.

REFERENCE

1. W. Haddon, Energy Damage and the Ten Counter-measure Strategies, *Hum. Factors J.*, (1973).

FURTHER READINGS

EG&G Idaho, Inc. *Barrier Analysis: DOE-76-451, SSDC-29, Safety Systems Development Center*, EG&G Idaho, Inc., July 1985.

Hocevar, C. J. and Orr, C. M., Hazard Analysis by the Energy Trace Method: Proceedings of the 9th International System Safety Conference, 1989, pp. H-59–H-70.

Stephenson, J., *System Safety 2000: A Practical Guide for Planning, Managing, and Conducting System Safety Programs*, John Wiley & Sons, pp. 147–152, 1991.

Chapter 29

Bent Pin Analysis

29.1 BPA INTRODUCTION

Bent pin analysis (BPA) is an analysis technique for identifying hazards caused by bent pins within cable connectors. It is possible to improperly attach two connectors together and have one or more pins in the male connector bend sideways and make contact with other pins within the connector. If this should occur, it is possible to cause open circuits and/or short circuits to +/− voltages, which may be hazardous in certain system designs. For example, a certain cable may contain a specific wire carrying the fire command signal (voltage) for a missile. This fire command wire may be a long wire that passes through many connectors. If a connector pin in the fire command wire should happen to bend and make a short circuit with another connector pin containing +28 VDC the missile fire command may be inadvertently generated. BPA is a tool for evaluating all of the potential bent pin combinations within a connector to determine if a potential safety hazard exists.

29.2 BPA BACKGROUND

Because the BPA technique is unique with a limited scope of coverage, it does not completely fulfill the requirements of any one of the seven basic hazard analyses types described in Chapter 4. However, BPA is often used to support the system design hazard analysis type (SD-HAT), detailed design hazard analysis type (DD-HAT), or preliminary design hazard analysis type (PD-HAT) analyses. An alternative name for the BPA technique is cable failure matrix analysis (CFMA).

The use of this technique is recommended for identification of system hazards resulting from potential bent connector pins. BPA should always be considered for systems

Hazard Analysis Techniques for System Safety, Second Edition. Clifton A Ericson, II.
© 2016 John Wiley & Sons, Inc. Published 2016 by John Wiley & Sons, Inc.

involving safety-critical circuits with connectors. The technique is uncomplicated and easily learned. Standard easily followed BPA worksheets and instructions are provided in this chapter. An understanding of electrical circuits is necessary, so is a good understanding of the system design and operation. BPA is often overlooked as a useful tool because it is not well known.

29.3 BPA HISTORY

The Boeing Company developed BPA circa 1965, on the Minuteman program, as a technique for identifying potential safety problems that might result from bent connector pins. The Minuteman program had been experiencing bent connector pins during system installation operations at many Minuteman sites. BPA proved to be successful in identifying potential safety problems that were subsequently eliminated through redesign.

29.4 BPA THEORY

The purpose of BPA is to determine the potential safety effect of one or more pins bending inside a connector and making contact with other pins or the casing. If a safety-critical circuit were to be short-circuited to another circuit containing positive or negative voltage, the overall effect might be catastrophic. BPA is a tool for evaluating all of the potential single pin-to-pin bent pin combinations within a connector to determine if a potential safety hazard exists, given that a bent pin occurs. Figure 29.1 is an illustration of the bent pin (BP) concept, whereby

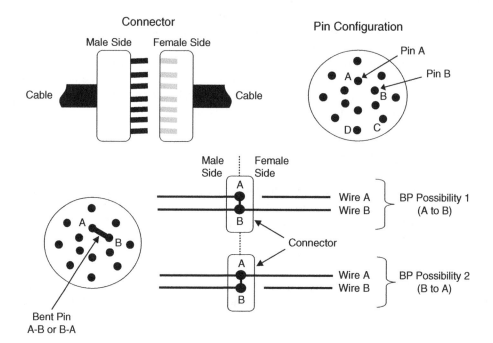

Figure 29.1 *BPA concept.*

a pin can bend over within the connector and make physical contact with another pin, thereby effectively causing a short circuit and an open circuit in the wiring.

Note in Figure 29.1 that pins A and B are close enough in proximity that they can make physical contact if one of them bends in the right direction, and that pins C and D are too far apart to make physical contact should one of them bend. BPA would evaluate the two scenarios of the pins A–B combination; pin A is bent such that it makes physical contact with pin B or pin B is bent such that it makes physical contact with pin A. Each of these two BP possibilities presents two different possible outcomes, for a total of four possible outcomes.

In this A–B and B–A scenario, there are four possible outcomes:

1. Pin A makes contact with pin B; wires A and B become a short circuit and the content of wire A has an upstream or downstream effect on wire B (depending upon the contents of both A and B).
2. Pin A makes contact with pin B; wire A becomes an open circuit after the short (downstream).
3. Pin B makes contact with pin A; wires A and B become a short circuit and the content of wire B has an upstream or downstream effect on wire A (depending upon the contents of both A and B).
4. Pin B makes contact with pin A; wire B becomes an open circuit after the short (downstream).

Figure 29.2 demonstrates the two possible outcomes from pin A bending and making contact with pin B. The overall system effect of these possibilities must be evaluated by the BPA. The upstream/downstream effect of the short can only be determined from the circuitry involved. Also, the overall effect of the open circuit can be determined only from the circuitry involved.

When considering connector failures, analysts sometimes omit the following open-circuit failure considerations:

- Dislodged contacts (those not properly seated and backing out of their working positions due to pulling from attached wires, or those pushed out by the mating contacts)
- Bad wiring connections at the contacts (improper crimps, improperly soldered joints, or corrosion)
- Corrosion at mating surfaces.

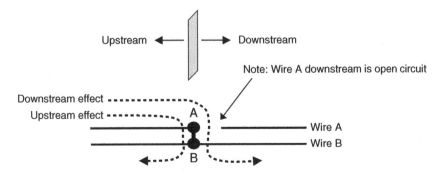

Figure 29.2 *Pin A to B short.*

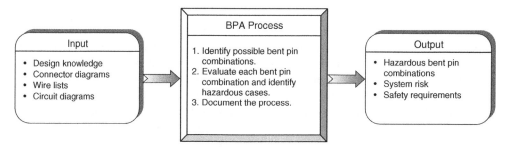

Figure 29.3   *BPA overview.*

- Bending to nothing; a pin can be flattened on the connector face while lying between neighboring pins.
- Bending to a grounded shell ignoring open circuits leads to a very incomplete analysis.

Multiple shorts can result when a pin is bent and it shorts to two other pins simultaneously. In some connector contact arrangements, a bent pin can be wedged between two nearby pins, or between one pin and the shell. In other arrangements, a pin can touch two other pins (or a pin and the shell) on the same side, rather than being wedged between them. In either case, this creates an entirely different failure mode: two signals shorting without one of them opening, plus a third signal shorting to both. Even if the bent pin is an unwired spare, it can still short two intact signal paths without introducing a third signal. Ignoring these possibilities can lead to an analysis that overlooks a large number of possible hazards.

29.5   BPA METHODOLOGY

The BPA technique evaluates connector pin layouts to predict worst-case effects of failure and ultimately system consequences. BPA generally considers and evaluates only the effect of a single bent pin contacting another pin within its radius, or with the connector case. BPA does not consider two pins bending and making contact with each other, or with a third pin. The probability of multiple bent pins occurring is extremely small, and would be considered only for high-consequence circuits.

Figure 29.3 summarizes the important relationships involved in BPA. This process consists of utilizing system design information and connector information to identify and mitigate hazards associated with potential bent connector pins.

Table 29.1 lists the basic steps in the BPA process, which involves performing a detailed analysis of all system electrical connectors.

It should be noted that BPA applies to both circular and rectangular shaped connectors. The examples in this chapter show round connectors, but the same methodology applies to rectangular connectors.

29.6   BPA WORKSHEET

BPA is a detailed hazard analysis of wire connectors, utilizing structure and rigor. It is desirable to perform the BPA using a form or worksheet to provide analysis structure and consistency.

TABLE 29.1   BPA Process

Step	Task	Description
1	Collect data	Identify and collect data on all system wire connectors. This includes connector pin layouts and wire content descriptions
2	Identify bent pin combinations	Review connector pin layouts and identify all BP combinations that are physically possible, and include contact with the case. Exclude BP combinations that are not possible
3	Evaluate combinations	Review the system design and operation to determine the effect of the potential BP combinations, should they occur. Analyze both A–B and B–A. For both A–B and B–A consider the possibilities: a. A short to B, with upstream/downstream effect on B b. A short to B, with A open from downstream side of connector c. B short to A, with upstream/downstream effect on A d. B short to B, with B open from downstream side of connector
4	Identify bent pin hazards	Identify those BP combinations where the system effect can result in a system hazard
5	Evaluate system risk	Identify the level of mishap risk presented by the BP hazard
6	Recommend corrective action	Establish design safety requirements to mitigate the identified hazard, such as changing pin locations or changing the system design to safely accommodate the failure
7	Track hazards	Transfer identified hazards into the hazard tracking system
8	Document BPA	Document the entire BPA process on the worksheets. Update for new information as necessary

The format of the analysis worksheet is not critical and can be modified to suit program requirements.

Typically, columnar-type worksheets are utilized to help maintain focus and structure in the analysis. As a minimum, the following basic information must be obtained from the analysis worksheets:

1. Possible bent pin combinations
2. The system effect on specific short or open circuits resulting from bent pins
3. The identification of bent pin combinations resulting in a hazard
4. System risk for the bent pin hazard

The recommended BPA matrix columnar type worksheet is shown in Figure 29.4. This particular BPA worksheet utilizes a columnar-type format. Other worksheet formats may exist because different organizations often tailor their analysis worksheet to fit their particular needs. The specific worksheet to be used may be determined by the system safety program (SSP), system safety working group, or the safety analysis customer.

The following instructions describe the information required under each column entry of the BPA worksheet:

1. *No.*   This column increments each BP combination in the analysis for reference purposes.
2. *Bent Pin*   This column identifies the particular pin-to-pin short combination that is being analyzed. The analysis considers only a single bent pin making contact with another pin within its bend radius, or a bent pin to the connector case.
3. *Pin Data*   This column identifies the specific electrical or data content on each of the pins involved.

Bent Pin Analysis							
No.	Bent Pin	Pin Data	Circuit State	Effect	Hazard	MRI	Comments
①	②	③	④	⑤	⑥	⑦	⑧

Figure 29.4   Recommended BPA worksheet.

4. *Circuit State*   This column identifies the two possible cases for BP analysis:
   a. upstream and downstream effect of pin-to-pin short
   b. downstream effect open circuit resulting from BP

5. *Effect*   This column identifies the effect of the bent pin (short or open circuit), assuming it occurs and includes the specific components impacted by the bent pin.

6. *Hazard*   This column identifies the hazard that may result from the bent pin. Generally, the worst-case system effect is stated in this column.

7. *Mishap Risk Index (MRI)*   This column provides a qualitative measure of mishap risk for the potential effect of the identified hazard, given that no mitigation techniques are applied to the hazard. Risk measures are a combination of mishap severity and probability, and the recommended values are shown the folloiwng table.

Severity	Probability
1 – Catastrophic	A – Frequent
2 – Critical	B – Probable
3 – Marginal	C – Occasional
4 – Negligible	D – Remote
	E – Improbable

8. *Comments*   This column provides a place to record useful information regarding the hazard or the analysis process.

29.7   BPA EXAMPLE

In order to demonstrate the BPA technique, the same hypothetical small missile system from Chapters 7 and 8 will be used. Figure 29.5 illustrates connector J1 for this example system, along with the connector pin layout and the pin content table. The pin content table contains a description of the electrical content of the pin (and associated wire).

Connector J1

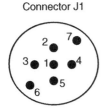

Pin	Content Description
1	Missile Fire Command (+5 VDC)
2	+28 VDC Missile Electronics Power
3	Ground
4	+5 VDC Missile Electronics Power
5	Ground
6	Ground
7	-28 VDC Missile Electronics Power
Case	Metallic case; grounded

Figure 29.5 *Missile connector diagram and pin content table.*

TABLE 29.2   Missile Connector Bent Pin Matrix

Pin	1	2	3	4	5	6	7	Case
1	S	X	X	X	X			
2	X	S						
3	X		S					
4	X			S				
5	X				S			
6						S		X
7							S	X

Code: S – self to self (ignore); X – possible bent pin contact.

Table 29.2 contains the bent pin matrix for connector J1. The bent pin matrix identifies which pins can physically make contact when one of the pins is bent. Also, a bent pin resulting in pin-to-case contact is considered.

The BP matrix is filled out by taking each of the pins in the left-hand column and identifying in the right-hand columns which pins it can physically contact when bent. In this BP matrix, an "S" indicates the pin itself listed both horizontally and vertically; it is not relevant and is ignored by the analysis. An "X" in the matrix indicates two pins are within contact radius of one another if one of the pins is bent. These are the pin-to-pin combinations, which must then be analyzed, in further detail. Connector pin layout, as well as pin length and position, is required to build this BP matrix.

The BP matrix and the pin content table are insufficient to determine the exact effect of pin-to-pin shorts. In order to determine the complete system effect of potential bent pins, it is necessary to have the complete circuit information available. Figure 29.6 contains an electrical circuit diagram for this example.

Tables 29.3, 29.4 and 29.5 contain the BPA worksheets that evaluate each of the possible bent pin combinations generated from the bent pin matrix.

The following conclusions can be drawn from the BPA of this example missile system:

1. If pin 2 is bent and contacts pin 1 (case 2a), immediate missile launch will occur when the system is turned on and +28 VDC is applied to the launch initiator.
2. If pin 4 is bent and contacts pin 1 (case 6a), immediate missile launch will occur when the system is turned on and +5 VDC is applied to the launch initiator
3. The FIRE command on pin 2 is a safety-critical function, and pin 2 should be isolated from other pins to avoid BP contact. If this is not possible, then pin 2 should only be in close proximity to safe short-circuit values, such as ground.

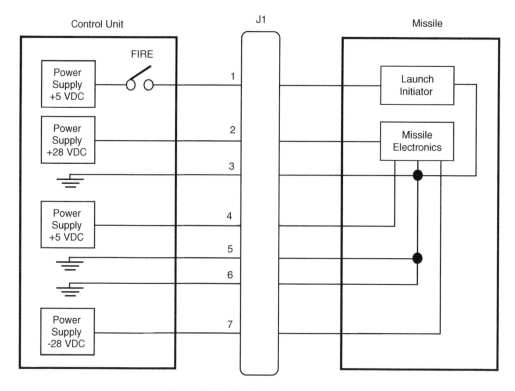

Figure 29.6 *Missile system schematic.*

29.8 BPA ADVANTAGES AND DISADVANTAGES

The following are advantages of the BPA technique:

1. BPA provides a pictorial aid for analysts in visualizing hazards.
2. BPA identifies hazards that might be overlooked by other techniques.

The disadvantage of the BPA technique is that it may require considerable time for obtaining detailed wire lists, electrical diagrams, and connector data.

29.9 COMMON BPA MISTAKES TO AVOID

When first learning how to perform a BPA, it is commonplace to commit some traditional errors. The following is a list of typical errors made during the conduct of a BPA:

1. Not completely analyzing the system effect of both open and short circuits resulting from a bent pin.
2. Not adequately determining pin length and bent pin contact radius.
3. Not fully documenting the entire analysis in detail.

TABLE 29.3   Example Bent Pin Analysis: Worksheet 1

Bent Pin Analysis

No.	Bent Pin	Pin Data	Circuit State	Effect	Hazard	MRI	Comments
1	1 to 2	1 – +5 VDC fire command	a. 1–2 short	+5 VDC short to +28 VDC when FIRE switch is closed	None		
		2 – +28 VDC	b. 1 open	Unable to fire missile	None		
2	2 to 1	Same as 1	a. 2–1 short	+28 VDC short to missile launch initiator	Inadvertent missile launch	1C	Change pin layout
			b. 2 open	No +28 VDC power to missile electronics	Missile state unknown with loss of power	2C	Further study required
3	1 to 3	1 – +5 VDC fire command	a. 1–3 short	+5 VDC short to ground when FIRE switch is closed	None		Unable to launch missile
		3 – Ground	b. 1 open	Same as 1b	None		
4	3 to 1	Same as 3	a. 3–1 short	Same as 3a	None		
			b. 3 open	Loss of ground to missile electronics and initiator	None		Has backup ground

TABLE 29.4   Example Bent Pin Analysis: Worksheet 2

Bent Pin Analysis

No.	Bent Pin	Pin Data	Circuit State	Effect	Hazard	MRI	Comments
5	1 to 4	1 – +5 VDC fire command	a. 1–4 short	+5 VDC short to +5 VDC when FIRE switch is closed	None		
		4 – +5 VDC	b. 1 open	Same as 1b	None		
6	4 to 1	Same as 5	a. 4–1 short	+5 VDC short to missile initiator	Inadvertent missile launch	1C	Change pin Layout
			b. 4 open	No +5 VDC power to missile electronics	Missile state unknown with loss of power	2C	Further study required
7	1 to 5	1 – +5 VDC fire command	a. 1–5 short	Same as 3a	None		Unable to launch missile
		5 – Ground	b. 1 open	Same as 1b	None		
8	5 to 1	Same as 7	a. 5–1 short	Same as 3a	None		Has backup ground
			b. 5 open	Same as 4b	None		

480

TABLE 29.5   Example Bent Pin Analysis: Worksheet 3

				Bent Pin Analysis			
No.	Bent Pin	Pin Data	Circuit State	Effect	Hazard	MRI	Comments
9	6 to case	6 – ground	a. 6 – case short	Ground to ground short	None		
		Case – ground	b. 1 open	Same as 4b	None		
10	7 to case	7 – –28 VDC case – ground	a. 7 – case short	–28 VDC short to ground	Arcs/sparks	2C	Change pin layout
			b. 7 open	No –28 VDC power to missile electronics	Missile state unknown with loss of power	2C	

29.10 SUMMARY

This chapter discussed the BPA technique. The following basic principles help summarize the discussion in this chapter:

1. The purpose of the BPA is to identify hazards caused by electrical circuits that are open or short-circuited from bent connector pins.
2. BPA should be a supplement to the PD-HAT, DD-HAT, and SD-HAT analyses.
3. BPA should always be considered whenever connectors are used to carry safety-critical signals.
4. BPA generally assumes a single bent pin-to-pin contact, since the probability of multiple bent pins contacting together is significantly less than the probability of a single bent pin.
5. BPA must address pin-to-case contact.
6. BPA is used to identify hazards that might be overlooked by other analyses, particularly those types of hazards involving bent connector pins causing electrical short circuits or open circuits.
7. The use of BPA worksheet forms and connector diagrams provides structure and rigor to the BPA process.

FURTHER READINGS

FAA System Safety Handbook, Chapter 9, Page 16, December 30, 2000.

Ozarin, N., What's Wrong with Bent Pin Analysis and What to Do about It, *J. Syst. Safety*, Sept–Oct, 2008, 19–28.

Ozarin, N. W., Rules for Automating Failure Modes and Effects Analysis for Electrical Connectors and Wiring, *SAE Aerospace Congress & Exhibition, Advances in Aviation Safety*, September, 2003.

Chapter 30

Management Oversight Risk Tree Analysis

30.1 INTRODUCTION TO MORT ANALYSIS

Management oversight and risk tree (MORT) is an analysis technique for identifying safety related oversights, errors, and/or omissions that lead to the occurrence of a mishap. MORT is primarily a reactive analysis tool for accident/mishap investigation, but it can also be used for the proactive evaluation and control of hazards. MORT analysis is used to trace out and identify all of the causal factors leading to a mishap or undesired event.

MORT utilizes the logic tree structure and rules of fault tree analysis (FTA), with the incorporation of some new symbols. Although MORT can be used to generate risk probability calculations like FTA, quantification is typically not employed. MORT analysis provides decision points in a safety program evaluation where design or program change is needed. MORT attempts to combine design safety with management safety.

30.2 MORT BACKGROUND

This analysis technique falls under the system design hazard analysis type (SD-HAT). Refer to Chapter 4 for a description of the analysis types. A smaller and less complex form of MORT has been developed that is referred to as mini-MORT.

MORT is a root cause analysis tool that provides a systematic methodology for planning, organizing, and conducting a detailed and comprehensive mishap investigation. It is used to identify those specific design control measures and management system factors that are less than adequate (LTA) and need to be corrected to prevent the reoccurrence of the mishap, or prevent the undesired event. The primary focus of MORT is on oversights, errors, and/or omissions and to determine what failed in the management system.

Hazard Analysis Techniques for System Safety, Second Edition. Clifton A Ericson, II.
© 2016 John Wiley & Sons, Inc. Published 2016 by John Wiley & Sons, Inc.

MORT analysis is applicable to all types of systems and equipment, with analysis coverage given to systems, subsystems, procedures, environment, and human error. The primary application of MORT is in mishap investigation to identify all root causal factors and to ensure that corrective action is adequate.

MORT analysis is capable of producing detailed analyses of root causes leading to an undesired event or mishap. By meticulously and logically tracking energy flows within and out of a system, MORT analysis compels a thorough analysis for each specific energy type. The degree of thoroughness depends on the self-discipline and ability of the analyst to track logically the flows and barriers in the system.

The analyst can master MORT analysis with appropriate training. The analyst must have the ability to understand energy flow concepts, for which at least a rudimentary knowledge of the behaviors of each of the basic energy types is necessary. Ability to logically identify energy sources and track flows in systems is an essential skill. Ability to visualize energy releases, energy exchange, or transformation effects is another helpful skill. Since MORT analysis is based on an extended form of FTA, the FTA technique itself could be used as a replacement for MORT analysis. A condensed version of MORT called mini-MORT could also be used.

The use of this technique is not recommended for the general system safety program, since it is complex, time-consuming, unwieldy in size, and difficult to understand. Other hazard analysis techniques are available that provide results more effectively. MORT could be used for mishap investigation, but FTA is more easily understood and just as effective.

30.3   MORT HISTORY

The MORT analysis technique was developed in circa 1970 by W. G. Johnson of the Aerojet Nuclear Company. The development work was sponsored by the Energy Research and Development Administration (Department of Energy, formerly the Atomic Energy Commission) at the Idaho National Engineering Laboratory (INEL). MORT analysis is predicated upon hazardous energy flows and safety barriers mitigating these flows.

30.4   MORT THEORY

The theory behind MORT analysis is fairly simple and straightforward. The analyst starts with a predefined MORT graphical tree that was developed by the original MORT developers. The analyst works through this predefined tree comparing the management and operations structure of his/her program to the ideal MORT structure, and develops a MORT diagram modeling the program or project. MORT and FTA logic and symbols are used to build the program MORT diagram. The predefined tree consists of 1500 basic events, 100 generic problem areas and a large number of judging criteria. This diagram can be obtained from The MORT User's Manual, DOE SSDC-4, 1983.

The concept emphasizes energy-related hazards in the system design and management errors. MORT analysis is based on energy transfer and barriers to prevent or mitigate mishaps. Consideration is given to management structure, system design, potential human error, and environmental factors.

Common terminology used in MORT analysis charts includes the following acronyms:

- LTA: less than adequate
- DN: did not

- FT: failed to
- HAP: hazard analysis process
- JSA: job safety analysis
- CS&R: code standards and regulations

The generic MORT diagram has many redundancies in it due to the philosophy that it is better to ask a question twice rather than fail to ask it at all.

MORT analysis is based on the following definitions:

Accepted or Assumed Risk This is very specific risk that has been identified, analyzed, quantified to the maximum practical degree, and accepted by the appropriate level of management after proper thought and evaluation. Losses from assumed risks are normally those associated with earthquakes, tornadoes, hurricanes, and other acts of nature.

Amelioration Postaccident actions such as medical services, fire fighting, rescue efforts, and public relations.

30.5 MORT METHODOLOGY

Table 30.1 shows an overview of the basic MORT analysis process and summarizes the important steps and relationships involved. This process consists of utilizing design information and known hazardous energy source information to verify complete safety coverage and control of hazards.

TABLE 30.1 MORT Analysis Process

Step	Task	Description
1	Define system	Define scope and the boundaries of the system. Define the mission, mission phases, and mission environments. Understand the system design and operation
2	Plan MORT analysis	Establish MORT analysis goals, definitions, worksheets, schedule, and process. Divide the system under analysis into the smallest segments desired for the analysis. Identify items to be analyzed and establish indenture levels for items/functions to be analyzed
4	Acquire data	Acquire all of the necessary design and process data needed (e.g., functional diagrams, code, schematics, and drawings) for the system, subsystems, and functions. Refine the system information and design representation for MORT analysis
5	Conduct MORT analysis	a. Using the predefined tree, draw a new diagram for the system under review. b. Color code events on tree diagram. c. Continue analysis until all events are sufficiently analyzed with supporting data.
6	Recommend corrective action	Recommend corrective action for hazards with unacceptable risk. Assign responsibility and schedule for implementing corrective action
7	Monitor corrective action	Review the MORT diagram at scheduled intervals to ensure that corrective action is being implemented
8	Track hazards	Transfer identified hazards into the hazard tracking system using the HAR format
9	Document MORT analysis	Document the entire MORT process on the worksheets. Update for new information and closure of assigned corrective actions

SYMBOL	NAME	DESCRIPTION
 ▭	General Event	Describes general event.
 ◯	Basic Event	A basic component failure; the primary, inherent, failure mode of a component. A random failure event.
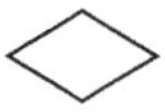 ◇	Undeveloped Event	An event that could be further developed if desired.
▢	Satisfactory Event	Used to show completion of logical analysis .
 ⌁	Normally Expected Event	An event that is expected to occur as part of normal system operation.
 ⬭	Assumed Risk Transfer	A risk that has been identified, analyzed, and quantified to the maximum practical degree, and accepted.
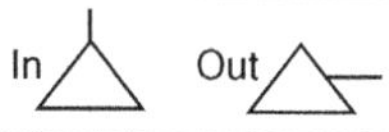 In △ Out △—	Transfer	Indicates where a branch or subtree is marked for the same usage elsewhere in the tree. In and Out or To/From symbols.
 ⌂	OR Gate	The output occurs only if at least one of the inputs occurs.
 ⌂	AND Gate	The output occurs only if all of the inputs occur together.
 —⬭	Constraint	Constraint on Gate Event or General Event.

Figure 30.1 MORT symbols.

30.6 MORT ANALYSIS WORKSHEET

The MORT analysis worksheet is essentially a slightly modified fault tree with some added symbols and color coding. All of the symbols, rules, and logic of FTA (see Chapter 15 on FTA) apply to MORT analysis. New symbols added specifically for MORT are shown in Figure 30.1.

Events on the MORT diagram are color coded according to the criteria in Table 30.2.

TABLE 30.2 MORT Color Coding

Color	Meaning
Red	Any factor or event found to be LTA is colored red on the chart. Should be addressed in the final report with appropriate recommendations to correct the deficiency. Use judiciously; must be supported by facts
Green	Any factor or event found to be adequate is colored green on the chart. Use judiciously; must be supported by facts
Black	Any factor or event found to be not applicable is color coded black (or simply crossed out) on the chart
Blue	Indicates that the block has been examined, but insufficient evidence or information is available to evaluate the block. All blue blocks should be replaced with another color by the time the investigation is complete

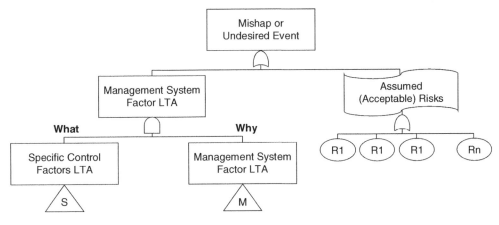

Figure 30.2 MORT top tiers.

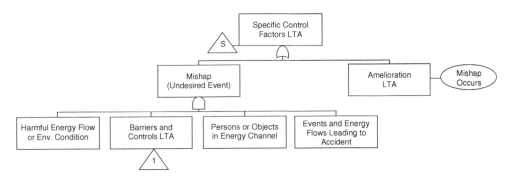

Figure 30.3 MORT-specific control factors.

MORT analysis is essentially an FTA that asks *what* oversights and omissions could have occurred to cause the undesired event or mishap, and *why* in terms of the management system. In some ways, MORT analysis is like using the basic MORT diagram as a checklist to ensure everything pertinent is considered.

Figure 30.2 shows the top level of the ideal MORT analysis from the MORT User's Manual.

Figure 30.3 expands the *S branch* of the MORT shown in 30.2.

Figure 30.4 expands the *M branch* of the MORT shown in 30.2.

Figure 30.5 expands the 1 branch of the MORT shown in 30.3.

Figure 30.6 expands the *2 branch* of the MORT shown in 30.5.

30.7 MORT ADVANTAGES AND DISADVANTAGES

The following are advantages of the MORT analysis technique:

1. Has a pictorial benefit that aids analysts in visualizing hazards.
2. Can be quantified (but usually is not).

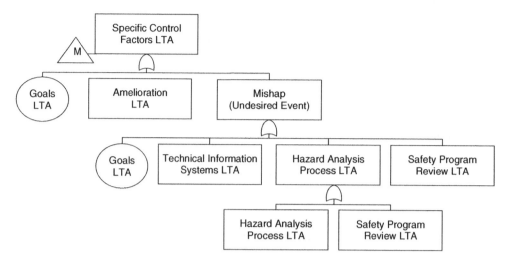

Figure 30.4 *MORT management system factors.*

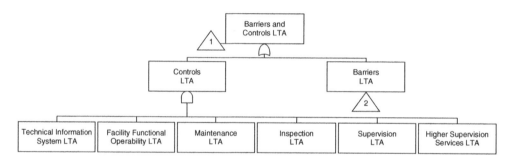

Figure 30.5 *MORT barriers and controls diagram.*

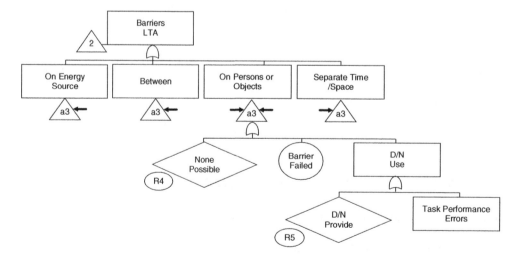

Figure 30.6 *MORT barriers diagram.*

3. Is simple to perform (once understood).
4. Commercial software is available to assist the analyst.

The following are disadvantages of the MORT analysis technique:

1. Though simple in concept, the process is labor intensive and requires significant training.
2. Is limited by the ability of the analyst to identify all the hazardous energy sources.
3. Tree size can become too large for effective comprehension by the novice.

30.8   COMMON MORT ANALYSIS MISTAKES TO AVOID

When first learning how to perform a MORT analysis it is commonplace to commit some traditional errors. The following is a list of typical errors made during the conduct of a MORT analysis:

1. Not obtaining the necessary training
2. Not thoroughly investigating all causal factor paths
3. Using the MORT analysis as a primary hazard analysis rather than a secondary hazard analysis.

30.9   MORT SUMMARY

This chapter discussed the MORT hazard analysis technique. The following are basic principles that help summarize the discussion in this chapter:

1. MORT analysis is a root cause analysis tool similar to FTA.
2. The primary purpose of MORT analysis is for mishap investigation analysis.
3. MORT analysis should be a supplement to the SHA.
4. MORT analysis involves a focus on hazardous energy sources and barriers.
5. MORT analysis is based on an existing pre-defined tree diagram.
6. MORT provides analysis of the management system for a project.

FURTHER READINGS

Clark, J. L., The Management Oversight and Risk Tree (MORT): A New System Safety Program, Proceedings of the 2nd International System Safety Conference, 1975, pp. 334–350.

Johnson, W. G., *MORT, the Management Oversight and Risk Tree, U.S. Atomic Energy Commission*, SAN-821-2, 1973, US Government Printing Office, Washington DC.

Johnson, W. G., *MORT Safety Assurance Systems*, Marcel Dekker, New York, 1980.

Knox, N. W. and Eicher, R. W., *MORT User's Manual*, SSDC-4 (revision 2), US Dept. of Energy, Idaho Falls, Idaho, 1983.

Stephenson, J., *System Safety 2000: A Practical Guide for Planning, Managing, and Conducting System Safety Programs*, Chapter 18 on MORT, John Wiley & Sons, pp. 218–255, 1991.

Chapter *31*

Job Hazard Analysis

31.1 JHA INTRODUCTION

In industry many different types of jobs are performed by workers on a daily basis. For example, milling machine operations, laying underground piping, performing maintenance on an aircraft, or building a new skyscraper, to name a few. Job hazard analysis (JHA) is an analysis technique for identifying hazards in a proposed job prior to the job being performed, where a job involves performing a specified set of tasks to accomplish a specific objective.

JHA is an analysis tool for specifically assessing the safety of operations (jobs or tasks) by integrally evaluating operational procedures, the system design, and the potential for human errors. Considerations include human error, task design error, ergonomics, noise, vibration, temperature, chemicals, hazardous materials, and so on. The intent is to identify and mitigate hazards prior to performing the job. JHA involves identifying hazards, hazard causal factors, hazard effects, hazard risk, and mitigating methods.

The scope of the JHA tasks includes normal operation, test, installation, maintenance, repair, training, storage, handling, transportation, and emergency/rescue operations. Consideration is given to task design, hardware failure modes, human error, and overall system design. Human factors and human system integration (HSI) considerations are a large factor in performing jobs and are, therefore, a major factor in JHA. A JHA is typically conducted prior to a scheduled job in order to affect a safe operation.

On the surface, the JHA methodology appears to be very similar in nature to the operating and support hazard analysis (O&SHA), and the question often arises as to whether they both accomplish the same objectives. The O&SHA evaluates operator tasks and activities for the identification of task hazards, which is effectively the same goal as for a JHA. The primary

Hazard Analysis Techniques for System Safety, Second Edition. Clifton A Ericson, II.
© 2016 John Wiley & Sons, Inc. Published 2016 by John Wiley & Sons, Inc.

difference is that the O&SHA originates from military standards, whereas the JHA originates from consumer industries, such as construction and manufacturing.

It should be noted that job hazard analysis is also sometimes referred to as job safety analysis (JSA), activity hazard analysis (AHA), or task hazard analysis (THA), depending on the industry and analysts involved.

31.2 JHA BACKGROUND

This analysis technique falls under the operations design hazard analysis type (OD-HAT) because it evaluates procedures and tasks performed by humans. It also includes the system design involved in the operational tasks. The basic analysis types are described in Chapter 4.

The JHA is a technique that focuses on job tasks as a way to identify potential worker hazards and mitigate them in order that they do not occur during performance of the job. JHA focuses on the relationship between the worker, the task, the tools, and the work environment. Ideally, after all the hazards have been identified, steps will be taken to eliminate or reduce job task hazards to an acceptable risk level.

The purpose of a JSA is to identify workplace hazards. The JHA helps to ensure the safety of personnel in the performance of a job comprised of a set of tasks. Hazards can be introduced by the system design, the task design, human error, and/or the environment.

The JHA is conducted prior to performing a scheduled task or job and is directed toward developing a safe set of procedures. The JHA identifies the functions and procedures that could be hazardous to personnel or, through personnel errors, could create hazards to equipment, personnel, or both. Corrective action resulting from this analysis is usually in the form of design requirements and procedural inputs to operating, maintenance, and training manuals. Many of the procedural inputs from system safety are in the form of caution and warning notes.

The JHA is applicable to the analysis of all types of operations, procedures, tasks, and functions. It can be performed on draft procedural instructions or detailed instruction manuals. The JHA technique provides sufficient thoroughness in identifying and mitigating operations and support-type hazards when applied to a given system/subsystem by experienced safety personnel. A basic understanding of the hazard analysis theory and the knowledge of system safety concepts is essential. Experience with or a good working knowledge of the particular type of system and subsystem is necessary in order to identify and analyze hazards that may exist within procedures and instructions. The methodology is uncomplicated and easily learned. Standard JHA forms and instructions have been developed that are included as part of this chapter.

JHA can provide insight into design or task changes that might adversely affect operational tasks and procedures. JHA effort should be performed sufficiently in advance of the scheduled job in order to enhance understanding and to allow implementation for any changes that may be necessary.

JHA is typically performed using a standard analysis worksheet, which provides a format for entering the sequence of operations, procedures, tasks, and steps necessary for task accomplishment. The worksheet also provides a format for analyzing this sequence in a structured process that produces a consistent and logically reasoned evaluation of hazards and controls. The JHA technique is uncomplicated and easily learned. A typical JHA worksheet, with instructions, is provided in this chapter.

31.3   JHA HISTORY

The history of JHA has been researched and explained by David Glenn [1]. JSA appears to have evolved from the scientific management practice of job analysis (JA). In fact, the first safety analyst to use the term *job safety analysis* was Heinrich (see Ref. [2], p. 96). The safety connection to scientific management is explicit in the subtitle to Heinrich's *Industrial Accident Prevention: A Scientific Approach*. Scientific management began with Frederick Taylor's proposal to improve wage-setting methods (see Ref. [3], p. 75). The time studies involved in this process consisted of "an analysis of a job as a whole into the elementary movements of man and machine" (see Ref. [3], p. 77).

The process of JSA actually preceded Heinrich's use of the term. A safety engineer from General Electric wrote in 1930 that "job analysis should bring out the hazards of the operations" so that standard procedures could be established (see Ref. [2], p. 32). A 1927 NSC magazine published "Job analysis for safety," which described a process of subdividing the operations, listing related hazards, and adopting standard methods for streetcar operators ([4], p. 80).

Today, JHA is a standard safety methodology applied by many industrial companies. It is also often required by the Occupational Safety and Health Administration (OSHA). The Federal OSHA Act states:

> Employers must furnish a place of employment free of recognized hazards that are causing or are likely to cause death or serious physical harm to employees (OSHA Act of 1970, 29 CFR 1903.2).

This OSHA requirement provides motivation for employers to perform JHA. All employers should be conducting JHA and JHA training in order to keep the workplace safe for employees. JHA covers all industries and all type of jobs and tasks. It should go without saying that JHA be extensively conducted on hazardous jobs and job sites, such as welding, machining, construction, and so on.

31.4   JHA THEORY

JHA is an analysis technique to identify the dangers of specific tasks in order to reduce the risk of injury to workers. JHA focuses on the identification of potential worker hazards resulting from a workers exposure to many different hazard sources found in the job site. In general terms, these hazard sources stem from tasks, processes, environments, chemicals, and materials. Specific health hazards and their impact on the human are assessed during the JHA.

Figure 31.1 shows an overview of the basic JHA process and summarizes the important relationships involved in the JHA process. This process consists of utilizing both design information and known hazard information to identify hazards. Known hazardous elements and mishap lessons learned are compared with the system design to determine if the design concept contains any of these potential hazard elements.

The JHA process involves the following:

1. Providing a safety focus from an operations and operational task viewpoint.
2. Identifying task or operational oriented hazards caused by design, hardware failures, software errors, human error, timing, and the like.

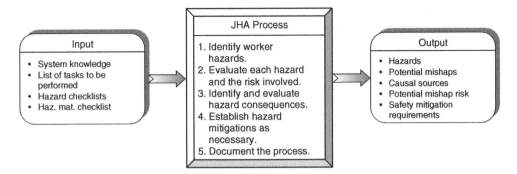

Figure 31.1 *JHA overview.*

3. Assessing the potential operations mishap risk.
4. Identifying design system safety requirements (SSRs) to mitigate operational task hazards.
5. Ensuring all operational procedures and tasks are safe.

An important and defining facet of a job-related hazard is its "mechanism of injury" (MoI). This is important because in the absence of an MoI there, it is not a hazard. The MoI is the method in which an injury occurs. This falls in line with the source, mechanism, outcome (S-M-O) theory of hazards and the MoI is the hazard mechanism. All hazards have a mechanism of occurrence or MoI.

For example, consider the hazard "worker trips over tool bag left in walkway, resulting in a fall that breaks the workers writs." In this case, the hazard source is tool bag in walkway, the MoI is worker trips over tool bag, and the hazard outcome is worker's broken wrist.

Common mechanisms of injury include the following:

• Slips
• Trips
• Fall to same/lower level falls
• Struck against/by
• Contact with/by
• Caught in/on/by/between
• Exposure to

31.5 JHA METHODOLOGY

Table 31.1 lists and describes the basic steps of the JHA process, which involves performing a detailed analysis of all potentially hazardous human health hazard sources.

The thought process behind the JHA methodology is shown in Figure 31.2. The idea supporting this process is utilize task information, system information, hazard checklists, and hazard knowledge to facilitate hazard identification. When using hazard checklists, of particular interest are hazard checklists dealing with human health issues and human failure modes. Also, data on human limitations and regulatory requirements are used to identify human hazards.

TABLE 31.1   JHA Process

Step	Task	Description
1	Acquire job information	Acquire information on the job, along with a complete list of the specific tasks to be performed
2	Acquire design information	Acquire all of the design and operational data for the system, particularly for the job being performed
3	Acquire hazard checklists	Acquire checklists of known hazard sources, such as chemicals, materials, processes, and so on. Also, acquire checklists of known human limitations in the operation of systems, such as noise, vibration, heat, and so on.
4	Acquire regulatory information.	Acquire all regulatory data and information that are applicable to the particular job under review
5	Identify job hazard sources	Examine the system and the tasks to be performed and identify all potentially hazardous sources and processes within the system
6	Identify hazards	Identify and list potential hazards created by the tasks to be performed using hazard knowledge, hazard checklists, and potential human errors and equipment failure modes
7	Evaluate system risk	Identify the level of mishap risk presented to personnel, both with and without design controls in the system design, in order to understand risk changes
8	Recommend corrective action	Determine if the hazard controls present are adequate, and if not, recommend controls that should be added to reduce the mishap risk
9	Track hazards	Transfer newly identified hazards into the HTS. Update the HTS as hazards, hazard causal factors, and risk are identified in the JHA.
10	Document JHA	Document the entire JHA process on the worksheets. Update for new information as necessary

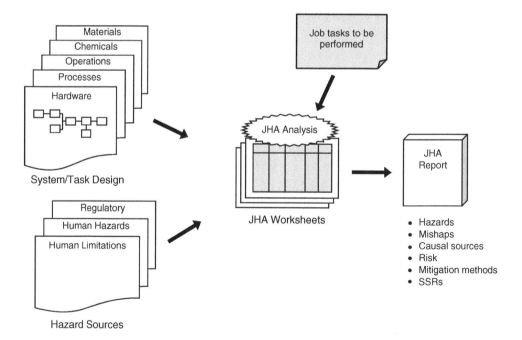

Figure 31.2   JHA methodology.

The JHA strongly depends upon the use of health hazard checklists. Hazard checklists are generic lists of known hazardous items and potentially hazardous designs or situations and should not be considered complete or all-inclusive. Checklists are intended as a starting point to help trigger the analyst's recognition of potential hazard sources from past lessons learned.

Typical personnel injury hazard checklist categories include the following:

1. Ergonomic
2. Noise
3. Vibration
4. Temperature
5. Chemicals
6. Biological
7. Hazardous materials
8. Physical stress
9. Pinch points
10. Slippery surfaces
11. Impact
12. Penetration

When performing the JHA, the following factors should be given consideration:

1. Toxicity, quantity, and physical state of materials.
2. Routine or planned uses and releases of hazardous materials or physical agents.
3. Accidental exposure potentials.
4. Hazardous waste generated.
5. Hazardous material handling, transfer, and transportation requirements.
6. Protective clothing/equipment needs.
7. Detection and measurement devices required to quantify exposure levels.
8. Number of personnel potentially at risk.
9. Design controls that could be used, such as isolation, enclosure, ventilation, noise or radiation barriers, and so on.
10. Potential alternative materials to reduce the associated risk to users/operators.
11. The degree of personnel exposure to the health hazard.
12. System, facility, and personnel protective equipment design requirements (e.g., ventilation, noise attenuation, radiation barriers, etc.) to allow safe operation and maintenance.
13. Hazardous material and long-term effects (such as potential for personnel and environmental exposure, handling and disposal issues/requirements, protection/control measures, and life cycle costs).
14. Means for identifying and tracking information for each hazardous material.
15. Environmental factors that affect exposure (wind, temperature, humidity, etc.)

When feasible engineering designs are not available to reduce hazards to acceptable levels, alternative protective measures must be specified (e.g., protective clothing, specific operation, or maintenance practices to reduce risk to an acceptable level). Identify potential nonhazardous or less-hazardous alternatives to hazardous materials if they exist, or provide a justification why an alternative cannot be used.

Additional example hazard sources to consider when performing a JHA include the following items:

- Heat
- Impact
- Sharp edges
- Penetration
- Compression
- Chemical exposures
- Repetitive motions
- Optical radiation
- Employee jewelry
- Harmful airborne contaminants
- Potential for being caught between pinch points
- Worker posture/balance
- Hazardous movements
- "Struck by" hazards
- Suspended loads
- Environmental hazards

When performing a JHA, the following parameters should be considered:

- Is personnel protective equipment (PPE) required/available?
- Has the worker been properly trained?
- Is the worker position/posture proper?
- Is lockout/tagout required?
- What is the flow of work?
- What are the sources of chemicals, noise, and so on?
- Are slips, trips, and falls a possibility?

When controlling or mitigating JHA hazards, the preferred control hierarchy is as follows:

- Engineering controls
- Administrative controls
- Procedures
- Personal protective equipment

When conducting a JHA, the following standards of performance should be considered and maintained:

- Maintain safety guards on all devices.
- No horseplay.

- Stick to assigned tasks.
- Maintain good housekeeping.
- Inspect all equipment prior to use.
- Qualified operators only.
- Follow PPE compliance (hard hats, gloves, eye protection, face protection, hearing protection, and respiratory protection).
- Follow safe lifting limits (never exceed 50 pounds by yourself).

All employees should be empowered to

- stop or not start work if they feel it is unsafe.
- actively participate in safety programs.
- comply with safety procedures.
- identify, report, and help to correct safety hazards.
- develop effective hazard controls through pretask planning and analysis.
- support and be informed about safety assessments.
- support investigations and cause analysis to prevent recurrence of unwanted events.

It is important to understand that it is not the JHA form or worksheet that will keep workers safe on the job, but rather the process it represents. It is of little value to identify hazards and devise controls if the controls are not put in place. Everyone in the workforce should be involved in creating the JHA. The more minds and the more years of experience applied to analyzing job hazards, the more successful the work group will be in controlling them. Remember, a JHA can become a quasi-legal document, and can be used in incident investigations, contractual disputes, and court cases.

31.6 JHA WORKSHEET

The JHA is a detailed hazard analysis utilizing structure and rigor. It is desirable to perform the JHA using a specialized worksheet. Although the format of the analysis worksheet is not critical, typically, matrix- or columnar-type worksheets are used to help maintain focus and structure in the analysis. Sometimes, a textual document layout worksheet is utilized. As a minimum, the following basic information is required from the JHA:

1. Worker job hazards
2. Hazard effects (mishaps)
3. Hazard causal factors (materials, processes, excessive exposures, etc.)
4. Risk assessment (before and after design safety features are implemented)
5. Derived safety requirements for eliminating or mitigating the hazards

The JHA worksheet can be simple or complex; however, simplicity and effectiveness are the key goals of the worksheet. A suggested JHA worksheet is shown in Figures 31.3. This particular JHA worksheet utilizes a columnar-type format and has been proven effective in many applications. The worksheet can be modified as found necessary by the safety team.

Job Hazard Analysis							
Task	Hazard	Causes	Effects	IMRI	Recommended Action	FMRI	Comments
①	②	③	④	⑤	⑥	⑦	⑧

Figure 31.3   *JHA worksheet.*

The information required under each column entry of the worksheet is described as follows:

1. *Task*   This column identifies the type of human health concern being analyzed, such as vibration, noise, thermal, chemical, and so on.

2. *Hazard*   This column identifies what can go wrong and the resulting undesirable consequences. This may require inspection of the job site. It may also require a team effort.

3. *Cause*   This column identifies conditions, events, or faults that could cause the hazard to exist, and the events that can trigger the hazardous elements to become a mishap or accident.

4. *Effect/Mishap*   This column identifies the effect and consequences of the hazard, should it occur. Generally, the worst-case result is the stated effect.

5. *Initial Mishap Risk Index (IMRI)*   This column provides a qualitative measure of mishap risk for the potential effect of the identified hazard, given that no mitigation techniques are applied to the hazard. Risk measures are a combination of mishap severity and probability, and the recommended values are shown in the following table. Risk knowledge helps in the prioritization of mitigation measures.

Severity	Probability
1 – Catastrophic	A – Frequent
2 – Critical	B – Probable
3 – Marginal	C – Occasional
4 – Negligible	D – Remote
	E – Improbable

6. *Recommended Action*   This column establishes recommended preventive measures to eliminate or control identified hazards. Safety requirements in this situation generally involve the addition of one or more barriers to keep the energy source away from the target. The preferred order of precedence for design safety requirements is as shown in the following table.

Order of Precedence

1 – Eliminate hazard through design selection
2 – Control hazard through design methods
3 – Control hazard through safety devices
4 – Control hazard through warning devices
5 – Control hazard through procedures and training

7. *Final Mishap Risk Index (FMRI)* This column provides a qualitative measure of mishap risk significance for the potential effect of the identified hazard, given that mitigation techniques and safety requirements are applied to the hazard. The same values used in column 5 are also used here. The initial and final risk values are important because they show the amount of risk improvement.

8. *Comments* This column provides a place to record useful information regarding the hazard or the analysis process that are not noted elsewhere.

31.7 EXAMPLE HAZARD CHECKLIST

Table 31.2 provides a list of typical health hazard sources that should be considered when performing a JHA.

Table 31.3 provides a list of example hazard sources that should be considered when performing a JHA.

TABLE 31.2 Typical Human Health Hazard Sources

JHA CATEGORY	EXAMPLES
Acoustic Energy Potential energy existing in a pressure wave transmitted through the air may interact with the body to cause loss of hearing or internal organ damage	• Steady-state noise from engines • Impulse noise from shoulder-fired weapons
Biological Substances Exposures to microorganisms, their toxins, and enzymes.	• Sanitation concerns related to waste disposal
Chemical Substances Exposure to toxic liquids, mists, gases, vapors, fumes, or dusts	• Combustion products from weapon firing • Engine exhaust products • Degreasing solvents
Oxygen Deficiency Hazard may occur when atmospheric oxygen is displaced in a confined/enclosed space and falls below 21% by volume; also used to describe the hazard associated with the lack of adequate ventilation in crew spaces	• Enclosed or confined spaces associated with shelters, storage tanks, and armored vehicles • Lack of sufficient oxygen and pressure in aircraft cockpit cabins • Carbon monoxide in armored tracked vehicles

(continued)

TABLE 31.2.   (*Continued*)

JHA CATEGORY	EXAMPLES
Ionizing Radiation Energy Any form of radiation sufficiently energetic to cause ionization when interacting with living matter	• Radioactive chemicals used in light sources for optical sights and instrumented panels
Nonionizing Radiation Emissions from the electromagnetic spectrum that has insufficient energy to produce ionization, such as lasers, ultraviolet, and radio frequency radiation sources	• Laser range finders used in weapons systems; microwaves used with radar and communication equipment
Shock Delivery of a mechanical impulse or impact to the body. Expressed as a rapid acceleration or deceleration	• Opening forces of a parachute harness • Back kick of firing a handheld weapon
Temperature Extremes Human health effects associated with hot or cold temperatures	• Increase to the body's heat burden from wearing total encapsulating protective chemical garments • Heat stress from insufficient ventilation to aircraft or armored vehicle crew spaces
Trauma Injury to the eyes or body from impact or strain	• Physical injury caused by blunt or sharp impacts • Musculoskeletal trauma caused by excessive lifting
Vibration Adverse health effects (e.g., back pain, hand-arm vibration syndrome (HAVS), carpel tunnel syndrome, etc.) caused by contact of a mechanically oscillating surface with the human body	• Riding in and/or driving/piloting armored vehicles or aircraft • Power hand tools • Heavy industrial equipment
Human–Machine Interface Various injuries such as musculoskeletal strain, disk hernia, carpel tunnel syndrome, and so on resulting from physical interaction with system components	• Repetitive ergonomic motion • Manual material handling – lifting assemblies or subassemblies • Acceleration, pressure, velocity, and force
Hazardous Materials Exposures to toxic materials hazardous to humans	• Lead • Mercury

TABLE 31.3   Example Job Hazards

Job Steps	Potential Hazard Types	Recommended Controls
General office work	Electrical shock and fire hazards – office equipment, electric heaters, appliances, and outlets Slips, trip, and falls – wet floors, workplace obstacles, and so on Chemical exposure – cleaning products, toner, and so on.	General office safety awareness training; office/workplace evaluations for hazard recognition Slip, trip, and fall prevention training Follow good housekeeping practices – including closing draws, properly disposing of waste, and proper storage of materials and equipment not in use Training; storage controls

TABLE 31.3. *(Continued)*

Job Steps	Potential Hazard Types	Recommended Controls
Data entry/ computer use/ typing	Repetitive motion disorders and musculoskeletal disorders	Ergonomics training, work breaks, and motion exercises
	Eye strain	Ensure work area is properly illuminated and eliminate glare on screen. Individuals requiring prescription eye wear should ensure proper focal distance is achieved without having to tilt the head. Adjust screen refresh rate as needed to eliminate screen flicker
Filing and materials handling	Musculosketal disorders – back strain	Ergonomics and lifting techniques training
Office equipment use – fax and copy machines	Potential exposure to ozone and other sensitizers	Ensure work area is adequately ventilated
Paper cutters, staplers, and so on	Lacerations; punctures	Do not use paper cutters if the guard is removed or blade is dull; keep fingers and hands away from point of operation
Material storage	Slips, trip, and falls: lacerations, bruises, or abrasions	Slip, trip, and fall prevention training
		Follow good housekeeping practices – including closing draws, properly disposing of waste, and proper storage of materials and equipment not in use. Do not store materials overhead height if possible. Use appropriate ladders or stepping devices (do not use chairs as step ladders)
General worker safety and reporting	Potential life-threatening situations – fire, explosion, chemical release, and so on	Know where fire or emergency pull stations are located in the building. Know the emergency evacuation routes for the building and the rally point once outside
General work site	Potential injury due to the presents of workplace hazards such as broken handrail, broken chair, and torn carpet presenting a trip hazard	Perform periodic workplace assessments for potential hazards; know your building contacts and report hazards to them for correction
		Know procedures for reporting work-related injuries

31.8 JHA TOOL

Occasionally, the steps in a job are not nicely listed for the worker or the JHA analyst. The task then becomes one of laying out the task steps in a complete and orderly manner. A tool for aiding in this effort is the "Ishikawa diagram" or as otherwise known the "fishbone diagram." This type of diagram helps the analyst in identifying all the necessary elements involved in a job. It can be done at both a macrolevel and a microlevel. Figure 31.4 demonstrates a fishbone diagram for changing the tire on a car. This example is at the macrolevel. Each of the items under "steps" could be broken down into more detail, or at the microlevel, using a fishbone diagram.

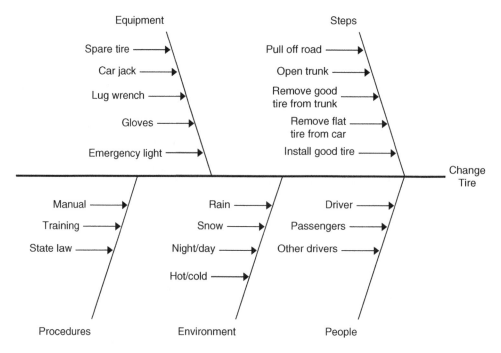

Figure 31.4 *Fishbone diagram for changing car tire.*

31.9   JHA EXAMPLE

In order to demonstrate the JHA methodology, a JHA is performed on two example jobs. Table 31.4 shows a JHA worksheet for an example job of using a ladder on a work site while repairing a light fixture. Table 31.5 shows a JHA worksheet for an example job of operating a bulldozer.

31.10   JHA ADVANTAGES AND DISADVANTAGES

The JHA technique the following advantages:

- Easily and quickly performed.
- Does not require considerable expertise for technique application.
- Relatively inexpensive, yet provides meaningful results.
- Provides rigor for focusing on job and task hazards.
- Quickly provides an indication of where hazards exist.
- Improves safety performance.
- Reduces worker absences.
- Increases productivity.
- Improves worker morale.
- Assists in OSHA compliance.

There are no notable disadvantages of the JHA technique.

TABLE 31.4 Example JHA for Ladder Task

Job Hazard Analysis

Task	Hazard	Causes	Effects	IMRI	Recommended Action	FMRI	Comments
Select ladder for location	Wrong type of ladder is selected, worker falls and is injured	Wrong ladder for location and/or use	Fall injury	3C	Training on ladder selection	3D	
	Damaged ladder is selected, worker falls and is injured	Ladder is damaged	Fall injury	3C	Inspect ladder prior to use
Ladder inspection program	3D	
Transport ladder to location	Worker is injured lifting the ladder	Ladder weight exceeds one man lift limit	Fall injury	3C	Use two people to carry the ladder	3D	
	Ladder collides with facility walls causing damage	Ladder hits wall	Facility damage	3C	Use two people, one as spotter	3D	
Set up ladder at task location	Ladder tips over due to unstable base causing worker injury	Uneven base for ladder	Fall injury	3C	Training on ladder setup	3D	
Ascend ladder to perform task	Worker falls from ladder during ascent, resulting in injury	Slippery footing	Fall injury	3C	Ensure ladder and feet are dry	3D	
		Tools snag on ladder			Use tool belt or bucket		

TABLE 31.5 Example JHA for Bulldozer Operation

Job Hazard Analysis

Task	Hazard	Causes	Effects	IMRI	Recommended Action	FMRI	Comments
Worker is operating a bulldozer at construction site	Operator thrown from cab, worker falls and is injured	Slanted terrain or bumps	Fall injury	3C	Operator training Seat belts usage Plan ahead for slopes	3D	
	Operator thrown from cab, worker is run over by dozer	Slanted terrain or bumps	Worker death	1C	Operator training Seat belt usage Plan ahead for slopes	1E	
	Bulldozer strikes other workers	Poor visibility Carelessness	Worker death	1C	Operator training	1E	
	Bulldozer strikes facility causing damage	Operator error Dozer control failure	Facility damage	3C	Operator training Dozer inspection prior to use Dozer maintenance program	3D	
	Bulldozer overturns and worker is run over	Slanted terrain Operator error	Worker death	1C	Operator training Plan ahead for slopes Seat belt usage and protective frame	1E	
	Operator exposed to excessive dust during operation resulting in lung injury	Dust environment	Lung injury	3C	Personnel protective equipment usage	3D	

Page: 1 of 1

504

31.11 COMMON JHA MISTAKES TO AVOID

When first learning how to perform a JHA, it is commonplace to commit some traditional errors. The following is a list of typical errors made during the conduct of a JHA:

- Assuming the analysis is limited to only human errors and does not consider system design or task design incongruities.
- The hazard description is not detailed enough.
- Design mitigation factors not stated or provided.
- Design mitigation factors do not address the actual causal factor(s).
- Overuse of project-specific terms and abbreviations.

31.12 SUMMARY

This chapter discussed the JHA technique. The following basic principles help summarize the discussion in this chapter:

- The primary purpose of JHA is to identify and mitigate worker hazards associated with a specific set of tasks to be performed in the accomplishment of a specific job.
- The use of hazard checklist greatly aids and simplifies the JHA process.
- The use of the recommended JHA worksheet aids the analysis process and provides documentation of the analysis.

REFERENCES

The following are some JHA references:

1. D. D. Glenn, *Job Safety Analysis: Its Role Today*, Professional Safety (ASSE), March 2011, pp. 48–57.
2. Job Analysis Reveals the Accident Causes, Goodspeed, M.C. (1930, July), National Safety News, 32, pp. 112–113.
3. Job Analysis for Safety, 1927, April, National Safety News, pp. 80.
4. H. Drury, *Scientific Management: A History and Criticism*, AMS Press, New York, 1922.

FURTHER READINGS

George, S., *Job Hazard Analysis: A Guide to Identifying Risks in the Workplace*, Government Institutes, an imprint of the Scarecrow Press, 2001.

Job Hazard Analysis, OSHA 3071, U.S. Department of Labor, Occupational Safety and Health Administration, 2002.

Heinrich, H.W., *Industrial Accident Prevention: A Scientific Approach*, 1st edition, McGraw-Hill Book Co., New York, 1931.

Job Safety Analysis, DOE 76-45/19, SSDC-19, EG&G Idaho Falls and DOE, 1979.

Roughton, J. E. and Crutchfield, N., *Job Hazard Analysis*, Buttersworth-Heineman an imprint of Elsevier, 2008.

Swartz, G., *JHA: A Guide to Identifying Risk in the Workplace*, Government Institute, 2001.

Chapter *32*

Threat Hazard Analysis

32.1 THA INTRODUCTION

The Threat hazard assessment (THA) is an evaluation of the munition life cycle environmental profile to determine the threats and hazards to which the munition may be exposed. The assessment includes threats posed by friendly munitions, enemy munitions, accidents, handling, transportation, storage, and so on. The assessment shall be based on analytical or empirical data to the extent possible. A THA is a mandatory requirement specified in MIL-STD-2105C, Hazard Assessment Tests for Nonnuclear Munitions. A THA covers the life cycle of the munition item, including friendly and hostile environments starting with production delivery and extending until the item is expended, or properly disposed. The THA identifies threats and hazards, both qualitatively and quantitatively, along with their causes and effects.

The THA evaluates potential threats and hazards throughout all the weapon system's life cycle scenarios, including combat threats and normal operational threats. Scenarios include transportation, handling, storage, and operational use. Potential threats are evaluated as hazards, and design action is taken to eliminate or mitigate these hazards. Identified hazardous scenarios are matched with insensitive munitions (IM) design and testing. Testing required by MIL-STD-2105C should be modified to address the hazards identified in the THA.

32.2 THA BACKGROUND

This analysis technique falls under the system design hazard analysis type (SD-HAT). The analysis types are described in Chapter 4. There are no alternative names for this technique, although there are other safety analyses approaches that may be utilized. The THA is primarily applicable to the analysis of weapon systems utilizing explosives.

Hazard Analysis Techniques for System Safety, Second Edition. Clifton A Ericson, II.
© 2016 John Wiley & Sons, Inc. Published 2016 by John Wiley & Sons, Inc.

The original purpose of the THA is to identify and evaluate hazards and threats to a weapon system containing nonnuclear munitions. Potential threats can result from enemy action or self-induced hazards. The identified hazards and threats are used as input into the munitions test program. The THA provides a framework for the development of a consolidated safety and IM assessment test program for nonnuclear munitions. It should be noted that the THA could also be applied to nuclear weapons and to other types of systems.

Note that the acronym *THA* is the same for both threat hazard analysis and test hazard analysis; use the acronym carefully to avoid confusion.

32.3 THA HISTORY

The THA originated as a requirement from MIL-STD-2105 (Series) to support munitions safety and in developing test requirements for munitions. In 1984, the Chief of Naval Operations introduced a plan for the US Navy's IM Program that would establish new guidelines for the sensitivity assessment of all naval munitions. The IM program proposed to increase the survivability of naval ships and aircraft by making all naval munitions less sensitive to unplanned stimuli. IM testing, in accordance with the requirements of NAV-SEAINST 8010.5B and MIL-STD-2105 (Series), provides a series of tests to assess the reaction of energetic materials to external stimuli representative of credible exposures in the life cycle of a weapon and requires the use of a THA in developing test plans. The IM program evolved to encompass all DoD munitions under development and procurement, its mandate now addressed in Public Law (Title 10, United States Code, Chapter 141, § 2389), and its implementation effected by DoDD 5000.1 of 12 May 2003 and CJCSM 3170.01 of 24 June 2003).

32.4 THA THEORY

The intent of the THA is to aid in designing a suitable hazard assessment test program for system containing munitions that will influence design early and give management the information necessary to determine the risk associated with the weapon system. The final product should be a list of hazards that prescribe safety tests and IM tests. The THA should provide inputs to other program safety analyses, such as the PHA, SSHA, and SHA. Some THA identified hazards may require more detailed analysis by another technique (e.g., fault tree analysis) to ensure that all causal factors are identified and mitigated.

The THA is capable of producing highly disciplined, thoroughly detailed analyses of hazards in new or existing systems. By meticulously and logically evaluating each portion of the munitions system life cycle for potential threats, the THA provides a thorough analysis. A technically trained safety analyst who is knowledgeable with the weapons system under study and explosives can easily master and accomplish the THA methodology. The technique is uncomplicated and easily learned. Standard, easily followed THA forms and instructions are provided in this chapter.

Figure 32.1 shows an overview of the basic THA process and summarizes the important relationships involved. This process consists of utilizing design information and the THA process to identify safety and IM hazards and threats. These threats are then mitigated through design and test.

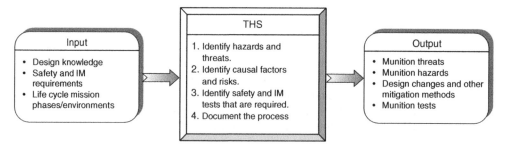

Figure 32.1 *THA overview.*

Summaries from the THA provide management information on IM testing requirements, as follows:

a. A list of potential threats to the weapon system munitions.

b. A list of potential hazards resulting when weapon system munitions are subjected to unplanned stimuli.

c. A list of design changes and other means of mitigating hazards.

d. A list of recommended safety and IM tests to evaluate the risk of the identified threat(s).

THA is based on the following definitions:

Explosive A solid or liquid energetic substance (or a mixture of substances) that is in itself capable, by chemical reaction, of producing gas at such temperature, pressure, and speed as to cause damage to the surroundings. Included are pyrotechnic substances even when they do not evolve gases. The term explosive includes all solid and liquid energetic materials variously known as high explosives and propellants, together with igniter, primer, initiation, and pyrotechnic (e.g., illuminant, smoke, delay, decoy, flare, and incendiary) compositions (refer to MIL-STD-2105C).

Explosive Device An item that contains explosive material(s) and is configured to provide quantities of gas, heat, or light by a rapid chemical reaction initiated by an energy source usually electrical or mechanical in nature (refer to MIL-STD-2105C).

Munition An assembled ordnance item that contains explosive material(s) and is configured to accomplish its intended mission (refer to MIL-STD-2105C).

Insensitive Munitions Munitions that reliably fulfill (specified) performance, readiness, and operational requirements on demand but that minimize the probability of inadvertent initiation and severity of subsequent collateral damage to the weapon platforms, logistic systems, and personnel when subjected to unplanned stimuli (refer to MIL-STD-2105C).

Stimulus Active threats, which transform a hazard from its potential state to one that causes harm to the system, related property, or personnel. Threats can stem from operational, combat, and logistical sources.

Threat An event or active condition that constitutes a stimulus on the weapon system munitions. For example, bullet strike is a munitions threat.

All Up Round (AUR) A completely assembled munitions as intended for delivery to a target or configured to accomplish its intended mission (refer to MIL-STD-2105C). The term is identical to all up weapon.

TABLE 32.1   THA Process

Step	Task	Description
1	Define system	Define scope and boundaries of the system. Define the mission, life cycle phases, life cycle scenarios and life cycle environments. Establish equipment and operational phases to be analyzed
2	Plan THA	Establish THA goals, definitions, worksheets, schedule, and process. Define credible threats and threat scenarios for the analysis
3	Select team	Select all team members to participate in THA and establish responsibilities
4	Acquire data	Acquire all necessary design and process data needed for THA. Refine the item indenture levels for analysis. Data can include functional diagrams, schematics, and drawings for the system and subsystems. Data also includes operational scenarios, tasks, and procedures.
5	Conduct THA	Perform the THA on each item in the established equipment and operational phases. A team and/or individual analyst performs the analysis. Utilize team analysis sessions when possible. Utilize the THA worksheets and have them reviewed by other system engineers
6	Recommend corrective action	Recommend corrective action for threats with unacceptable risk or criticality. Recommend additional IM testing that will be required
7	Monitor corrective action	Review the THA at scheduled intervals to ensure that corrective action is being implemented
8	Track hazards	Transfer identified hazards into the hazard tracking system using the HAR format
9	Document THA	Document the entire THA process on the worksheets. Update for new information and closure of assigned corrective actions

32.5   THA METHODOLOGY

Table 32.1 describes the basic steps of the THA process. This process involves performing a detailed analysis of all potential threats to the system IM that can result in explosives detonation.

The THA analyst must ensure that all system configurations are adequately addressed. In most cases, the analyst is ultimately interested only in the AUR configuration. It is conceivable, however, that major subassemblies or components may be separately considered during the early phases. It is also possible that the latter phases may involve more than one configuration. If the system is enclosed in a shipping container, then that container must be identified.

The THA analyst must identify all major energy sources internal to the system. Energy sources include propellants, explosives, gas generators, compressed gases, and mitigating devices. An uncontrolled release of a small amount of energy may be an acceptable risk, whereas the uncontrolled release of a large amount of energy could be disastrous. The approximate quantity should be listed for each identified energy source. Though toxic materials are not energy sources in the true sense, they should also be included in the THA. The analysis should list all devices that restrain the AUR, such as tie-downs, pallets, and containers.

32.5.1   Cradle-to-Grave Sequences

The THA should consider all aspects of the cradle-to-grave sequence (CGS) for the system life cycle missions and phases, such as follows:

1. On/offload to truck
2. Transport by truck

3. Weapon station storage
4. On/offload to ship
5. Shipboard stowage/transport by ammo/resupply ship
6. On/offload to railcar
7. Transport by rail
8. Battlefield storage
9. Airfield storage
10. On/offload on air transport (fixed wing or helo)
11. Airplane takeoff/landing
12. Transport by air
13. Fixed wing air drop
14. Low altitude extraction
15. Helo airdrop
16. Transport by troops (hand carry)
17. Amphibious flight deck
18. Amphibious magazine storage
19. Remove from amphibious magazine storage
20. Load to gun or launcher
21. Firing/launch
22. Flight to target
23. Target – end of mission

The first phase in analyzing a weapon system would be a palletized, containerized, or AUR (or energetic component) on the manufacturer's shipping dock (or storage bunker). This situation is normal when the government takes possession. At this point, the item may be subjected to a fire, handling accident, terrorist-type activities, or any other external energy (stimuli). Every situation that can be anticipated must be identified, such as transportation, handling, transfers, storage (land and shipboard), and any other activity right up to the item's final usage. An excellent guide in defining the CGS phases is the Integrated Logistics Support Plan.

Scenarios that are similar in nature may be combined. Transportation to a weapons station is similar to transportation from a weapons station. Handling operations at a Naval Weapons Station may be similar enough to combine with handling operations at a Naval Air Station. Storage/handling operations onboard different types of ammunition ships might also be combined. Differences that must be considered include personnel and equipment variances that could result in different types or amounts of losses.

Scenarios that include identical operations and even onboard the same platform but that could result in different losses must be treated as different scenarios. For example, handling operations onboard an aircraft carrier may have different potential losses when performed at different locations, such as the hanger deck, magazine area, and populated spaces like the Mess Decks that are sometimes used to break out and assemble weapons.

32.5.2 Threat Scenarios

Table 32.2 contains examples of major munitions threat scenarios for various CGSs.

TABLE 32.2 Example CGS Threat Scenarios

CGS	Possible Threat
Storage	During storage, the main threats is a fire in the storage facility (initiated from sources other than the stored munitions); actions from an accident, sabotage, terrorists, enemy attack, and the sympathetic impact of fragments from the detonation of adjacent rounds
Handling	During handling operations, example threats include the accidental drop of the round or container, fire in the vicinity from external sources (e.g., the transport vehicle being loaded or unloaded), and actions from sabotage, terrorists, and enemy attack. The drop distance for the uploading and downloading of trucks and rail transport can be assumed to be approximately 6 feet. For other uploading/downloading operations, such as aboard ship, the drop distance can be assumed to be 40 feet
Truck transport	When transporting by truck, the most serious threat is a vehicle accident. Accidents can result in a vehicle fire or violent impact; both outcomes pose serious threats to the munitions cargo. Vehicle fires can also result from other sources, for example, fuel leaks or electrical problems; these fires can then lead to combustion of the gasoline or diesel fuel. Actions from sabotage, terrorists, and enemy attacks are also possible
Rail transport	When transporting by rail, the most serious threat is a vehicle accident, such as derailment, resulting in a violent impact. Fires from other sources could pose a threat. Actions from sabotage, terrorists, and enemy attacks are also possible
Aircraft transport	Threats considered in aircraft transport are aircraft crash, an aircraft fire from sources other than the munitions, an accident during takeoff or landing, and actions from sabotage, terrorists, and enemy attacks are also possible
Ship transport	During transport by ship, fire inside the weapons locker is a serious threat. Also, fire from an external source such as sabotage, terrorist, and enemy attacks are threats

32.5.3 Characterization of Environments

In the course of the system CGS, the munitions will potentially be exposed to a variety of hazardous environments. The expected environments are characterized in Table 32.3.

32.5.4 Threats

Table 32.4 lists some of the potential threats that can cause serious munitions safety concerns. This list of threats should be considered in the THA.

32.6 THA WORKSHEET

The THA should be performed using a worksheet to provide the analysis structure and consistency. The format of the analysis worksheet is not critical. The worksheets assist in the development of scenarios and the presentation of data for the THA. As a minimum, the following basic information should be obtained from the analysis worksheet:

1. The threat (including CGS, source, stimuli, and energy release)
2. The potential worst-case effect of the threat
3. Risk assessment
4. Mitigation methods
5. IM tests that are recommended

TABLE 32.3 Threat Environments

Environment	Possible Threat
Fire fast cook-off (FCO)	Fast cook-off occurs when the ammunition is exposed to a rapid increase in temperature. This situation usually results from being engulfed in flames. The combustion of hydrocarbon fuels (such as diesel fuel or jet fuel) will usually result in a fast cook-off environment. FCO is most likely to occur during transport, storage, or aircraft carrier deployment
	The standard FCO test exposes a weapon to an enveloping hydrocarbon fuel fire. Such fires typically deposit 10 BTU/square feet/second on the item under test. The shipping container can be expected to melt rapidly in a fuel fire and will delay the time to initial reaction only a short time
	Fast Cook-off During Rail Transport
	Rail transport is normally a controlled condition. The threat of fire from nonmunition sources in an adjacent railcar should be limited. However, a second source that must be considered is a stuck cast-iron brake shoe. A frozen brake shoe could ignite, and has ignited, the floor of the boxcar. An FCO during rail transport is a possible scenario and should be addressed
	Fast Cook-off During Truck Transport
	A fuel fire is the most likely cause of fast cook-off during truck transport. The combustion characteristics of diesel fuel and gasoline are similar to jet aircraft fuel fires, which produce intense heat between 1600 °F and 2200 °F. An FCO during truck transport is a possible scenario and should be addressed
	Fast Cook-off During Storage
	The most likely source of an FCO scenario during storage would be direct exposure to a fire while stored in an ammunition dump during combat. Such a fire could subject the ammunition to an FCO environment
Slow cook-off (SCO)	The slow cook-off environment is a gradual increase in temperature where the ammunition is not directly exposed to flames. It is normally caused by a fire in an adjacent magazine or exposed to a heat source. Slow, continual heating could occur during a shipboard fire while stowed in a magazine. The standard SCO test consists of exposing a weapon to a 6 °F per hour temperature increase until a reaction occurs
Bullet impact (BI)	Munitions may be exposed to bullet impact during land transport and at forward ammunition dumps. In these sequences, the munitions are stored in their shipping containers. Terrorist attack, hunting accidents, and aircraft strafing of ammunition storage areas are the primary threats. The 7.62 mm, 20 mm, and 30 mm rounds are common threats. The 7.62 mm round is representative of both a hunter and terrorist threat. The 20 mm and 30 mm round are representative of large diameter, high rate of fire threats, which could impact the ammunition container as a result of an accidental "friendly fire" or from enemy action
Fragment impact (FI)	Impacts from high-energy fragments may have a myriad of sources, both accidental and intentional, such as terrorist attack, accidents, or combat activities. Hand grenades may be used by terrorists or enemy personnel. Fratricide accidents caused by the detonation of an ordnance item adjacent to or near the containerized munition may occur during shipping or storage. The standard fragment impact test utilizes multiple (5–25) one-half inch mild steel cubes (AISI 1020, ~250 grains) impacting the test item at 8300 ± 300 feet per second
Sympathetic detonation	IM sympathetic detonation is the interaction between identical weapons in their storage or transport configuration. The first round(s) (donor) can be detonated by any of the other IM stimuli (FCO, SCO, FI, or BI).
Shaped charge jet impact	IM detonation due to a shaped charge jet impact
Spall impact	IM detonation due to the impact of hot spall fragments

TABLE 32.4 Potential Threats

Threat	Possible Sources
Fire	Fuel fire, fire from crash during transportation
Bullet impact	Terrorism, war activity, and accidental gun fire
Shrapnel impact	Bullet impact, shrapnel impact; damaged container; handling equipment
Impact (general)	Damaged container, handling equipment, and forklift
Shock	Vehicle crash, dropped
Energy	RF energy, ESD energy; electrical energy; lightning
Heat	Desert activity, near a fire
Vibration	Vibration from ship, train, truck, or aircraft
Jet impact	Shaped charge jet impact
Spall impact	Hot spall fragments from molten metal
RF Energy	RF energy from radar equipment
ESD Energy	ESD energy from human handlers
Water	Dropped in water, water deluge from fire extinguisher

Some example THA worksheets are shown in Figures 32.2 and 32.3. Note that the same basic information is derived from both types of worksheets. The specific worksheet to be used may be determined by the managing authority, the safety working group, the safety team, or the safety analyst performing the analysis. Other worksheet formats may exist because different organizations often tailor their THA worksheet to fit their particular needs.

The recommended THA worksheet for system safety usage is shown in Figure 32.3. The THA steps for this worksheet are as follows:

1. *System* This column identifies the system under analysis.
2. *Cradle-to-Grave Sequence* This column identifies the particular CGS life cycle being assessed.

System:					THREAT HAZARD ASSESSMENT WORKSHEET					Date:	
CGS:										Prepared By:	
THREAT			ENERGY		RESULTS		RISK			MITIGATION METHOD	NOTES
No.	Stimuli	Eq. Env.	Type	Release	Allowed	Expected	Sev.	Prob.	Basis		

Energy Release Probability:
1 – Possible for all weapons
2 – Possible for several weapons
3 – Not likely for any weapon
x – Undetermined

Risk Severity:
I – Catastrophic
II – Critical
III – Marginal
IV – Negligible

Risk Probability:
A – Frequent
B – Probable
C – Occasional
D – Remote
E – Improbable

Risk Basis:
H – Documented experience
M – Test data
L – Engineering judgement

Figure 32.2 *Example THA worksheet 1.*

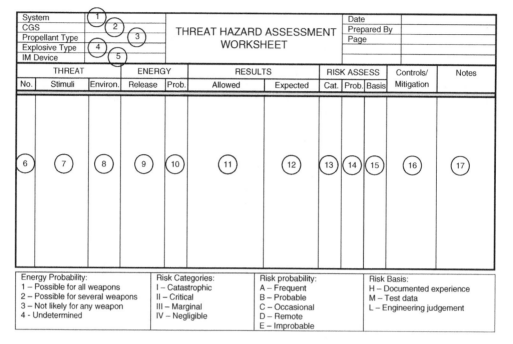

Figure 32.3 *Example THA worksheet 2.*

3. *Propellant Type*   This column identifies the propellant type.

4. *Explosive Type*   This column identifies the explosive type.

5. *IM Device*   This column identifies the IM device involved.

6. *Threat No.*   This column sequentially identifies each threat listed in the analysis. This number is for reference purposes.

7. *Threat Stimuli*   This column identifies the external stimuli (threats) that could cause the weapon to become damaged. This could include fire, bullet impact, transportation crash, terrorist act, and so on.

8. *Threat Environment*   This column identifies a realistic life cycle threat environment to the weapon, and the type of simulation test required to assess this threat event. Environments include FCO, SCO, bullet impact, and so on.

9. *Energy Release*   This column identifies the type of energy release that will occur from the stimuli and environment, such as quiescent burn, active burn, deflagration, explosion, detonation, and so on. The explosive reaction terms from MIL-STD-2105 (Series) include the following:

Type I – Detonation

Type II – Partial detonation

Type III – Explosion

Type IV – Deflagration

Type V – Burning

NR – No reaction

10. *Energy Probability*   This provides a qualitative determination of the probability that the energy release will be of the type stated.

11. *Results Allowed*   This column identifies the hazard results (i.e., mishap) that can be tolerated by the system. For example, a small quantity of propellant may be allowed to react as an active burn, whereas a large quantity might not be tolerated. For many situations there may be a "None" in this column. This column establishes the maximum allowable reaction requirement.

12. *Results Expected*   This column identifies the worst-case hazard results (i.e., mishap) that can be expected by the system. For example, fire in a ship's magazine may result in the loss of weapons in an adjacent magazine. Types of results include loss of weapons, loss of adjacent weapons, loss of vehicle, loss of facility, loss of life, and so on.

13. *Risk Category*   This column provides a qualitative determination of the mishap risk severity for the expected mishap outcome. The following qualitative ratings are from MIL-STD-882:

    I – Catastrophic

    II – Critical

    III – Marginal

    IV – Negligible

14. *Risk Probability*   This column provides a qualitative determination of the mishap probability for the expected result. The following qualitative ratings are from MIL-STD-882:

    A – Frequent

    B – Probable

    C – Occasional

    D – Remote

    E – Improbable

15. *Risk Basis*   This column provides a qualitative determination of the rational behind the severity and probability judgements. The qualitative ratings used are as follows:

    High (H) – Based on documented experienceMedium (M) – Based on test data

    Low (L) – Based on engineering judgement

16. *Controls/Mitigation*   This column identifies any requirements necessary to reduce the mishap severity and/or the probability of the mishap occurring. This column also identifies design features already in effect to reduce the mishap risk.

17. *Notes*   This column contains additional information the analyst wants to add as supplemental information. Requirements for special IM test can be placed in this column.

32.7   THA EXAMPLE

In order to demonstrate the THA methodology, the following example is provided. A partial THA is performed on a hypothetical 8-inch projectile for a shipboard gun. The projectile contains a high explosive (HE) known as cyclotrimetholenetrinitramine (RDX).

Tables 32.5 and 32.6 contain THA worksheets for one CGS phase of shipboard storage. Note that a complete THA would analyze all of the system CGS phases for the system.

TABLE 32.5 THA Worksheet for Example System (Page 1)

System	8 Inch Artillery Round		Date	
CGS	Shipboard Storage		Prepared By	
Propellant Type			Page	1 / 2
Explosive Type				
IM Device				

THREAT HAZARD ASSESSMENT WORKSHEET

THREAT			ENERGY		RESULTS		RISK ASSESS			Controls/ Mitigation	Notes
No.	Stimuli	Environ	Release	Prob	Allowed	Expected	Cat	Prob	Basis		
1	Dropped from truck during handling	6-foot drop test	Detonation	1	Minor round damage	Explosion of single round resulting in death/injury	I	D	M	Proper handling equipment and procedures	6-foot drop test
2	Fire; loaded truck catches on fire	FCO	Burn	1	None	Explosion of multiple rounds resulting in death/injury	I	D	M	Proper handling equipment and procedures; fire fighting equipment	FCO test
		SCO	Deflagration	1		Loss of facility	II	D	M		SCO test
3	Sympathetic detonation	SD	Explosion	1	None	Explosion of multiple rounds resulting in death/injury	I	D	M	IM material Protective containers Quantity limit	SD test
4	Bullet/grenade strike from terrorist activity	BI FI	Explosion	1	None	Explosion of single round resulting in sympathetic detonation and death/injury	I	D	M	Security Barriers Protective containers	BI test FI test SD test

TABLE 32.6 THA Worksheet for Example System (Page 2)

System	8 Inch Artillery Round									Date	
CGS	Shipboard Storage				THREAT HAZARD ASSESSMENT WORKSHEET					Prepared By	
Propellant Type										Page	2 / 2
Explosive Type											
IM Device											
THREAT			ENERGY			RESULTS		RISK ASSESS			Controls/	Notes
No.	Stimuli	Environ	Release	Prob	Allowed	Expected	Cat	Prob	Basis	Mitigation	
5	Dropped from storage rack	6-foot drop test	Detonation	1	Minor round damage	Explosion of single round resulting in death/injury	I	D	M	Proper handling equipment and procedures	6-foot drop test
6	Fire in storage bay	FCO SCO	Burn Deflagration	1 1	None	Explosion of multiple rounds resulting in death/injury Loss of facility	I II	D D	M M	Proper handling equipment and procedures; fire fighting equipment	FCO test SCO test
7	Sympathetic detonation	SD	Explosion	1	None	Explosion of multiple rounds resulting in death/injury	I	D	M	IM material Protective containers Quantity limit	SD test
8	Bullet or shrapnel strike from enemy activity	BI FI	Explosion	1	None	Explosion of single round resulting in sympathetic detonation and death/injury	I	D	M	Barriers Protective containers	BI test FI test SD test

517

32.8 THA ADVANTAGES AND DISADVANTAGES

The THA technique offers the following significant advantages:

1. THA provides a consistent methodology for identifying and evaluating system munitions threats.
2. THA threats can be qualitatively evaluated for risk and quantitatively evaluated through further analysis and/or testing.
3. THA is a relatively inexpensive analysis tool compared to some other tools.

The following are significant disadvantages of the THA technique:

1. THA is limited by the ability of the analyst to identify all system threats.
2. THA does not identify all system hazards, only those associated with munitions threats.

32.9 COMMON THA MISTAKES TO AVOID

When first learning how to perform a THA, it is commonplace to commit some traditional errors. The following is a list of typical errors made during the conduct of a THA:

1. The entire Cradle to Grave Sequence (CGS) is not considered
2. All potential threats are not thoroughly considered and/or investigated
3. Use of design mitigation methods for identified hazards is inadequate
4. Hazards are not adequately and thoroughly described

32.10 SUMMARY

In summary, the following key points are reiterated to help emphasize the concepts discussed in this chapter:

1. A THA must be performed on all nonnuclear munitions systems in accordance with MIL-STD-2105C.
2. The primary purpose of the THA is to identify munitions threats and hazards, which are then used to influence designs.
3. THA provides inputs to other program safety analyses, such as the PHA, SSHA, and SHA.
4. THA involves a detailed focus on system life cycle scenarios and possible munitions threats.
5. Some THA-identified hazards may require more detailed analysis by another technique (e.g., FTA) to ensure that all causal factors are identified and mitigated.
6. The use of worksheets provides structure and rigor to the THA process.

FURTHER READINGS

DI-SAFT-81124, Threat Hazard Assessment Format and Content Preparation Instructions.

Mathre, J. K., Chirkis, K. R., and Banister, J. D. Insensitive Munitions Threat Hazard Assessment Methodology, presented at the 15th International System Safety Conference, Albuquerque, NM, 1996.

MIL-STD-2105C, Notice 1, Hazard Assessment Tests for Non-Nuclear Munitions, 23 July 2003.

NAVSEAINST 8010.5B, Insensitive Munitions Program Planning and Execution.

NAVSEAINST 8020.5C, Qualification and Final (TYPE) Qualification Procedures for Navy Explosives (High Explosives, Propellants, Pyrotechnics and Blasting Agents).

OPNAVINST 8010.13B, Department of the Navy Policy on Insensitive Munitions.

Chapter *33*

System of Systems Hazard Analysis

33.1 SoSHA INTRODUCTION

Systems of systems (SoS) is a new category of systems that has emerged in the last decade. This type of system is very large, complex, and comprised of multiple legacy systems. These legacy systems can operate independently to achieve their own goals, but they also contribute to the goals of the SoS. The SoS concept has been gaining traction, with more of these type systems in development and operation.

The SoS is an integrated set of independent systems, plus a central SoS hub. It is akin to a large system federation comprised of smaller independent systems. The SoS has a higher goal than the individual systems supporting it, but the SoS requires the combined effort of all of the independent systems to achieve that goal.

A SoS is a collection of task-oriented or dedicated systems that pool their resources and capabilities together to create a new, more complex system that offers more functionality and performance than simply the sum of the constituent systems. One could argue that SoS have really been around for many years, they were just not classified as such, mainly because the technology available and the limited interconnectivity of resources kept them at a smaller scale. The latest technology in distributed processing, the Internet, and satellite capability has made these types of systems more viable and universal. Because SoS are becoming more pervasive, and since they present new safety challenges, the most recent version of MIL-STD-882(version E, 2012) has included a new requirement for a SoS hazard analysis (SoSHA).

There ae many different types of SoS, for example:

- An aircraft is a system, whereas an airport comprised of many planes, radars, communications, and a control tower is a SoS.

Hazard Analysis Techniques for System Safety, Second Edition. Clifton A Ericson, II.
© 2016 John Wiley & Sons, Inc. Published 2016 by John Wiley & Sons, Inc.

- An automobile is a system, whereas an interstate traffic system comprised of many cars, trucks, traffic lights, road signs, road lanes, emergency vehicles, and monitoring systems is a SoS.
- A ship is a system, whereas a battle force is comprised of ships, radars, satellites, communications, missiles, and a command and control center is a SoS.

33.2 SoSHA BACKGROUND

This analysis falls under the system design hazard analysis type (SD-HAT). The analysis types are described in Chapter 4. There are no alternative names for this technique, although there are other safety analysis approaches that may be utilized to fulfill the need for a SoSHA.

In reality, a SoS is just another system, except it is larger and is comprised of other systems developed and operating independently. For this reason they tend to be more complex and tend to present a larger problem in the area of complete understanding, particularly in the area of safety. SoS hazard analysis does not really require a new analysis technique or methodology, it requires a larger focus using existing hazard analysis techniques. Part of this larger focus includes management responsibility for the SoS along with the appropriate funding needed for an adequate system safety program.

The overarching purpose of a SoSHA is to evaluate a SoS for the following:

a. Compliance with specified safety criteria and requirements
b. The identification of hazards created by the SoS hub architecture, both within the hub and if they should impact any of the linked systems
c. The identification of hazards in the independent systems created by their link into the SoS, both with the system and if they should impact another system or the SoS hub

A basic understanding of hazard analysis theory is essential, so is knowledge of system safety concepts in order to conduct a SoSHA. Experience with and/or a good working knowledge of the particular SoS, system, and subsystems is necessary in order to identify and analyze all hazards.

SoS share some characteristics with systems, but there are a number of characteristics that distinguish them and that can lead to an increased system safety and hazard analysis process. Some of the primary characteristics of a SoS include the following:

Complexity Systems by themselves are complex; it logically follows that an integrated set of systems will be even more so. SoS employ complex interactions and dependencies between complex systems and as such, the complexity of a SoS is significantly greater than the complexity of any single system. Systems may enter and leave the SoS, may perform different roles, or may be connected to another system. This complexity means that a fully funded SoS safety program must be applied.

States The states of a SoS can become very large when combining independent system states, each having a large number of system states and modes. This complexity means that a fully funded SoS safety program must be applied.

Size The sheer size of a SoS can become very large when combining independent system states. The size alone can be overwhelming, requiring a large dedicated SoS safety program.

Emergent Behavior A SoS can perform functions that the component systems alone cannot achieve. The difference between what the systems can achieve individually and what the SoS can achieve is referred to as emergent behavior; the special behavior emerges from the unique design. While some emergent behaviors are intentional, others are not. Unintended emergent behaviors can have significant safety consequences and efforts must be made to ensure that the consequences are recognized and evaluated for hazards.

Autonomy A system is constructed of components and subsystems that are unable to operate on their own to achieve a useful purpose. In contrast, a SoS is constructed of systems that operate on their own on a regular basis and are able to perform functions independently that may be unrelated to the functions of the SoS.

33.3 SoSHA HISTORY

SoS theory has been growing and gaining traction over the past decade, along with a growing list of SoS applications. The DoD finally recognized the need for a SoSHA requirement when MIL-STD-882E was released in 2012. However, the bottom line is that SoSHA is a need or requirement but not a specific technique. SoSHA is actually achieved through the application of existing HA techniques to the SoS design and architecture. To some degree, SoS have been around for many years, yet they were not known or recognized as such.

33.4 SoS THEORY

The concept of SoS is that it is a large system comprised of an integrated set of independent systems. It is akin to a large system federation comprised of smaller independent systems. The SoS has a higher goal than the individual systems supporting it, but the SoS requires the combined effort of all of the independent systems to achieve that goal.

Typically, each of the independent systems is developed independently of the SoS. These independent systems are sometimes referred to as legacy systems; they were previously developed and in use prior to the SoS. The SoS goal is achieved through the combined effort of the individual systems capabilities. The SoS may use the full capabilities of the individual systems or it may utilize only partial capabilities.

The SoS concept can be confusing and misunderstood; definitions are not firmly agreed upon and they can be rather vague and generic. Some common SoS definitions are provided below.

Capability A capability is the ability to achieve a desired effect under specified standards and conditions through combinations of ways and means to perform a set of tasks [1].

System A system is the organization of hardware, software, material, facilities, personnel, data, and services needed to perform a designated function within a stated environment with specified results [2].

Systems of Systems

1. A SoS is defined as a set or arrangement of systems that results when independent and useful systems are integrated into a larger system that delivers unique capabilities. Both individual systems and SoS conform to the accepted definition of a system in that each consists of parts, relationships, and a whole that is greater than the sum of the parts; however, although an SoS is a system, not all systems are SoS [1].

2. A SoS is a set or arrangement of interdependent systems that are related or connected to provide a given capability [2].

3. A SoS is an amalgamation of legacy systems and developing systems that provide an enhanced military capability greater than any of the individual systems within the system of systems [3].

Family of Systems   An FoS is defined as a set of systems that provide similar capabilities through different approaches to achieve similar or complementary effects. For instance, the war fighter may need the capability to track moving targets. The FoS that provides this capability could include unmanned or manned aerial vehicles with appropriate sensors, a space-based sensor platform, or a special operations capability. Each can provide the ability to track moving targets but with differing characteristics of persistence, accuracy, timeliness, and so on [1]. FoS are fundamentally different from SoS because a family of systems (FoS) lacks the synergy of a SoS. The FoS does not acquire new properties as a result of the grouping. In fact, the member systems may not be connected into a whole.

Net-Centric   A large set of disperate pieces of data provided by differing systems and/ or sensors littered across a network awaiting multiple users, service providers, or applications to use them to create information to be used by humans to solve problems

A system is a collection of components and subsystems organized to accomplish a specific function or set of functions. A SoS is a "super system" comprised of elements that are themselves complex, independent systems that interact to achieve a common goal. SoS systems engineering deals with planning, analyzing, organizing, and integrating the capabilities of a mix of existing and new systems into a SoS capability greater than the sum of the capabilities of the constituent parts.

Common SoS characteristics include the following:

- The component systems achieve well-substantiated purposes in their own right even if detached from the overall SoS; they are expected to cooperate within the SoS.
- The component systems are managed in large part for their own purposes rather than the purposes of the whole SoS.
- The component systems are typically geographically dispersed.
- The SoS exhibits behavior, including emergent behavior, not achievable by the component systems acting independently.0
- Constituent systems and functions may be added or removed from the SoS during its use.
- The SoS typically has a mission focus, which can evolve or change over time.
- SoS mishap risk is spread among the component systems and the SoS.
- The SoS typically has a heavy dependency on network communications (i.e., net-centric).
- The Sos is a dynamic collaboration of component systems to achieve common objectives.

SoS can take different forms. Based on a recognized taxonomy of SoS, there are four types of SoS that are found in the DoD [1]. These are as follows:

Virtual SoS   Virtual SoS lack a central management authority and a centrally agreed upon purpose for the system-of-systems. Large-scale behavior emerges—and may be desirable—but this type of SoS must rely upon relatively invisible mechanisms to maintain it.

Collaborative SoS   In a collaborative SoS, the component systems interact more or less voluntarily to fulfill agreed upon central purposes. The Internet is a collaborative system. The Internet Engineering Task Force works out standards but has no power to enforce them. The central players collectively decide how to provide or deny service, thereby providing some means of enforcing and maintaining standards.

Acknowledged SoS   Acknowledged SoS have recognized objectives, a designated manager, and resources for the SoS; however, the constituent systems retain their independent ownership, objectives, funding, development, and sustainment approaches. Changes in the systems are based on collaboration between the SoS and the system.

Directed SoS   Directed SoS are those in which the integrated system-of-systems is built and managed to fulfill specific purposes. It is centrally managed during long-term operations to continue to fulfill those purposes as well as any new ones the system owners might wish to address. The component systems maintain an ability to operate independently, but their normal operational mode is subordinated to the central managed purpose.

Management issues for SoS are substantial; successful SoS management requires reaching across organizational boundaries to establish an end-in-mind set of objectives and the resourced plan. Experienced managers are needed in a SoS environment, which requires considerable flexibility to negotiate among the competing interests in a SoS environment. SoS increases the complexity, scope, and cost of both the planning process and the systems engineering, and introduces the need to coordinate interprogram activities and manage agreements among multiple system managers as stakeholders who may not have a vested interest in the SoS.

SoS have an emergent behavior, which is considered to be behavior that is unexpected or cannot be predicted by knowledge of the system's constituent parts. For the purposes of an SoS, "unexpected" means unintentional, not purposely or consciously designed-in, not known in advance, or surprising to the developers and users of the SoS. In a SoS context, "not predictable by knowledge of its constituent parts" means the impossibility or impracticability (in time and resources) of subjecting all possible logical threads across the myriad functions, capabilities, and data of the systems to a comprehensive SE process. The emergent behavior of a SoS can result from either the internal relationships among the parts of the SoS or as a response to its external environment. Consequences of the emergent behavior may be viewed as negative/harmful, positive/beneficial, or neutral/unimportant by stakeholders of the SoS. This emergent behavior is of prime importance and consideraion to system safety engineering.

SoS design considerations include the following:

- The SoS must balance SoS needs with individual system needs.
- SoS planning and implementation must consider and leverage the development plans of the individual systems.
- SoS must address the end-to-end behavior of the ensemble of systems, addressing the key issues that affect that behavior.
- The SoS focuses primarily on the end-to-end behavior of the SoS and addresses the constituent systems only from that perspective.

As would be expected, a SoS can easily become large, complex, and overwhelming. However, the SoS development process must follow the standard engineering development process. The core SOS elements include the following:

- Translating SoS capability objectives into requirements
- Understanding systems and relationships
- Assessing SoS performance toward capability objectives
- Developing and evolving the SoS architecture
- Monitoring and assessing SoS changes
- Addressing SoS requirements and solution options
- SoS risk management
- SoS requirements management
- SoS configuration management
- SoS data management
- SoS interface management
- SoS safety program

It should be obvious by now that the actual SoS design and architecture can be very complex and take many different forms. Figure 33.1 depicts a generic picture of the SoS concept.

Independent systems are linked together via a central SoS "hub." Each system performs its own objectives and also contributes to the SoS goals demanded by the hub. This may be achieved through any of several different SoS architectures. Also, some of the systems may be interconnected as part of their normal operation. In this example, a hypothetical military SoS on a ship may detect target threats to the ship and fire its missiles at the target independent of the SoS. In another scenario, the AWACS aircraft may detect a different threat and send the threat information to the SoS hub, where the hub determines that system should fire missiles or drop bombs at the threat.

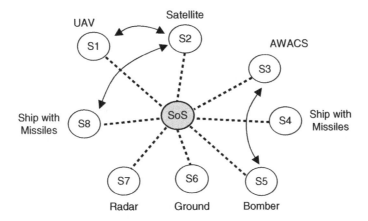

Figure 33.1 *SoS concept overview.*

33.5 SoS SAFETY AND HAZARDS

A SoS safety program must address both the safety of the larger SoS element and the safety of the individual system elements as they exist within the SoS. One advantage here is that, typically, each of the individual systems was developed with its own safety program, including hazard analyses. In optimum situations, this means that hazard analyses already exist for the individual systems. However, in the real world, one of the problems in SoS safety is that sometimes this data is not available, the analysis was not performed or it is of poor quality. This situation makes the safety and hazard analysis job more difficult.

The objective of the SoS system safety program is to ensure safety of the entire SoS. This breaks down into ensuring that all SoS-related hazards are identified and have been eliminated or mitigated to an acceptable level of risk. In order to meet this objective, a SoS hazard analysis program must be conducted.

Many analysts think that SoS hazards are typically separated into two distinct categories: single legacy system hazards and emergent SoS hazards. A single system hazard is a hazard that is attributable to a single system alone, an emergent hazard is a hazard that results from the integration of several systems into a SoS and hence cannot be attributed to a single system. Some researchers further subdivide emergent hazards into reconfiguration hazards, integration hazards, and interoperability hazards. These subdivisions are, however, tricky and vague, possibly causing more confusion than understanding when conducting a SoS hazard analysis.

A more thorough and pragmatic SoS hazard category breakdown would include the following considerations:

- Hazards within the independent system that only affect that system. This should have been done during development of the legacy system.
- Hazards within the independent system that affect the SoS. The SoS hazard analysis will have to cover this area.
- Hazards within the independent system that affect other independent systems when linked into the SoS. The SoS hazard analysis will have to cover this area.
- Hazards that arise from the SoS (i.e., emergent hazards) federation architecture. The SoS hazard analysis will have to cover this area.
  - Hazards solely within the SoS scope
  - Hazards from the SoS architecture that affect the SoS
  - Hazards from the SoS architecture that direct back to affect an individual system

The SoSHA is initiated as early as possible in the design process. However, it cannot be completed until the detailed design data is available. One of the major roadblocks is collecting adequate design data and hazard analyses from the individual systems. If this information is not available or is inadequate, then the SoS safety team must decide whether or not to obtain the information or accept the unknown safety risk. To obtain the information, they may have to pay the original system development organization to perform the HAs or they must perform the HAs themselves.

SoS hazards can be broken down into two main categories. Single system hazards are those that are attributable to a single system only and require no system interaction in order to occur, and emergent hazards are those that result from the interaction systems and the SoS hub.

There are three subcategories of emergent hazards. Reconfiguration hazards are unique to systems of systems and do not have a parallel within the system space. They are caused by the transition of a system of systems from one state to another. Interoperability hazards result from miscommunications. A receiving system interprets a message or signal in a manner that conflicts with the intent of the transmitting system. The third subcategory is integration hazards. These hazards can best be described by the subtypes of the category. There are three subtypes, interface hazards, resource hazards, and proximity hazards. An interface hazard results from a failure or partial performance in one system causing a mishap in a second system. A resource hazard occurs when there are insufficient levels of shared resources to support the systems present, or when there is a conflict over a certain part of the shared resource. A proximity hazard occurs when one system causes a mishap in another without using a defined interface.

Single System Hazards   A single system hazard is a hazard that is attributable to a single system alone. These hazards are identified by the system hazard analysis (SHA) process and are outside the scope of a system of systems hazard analysis. The purpose of a SoS hazard analysis is to identify all SoS hazards except for single-system hazards. A single-system hazard is any hazard that may occur within a SoS that is attributable to a single system and may occur whether or not that system is operating within the system of systems context.

Integration Hazards   Integration hazards are a type of emergent hazard that result from the integration of systems into a system of systems. The vast majority of system of systems hazards are integration hazards, which can be further subcategorized into three types, interface hazards, proximity hazards, and resource hazards.

Interface Hazards An interface hazard is a hazard in which one system causes a mishap in another system by transferring a failure or partial performance over a defined interface, possibly through another system. The dependency between systems within a system of systems can result in a failure in one system causing a mishap in another. What may be a benign failure in one system may be catastrophic when transferred to another.

Proximity Hazards   A proximity hazard is a hazard in one system that is caused by the operation, failure or partial performance of another system that is transferred to the victim system by a means other than a defined interface. A system of systems utilizes a network of systems that is, more often than not, geographically distributed. However, the systems of a system of systems operate with a certain amount of autonomy. As such, it is possible that systems may come within close physical proximity of one another. In some systems of systems, systems may come in close proximity as a part of regular operations. An interface hazard occurs when one system adversely affects another system via a defined interface. When systems come within close proximity of each other, it is possible for one system to adversely affect another system without using a defined interface. This is a proximity hazard.

Resource Hazards   A resource hazard is a hazard that results from insufficient shared resources or resource conflicts. Systems within a system of systems share resources. When systems, subsystems, or components are integrated, unless it is through an outdated point-to-point wiring, then it is through shared resources. If the systems are dependent upon the resource, or dependent upon the integration that the resource provides, then mishaps may result from a compromise of that resource. Examples of resources that may be shared are bandwidth, airspace, network addresses, or memory. There are two types of resource hazards, insufficient resources or resource conflicts.

Reconfiguration Hazards A reconfiguration hazard is a hazard that results from the transfer of a system of systems from one state to another. One aspect of system of systems that makes them so capable is their ability to dynamically reconfigure as operational needs demand. A system of systems evolves and morphs in both short- and long-time frames. In the long-time frame, new systems are developed and added, old systems are retired. In the short-time frame, systems are added or removed from the operating network dynamically depending upon the demands of the task at hand.

Interoperability Hazards An interoperability hazard is a hazard that occurs when the command, response or data of one system is interpreted by a second system in a manner that is inconsistent with the intent of the first system. Interoperability mishaps happen with alarming frequency, in particular when armed forces of different nations are operating in a combined environment. The mishaps generally occur when the intent of one party is not clearly communicated to the other. Neither party is solely at fault. The party that gave the command may have done so in accordance with their rules, as may have the party who responds to the command, and yet a mishap still occurs.

33.6 SoSHA TOOLS

There are some tools available that can be utilized in the SoS hazard analysis process. The first tool is the system mishap model (SMM). The second tool is the SoS component system matrix.

33.6.1 SMM

These tools help in the development of a map that delineates the "hazard space" in a SoS. In this concept, hazard space is defined as the total set of hazards and their reference points in a system or a SoS. The hazard reference points are the TLMs that establish the main category of hazards in a system. An SMM displays the TLMs for a system in an organized manner that is easily understandable. The TLMs can then be broken down into the major system causal factor areas contributing to the TLM. The SMM is not an HA, but when complete it provides a hazard space map that aids the analyst in determining where to look and what to look for during the HA. The TLM and SMM concepts are explained in detail in Chapter 5.

Figures 33.2–33.4 depict the steps in developing a SMM for a hypothetical SoS. The first step is to establish the SoS linkage with all of the system components. Figure 33.3 shows the

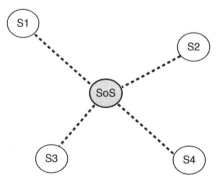

Figure 33.2 *Example SoSHA architecture.*

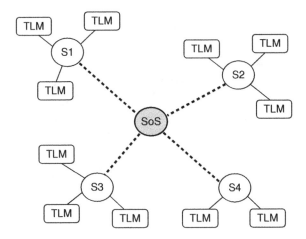

Figure 33.3 *TLMs for systems in example SoSHA.*

SoS architecture comprised of systems 1–4, which are independent systems linked together to form a SoS around a SoS "hub".

The second step is to lay out the component system TLMs, typically obtained from the system developers. Each of the systems has their own unique set of TLMs derived from their individual system hazard analyses, as depicted in Figure 33.4.

The third step is to add TLMs unique to the SoS hub. The SoS has its own unique set of TLMs derived from the SoS hazard analysis. In addition, SoS faults may contribute to, or cause new hazards, in any of the systems. A system hazard could affect another system TLM (see A). A SoS hazard could affect another system TLM (see B).

The fourth step is to identify SoS hazards using the SoS TLMs as a guideline. The TLMs provide a good indication of where the SoS hazards are located within the SoS architecture. This step considers hazards generated by the SoS hub and also by hazard in each of the

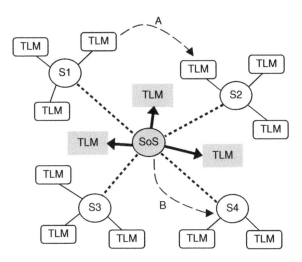

Figure 33.4 *SMM for example SoSHA.*

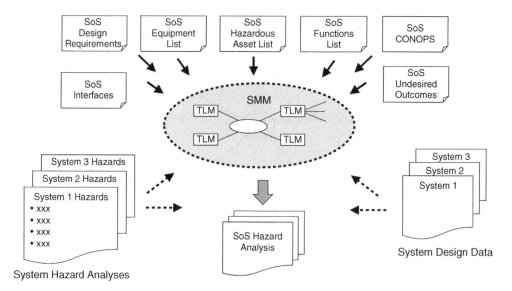

Figure 33.5   *SoSHA methodology.*

component systems. This step also includes looking at SoS causal factors that may contribute back to a component system hazard, as indicated by path B in Figure 33.5.

The last step involves looking at each component system to determine if causal factors exist that could contribute to a hazard in another component system, as indicated by path A in Figure 33.5.

33.6.2   SoS Component System Matrix

Another very useful tool to aid in the SoSHA is component system matrix. This is a matrix of all the component systems in a SoS that delineates which systems can have a potential hazardous impact on any other system within the SoS. Table 33.1 demonstrates this matrix.

This table is read by starting in the left column. Compare the system of interest in the left column to each of the other systems under the "System of interest can Impact" column. If the system of interest can adversely impact the other system safety-wise, indicate with a yes or no.

TABLE 33.1   SoS Component System Interface Matrix

System of Interest	System of Interest can Impact						
	System 1	System 2	System 3	System 4	System 5	System 6	System 7
System 1	—	Yes	No	No	No	Yes	Yes
System 2	Yes	—	No	Yes	No	No	Yes
System 3	No	No	—	No	No	Yes	No
System 4	No	Yes	No	—	No	No	Yes
System 5	Yes	Yes	Yes	Yes	—	Yes	Yes
System 6	No	No	No	No	No	—	No
System 7	Yes	No	No	No	No	Yes	—

The value of the component system matrix is that when you have finished establishing which systems can adversely affect another system; this provides a starting point to look for hazards involving those SoS interfaces.

33.7 SoSHA METHODOLOGY

A SoSHA is slightly different from other types of HA. First, the SoSHA is really more of an objective rather than a specific technique; the goal is to identify hazards created by the development of a SoS from existing systems. Second, because of item 1, the SoSHA does not require a special methodology; any of several existing techniques can be used. Third, a SoSHA is best applied when it is based upon an SMM that lays out the entire hazard space of the SoS. The SMM concept is explained in Chapter 5.

As part of the SoSHA, it is beneficial if all identified hazards are combined under top-level mishaps (TLMs). The TLM concept is explained in Chapter 5. The SoSHA then evaluates each TLM to determine if all causal factors are identified and adequately mitigated to an acceptable level of system risk. A review of the TLMs in the SoSHA will indicate if additional in-depth analysis of any sort is necessary, such as for a safety-critical hazard or an interface concern.

An overview of the SoSHA methodology is shown in Figure 33.5. This figure demonstrates that significant amount of information and data must be input to and utilized by the SoSHA process.

Basic design data is part of the data input, which includes the following as a minimum:

- SoS design requirements
- SoS equipment and hardware list
- SoS list of functions
- SoS concept of operations (CONOPS)
- SoS list of hazardous assets (e.g., fuel explosives, radar energy, etc.)
- SoS interfaces
- SoS undesired outcomes
- Design data and information for each independent system
- HA data for each independent system

The SoSHA process involves reviewing and utilizing the results of previously identified hazards. This review is primarily focused on the evaluation of subsystem interfaces for potential hazards not yet identified. System and subsystem design safety requirements are utilized by the SoSHA to evaluate system compliance. Subsystem interface information, primarily from the SSoSHA and interface specifications, is also utilized by the SoSHA to assist in the identification of interface-related hazards.

Every hazard analysis should always consider the following:

- The functional design
- The physical design
- The interface design
- The operational design
- The software design
- The human interface design

SoS hazard categories include the following:

- Conditions within an independent system that cause hazards to that system
- Conditions within an independent system that cause hazards to another system within the SoS
- Conditions within an independent system that cause hazards within the SoS
- Conditions within the SoS that cause hazards to a system linked to the SoS
- Conditions within the SoS that cause SoS hazards

Table 33.2 lists and describes the basic steps in the SoSHA process.

TABLE 33.2   SoSHA Process

Step	Task	Description
1	Define the SoS	Define the SoS scope, boundaries, and component systems. Define the mission, mission phases, and mission environments. Understand the SoS design, architecture, and operation. Understand the SoS requirements
2	Plan SoSHA	Establish SoSHA goals, definitions, methodologies, schedule, and process. Identify the subsystems and interfaces to be analyzed. It is very likely that SCFs and TLMs will provide the interfaces of most interest
3	Establish safety criteria	Identify applicable design safety criteria, safety precepts/principles, safety guidelines, and safety-critical factors
4	Obtain data on independent systems	Obtain design data and HA data on existing systems being linked to the SoS architecture. Ensure there is sufficient data to support a SoSHA. If sufficient data is not available, take action to obtain the needed information
5	Establish SMM of SoS hazard space	Establish the TLMs from the identified hazards and the system safety-critical functions (SCFs). Also, begin establishing the SoS hazard space using an SMM
6	Identify hazards	Identify SoS hazard using the most applicable HA techniques already available. Utilize the SMM to aid in recognizing critical safety areas and hazards in those areas
7	Perform supporting analyses	Certain safety-critical (SC) hazards may require a more detailed and/or. Perform more detailed analyses as found necessary during the SoSHA, such as quantitative analysis, FTA, CCFA, and so on
8	Evaluate risk	Identify the level of SoSHA risk presented to the system by each identified hazard, both before and after recommended hazard mitigations have been established for the system design
9	Recommend corrective action	Recommend any corrective action as found necessary during the SoSHA. Place recommendations into system safety requirements (SSRs) and incorporate SSRs into design and operational requirements
10	Monitor corrective action	Monitor SoS test activities and review test results to ensure that safety recommendations and SSRs are effective in mitigating hazards as anticipated. Also, use test activities to help identify SoS hazards
11	Track hazards	Transfer newly identified hazards into the HTS. Update the HTS as hazards, hazard causal factors, and risk are identified in the SoSHA
12	Document SoSHA	Document the entire SoSHA process on the worksheets. Update for new information and closure of assigned corrective actions

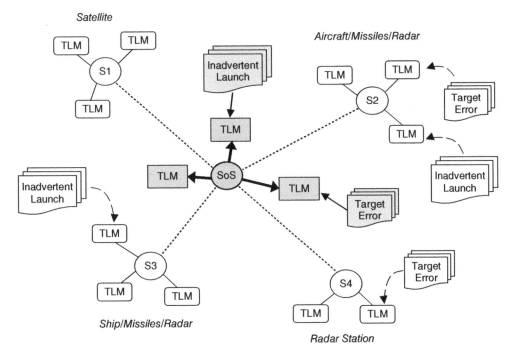

Figure 33.6   SMM for hypothetical SoS.

33.8   SoSHA EXAMPLE

Figure 33.6 contains a hypothetical SoS consisting of the SoS hub and the following component systems: satellite, aircraft with missiles, ground radar station, and a ship with missiles. The objective of this SoS is to detect threats and destroy them with missiles. Threats can be detected by any of the three radar systems, the information sent to the satellite that then relays the information to the SoS hub, whereby the hub determines whether to use missiles from the ship or the aircraft.

The first step in establishing the SMM is to link the component systems to the SoS hub. Then, the TLMs for each component system are taken from the system's hazard analyses and placed around the component system's bubble. Of the many possible TLMs for the component systems, only two are shown in Figure 33.3, in order to keep the diagram uncluttered. The two TLMs are (1) inadvertent missile launch and (2) targeting error. Each TLM symbol shows multiple pages in order to depict that each TLM has multiple hazards below it.

Now, a system impact table is established, which is shown in Table 33.3 for this hypothetical example. This table shows, for example, that ship faults/hazards cannot impact the satellite. It also shows that radar faults/hazards can impact the satellite, primarily by sending incorrect target data to the satellite, which the satellite then sends to the ship and aircraft. Note that these interface conditions are for example only and may not apply to real systems.

From the SMM and the system impact table, TLMs can be derived for the SoS hub. The two TLMs for the systems also apply directly to the SoS hub. Again, these TLMs are (1) inadvertent missile launch and (2) targeting error. Additional analysis must be performed to determine if additional TLMs exist for the SoS hub.

TABLE 33.3   System Impact Table

System of Interest	System of Interest can Impact			
	System 1	System 2	System 3	System 4
1. Satellite	—	Yes	Yes	No
2. Ship	No	—	Yes	No
3. Aircraft	No	Yes	—	No
4. Radar	Yes	Yes	Yes	—

TABLE 33.4   Missile System TLMs

No.	TLM
1	Inadvertent missile launch
2	Inadvertent missile destruct
3	Inadvertent warhead explosives initiation
4	Incorrect missile target
5	Fire occurs within the missile
6	Unknown missile state
7	Inadvertent explosives detonation
8	Unable to safe warhead

TLMs that are connected with the missiles aboard the ship are shown in Table 33.4. These TLMs are just for the missiles, many other ship TLMs will also exist.

The following are some example hazards that could result from the above information when performing the SoSHA:

1. The satellite receives threat target data from the ground radar, but faults in the satellite result in incorrect target information sent to the SoS, which then forwards the information to the aircraft for missile release. This causes the missile to hit an incorrect target resulting in death/injury to friendly force and/or civilians.
2. After a threat is detected the SoS hub selects the wrong component system to launch missiles at the threat. The threat is not neutralized resulting in it causing death/injury to ground forces or one of the component systems.
3. A fault/error in the SoS hub causes an inadvertent launch of a missile in one of the component systems, which results in death/injury.

33.9   SoSHA WORKSHEET

There is no unique HA worksheet for a SoSHA analysis; a SoSHA is more of an objective than a technique. A SoSHA can be performed using any one of a number of HA worksheets that already exist. For example, the following hazard analysis techniques can be utilized to achieve a SOSHA: preliminary hazard analysis (PHA), subsystem hazard analysis (SSHA), SHA, operations and support hazard analysis (O&SHA), and/or hazard and operability analysis (HAZOP). Other HA techniques can be used as found necessary and applicable.

As a minimum, the following basic information should be obtained from the SoSHA:

1. SoS-related hazards
2. Hazard causal factors

3. Hazard risk
4. Information to determine if further causal factor analysis is necessary for a particular hazard
5. Recommended risk mitigation methods

33.10 SoSHA GUIDELINES

The following are some basic guidelines that should be followed when completing the SoSHA worksheet:

1. The SoSHA identifies hazards caused by subsystem interface factors, environmental factors, or common cause factors.
2. The SoSHA is more than a collection of the HAs performed on each individual system. It must also consider the SoS hub and all of the component system interfaces with the hub and with each other.
3. Start the SoSHA by considering TLMs, SCFs, and the SoS hazard space. The hazard space should be defined and managed via an SMM.
4. Many of the component system TLMs will also become TLMs for the SoS hub.
5. When reviewing the SMM it may become apparent that one of the component systems did not have a TLM that it should have, thus requiring more SoS effort to have the system TLM evaluated.
6. Perform supporting analyses as determined necessary by SoS hazards and risk. Supporting analyses may include FTA, bent pin analysis, CCFA, and so on.

33.11 SoSHA ADVANTAGES AND DISADVANTAGES

The following are advantages of performing a SoSHA:

1. The SoSHA identifies SoS hazards, both emergent and system.
2. The SoSHA consolidates hazards to ensure that all causal factors are thoroughly investigated and mitigated.
3. The SoSHA identifies critical SoS-level hazards that must be evaluated in more detail through the use of other analysis techniques.
4. The SoSHA provides the basis for making an assessment of overall SoS risk.

There are no disadvantages to performing a SoSHA.

33.12 COMMON SoSHA MISTAKES TO AVOID

When first learning how to perform a SoSHA, it is commonplace to commit some typical mistakes. The following is a list of errors often encountered during the conduct of a SoSHA:

1. The SOSHA focus is not on emergent hazards.
2. Failure conditions in one system that adversely affect other linked systems are ignored.

3. The hazard space in an SoS is not managed via an SMM.

4. The causal factors for a hazard are not thoroughly investigated.

5. The SoSHA mishap risk index (MRI) risk severity level does not appropriately support the identified hazardous effects.

6. Hazards are closed prematurely without complete causal factor analysis and test verification.

7. Failure to consider common cause events and dependent events.

8. Failure to include human errors and software functional failures.

9. Incompleteness in the entire SoSHA, including the hub and component systems.

33.13   SUMMARY

This chapter discussed the SoSHA Technique. The following are basic principles that help summarize the discussion in this chapter:

1. The primary purpose of the SoSHA is to ensure safety of the total SoS, which involves ensuring that the system risk is acceptable.

2. The SoSHA assesses system compliance with safety requirements and criteria, through traceability of hazards and verification of SSRs.

3. The SoSHA identifies emergent hazards resulting from the SoS architecture.

4. The use of an SMM helps organize and manage hazards in the SoS; it also helps guide the SoSHA.

REFERENCES

The following are some references on SoS and SoS hazard analysis:

1. Systems Engineering Guide for Systems of Systems, Office of the Deputy Under Secretary of Defense for Acquisition and Technology, Version 1.0, August, 2008.
2. MIL-STD-882E, System Safety, DoD Standard Practice, 11 May 2012.
3. D. S. P. Caffal and J. B. P. Michael, Architectural Framework for a System-of-Systems, OT IEEE International Conference on Systems, Man and Cybernetics, 2005, pp. 1876–1881.

FURTHER READINGS

A System of Systems Interface Hazard Analysis Technique, Patrick Redmond, Naval Postgraduate School, Master's Thesis, March, 2007.

Boxer, P., Morris, E., Anderson, W., and Cohen, B. Systems-of-Systems Engineering and the Pragmatics of Demand, IEEE Systems Conference, 7–10 April, Montreal, Canada, 2008.

Brownsword, L., Fisher, D., Morris, E., Smith, J., and Kirwan, P. *System of Systems Navigator: An Approach for Managing System of Systems Interoperability*, Integration of Software-Intensive Systems Initiative, Software Engineering Institute, 2006.

Maier, M. Architecting Principles for Systems-of-Systems, *Syst. Eng.*, **1** (4):267–284 (1998).

Chapter 34

Summary

34.1 TENETS OF HAZARD ANALYSIS

This book has focused on two main objectives. The first objective was to provide an understanding of hazard theory, in order that hazards can be better understood and therefore more easily identified and described. The second objective was to explain in detail how to perform the 28 most used hazard analysis techniques in system safety. In order to be truly professional, the system safety analyst must be able to correctly apply the appropriate techniques in order to identify and mitigate hazards.

Overall, the hazard concepts presented in this book can be briefly summarized by the following key tenets:

- *Principle 1*   Hazards and mishaps are not chance events; hazards lead to mishaps if left unchecked.
- *Principle 2*   Hazards are created during system design and exist with the design.
- *Principle 3* Hazards are comprised of three components: HA, IMs, and TTO.
- *Principle 4*   Many hazards cannot be eliminated due to the hazard sources that are required by the system.
- *Principle 5*   Hazards present risk; risk is the metric for measuring the criticality or danger level of a hazard.
- *Principle 6*   Hazards can be modified via design methods, which in turn can reduce risk.
- *Principle 7*   Hazard analysis is the key to preventing mishaps; hazard identification and mitigation reduces mishap risk.
- *Principle 8*   The system mishap model (SMM) is an effective hazard analysis tool.

Hazard Analysis Techniques for System Safety, Second Edition. Clifton A Ericson, II.
© 2016 John Wiley & Sons, Inc. Published 2016 by John Wiley & Sons, Inc.

- *Principle 9* Hazard analysis and hazard descriptions can easily become abused, confused, and/or misused.
- *Principle 10* Utilizing more than one hazard analysis technique is recommended.
- *Principle 11* Hazard mitigation is not hazard elimination.
- *Principle 12* Hazard risk is the same as mishap risk.
- *Principle 13* There are both primary and secondary hazard analysis techniques.
- *Principle 14* There are pseudo-hazards and real hazards.

34.2 DESCRIPTION OF TENETS

34.2.1 Hazards and Mishaps are Not Chance Events; Hazards Lead to Mishaps If Left Unchecked

A mishap is not a random chance event, but instead it is a deterministic event. Mishaps and accidents do not just happen; they are the result of a unique set of conditions (i.e., hazards). A hazard is a potential condition that can result in a mishap or accident. This means that mishaps can be predicted via hazard identification. And, mishaps can be prevented or controlled via hazard elimination, control, or mitigation.

A hazard is the precursor to a mishap; a hazard is a condition that defines a potential event (i.e., mishap), while a mishap is the occurred event. This results in a direct relationship between a hazard and a mishap, whereby a hazard and a mishap are two separate states of the same phenomenon, linked by a state transition that must occur. A hazard is a "potential event" at one end of the spectrum, which may be transformed into an "actual event" (the mishap) at the other end of the spectrum. The transition from the hazard state to the risk state is based upon the risk involved.

The hazard–mishap relationship is shown in Figure 34.1.

Hazards are a reality and it should go without saying that if a hazard is left unchecked (i.e., not eliminated or mitigated), then it will eventually result in a mishap. The mishap will occur with a probability based on the hazard components. The significance of hazard analysis is to identify hazards and then implement design measures to check the hazards from becoming mishaps

34.2.2 Hazards are Created During System Design and Exist with the Design

How do hazards come into existence? Are they acts of nature or man-made? Mishaps do not just happen; they are the result of hazards. And, hazards do not just randomly happen either, they are the result of circumstance and/or design flaws inadvertently built into the system design.

Figure 34.1 *Hazard–mishap relationship.*

The basic reasons hazards exist are (1) they are unavoidable because hazardous elements must be used in the system and/or (2) they are the result of inadequate design safety consideration. Inadequate design safety consideration results from poor or insufficient design or the incorrect implementation of a good design. This includes inadequate consideration given to the potential effect of hardware failures, sneak paths, software glitches, human error, and so on.

Hazards are man-made, created during the design of a system; they are typically the by-product of man-made systems. Hazards are the result of flawed designs, hazardous assets, and potential component failures. Quite often, hazards are inadvertently injected into a system design through design "blunders," design errors, or lack of foresight. For example, two subsystems may have been independently designed and when combined in a system have unforeseen interface problems that result in a hazard. Or, an electrical system may have been designed with unforeseen sneak paths that create hazards. System safety practitioners must use hazard analysis to identify these types of hazards, along with their causal factors.

Fortunately, since hazards are created through the design and development process, this makes them deterministic, predictable, and identifiable. This generates the need for hazard analysis since this is the method for identifying hazards in order that they can be changed.

34.2.3 Hazards are Comprised of Three Components: HA, IMs, and TTO

A hazard is a unique and discrete entity comprised of a unique set of causal factors and outcomes. Each and every hazard is unique, with a unique level of risk attached to it. A hazard is like a mini-system; it has a dormant life until a state transition transforms it into a mishap. A hazard defines the terms and conditions of a potential mishap; it is the wrapper containing the entire potential mishap description. The mishap that results is the product of the hazard components.

A hazard is comprised of the following three basic components, each of which must be present in order for the hazard to exist:

1. *Hazard source (HS)*   This is the basic hazardous resource creating the impetus for the hazard, such as a hazardous energy source like explosives being used in the system.
2. *Initiating mechanism (IM)*   This is the trigger or initiator event(s) causing the hazard to occur. This is the mechanism(s) that causes actualization of the hazard from a dormant state to an active mishap.
3. *Target/threat outcome (TTO)*   This is the person or thing that is vulnerable to injury and/or damage; it describes the severity of the mishap event. This is the mishap event outcome and the expected consequential damage and loss.

These three components form the hazard triangle, as shown in Figure 34.2.

Figure 34.2   *Hazard triangle.*

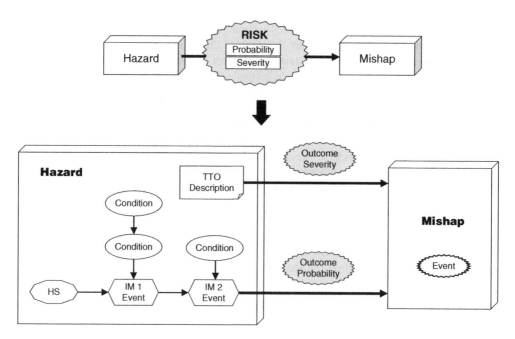

Figure 34.3 *Hazard component expansion.*

The hazard triangle illustrates that a hazard consists of *three necessary* and *coupled* components, each of which forms the side of a triangle. All three sides of the triangle are necessary and essential in order for a hazard to exist. Remove any one of the triangle sides and the hazard is eliminated because it is no longer able to produce a mishap (i.e., the triangle is incomplete). Reduce the probability of the IM triangle side and the mishap probability is reduced. Reduce an element in the HS or the TTO side of the triangle and the mishap severity is reduced.

It should be noted that sometimes the three hazard components are referred to as source, mechanism, and outcome. These terms are just as acceptable and provide the same definitions as their HS, IM, and TTO counterparts.

Figure 34.3 expands on the hazard–mishap model shown in Figure 34.1. The expansion illustrates that the IMs are essentially events that trigger the mishap, and the IMs are a function of one more system conditions that must occur. For example, failure of a particular resistor may be a condition to cause an inadvertent missile launch, and when the resistor does fail it is noted as an IM event. The mishap probability is derived from the IM probabilities, and the mishap outcome is derived from the TTO information.

34.2.4   Many Hazards Cannot be Eliminated due to the Hazard Sources that are Required by the System

As already stated, a hazard is comprised of an HS, IM, and TTO. In order to eliminate a hazard, one of these three components must be eliminated, as a minimum. Unfortunately, in most systems, the HS cannot be eliminated because it is needed by the system in order for the system to perform its function. For example, an automobile requires gasoline in order to operate, and gasoline is a hazard source. For a gasoline-operated auto, this HS cannot be removed, thus the HS will spawn many gasoline-related hazards. These hazards cannot be eliminated, but they can be reduced in risk via design safety measures. If the auto can be redesigned to use batteries rather than gasoline, then the gasoline-related hazards would be eliminated.

By circumstance, systems with hazardous sources will always have hazards that cannot be eliminated. In these situations, the hazards are usually well known and the objective is to reduce the system mishap risk to an acceptable level. In order to reduce the risk, the hazard causal factors must first be identified, which is accomplished via hazard analysis.

For every system that contains a hazard source, there will be a commensurate set of inherent hazards associated with it, each of which presents its own potential mishap risk. A hazard source is the rudimentary element of a hazard that provides the impetus for the hazard to exist. A hazard source is an inherent source of danger. Without a hazard source, there would be no hazard. Hazard sources are generally energy sources or safety-critical functions. Example hazard energy sources include fuel, electricity, lightning, compressed gas, compressed fluid, rotating machinery, nuclear power, temperature, velocity, and so on. It is almost guaranteed that when these types of potentially dangerous entities exist within a system, they bring potential hazards with them, which cannot be eliminated.

34.2.5 Hazards Present Risk; Risk is the Metric for Measuring the Criticality or Danger Level of a Hazard

Risk is the measure that rates the safety significance, or criticality, of a hazard. Risk is a combined function of probability and severity, where the probability indicated how often the hazard/mishap will occur and the severity indicated how bad the hazard/mishap outcome will be when it does occur.

The risk of each hazard is a unique value based on the system design. The hazard components and the unique system design establish probability and severity. When the system design is modified, the risk is also modified.

Risk theory involves three parameters: a future event, a probability for the event, and an outcome for the event. Risk probability is the likelihood of the future event occurring. Risk outcome is the final expected result of the future event, given that it occurs. The risk outcome can actually be a threat or an opportunity; however, in system safety, it is a threat of loss and/or death/injury.

Risk is an intangible quality; it does not have physical or material substance (a mishap does, but not risk). It is a future value concept with some quantifiable metrics, probability, and severity, which characterize the future event. Risk can be thought of as the net present value of a future event. In system safety, risk is a quantitative measure of the future event, where the event is an expected mishap. Whether or not risk is expressed qualitatively or numerically, it is still a quantitative measure of some type.

34.2.6 Hazards can be Modified via Design Methods, which in Turn can Reduce Risk

When the risk presented by a hazard is determined to be unacceptable, it must be mitigated. Mitigation is typically done through the implementation of a design safety feature (DSF), such as a redundant component, fail–safe mechanism, interlock, and so on. System safety requirements (SSRs) are the vehicle that establishes the DSFs for the system design. By implementing the DSFs contained in the SSRs, the mishap risk is reduced to an acceptable level. But, it does

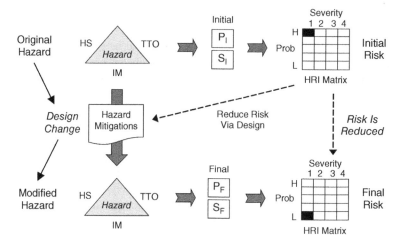

Figure 34.4 *Hazard risk reduction.*

not end there; the SSRs must be verified and validated (V&V) to ensure they are implemented and effectively eliminate the hazard or reduce it to an acceptable level of risk. Mitigation success is verified through appropriate analysis, testing, demonstration, or inspection.

A hazard risk index (HRI) table is recommended in MIL-STD-882E that is used to determine the relative level of risk a hazard presents. This table can be used to measure the initial risk the hazard presents and the final risk after the system design has been modified in order to change the probability or severity components of the hazard.

Figure 34.4 depicts how risk reduction is measured using hazard probability, severity, and a simplified HRI matrix.

In this example, the system is an electric coffee grinder, where the coffee bean cutting blades are at the bottom of a plastic cylinder. When the grinder lid is placed on top of the cylinder base, the blades automatically start to spin and grind the beans. There is a pin at the top of the cylinder that slides into the lid and acts as an on/off switch. The initial hazard is

> The operator accidentally depresses the on/off pin while his fingers are inside the cylinder, thereby accidentally turning the blades on and cutting a finger tip off.

The risk of this hazard is determined to have a severity of High (H on the HRI) and a probability of high (1 on the HRI) and therefore unacceptable. To reduce hazard risk, the design is modified to make the on/off pin part of the lid rather than the base. The pin must then slide down into a recessed hole in the base to activate the on/off switch. This new design prevents the operator from accidentally activating the blades by touching the exposed pin. The modified hazard is

> While the operator has his fingers inside the cylinder, internal on/off switch fails in the on position, thereby accidentally turning the blades on and cutting a finger tip off.

The modified system design has modified the hazard, which in turn has reduced the hazard risk from 1H to 1L (high to low) as shown in Figure 34.4. In this example, the hazard severity has not changed, but the hazard probability has been reduced.

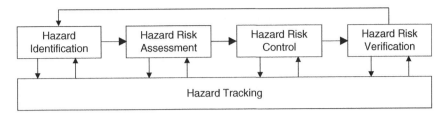

Figure 34.5 *System safety focuses on hazards.*

34.2.7 Hazard Analysis is the Key to Preventing Mishaps; Hazard Identification and Mitigation Reduce Mishap Risk

To reduce mishap risk, a project should implement a system safety program that focuses on hazards, because hazards create the risk. The identification of hazards and hazard causal factors is largely achieved through hazard analysis. Hazard analysis requires methodologies that are designed specifically for the purpose of identifying and evaluating hazards and mishap risk. Hazard analysis tools are a necessary ingredient in the system safety process. The system safety analyst should be familiar with each of the hazard analysis tools presented in this book. They form the basic building blocks for performing hazard and safety analysis on any type of system.

The purpose of a system safety program is to design-in safety and reduce mishap risk. From a historical perspective, it has been learned that a proactive preventive approach to safety during design is much more cost-effective than trying to implement safety into a system after the occurrence of an accident or mishap. This is accomplished via many different system safety tasks; however, the core system safety process revolves around hazards – hazard identification, mishap risk assessment from hazards, and then hazard elimination or mitigation. The core system safety process can be reduced to the closed loop process shown in Figure 34.5.

This is a hazard and mishap risk management process, whereby safety is achieved through the identification of hazards, the assessment of hazard mishap risk, and the control of hazards presenting unacceptable risk. It is a closed loop process because hazards are identified and continuously tracked and updated until acceptable closure action is implemented and verified. An automated hazard tracking system is generally used for collecting and storing hazard information.

34.2.8 The System Mishap Model is an Effective Hazard Analysis Tool

Sometimes, a system is so large and complex that it is difficult to get a handle on where to start the hazard analysis and how to correctly model hazards. The SMM provides a visual model of the hazard space in a system that allows the analyst to visualize all system hazard connections. A system mishap model is a model organizing the major potential mishaps that a system is susceptible to; it establishes and links the major mishap categories comprising a system. The purpose of a SMM is to better understand a system, its associated hazards, and potential mishaps. The SMM provides hazard analysis direction and visibility. It also provides a means for summing risk at the individual mishap level. The SMM is an a priori (before the fact) approach to understanding hazards and mishaps.

When conducting an HA, one of the hazard identification tools is the SMM. When potential system mishap vulnerability is known, hazards leading to that mishap category can be more easily identified. An established methodology for an SMM involves utilizing a mishap tree-linking system top-level mishaps. These concepts and methods are explained below.

34.2.9   Hazard Analysis and Hazard Descriptions can Easily Become Abused, Confused, and/or Misused

Hazard analysis appears to be a simple process, but in reality it is more complex than most people envision. This factor can easily lead to misuse and confusion in a hazard analysis. In addition, some projects try to cut costs by performing a rudimentary hazard analysis and/or using untrained analysts, which also leads to abuse and misuse of hazard analysis.

There seems to be the perennial question of hazards versus failures. Some analysts believe that hazards are only the result of failures and therefore they only concentrate on failures. A failure is not a hazard per se, but a failure (or failures) can be part of a hazard. And, not all failures even contribute to hazards. A hazard analysis is not an FMEA, or vice versa. Hazards are more than failure; hazards are an entity, which can involve many different factors, such as human error, design errors, interface flaws, environmental conditions, software errors, algorithm errors, and so on.

Hazard count often causes confusion and/or concern. Quite often, a large and complex system can have 2000 hazards, or more. This sometimes concerns system developers because they (erroneously) think this means their system design is unsafe. The number of hazards is not what makes an unsafe system; it is the residual risk presented by the hazards after hazard analysis and mitigation that reflect the level of safety.

In reality, the quantity of hazards is not an indication of a bad design, it is an indication that the system is dealing with a lot of hazard sources. The problem of safety is not that of having hazards in a system, the problem is in not fully recognizing all hazards and eliminating them, or reducing their risk to an acceptable level, and not being honest about their existence.

Sometimes, system developers will try to bundle several hazards into one hazard, in order to reduce the hazard count. This is not recommended because it hides hazard and risk visibility. It also violates the general hazard rule of hazard causal factors can only be ANDed together (an OR situation generally indicates multiple hazards). In reality, the hazards are being combined under a generic TLM outcome and not a single hazard.

Hazard risk can be easily manipulated by identifying hazards at too low a level in the system hierarchy or too high a level. For example, if a critical subsystem is analyzed at the nut and bolt level, a single bolt failure will not appear to present high risk because there might be three redundant bolts involved; however, if risk is evaluated at the subsystem level, then it will likely appear much more significant when all of the subsystem failures are summed together. The system risk levels of risk rating should be thought out carefully prior to the hazard analysis.

34.2.10   Utilizing More than One Hazard Analysis Technique is Recommended

Using more than one hazard analysis technique helps ensure complete hazard identification coverage because different techniques look at different system aspects. Several hazard analysis techniques work in conjunction with each other, one analysis building upon the information from the previous analysis.

There are many different hazard analysis techniques and one particular hazard analysis technique does not necessarily identify all the hazards within a system; it may take more than one type. When used together, their combined effect provides an optimum process for identifying hazards, mitigating hazards, and reducing system residual risk. A best practice system safety program includes using several analysis techniques to ensure complete hazard coverage and provide optimum safety assurance.

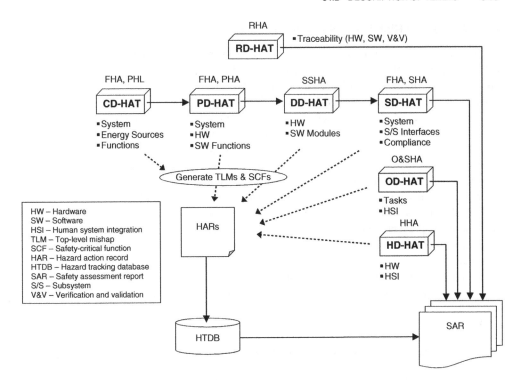

Figure 34.6 *Hazard analysis relationships.*

Figure 34.6 shows the relative relationship of several key hazard analysis types and techniques, along with their interdependencies. This figure shows how the output of one analysis type can provide input data for another analysis type, which together achieve the necessary system safety program tasks.

34.2.11 Hazard Mitigation is not Hazard Elimination

Every identified hazard should have at least one corresponding safety requirement, as a minimum, which either eliminates or mitigates the hazard. People not totally familiar with the system safety process sometimes become confused over the difference between elimination and mitigation, thinking that if a hazard has been mitigated, then it is eliminated and can be forgotten. This is a dangerous misbelief.

Hazard elimination means to completely remove the hazard from the system design. This generally involves completely removing the hazardous element (e.g., fuel, explosives, etc.) from the system design and replacing it with another nonhazardous item. In this case, the hazard no longer exists.

Hazard mitigation means reducing the mishap risk of a hazard because it cannot be eliminated. Mitigation reduces the hazard mishap probability and/or the hazard mishap outcome severity. Hazard mitigation involves taking specific design action to reduce the hazard mishap risk to an acceptable level. Design action is achieved through design measures, the use of safety devices, warning devices, training, or procedures. In this case, the hazard exists, but has been made relatively same through the incorporation of special design measures.

34.2.12 Hazard Risk is the Same as Mishap Risk

Safety risk is sometimes referred to as *hazard risk* and sometimes as *mishap risk*, with the idea that they are different. The idea that they represent different risks is incorrect and can lead to much confusion during a hazard analysis. It should be noted that hazard risk and mishap risk are really the same entity, just viewed from two different perspectives.

Hazard risk is the safety metric characterizing the amount of danger presented by a hazard, where the likelihood of a hazard occurring and transforming into a mishap is combined with the expected severity of the mishap outcome predicted by the hazard.

Mishap risk is the safety metric characterizing the amount of danger presented by a potential mishap, where the likelihood of the mishap's occurrence is combined with the resulting severity of the mishap. Mishap risk likelihood defines the likelihood of the mishap occurring, while mishap risk severity defines the expected final consequences and loss outcome expected from the mishap event. The mishap likelihood and severity can be computed only from the information contained in the hazard description, when the mishap is being predicted in advance.

Hazard risk is a view of risk in the future. Mishap risk is a view of risk as an event that has occurred. As the hazard–mishap model depicts, hazards are directly linked to mishaps, thus the risk is the same from each viewpoint.

34.2.13 There are Both Primary and Secondary Hazard Analysis Techniques

There is some confusion regarding which hazard analysis techniques are truly hazard analyses and which are not. Within the system safety discipline, there are over 100 different hazard analysis techniques that have been proposed, some of which are unique, some of which are variants of others, some of which are extremely useful and some of which are not useful at all. Some of the techniques are not true hazard analyses, and many are merely variations of other hazard analysis techniques. There are only about 15–20 hazard analysis techniques that are commonly used by system safety experts.

Essentially, there are *primary* and *secondary* hazard analysis techniques. The primary hazard analysis techniques are full-fledged, or complete, formal methodologies that are designed for identifying all, or most, system hazards. The secondary hazard analysis techniques are limited in their hazard identification ability; typically, they are not designed to identify all hazards. The secondary techniques essentially provide support for the primary techniques; many help identify the root causal factors of already identified hazards.

It should be noted that many analysis techniques are incorrectly used for hazard analysis, such as failure mode and effects analysis (FMEA). FMEA results can be used as resource information for a hazard analysis, but the FMEA does not suffice for a true hazard analysis because it does not thoroughly cover system hazard–mishap scenarios and it does not cover the combined effect of multiple simultaneous failures that often cause hazards. An FMEA is not a primary hazard analysis technique and should not be used in place of it either, but it can be used to supplement as a secondary supporting technique; it is an excellent source for failure modes and failure rates.

34.2.14 There are Pseudo-Hazards and Real Hazards

Although a hazard is a real entity that exists within the design of a system, it is somewhat invisible; that is, it is camouflaged or concealed by the overall system design and complexity.

TABLE 34.1   Pseudo-hazards versus true hazards

Pseudo Hazard	True Hazard
Electrocution	Worker makes contact with exposed high-voltage conductor
Hearing damage	Worker exposed to high noise environment without hearing protection
Fall injury	Worker trips and falls from platform because no barriers or harness protection are present

Some hazards are much easier to recognize or visualize than others. In this regard, care must be taken to differentiate between *pseudo-hazards* and true hazards. For example, consider the examples shown in Table 34.1.

Pseudo-hazard descriptions typically address only the mishap outcome, whereas a true hazard description describes all of the relevant factors involved. The problem with pseudo-hazards is that they do not provide sufficient information to fully evaluate the hazard; they do not identify the HS, IMs, or TTO. Without the HS, IMs, and TTO information, the hazard probability and severity cannot be determined.

Pseudo-hazards can cause confusion in hazard analysis, and obtaining a risk assessment for them can be difficult, if not impossible, due to the lack of concise information and knowledge regarding the hazard.

34.3   FINIS

Remember, absolute safety is not possible because complete freedom from all hazardous conditions is not possible, particularly when dealing with inherently hazardous systems. Hazards will exist, but their risk can and must be made acceptable. Safety is an optimized level of mishap risk that is managed and constrained by cost, time, and operational effectiveness (performance). System safety requires that risk be evaluated and the level of risk accepted or rejected by an appropriate authority within the organization. Mishap risk management is the basic origin of system safety's requirement for both engineering and management functions. System safety is a process of disciplines and controls employed from the initial design steps through system disposal or demilitarization.

System safety is an investment rather than an expense of doing business. By proactively applying system safety during system development, future potential mishaps are eliminated and reduced. The overall cost of a system safety program is significantly less than the cost of one or more mishaps that could be experienced during the life of a system. Mishap costs are calculated in terms of money, time, lives, environmental damage, publicity, and public relations.

Hazard analysis is the core of the system safety process, and understanding when and which hazard analysis techniques to utilize, and how to correctly apply the technique, is crucial in the system safety discipline of designing in safety.

Appendix *A*

List of Acronyms

ALARP	as low as reasonably practical
BA	barrier analysis
BPA	bent pin analysis
CCA	cause–consequence analysis
CCF	common cause failure
CCFA	common cause failure analysis
CDR	critical design review
CIL	critical item list
CM	configuration management
CMF	common mode failure
COTS	commercial off-the-shelf
CSI	critical safety item
CS	cut set
CSCI	computer software configuration item
CSU	computer software unit
EHA	environmental hazard analysis
ET	event tree
ETA	event tree analysis
FMEA	failure mode and effects analysis
FMECA	failure mode and effects and criticality analysis
FHA	fault hazard analysis

Hazard Analysis Techniques for System Safety, Second Edition. Clifton A Ericson, II.
© 2016 John Wiley & Sons, Inc. Published 2016 by John Wiley & Sons, Inc.

FHA	functional hazard analysis
FT	fault tree
FTA	fault tree analysis
HAR	hazard action record
HAZOP	hazard and operability
HCF	hazard causal factor
HCR	hazard control record
HF	human factors
HHA	health hazard assessment
HSI	human system integration
HTS	hazard tracking system
HWCI	hardware configuration item
IE	initiating event
LTA	less than adequate
MA	Markov analysis
MCS	minimal cut set
MORT	management and oversight risk tree
NDI	non-developmental item
O&SHA	operating and support hazard analysis
OS	operating system
PDR	preliminary design review
PNA	Petri net analysis
PHA	preliminary hazard analysis
PHA	process hazard analysis
PHL	preliminary hazard list
PPE	personnel protective equipment
PRA	probabilistic risk assessment
RCA	root cause analysis
RHA	requirements hazard analysis
RPN	risk priority number
SAE	Society of Automotive Engineers
SBP	software build plan
SC	safety-critical
SCA	sneak circuit analysis
SCF	safety-critical function
SDF	software development file
SDP	software development plan
SDR	system design review
SHA	system hazard analysis
SMM	system mishap model
SoS	system of systems

SoSHA	system of systems hazard analysis
SPR	software problem report
SRA	safety requirements analysis
SRCA	safety requirements/criteria analysis
SSHA	subsystem hazard analysis
SSP	system safety program
SSPP	system safety program plan
SSR	system safety requirement
SSVV	system safety verification and validation
STP	software test plan
STR	software trouble report
SUM	software user manual
SwSA	software safety assessment
SwHA	software hazard analysis
SWSSP	software system safety program
THA	test hazard analysis
THA	threat hazard analysis
TLM	top-level mishap
VDD	version description document

Glossary

The following is a glossary of the terms and concepts most used in system safety. This list is by no means exhaustive; for a more thorough and complete list of terms refer to the textbook "Concise Encyclopedia of System Safety: Definition of Terms and Concepts", C. A. Ericson II, John Wiley & Sons, 2011.

Acceptable risk   The part of identified mishap risk that is allowed to persist without taking further engineering or management action to eliminate or reduce the risk, based on knowledge and decision making. The system user is consciously exposed to this risk.

Accepted risk   Accepted risk has two parts: (1) risk that is knowingly understood and accepted by the system developer or user, and (2) risk that is not known or understood and is accepted by default.

Accident   An unexpected event that culminates in the death or injury of personnel, system loss, or damage to property, equipment, or the environment.

Accident scenario   A series of events that ultimately result in an accident. The sequence of events begins with an initiating event and is (usually) followed by one or more pivotal events that lead to the undesired end state.

As low as reasonably practical (ALARP)   A level of mishap risk that has been established and is considered as low as reasonably possible and still acceptable. It is based on a set of predefined ALARP conditions that is considered acceptable.

Barrier analysis (BA)   BA is an analysis technique for identifying hazards specifically associated with hazardous energy sources. BA provides a tool to evaluate the unwanted flow of (hazardous) energy to targets (personnel or equipment) through the evaluation of barriers preventing the hazardous energy flow. BA is based on the theory that when hazardous

Hazard Analysis Techniques for System Safety, Second Edition. Clifton A Ericson, II.
© 2016 John Wiley & Sons, Inc. Published 2016 by John Wiley & Sons, Inc.

energy sources exist within a system, they pose a hazardous threat to certain targets. Placing barriers between the energy source and the target can mitigate the threat to targets. BA is performed according to an established set of guidelines and rules.

Bent pin analysis (BPA) BPA is a specialized analysis of pins within a connector to determine the safety impact of potential bent pins within the connector. A bent pin is a pin inside a connector that is bent sideways, while two connectors are otherwise normally mated. The concern with a bent pin is that it makes electrical contact with another pin or the connector casing during system operation. If this should occur, it is possible to cause open circuits and/or short circuits to +/− voltages, which may be hazardous in certain system designs.

Cascading failure A failure event for which the probability of occurrence is substantially increased by the existence of a previous failure. Cascading failures are dependent events, where the failure of one component causes the failure of the next component in line, similar to the falling domino effect.

Combinatorial model A graphical representation of a system that logically combines system components together according to the rules of the particular model. Various types of combinatorial models that are available include reliability block diagrams (RBDs), fault trees (FTs) and success trees.

Commercial off-the-shelf (COTS) An item that can be purchased commercially from a vendor's catalog. No development or manufacturing is required.

Common cause component group (CCCG) A group of components that share a common coupling factor are referred to as a CCCG.

Common cause failure (CCF) The failure (or unavailable state) of more than one component due to a shared cause during system operation. Viewed in this fashion, CCFs are inseparable from the class of dependent failures. An event or failure, which bypasses or invalidates redundancy or independence (ARP-4761).

A CCF is the simultaneous failure of multiple components due to a common or shared cause. For example, when two electrical motors become inoperable simultaneously due to a common circuit breaker failure that provides power to both motors. CCFs include CMFs, but CCF is much larger in scope and coverage. Components that fail due to a shared cause normally fail in the same functional mode. CCFs deal with causes other than just design dependencies, such as environmental factors, human error, and so on. Ignoring the effects of dependency and CCFs can result in overestimation of the level of reliability and/or safety. For system safety, a CCF event consists of item/component failures that meet the following criteria:

1. Two or more individual components fail or are degraded such that they cannot be used when needed, or used safely if still operational.
2. The component failures result from a single *shared cause* and *coupling mechanism*.

Common cause failure coupling factor A coupling factor is a qualitative characteristic of a group of components or piece parts that identifies them as susceptible to the same causal mechanisms of failure. Such factors include similarity in design, location, environment, mission, and operational, maintenance, and test procedures. The coupling factor(s) is part of the root cause for a CCF. The identification of coupling factors enables the analyst to implement defenses against common root cause failure vulnerabilities.

Common cause failure root cause   The root cause is the most basic reason(s) for the component failure, which if corrected, would prevent recurrence. Example CCF root causes include events such as heat, vibration, moisture, and so on. The identification of a root cause enables the analyst to implement design defenses against CCFs.

Common mode failure (CMF)   A *CMF* is the failure of multiple components in the same mode. An event, which simultaneously affects a number of elements otherwise considered as being independent. For example, a set of identical resistors from the same manufacturer may all fail in the same mode (and exposure time) due to a common manufacturing flaw. The term CMF, which was used in the early literature and is still used by some practitioners, is more indicative of the most common symptom of the CCF, but it is not a precise term for describing all of the different dependency situations that can result in a CCF event. A CMF is a special case or a subset of a common cause failure.

Computer program   A combination of computer instructions and data definitions that enable computer hardware to perform computational or control functions. A computer program is also known as software.

Computer software configuration item (CSCI)   An aggregation of software that satisfies an end use function and is designated for separate configuration management by the developer or acquirer.

Computer software unit (CSU)   An element in the design of a CSCI; for example, a major subdivision of a CSCI, a component of that subdivision, a class, object, module, function, routine, or database. Software units may occur at different levels of a hierarchy and may consist of other software units.

Concurrent engineering   This method performs several of the development tasks concurrently, in an attempt to save development time. This method has a higher probability for technical risk problems since some items are in preproduction before full development and testing.

Critical item list (CIL)   A list of items that are considered critical for reliable and/or safe operation of the system. The list is usually generated from the failure mode and effects analysis (FMEA).

Deductive analysis   An analysis that reasons from the general to the specific to determine the causal factors for how an event actually occurred or how a suspected potential event might occur (e.g., fault tree analysis). Deduction tends to fill in the holes and gaps in a premise to validate the premise.

Dependence (in design)   A design whereby the failure of one item directly causes, or leads to, the failure of another item. This refers to when the functional status of one component is affected by the functional status of another component. CCF dependencies normally stem from the way the system is designed to perform its intended function. Dependent failures are those failures that defeat redundancy or diversity, which are intentionally employed to improve reliability and/or safety. In some system designs, dependency relationships can be very subtle, such as in the following cases:

a. *Standby Redundancy* When an operating component fails, a standby component is put into operation, and the system continues to function. Failure of an operating component causes a standby component to be more susceptible to failure because it is now under load.

b. *Common Loads* When failure of one component increases the load carried by other components. Since the other components are now more likely to fail, we cannot assume statistical independence.

c. *Mutually Exclusive Events* When the occurrence of one event precludes the occurrence of another event.

Dependent event   Events are dependent when the outcome of one event directly affects or influences the outcome of a second event (probability theory). To find the probability of two dependent events both occurring, multiply the probability of A and the probability of B after A occurs; $P(A \text{ and } B) = P(A) \cdot P(B \text{ given } A)$ or $P(A \text{ and } B) = P(A) \cdot P(B|A)$. This is known as conditional probability. For example, a box contains a nickel, a penny, and a dime. Find the probability of choosing first a dime and then, without replacing the dime, choosing a penny. These events are dependent. The first probability of choosing a dime is $P(A) = 1/3$. The probability of choosing a penny is $P(B|A) = 1/2$ since there are now only two coins left. The probability of both is $1/3 \cdot 1/2 = 1/6$. Keywords such as "not put back" and "not replace" suggest that events are dependent. Two failure events A and B are said to be dependent if $P(A \text{ and } B) \neq P(A)P(B)$. In the presence of dependencies, often, but not always, $P(A \text{ and } B) > P(A)P(B)$. This increased probability of two (or more) events is why CCFs are of concern.

Deterministic process   A deterministic process or model predicts a single outcome from a given set of circumstances. A deterministic process results in a sure or certain outcome and is repeatable with the same data. A deterministic model is sure or certain, and is the antonym of random.

Embedded software   Embedded systems are electronic devices that incorporate micro-processors within their implementations 18. The use of a microprocessor simplifies system design and provides flexibility. Embedded software is usually stored in a read-only memory (ROM) chip, meaning that modification requires replacing or reprogramming the chip.

Energy barrier   Any design or administrative method that prevents a hazardous energy source from reaching a potential target in sufficient magnitude to cause damage or injury. Barriers separate the target from the source by various means involving time or space. Barriers can take many forms, such as physical barriers, distance barriers, timing barriers, procedural barriers, and so on.

Energy path   The path of energy flow from source to target.

Energy source   Any material, mechanism, or process that contains potential energy that can be released. The safety concern is that the released energy may cause harm to a potential target. Energy sources generally provide the hazard source element leg of the hazard triangle.

Engineering development model   This is the traditional system development approach that has been in use for many years whereby each of the system life cycle phases is performed sequentially. The system life cycle is typically defined as the stages of conceptual design, preliminary design, detailed design, test, manufacture, operation, and disposal (demilitari-zation). The operational stage is usually the longest and can be 30–50 years or longer. Under this model, the development stages are conceptual design, preliminary design, detailed design, and test. Each development phase must be complete and successful before the next phase is entered. This development method normally takes the longest length of time because the system is developed in sequential stages. Three major design reviews are conducted for exit from one phase and entry into the next. These are the system design review (SDR), preliminary design review (PDR), and critical design review (CDR). These design reviews are an important aspect of the hazard analysis Types.

Error   (1) An occurrence arising as a result of an incorrect action or decision by personnel operating or maintaining a system and (2) a mistake in specification, design, or implementation (SAE ARP-4761).

Event tree (ET)   An ET is a graphical model of an accident scenario that yields multiple outcomes and outcome probabilities. ETs are one of the most used tools in a probabilistic risk assessment (PRA).

Fail–safe   A design feature that ensures the system remains safe, or in the event of a failure, causes the system to revert to a state that will not cause a mishap (MIL-STD-882D).

Failure   A failure is departure of an item from its required or intended operation, function, or behavior; problems that users encounter. The inability of a system, subsystem, or component to perform its required function. The inability of an item to perform within previously prescribed limits.

Failure cause   The failure cause is the process or mechanism responsible for initiating the failure mode. The possible processes that can cause component failure include physical failure, design defects, manufacturing defects, environmental forces, and so on.

Failure effect   The consequence(s) a failure mode has on the operation, function, or status of an item and on the system.

Failure mode   A failure mode is the manner by which an item fails; the mode or state the item is in after it fails. The way in which the failure of an item occurs. For example, a resistor has two primary modes of failure, failed open and failed shorted.

Failure mode and effects analysis (FMEA)   FMEA is a tool for evaluating the effect(s) of potential failure modes of subsystems, assemblies, components, or functions. It is primarily a reliability tool to identify failure modes that would adversely affect overall system reliability. FMEA has the capability to include failure rates for each failure mode, in order to achieve a quantitative probabilistic analysis. Additionally, the FMEA can be extended to evaluate failure modes that may result in an undesired system state, such as a system hazard, and thereby also be used for hazard analysis. FMEA is performed according to an established set of guidelines and rules.

Fault   An undesired anomaly in the functional operation of an equipment or system. The occurrence of an undesired state, which may be the result of a component failure.

Fault tree analysis (FTA)   FTA is a systems analysis technique used to determine the root causes and probability of occurrence of a specified undesired event. A fault tree (FT) is a model that logically and graphically represents the various combinations of possible events, faulty and normal, occurring in a system that leads to a previously identified hazard or undesired event. It is performed according to an established set of guidelines, rules, and logic gates to model cause–effect relationships.

Functional hazard analysis (FHA)   FHA is an analysis technique used to identify system hazards by the analysis of functions. Functions are the means by which a system operates to accomplish its mission or goals. System hazards are identified by evaluating the safety impact of a function failing to operate, operating incorrectly, or operating at the wrong time. When a function's failure can be determined hazardous, the casual factors of the malfunction should be investigated in greater detail via another root cause analysis. FHA is performed according to an established set of guidelines and rules.

Guide word   A guide word is a special word used in a hazard and operability (HAZOP) analysis to help guide or focus the analysis. HAZOP uses a set of guide words, such as more, less, early, late, and so on.

Hardware   An object that has physical being. Generally refers to line replaceable units (LRUs), circuit cards, power supplies, and so on (SAE ARP-4761).

Hardware configuration item (HWCI)   An aggregation of hardware that satisfies an end use function and is designated for separate configuration control by the acquirer.

Hazard   Any real or potential condition that can cause injury, illness, or death to personnel; damage to or loss of a system, equipment, or property; or damage to the environment (MIL-STD-882D). A potentially unsafe condition resulting from failures, malfunctions, external events, errors, or a combination thereof (SAE ARP-4761).

Hazard causal factor (HCF)   Hazard causal factors are the specific items responsible for the existence of a hazard. At a high level, the causal factors are the hazardous elements and initiating mechanisms of the hazard. At a more refined level, the causal factors are the result of poor or insufficient design, incorrect implementation of a good design, or potential or actual failures that would have to occur in order to result in the condition defined as a hazard.

Hazard control record (HCR)   A hazard record employed for hazard tracking; a HCR contains all information relevant to the identification, assessment, mitigation, and closing of a hazard.

Hazard triangle   The three components of a hazard form a hazard triangle because each of the sides must be present in order for the hazard to exist. If the triangle cannot be completed, then the hazard does not exist. The three sides of the triangle are comprised of the following components:

1. *Hazard Source (HS)* This is the basic hazardous resource creating the impetus for the hazard, such as a hazardous energy source like explosives being used in the system.
2. *Initiating Mechanism (IM)* This is the trigger or initiator event(s) causing the hazard to occur. The IMs causes actualization or transformation of the hazard from a dormant state to an active mishap state.
3. *Target and Threat Outcome (TTO)* This is the person or thing that is vulnerable to injury and/or damage, and it describes the severity of the mishap event. This is the mishap outcome and the expected consequential damage and loss.

The three components of a hazard, HS, IM, and TTO, are also often referred to as source (S), mechanism (M), and outcome (O). They are essentially equivalent in meaning.

Hazard and operability (HAZOP)   HAZOP analysis is a technique for identifying and analyzing hazards and operational concerns of a system. It is used primarily in the chemical process industry. HAZOP analysis looks for hazards resulting from identified potential deviations in design operational intent.

Health hazard analysis (HHA)   The HHA is an analysis that evaluates the system design and operational procedures to identify hazards strictly involving human health. For example, it would consider the system effect of noise, vibration, toxicity, heat, hazardous materials, and so on, on humans. It is performed according to an established set of guidelines and rules.

Human engineering   The application of knowledge about human capabilities and limitations to system or equipment design and development to achieve efficient, effective, and safe system performance at minimum cost and manpower, skill, and training demands. Human engineering assures that the system or equipment design, required human tasks, and work environment are compatible with the sensory, perceptual, mental, and physical attributes of the personnel who will operate, maintain, control, and support it (MIL-HDBK-1908A).

Human error   Unwanted actions or inactions that arise from problems in sequencing, timing, knowledge, interfaces, and/or procedures that result in deviations from expected standards or norms that places people, equipment, and systems at risk.

Human factors (HF)   HF is a body of scientific facts about human characteristics. The term covers all biomedical and psychosocial considerations; it includes, but is not limited to, principles and applications in the areas of human engineering, personnel selection, training, life support, job performance aids, and human performance evaluation (MIL-HDBK-1908A).

Human system integration (HSI)   HSI is the application of human factors and human engineering to system design to ensure the safe and reliable operation of the system throughout its life cycle. Since personnel are a major component of any system, special design consideration must be given to human performance. The human–machine interface, as well as the human influence on the system, must be part of all system design considerations.

Identified risk   Identified risk is a known risk that has been determined through the identification and evaluation of a hazard.

Incremental development   In the incremental development process, a desired capability is identified, an end-state requirement is known, and that requirement is met over time by developing several increments, each dependent on available mature technology. This method breaks the development process into incremental stages, in order to reduce development risk. Basic designs, technologies, and methods are developed and proven before more detailed designs are developed.

Indenture level   The item levels that identify or describe relative complexity of an assembly or function. The levels progress from the more complex (system) to the simpler (part) divisions (MIL-STD-1629A). Equipment indenture levels are used, for example, to develop equipment hierarchy lists that aid in system understanding. Example hierarchy levels are system, subsystem, unit, assembly, and component.

Independence (in design)   Independence is a design concept that ensures the failure of one item does not cause the failure of another item 1. This concept is very important in many safety and reliability analysis techniques due to the impact on logic and mathematics. Many models, such as FTA, assume event independence.

Independent event   Events are independent when the outcome of one event does not influence the outcome of a second event (probability theory). To find the probability of two independent events both occurring, multiply the probability of the first event by the probability of the second event; for example, $P(\text{A and B}) = P(\text{A}) \cdot P(\text{B})$. For example, find the probability of tossing two number cubes (dice) and getting a 3 on each one. These events are independent: $P(3) \cdot P(3) = (1/6) \cdot (1/6) = 1/36$. The probability is 1/36.

Inductive analysis   An analysis that reasons from the specific to the general to determine what overall system effect could result from a component failure (e.g., failure mode and effects analysis). Induction tends to establish a premise from data where the data is not complete enough to entirely validate the premise (more data is necessary).

Initiating event (IE)   A failure or undesired event that initiates the start of an accident sequence. The IE may result in a mishap, depending upon successful operation of the hazard countermeasure methods designed into the system. Refer to Chapter 2 on hazard theory for information on the components of a hazard.

Interlock   A safety interlock is a single device and/or functionality that is part of a larger system function. Its purpose is to prevent the overall system function from being performed until a specified set of safety parameters are satisfied. If a known hazardous state is about to be entered the interlock interrupts the system function, thereby preventing a mishap. For example, if a hazardous laser is being operated in a locked room with no personnel in the room, a sensor on the door would be a safety interlock that automatically removes power from the laser system when the door is opened by someone inadvertently entering the room.

A safety interlock is also used to prevent a function from performing unintentionally due to possible system failure modes. For example, power to launch a missile would not reach the missile until three separate and independent switches are closed. These switches are considered three independent interlocks that significantly reduce the probability of preventing inadvertent launch due to random failures. An interlock is like a temporary barrier that prevents a functional path from being completed until desired.

Latent failure   A latent failure is a failure that is not detected and/or enunciated when it occurs. A latent failure of a backup system means the user is not aware that the backup has failed.

Markov analysis (MA)   MA is the analysis and evaluation of systems using Markov chains and Markov processes. MA provides a combinatorial-type analysis of components that is useful for dependability and reliability studies.

Markov chain   These are sequences of random variables in which the future variable is determined by the present variable, but is independent of the way in which the present state arose from its predecessors (the future is independent of the past given the present). The Markov chain assumes discrete states and a discrete time parameter, such as a global clock.

Markov process   The Markov process assumes states are continuous. The Markov process evaluates the probability of jumping from one known state into the next logical state until the system has reached the final state. For example, the first state is everything in the system working, the next state is the first item failed, and this continues until the final system failed state is reached. The behavior of this process is that every state is memory-less, meaning that the future state of the system depends only on its present state. In a stationary system, the probabilities that govern the transitions from state to state remain constant, regardless of the point in time when the transition occurs.

Mishap   An unplanned event or series of events resulting in death, injury, occupational illness, damage to or loss of equipment or property, or damage to the environment (MIL-STD-882D).

Mishap risk   Mishap risk is an expression of the impact and possibility of a mishap in terms of potential mishap severity and probability of occurrence (MIL-STD-882D).

Mitigation   Mitigation is the action taken to reduce the risk presented by a hazard, by modifying the hazard in order to decrease the mishap probability and/or the mishap severity. Mitigation is generally accomplished through design measures, use of safety devices, warning devices, training, or procedures. It is also referred to as hazard mitigation and risk mitigation.

Multitasking software   A task is a job to be done by the software. Multitasking is the process of performing multiple tasks (or threads) concurrently and switching back and forth between them. Since most computers utilize a single CPU, they can perform only a single task at a time, multitasking only gives the appearance of concurrent operation of tasks (or threads).

Mutually exclusive events   Two events are mutually exclusive if the occurrence of one event precludes the occurrence of the other. For example, if the event "switch A fails closed" occurs, then the event "switch A fails open" cannot possibly occur. The switch cannot be in both states simultaneously.

Network trees   A diagram that represents a simplified version of the system circuitry, created by selective deletion of extraneous circuitry detail to reduce system complexity, while retaining all circuit elements pertinent. Network trees are used in sneak circuit analysis.

Nondevelopmental item (NDI)   An already developed item that is provided from another system or development program. No development or manufacturing is required for the current program using the NDI item.

Operations and support hazard analysis (O&SHA)   The O&SHA is an analysis that is performed to identify and evaluate operational type hazards. It is based upon detailed design information and is an evaluation of operational tasks and procedures. It considers human system integration factors such as human error, human task overload, cognitive mis-perception, the effect on humans of hardware failure, and so on. The O&SHA establishes the necessary cautions and warnings that are included in the operational procedures. Occasionally, the O&SHA necessitates design changes or workarounds. It is performed according to an established set of guidelines and rules.

Operating system (OS)   An OS is the overarching program that guides and controls the operation of a computer. The OS manages processor time, memory utilization, application programs, concurrent threads, and so on.

Operation   An operation is the performance of procedures to meet an overall objective. For example, a missile maintenance operation may be "replacing missile battery." The objective is to perform all the necessary procedures and tasks to replace the battery.

Qualitative analysis   An analysis or evaluation based on qualitative values. Mathematical calculations are generally not involved; however, qualitative indices may be combined. A qualitative result is produced, which is considered subjective and/or fuzzy.

Quantitative analysis   An analysis or evaluation based on numerical values and/or mathematical calculations. A quantitative result is produced, which is considered objective and concrete.

Pivotal events   Pivotal events are the intermediary events between an initiating event and the final mishap. These are the failure/success events of the design safety methods established to prevent the IE from resulting in a mishap. If a pivotal event works successfully, it stops the accident scenario and is referred to as a mitigating event. If a pivotal event fails to work, then the accident scenario is allowed to progress and is referred to as an aggravating event.

Preliminary hazard analysis (PHA)   The PHA is generally the first rigorous analysis that is performed to identify hazards, hazard causal factors, mishaps, and system risk. It is usually performed during the preliminary design phase and is, therefore, considered preliminary in nature. It is performed according to an established set of guidelines and rules. The PHA begins with hazards identified from the PHL and expands upon them. The PHA is system oriented and generally identifies system-level hazards.

Preliminary hazard list (PHL)   The PHL is an analysis that results in the generation of a list of hazards. This list is considered preliminary because it is the first hazard analysis performed, and it is generally performed early in the system development process when only conceptual information is available. The PHL analysis is more of a brainstorming type analysis that is intended to quickly focus on hazards that can be expected by the conceptual design.

Probabilistic risk assessment (PRA)   PRA is a comprehensive, structured, and logical analysis method for identifying and evaluating risk in a complex technological system. The detailed identification and assessment of accident scenarios with a quantitative analysis is the PRA goal.

Procedure   A procedure is a set of tasks that must be performed to accomplish an operation. Tasks within a procedure are designed to be followed sequentially to properly and safely accomplish the operation. For example, the above battery replacement operation may be comprised of two primary procedures, (1) battery removal and (2) battery replacement. Each of these procedures contains a specific set of tasks that must be performed.

Reachability   A system can have many different possible states; reachability refers to the systems capability to reach any or all of those states during operation. As designed, the system may not be able to reach some states.

Real-time kernel   Many embedded systems use a real-time kernel, which is a small segment of code that manages processor time and memory utilization among a number of concurrent threads.

Real-time system   A real-time system is one that controls an environment by receiving data, processing them, and returning results sufficiently fast to affect the environment at that time. In a real-time system, response time is a critical element and performance deadlines depend on many factors.

Redundancy   Redundancy is a design methodology using multiple identical components, such that if one component fails the next one will perform the function. This methodology creates a higher functional reliability. Multiple independent means are incorporated to accomplish a given function.

Reliability   The probability that an item will perform a required function under specified conditions, without failure, for a specified period of time. A built-in characteristic of an item that emerges as a function of design and manufacturing.

Repair   Repair refers to the capability to physically repair a failed component, item, sub-system, or system and restore it to an operational state.

Requirement   An identifiable element of a specification that can be validated and against which an implementation can be verified (SAE ARP-4761).

Residual risk   Residual risk is the overall risk remaining after system safety mitigation efforts have been fully implemented. It is, according to MIL-STD-882D, "The remaining mishap risk that exists after all mitigation techniques have been implemented or exhausted, in accordance with the system safety design order of precedence." Residual risk is the sum of all risk after mishap risk management has been applied. This is the total risk passed on to the user.

Reusable software   Software that has been separately developed, typically for some other application, but is considered usable for the present application, usually as-is without modification.

Risk   Exposure to possible loss or injury; danger (dictionary). Risk refers to the measure of expected loss presented by a potential event, such as a financial failure event, a schedule failure event, or a mishap event. In system safety it refers to mishap risk.

Risk analysis   Risk analysis is the process of identifying safety risk. This involves identifying hazards that present the mishap risk and assessing the level of risk.

Risk assessment Risk assessment is the process of determining the risk presented by the identified hazards. This involves evaluating the identified hazard causal factors and then characterizing the risk as the product of the hazard severity times the hazard probability.

Risk communication Risk communication is an interactive process of exchanging risk information and opinions among stakeholders.

Risk management Risk management is the process, by which assessed risks are mitigated, minimized, or controlled through engineering, management, or operational means. This involves the optimal allocation of available resources in support of safety, performance, cost, and schedule.

Risk priority number (RPN) A risk ranking index for reliability, where RPN = (probability of occurrence) × (severity rating) × (detection capability rating).

Root cause analysis (FCA) Root cause analysis is the process of identifying the basic lowest level causal factors for an event. Usually, the event is an undesired event, such as a hazard or mishap. There are different analysis techniques available for RCA.

Safety Freedom from those conditions that can cause death, injury, occupational illness, damage to or loss of equipment or property, or damage to the environment (MIL-STD-882). The ability of a system to exclude certain undesired events (i.e., mishaps) during stated operation under stated conditions for a stated time. The ability of a system or product to operate with a known and accepted level of mishap risk. A built-in system characteristic.

Safety case A safety case is a formal documented body of evidence that provides a convincing and valid argument that a system is adequately safe for a given application in a given environment. The safety case documents the safety requirements for a system, the evidence that the requirements have been met, and the argument linking the evidence to the requirements. Elements of the safety case include safety claims, evidence, arguments, and inferences. Claims are simply propositions about properties of the system supported by evidence. Evidence may either be factual findings from prior research or scientific literature or subclaims that are supported by lower level arguments. The safety argument is the set of inferences between claims and evidence that leads from the evidence forming the basis of the argument to the top-level claim, which is typically that the system is safe to operate in its intended environment. Producing a safety case does not require a specific process; any methodology is acceptable as long as it provides a compelling argument that the system meets its safety requirements. System safety is argued through satisfaction of the requirements, which are then broken down further into more specific goals that can be satisfied directly by evidence.

Safety-critical (SC) A term applied to any condition, event, operation, process, or item whose proper recognition, control, performance, or tolerance is essential to safe system operation and support (e.g., safety-critical function, safety-critical path, or safety-critical component; MIL-STD-882D).

Safety-critical function (SCF) A function comprised of hardware, software, and/or HSI, whose correct operation is necessary and essential for safe system operation. The definition of safe system operation may be program or system dependent, but generally it means operation precluding the occurrence of mishaps that will result in death/injury and/or system loss. Therefore, an SCF is any function whose failure or misbehavior could result in death/injury and/or system loss.

Safety-related A term applied to any condition, event, operation, process, or item that does not meet the definition of safety-critical, but causes an increase in risk.

Safety requirements/criteria analysis (SRCA)   The SRCA is an analysis for evaluating system safety requirements (SSRs) and the criteria behind them. SRCA has a twofold purpose: (1) to ensure that every identified hazard has at least one corresponding safety requirement and (2) to verify that all safety requirements are implemented and are validated successfully. The SRCA is essentially a traceability analysis to ensure that there are no holes or gaps in the safety requirements and that all identified hazards have adequate and proven design mitigation coverage. The SRCA applies to hardware, software, firmware, and test requirements.

Safety-critical function (SCF) thread   An SCF thread refers to the items (functions, components, etc.) comprising the SCF that are necessary for the successful performance of the SCF. Quite often, but not always, an SCF thread is the inverse of a significant top-level mishap. For example, the TLM "inadvertent missile launch" indirectly establishes the SCF "missile launch function." It becomes an SCF because everything in that SCF thread must work correctly for safe system operation. This thread contains all of the elements that might contribute to the TLM.

Semi-Markov process   The Semi-Markov process is similar to that of a pure Markov model, except the transition times and probabilities depend upon the time at which the system reached the present state. The semi-Markov model is useful in analyzing complex dynamical systems, and is frequently used in reliability calculations.

Sneak circuit   A latent path or condition in an electrical system that inhibits a desired condition or initiates an unintended or unwanted action through normal system operation, without failures involved.

Sneak clues   A checklist of items or clues that helps the analyst identify a sneak. The analyst compares the clues to the network tree and topograph to recognize sneaks. The clue list has been developed from past experience and research.

Software   Computer programs, procedures, rules, and any associated documentation pertaining to the operation of a computer system (ARP 4761).

Software build   A build is a version of software that meets a specified set of the requirements. A final system may be developed in several incremental builds. Builds are often necessary to correct anomalies and/or deficiencies, or to incorporate new enhancements.

Software build plan (SB)   A document describing the overall plan for building software in incremental builds. This plan includes a schedule and the build naming convention that will be used. It is useful in software safety analysis.

Software configuration management (CM) plan   This document defines the software configuration management process. It is critical that rigorous control be maintained over the various software builds and versions, and that software cannot be modified without following the appropriate authority and control measures. It is useful in software safety analysis.

Software development file (SDF)   A repository for material pertinent to the development of a particular body of software. Contents typically include (either directly or by reference) considerations, rationale, and constraints related to requirements analysis, design, and implementation; developer-internal test information; and schedule and status information. It is useful in software safety analysis.

Software development plan (SDP)   This document describes the software engineering tasks that must be performed in developing the software. It includes tools, definitions, constraints, and so on. It is useful in software safety analysis.

Software Engineering Standard   A standard delineating the software development method, practices, coding rules, and so on for the development of software within that organization. It is useful in software safety analysis.

Software problem report (SPR)   This is a report that documents software problems or anomalies identified during formal testing. Also known as a software trouble report, it is useful in software safety analysis.

Software test plan (STP)   This document contains the plan for software testing, particularly CSCI qualification testing. The STP includes test objectives, qualification methods, and provides traceability from requirements to specific tests. It is useful in software safety analysis.

Software threads   A thread is a sequence of instructions, designed to perform a single task. A single task may be partitioned into one or more threads running concurrently, but a single thread may also contribute to the objective of one or more tasks. Separating the code into threads simplifies software development and maintenance by allowing the programmer to concentrate on one task at a time.

Software trouble report (STR)   This report documents software problems or anomalies identified during formal testing. Also known as a software problem report, it is useful in software safety analysis.

Software user manual (SUM)   The SUM is a document that records information needed by hands-on users of the software (persons who will both operate the software and make use of its results). It is useful in software safety analysis.

Specification   A collection of requirements that, when taken together, constitute the criteria that define the functions and attributes of a system, or an item (SAE ARP-4761).

Spiral development   In the spiral development process, a desired capability is identified, but the end-state requirements are not known at program initiation. Requirements are refined through demonstration, risk management, and continuous user feedback. Each increment provides the best possible capability, but the requirements for future increments depend on user feedback and technology maturation.

State   The condition of a component or system at a particular point in time (i.e., operational state, failed state, degraded state, etc.).

State transition diagram   A state transition diagram is a directed graph representation of system states, transitions between states, and transition rates. These diagrams contain sufficient information for developing the state equations, which are used for probability calculations in Markov analysis.

Stochastic process   A stochastic process or model predicts a set of possible outcomes weighted by their likelihoods or probabilities. A stochastic process is a random or chance outcome.

Subsystem   A grouping of items satisfying a logical group of functions within a particular system (MIL-STD-882D).

Subsystem hazard analysis (SSHA)   The SSHA is generally the second rigorous analysis that is performed to identify hazards, hazard causal factors, mishaps, and system risk. It is usually performed during the detailed design phase, and is performed according to an established set of guidelines and rules. The SSHA begins with hazards identified by the PHA and expands upon their casual factors. The SSHA is limited to hazards within the subsystem under analysis.

System   A composite, at any level of complexity, of personnel, procedures, materials, tools, equipment, facilities, and software. The elements of this composite entity are used together in the intended operational or support environment to perform a given task or achieve a specific purpose, support, or mission requirement (MIL-STD-882C).

System hazard analysis (SHA)   The SHA is an analysis that is performed to identify and evaluate system-level hazards that are generally the result of subsystem interface issues. It is based upon detailed design information and is performed according to an established set of guidelines and rules. Often, a special purpose analysis is performed to support the SHA. For example, a fault tree analysis might be performed to quantitatively evaluate a system hazard of an inadvertent missile launch. The SHA is system focused and does not look into hazards that are strictly within a subsystem.

System life cycle   The system life cycle is typically defined as the stages of conceptual design, preliminary design, detailed design, test, manufacture, operation, and disposal (demilitarization). The operational stage is usually the longest and can be 30–50 years or longer.

System safety   The application of engineering and management principles, criteria, and techniques to achieve acceptable mishap risk, within the constraints of operational effectiveness and suitability, time, and cost, throughout all phases of the system life cycle (MIL-STD-882). The application of engineering and management processes during system development to intentionally design-in safety in order to prevent mishaps, or reduce the risk of a mishap to an acceptable level.

System safety program (SSP)   The combined tasks and activities of system safety management and system safety engineering implemented during system development to develop a safe system.

System safety program plan (SSPP)   A description of the planned tasks and activities to be used by a contractor to implement a system safety program. This description includes organizational responsibilities, resources, methods of accomplishment, milestones, depth of effort, and integration with other program engineering and management activities and related systems.

Task   A task is an element of work, which together with other elements of work comprises a procedure. For example, battery removal may consist of a series of sequential elements of work, such as power shutdown, removal of compartment cover, removal of electrical terminals, unbolting of battery hold down bolts, and removal of battery.

Top-level mishap (TLM)   A TLM is a generic mishap category for collecting various hazards that share the same general outcome or type of mishap. A TLM is a significant mishap that can be caused by multiple different hazards. Its purpose is to serve as a collection point for of all the potential hazards that can result in the same outcome, but have different causal factors.

Topograph   A topological pattern that appears in a network tree. Topographs are used in sneak circuit analysis.

Total risk   Total risk is the sum of individual identified risks (what is known) and unidentified risks (what is unknown).

Unacceptable risk   Unacceptable risk is the risk that cannot be tolerated.

Unidentified risk   Unidentified risk is an unknown risk that has not been detected. It is real, and it is important, but it has not been recognized. Some unidentified risks are only subsequently identified when a mishap occurs. Some risks are never known (i.e., the probability is so small that they never happen).

Validation   The determination that the requirements for a product are sufficiently correct and complete.

Verification   The evaluation of an implementation to determine that the applicable requirements are met.

Version description document (VDD)   The VDD is a description of a unique release or build of the software. This document is basically a "packing list" of what is included in the release. It is useful in software safety analysis.

Appendix *C*

Hazard Checklists

This chapter contains system safety hazard checklists from many different sources. Hazard checklists are an invaluable aid for assisting the system safety analyst in identifying hazards. For this reason, the more checklists that are available the greater the likelihood of identifying all hazards.

In performing a hazard analysis, the analyst compares design knowledge and information to hazard checklists. This allows the analyst to visualize or postulate possible hazards. For example, if the analyst discovers that the system design will be using jet fuel, he then compares jet fuel to a hazard checklist. From the hazard checklist, it will be obvious that jet fuel is a hazardous element and that a jet fuel fire/explosion is a potential mishap with many different ignition sources presenting many different hazards.

The hazard checklist should not be considered a complete, final, or all-inclusive list. Hazard checklists help trigger the analyst's recognition of potential hazardous sources, from past lessons learned. Hazards checklist are not a replacement for good engineering analysis and judgment. A checklist is merely a mechanism or catalyst for stimulating hazard recognition.

When using multiple hazard checklists, redundant entries may occur; however, this nuisance factor should be overlooked for the overall value provided by many different clues.

Table C-1 lists the different types of checklists included in Appendix C.

Hazard Analysis Techniques for System Safety, Second Edition. Clifton A Ericson, II.
© 2016 John Wiley & Sons, Inc. Published 2016 by John Wiley & Sons, Inc.

TABLE C-1   Hazard Checklist in Appendix C

Section	Title
C.1	General hazards checklist
C.2	Hazard checklist for energy sources
C.3	Hazard checklist for general sources
C.4	Hazard checklist for space functions
C.5	Hazard checklist for general operations
C.6	Operational hazards checklist
C.7	Hazard checklist for failure states
C.8	Hazard guide words for failure states
C.9	Work site hazards checklist
C.10	Electrical hazards checklists
C.11	Food industry hazards checklist

C.1   GENERAL HAZARDS CHECKLIST

This checklist is a general list of possible hazards sources. When performing a hazard analysis, each of these items should be considered for hazardous impact within the system. The source for this checklist is NASA Reference Publication 1358, System Engineering "Toolbox" for Design Oriented Engineers, 1994.

Acceleration/Deceleration/Gravity:

- Inadvertent motion
- Loose object translation
- Impacts
- Falling objects
- Fragments/missiles
- Sloshing liquids
- Slip/trip
- Falls

Chemical/Water Contamination:

- System cross-connection
- Leaks/spills
- Vessel/pipe/conduit rupture
- Backflow/siphon effect

Common Causes:

- Utility outages
- Moisture/humidity
- Temperature extremes
- Seismic disturbance/impact
- Vibration

- Flooding
- Dust/dirt
- Faulty calibration
- Fire
- Single-operator coupling
- Location
- Radiation
- Wear out
- Maintenance error
- Vermin/varmints/mud daubers
- Manufacturing defect

Contingencies (Emergency Responses by System/Operators to "Unusual" Events):

- "Hard" shutdowns/failures
- Freezing
- Fire
- Windstorm
- Hailstorm
- Utility outrages
- Flooding
- Earthquake
- Snow/ice load

Control Systems:

- Power outage
- Interferences (EMI/RFI)
- Moisture
- Sneak circuit
- Sneak software
- Lightning strike
- Grounding failure
- Inadvertent activation

Electrical:

- Shock
- Burns
- Overheating
- Ignition of combustibles
- Inadvertent activation
- Power outage

- Distribution back-feed
- Unsafe failure to operate
- Explosion/electrical (electrostatic)
- Explosion/electrical (arc)

Mechanical:

- Sharp edges/points
- Rotating equipment
- Reciprocating equipment
- Pinch points
- Lifting weights
- Stability/topping potential
- Ejected parts/fragments
- Crushing surfaces

Pneumatic/Hydraulic Pressure:

- Overpressurization
- Pipe/vessel/duct rupture
- Implosion
- Mislocated relief device
- Dynamic pressure loading
- Relief pressure improperly set
- Backflow
- Cross-flow
- Hydraulic ram
- Inadvertent release
- Miscalibrated relief device
- Blown objects
- Pipe/hose whip
- Blast

Temperature Extremes:

- Heat source/sink
- Hot/cold surface burns
- Pressure evaluation
- Confined gas/liquid
- Elevated flammability
- Elevated volatility
- Elevated reactivity
- Freezing
- Humidity/moisture

- Reduced reliability
- Altered structural properties (e.g., embrittlement)

Radiation (Ionizing):

- Alpha
- Beta
- Neutron
- Gamma
- X-Ray

Radiation (Nonionizing):

- Laser
- Infrared
- Microwave
- Ultraviolet

Fire/Flammability: Presence of

- Fuel
- Ignition source
- Oxidizer
- Propellant

Explosives (Initiators):

- Heat
- Friction
- Impact/shock
- Vibration
- Electrostatic discharge
- Chemical contamination
- Lightning
- Welding (stray current/sparks)

Explosives (Effects):

- Mass fire
- Blast overpressure
- Thrown fragments
- Seismic ground wave
- Meteorological reinforcement

Explosives (Sensitizes)

- Heat/cold
- Vibration

□ Impact/shock
□ Low humidity
□ Chemical contamination

Explosives (Conditions)

□ Explosive propellant present
□ Explosive gas present
□ Explosive liquid present
□ Explosive vapor present
□ Explosive dust present

Leaks/Spills (Material Conditions)

□ Liquid/cryogens
□ Gases/vapors
□ Dusts – irritating
□ Radiation sources
□ Flammable
□ Toxic
□ Reactive
□ Corrosive
□ Slippery
□ Odorous
□ Pathogenic
□ Asphyxiating
□ Flooding
□ Run off
□ Vapor propagation

Physiological (see Ergonomic)

□ Temperature extremes
□ Nuisance dusts/odors
□ Baropressure extremes
□ Fatigue
□ Lifted weights
□ Noise
□ Vibration (Raynaud's syndrome)
□ Mutagens
□ Asphyxiants
□ Allergens
□ Pathogens
□ Radiation (see radiation)

- Cryogens
- Carcinogens
- Teratogens
- Toxins
- Irritants

Human Factors (see Ergonomic):

- Operator error
- Inadvertent operation
- Failure to operate
- Operation early/late
- Operation out of sequence
- Right operation/wrong control
- Operated too long
- Operated too briefly

Ergonomic (see Human Factors)

- Fatigue
- Inaccessibility
- Nonexistent/inadequate "Kill" switches
- Glare
- Inadequate control/readout differentiation
- Inappropriate control/readout location
- Faulty/inadequate control/readout labeling
- Faulty work station design
- Inadequate/improper illumination

Utility Outages:

- Electricity
- Steam
- Heating/cooling
- Ventilation
- Air conditioning
- Compressed air/gas
- Lubrication drains/slumps
- Fuel
- Exhaust

Mission Phasing:

- Transport
- Delivery

- Installation
- Calibration
- Checkout
- Shake down
- Activation
- Standard start
- Emergency start
- Normal operation
- Load change
- Coupling/uncoupling
- Stressed operation
- Standard shutdown
- Shutdown emergency
- Diagnosis/trouble shooting
- Maintenance

C.2 HAZARD CHECKLIST FOR ENERGY SOURCES

This checklist is a general list of potentially hazardous energy sources. A system that uses any of these energy sources will very likely have various associated hazards. This checklist was collected by C. Ericson.

1. Fuels
2. Propellants
3. Initiators
4. Explosive charges
5. Charged electrical capacitors
6. Storage batteries
7. Static electrical charges
8. Pressure containers
9. Spring-loaded devices
10. Suspension systems
11. Gas generators
12. Electrical generators
13. RF energy sources
14. Radioactive energy sources
15. Falling objects
16. Catapulted objects
17. Heating devices
18. Pumps, blowers, and fans
19. Rotating machinery

20. Actuating devices
21. Nuclear

C.3 HAZARD CHECKLIST FOR GENERAL SOURCES

This is another checklist of general items that often generates hazards within a system. When performing a hazard analysis each of these items should be considered for hazardous impact within the system. This checklist was collected by C. Ericson.

1. Acceleration
2. Contamination
3. Corrosion
4. Chemical dissociation
5. Electrical
   - Shock
   - Thermal
   - Inadvertent activation
   - Power source failure
   - Electromagnetic radiation
6. Explosion
7. Fire
8. Heat and temperature
   - High temp.
   - Low temp.
   - Temp variations
9. Leakage
10. Moisture
    - High humidity
    - Low humidity
11. Oxidation
12. Pressure
    - High
    - Low
    - Rapid change
13. Radiation
    - Thermal
    - Electromagnetic
    - Ionizing
    - Ultraviolet
14. Chemical replacement
15. Shock (mechanical)

16. Stress concentrations
17. Stress reversals
18. Structural damage or failure
19. Toxicity
20. Vibration and noise
21. Weather and environment

C.4 HAZARD CHECKLIST FOR SPACE FUNCTIONS

This is a checklist of general space-related functions that mostly generates hazards within a system. When performing a hazard analysis, each of these items should be considered for hazardous impact within the system. This checklist was collected by C. Ericson.

1. Crew egress/ingress
2. Ground to stage power transfer
3. Launch escape
4. Stage firing and separation
5. Ground control communication transfer
6. Rendezvous and docking
7. Ground control of crew
8. Ground data communication to crew
9. Extra vehicular activity
10. In-flight tests by crew
11. In-flight emergencies
    - Loss of communications
    - Loss of power/control
    - Fire toxicity
    - Explosion
    - Life support
12. Reentry
13. Parachute deployment and descent
14. Crew recovery
15. Vehicle safing and recovery
16. Vehicle inerting and decontamination
17. Payload mating
18. Fairing separation
19. Orbital injection
20. Solar panel deployment
21. Orbit positioning
22. Orbit correction
23. Data acquisition

24. Mid-course correction
25. Star acquisition (navigation)
26. On-orbit performance
27. Retro-thrust

C.5   HAZARD CHECKLIST FOR GENERAL OPERATIONS

This is a checklist of general operations that often generates hazards within a system. When performing a hazard analysis, each of these items should be considered for hazardous impact within the system. This checklist was collected by C. Ericson.

1. Welding
2. Cleaning
3. Extreme temperature operations
4. Extreme weight operations
5. Hoisting, handling, and assembly operations
6. Test chamber operations
7. Proof test of major components/subsystems/systems
8. Propellant loading/transfer/handling
9. High-energy pressurization/hydrostatic–pneumostatic testing
10. Nuclear component handling/checkout
11. Ordnance installation/checkout/test
12. Tank entry/confined space entry
13. Transport and handling of end item
14. Manned vehicle tests
15. Static firing
16. Systems operational validations

C.6   OPERATIONAL HAZARDS CHECKLIST

This is a checklist of general operational considerations that often generates hazards within a system. When performing a hazard analysis, each of these items should be considered for hazardous impact within the system. This checklist was collected by C. Ericson.

1. Work Area
   Tripping, slipping, corners
   Illumination
   Floor load, piling
   Ventilation
   Moving objects
   Exposed surfaces – hot, electric

Cramped quarters
Emergency exits
2. Materials Handling
Heavy, rough, and sharp
Explosives
Flammable
Awkward, fragile
3. Clothing
Loose, ragged, and soiled
Necktie, jewelry
Shoes, high heels
Protective
4. Machines
Cutting, punching, and forming
Rotating shafts
Pinch points
Flying pieces
Projections
Protective equipment
5. Tools
No tools
Incorrect tools
Damaged tools
Out-of-tolerance tools
6. Emergency
Plans, procedures, and numbers
Equipment
Personnel
Training
7. Safety Devices
Fails to function
Inadequate

C.7   HAZARD CHECKLIST FOR FAILURE STATES

This is a checklist of failure modes or failure states that can generate hazards within a system. When performing a hazard analysis, each of these items should be considered for hazardous impact within the system. This checklist was collected by C. Ericson.

1. Fails to operate
2. Fails to function

3. Malfunction

4. Degraded function

5. Operates/functions incorrectly/erroneously

6. Operates/functions inadvertently

7. Operates/functions at incorrect time (early, late)

8. Unable to stop operation

9. Receives erroneous data

10. Sends erroneous data

11. Interprets data incorrectly

12. Operator confusion

13. Operator misuse

14. Loss of control capability

C.8   HAZARD GUIDE WORDS FOR FAILURE STATES

These are guide words typically used in a HAZOP analysis. The guide words are applied to every component, subsystem, and/or function in the system. When the guide word question is asked, the answer will help lead to the identification of a hazard when the design conditions are not adequately safe.

Guide Word	Meaning
No	The design intent does not occur (e.g. Flow/No), or the operational aspect is not achievable (Isolate/No)
Less	A quantitative decrease in the design intent occurs (e.g. Pressure/Less)
More	A quantitative increase in the design intent occurs (e.g. Temperature/More)
Reverse	The opposite of the design intent occurs (e.g. Flow/Reverse)
Also	The design intent is completely fulfilled, but in addition some other related activity occurs (e.g. Flow/Also indicating contamination in a product stream, or Level/Also meaning material in a tank or vessel that should not be there)
Other	The activity occurs, but not in the way intended (e.g. Flow/Other could indicate a leak or product flowing where it should not, or Composition/Other might suggest unexpected proportions in a feedstock)
Other than	Complete substitution
Fluctuation	The design intention is achieved only part of the time (e.g. an air lock in a pipeline might result in Flow/Fluctuation)
Early	The timing is different from the intention. Usually used when studying sequential operations, this would indicate that a step is started at the wrong time or done out of sequence
Late	Timing concern similar to Early
As well as	Qualitative modification/increase
More than	Exceeds a specified value
Part of	Only some of the design intention is achieved
Reverse	Logical opposite of the design intention occurs

(Continued)

Guide Word	Meaning
Where else	Applicable for flows, transfers, sources, and destinations
Before/after	The step (or some part of it) is effected out of sequence
Faster/slower	The step is done/not done with the right timing
Fails	Fails to operate or perform its intended purpose
Inadvertent	Function occurs inadvertently or prematurely (i.e., unintentionally)

C.9 WORKSITE HAZARDS CHECKLIST

This is a checklist of typical hazards encountered at a work site. When performing a hazard analysis of work site operations, each of these items should be taken into account for hazardous impact, considering both system design and task operation.

General Work Environment

- Are all work sites clean and orderly?
- Are work surfaces kept dry or appropriate means taken to assure the surfaces are slip resistant?
- Are all spilled materials or liquids cleaned up immediately?
- Is combustible scrap, debris, and waste stored safely and removed from the work site promptly?
- Is accumulated combustible dust routinely removed from elevated surfaces, including the overhead structure of buildings?
- Is combustible dust cleaned up with a vacuum system to prevent the dust going into suspension?
- Is metallic or conductive dust prevented from entering or accumulation on or around electrical enclosures or equipment?
- Are covered metal waste cans used for oily and paint-soaked waste?
- Are all oil- and gas-fired devices equipped with flame failure controls that will prevent the flow of fuel if pilots or main burners are not working?
- Are paint spray booths, dip tanks, and the like cleaned regularly?
- Are the minimum number of toilets and washing facilities provided?
- Are all toilets and washing facilities clean and sanitary?
- Are all work areas adequately illuminated?
- Are pits and floor openings covered or otherwise guarded?

Personal Protective Equipment (PPE)

- Are protective goggles or face shields provided and worn where there is any danger of flying particles or corrosive materials?
- Are approved safety glasses required to be worn at all times in areas where there is a risk of eye injuries such as punctures, abrasions, contusions, or burns?

- Are employees who need corrective lenses (glasses or contact lenses) in working environments with harmful exposures required to wear only approved safety glasses, wear protective goggles, or use other medically approved precautionary procedures?
- Are protective gloves, aprons, shields, or other means provided against cuts, corrosive liquids, and chemicals?
- Are hard hats provided and worn where danger of falling objects exists?
- Are hard hats inspected periodically for damage to the shell and suspension system?
- Is appropriate foot protection required where there is the risk of foot injuries from hot, corrosive, poisonous substances, falling objects, and crushing or penetrating actions?
- Are approved respirators provided for regular or emergency use where needed?
- Is all protective equipment maintained under a sanitary condition and ready for use?
- Do you have eye wash facilities and a quick drench shower within the work area where employees are exposed to injurious corrosive materials?
- Where special equipment is needed for electrical workers, is it available?
- When lunches are eaten on the premises, are they eaten in areas where there is no exposure to toxic materials or other health hazards?
- Is protection against the effects of occupational noise exposure provided when sound levels exceed those of the Cal/OSHA noise standard?

Walkways

- Are aisles and passageways kept clear?
- Are aisles and walkways marked as appropriate?
- Are wet surfaces covered with nonslip materials?
- Are holes in the floor, sidewalk, or other walking surface repaired properly, covered, or otherwise made safe?
- Is there safe clearance for walking in aisles where motorized or mechanical handling equipment is operating?
- Are spilled materials cleaned up immediately?
- Are materials or equipment stored in such a way that sharp projectiles will not interfere with the walkway?
- Are changes of direction or elevations readily identifiable?
- Are aisles or walkways that pass near moving or operating machinery, welding operations, or similar operations arranged so employees will not be subjected to potential hazards?
- Is adequate headroom provided for the entire length of any aisle or walkway?
- Are standard guardrails provided wherever aisle or walkway surfaces are elevated more than 30 inches above any adjacent floor or the ground?
- Are bridges provided over conveyors and similar hazardous equipment?

Floor and Wall Stairways

- Are floor openings guarded by a cover, guardrail, or equivalent on all sides (except at entrance to stairways or ladders)?

▫ Are toe boards installed around the edges of a permanent floor opening (where persons may pass below the opening)?

▫ Are skylight screens of such construction and mounting that they will withstand a load of at least 200 pounds?

▫ Is the glass in windows, doors, and glass walls that are subject to human impact of sufficient thickness and type for the condition of use?

▫ Are grates or similar type covers over floor openings, such as floor drains, of such design that foot traffic or rolling equipment will not be affected by the grate spacing?

▫ Are unused portions of service pits and pits not actually in use either covered or protected by guardrails or equivalent?

▫ Are manhole covers, trench covers, and similar covers, plus their supports, designed to carry a truck rear axle load of at least 20,000 pounds when located in roadways and subject to vehicle traffic?

▫ Are floor or wall openings in fire-resistive construction provided with doors or covers compatible with the fire rating of the structure and provided with self-closing feature when appropriate?

Stairs and Stairways

▫ Do standard stair rails or handrails on all stairways have four or more risers?

▫ Are all stairways at least 22 inches wide?

▫ Do stairs have at least a 6′6″ overhead clearance?

▫ Do stairs angle no more than 50 and no less than 30 degrees?

▫ Are stairs of hollow pan-type treads and landings filled to noising level with solid material?

▫ Are step risers on stairs uniform from top to bottom, with no riser spacing greater than 7.5 inches?

▫ Are steps on stairs and stairways designed or provided with a surface that renders them slip resistant?

▫ Are stairway handrails located between 30 and 34 inches above the leading edge of stair treads?

▫ Do stairway handrails have a least 1.5 inches of clearance between the handrails and the wall or surface they are mounted on?

▫ Are stairway handrails capable of withstanding a load of 200 pounds, applied in any direction?

▫ Where stairs or stairways exit directly into any area where vehicles may be operated, are adequate barriers and warnings provided to prevent employees stepping into the path of traffic?

▫ Do stairway landings have a dimension measured in the direction of travel, at least equal to width of the stairway?

▫ Is the vertical distance between stairway landings limited to 12 feet or less?

Elevated Surfaces

▫ Are signs posted, when appropriate, showing the elevated surface load capacity?

▫ Are surfaces elevated more than 30 inches above the floor or ground provided with standard guardrails?

▫ Are all elevated surfaces (beneath which people or machinery could be exposed to falling objects) provided with standard 4-inch toe boards?

▫ Is a permanent means of access and egress provided to elevated storage and work surfaces?

▫ Is required headroom provided where necessary?

▫ Is material on elevated surfaces piled, stacked, or racked in a manner to prevent it from tipping, falling, collapsing, rolling, or spreading?

▫ Are dock boards or bridge plates used when transferring materials between docks and trucks or rail cars?

Exiting or Egress

▫ Are all exits marked with an exit sign and illuminated by a reliable light source?

▫ Are the directions to exits, when not immediately apparent, marked with visible signs?

▫ Are doors, passageways, or stairways, which are neither exits nor access to exits and which could be mistaken for exits, appropriately marked "NOT AN EXIT," "TO BASEMENT," "STOREROOM", and the like?

▫ Are exit signs provided with the word "EXIT" in lettering at least 5 inches high and the stroke of the lettering at least 0.5 inch wide?

▫ Are exit doors side-hinged?

▫ Are all exits kept free of obstructions?

▫ Are at least two means of egress provided from elevated platforms, pits, or rooms where the absence of a second exit would increase the risk of injury from hot, poisonous, corrosive, suffocating, flammable, or explosive substances?

▫ Are there sufficient exits to permit prompt escape in case of emergency?

▫ Are special precautions taken to protect employees during construction and repair operations?

▫ Are the number of exits from each floor of a building and the number of exits from the building itself appropriate for the building occupancy load?

▫ Are exit stairways that are required to be separated from other parts of a building enclosed by at least 2 h fire-resistive construction in buildings more than four stories in height, and not less than 1 h fire-resistive construction elsewhere?

▫ When ramps are used as part of required exiting from a building, is the ramp slope limited to 1 foot vertical and 12 feet horizontal?

▫ Where exiting will be through frameless glass doors, glass exit doors, storm doors, and the like, are the doors fully tempered and meet the safety requirements for human impact?

Exit Doors

▫ Are doors that are required to serve as exits designed and constructed so that the way of exit travel is obvious and direct?

▫ Are windows that could be mistaken for exit doors, made inaccessible by means of barriers or railings?

▫ Are exit doors openable from the direction of exit travel without the use of a key or any special knowledge or effort?

- Is a revolving, sliding, or overhead door prohibited from serving as a required exit door?
- Where panic hardware is installed on a required exit door, will it allow the door to open by applying a force of 15 pounds or less in the direction of the exit traffic?
- Are doors on cold storage rooms provided with an inside release mechanism that will release the latch and open the door even if it is padlocked or otherwise locked on the outside?
- Where exit doors open directly onto any street, alley, or other area where vehicles may be operated, are adequate barriers and warnings provided to prevent employees stepping into the path of traffic?
- Are doors that swing in both directions and are located between rooms where there is frequent traffic provided with viewing panels in each door?

Portable Ladders

- Are all ladders maintained in good condition, joints between steps and side rails tight, all hardware and fittings securely attached, and moveable parts operating freely without binding or undue play?
- Are nonslip safety feet provided on each ladder?
- Is nonslip safety material provided on each ladder rung?
- Are ladder rungs and steps free of grease and oil?
- Is it prohibited to place a ladder in front of doors opening toward the ladder except when the door is blocked open, locked, or guarded?
- Is it prohibited to place ladders on boxes, barrels, or other unstable bases to obtain additional height?
- Are employees instructed to face the ladder when ascending or descending?
- Are employees prohibited from using ladders that are broken, missing steps, rungs, or cleats, broken side rails, or other faulty equipment?
- Are employees instructed not to use the top 2 steps of ordinary stepladders as a step?
- When portable rung ladders are used to gain access to elevated platforms, roofs, and the like, does the ladder always extend at least 3 feet above the elevated surface?
- Is it required that when portable rung or cleat-type ladders are used, the base is so placed that slipping will not occur, or it is lashed or otherwise held in place?
- Are portable metal ladders legibly marked with signs reading "CAUTION" "Do Not Use Around Electrical Equipment" or equivalent wording?
- Are employees prohibited from using ladders as guys, braces, skids, gin poles, or for other than their intended purposes?
- Are employees instructed to only adjust extension ladders while standing at a base (not while standing on the ladder or from a position above the ladder)?
- Are ladders inspected for damage?
- Are the rungs of ladders uniformly spaced at 12 inches, center to center?

Hand Tools and Equipment

- Are all tools and equipment (both, company and employee-owned) used by employees at their workplace in good condition?

▫ Are hand tools such as chisels and punches, which develop mushroomed heads during use, reconditioned or replaced as necessary?

▫ Are broken or fractured handles on hammers, axes, and similar equipment replaced promptly?

▫ Are worn or bent wrenches replaced regularly?

▫ Are appropriate handles used on files and similar tools?

▫ Are employees made aware of the hazards caused by faulty or improperly used hand tools?

▫ Are appropriate safety glasses, face shields, and similar equipment used while using hand tools or equipment that might produce flying materials or be subject to breakage?

▫ Are jacks checked periodically to assure they are in good operating condition?

▫ Are tool handles wedged tightly in the head of all tools?

▫ Are tool cutting edges kept sharp so the tool will move smoothly without binding or skipping?

▫ Are tools stored in dry, secure location where they would not be tampered with?

▫ Is eye and face protection used when driving hardened or tempered spuds or nails?

Portable (Power Operated) Tools and Equipment

▫ Are grinders, saws, and similar equipment provided with appropriate safety guards?

▫ Are power tools used with the correct shield, guard, or attachment recommended by the manufacturer?

▫ Are portable circular saws equipped with guards above and below the base shoe?

▫ Are circular saw guards checked to assure they are not wedged up, thus leaving the lower portion of the blade unguarded?

▫ Are rotating or moving parts of equipment guarded to prevent physical contact?

▫ Are all cord-connected, electrically-operated tools and equipment effectively grounded or of the approved double insulated type?

▫ Are effective guards in place over belts, pulleys, chains, and sprockets, on equipment such as concrete mixers, air compressors, and the like?

▫ Are portable fans provided with full guards or screens having openings 0.5 inch or less?

▫ Is hoisting equipment available and used for lifting heavy objects, and are hoist ratings and characteristics appropriate for the task?

▫ Are ground-fault circuit interrupters provided on all temporary electrical 15 and 20 ampere circuits, used during periods of construction?

▫ Are pneumatic and hydraulic hoses on power-operated tools checked regularly for deterioration or damage?

Abrasive Wheel Equipment Grinders

▫ Is the work rest used and kept adjusted to within 1/8 inch of the wheel?

▫ Is the adjustable tongue on the top side of the grinder used and kept adjusted to within 1/4 inch of the wheel?

▫ Do side guards cover the spindle, nut, and flange and 75% of the wheel diameter?

▫ Are bench and pedestal grinders permanently mounted?

- Are goggles or face shields always worn when grinding?
- Is the maximum RPM rating of each abrasive wheel compatible with the RPM rating of the grinder motor?
- Are fixed or permanently mounted grinders connected to their electrical supply system with metallic conduit or other permanent wiring method?
- Does each grinder have an individual on and off control switch?
- Is each electrically operated grinder effectively grounded?
- Before new abrasive wheels are mounted, are they visually inspected and ring tested?
- Are dust collectors and powered exhausts provided on grinders used in operations that produce large amounts of dust?
- Are splashguards mounted on grinders that use coolant, to prevent the coolant reaching employees?
- Is cleanliness maintained around grinder?

Powder-Actuated Tools

- Powder-actuated tool cartridges work similar to blank firearm cartridges for ramming force.
- Are employees who operate powder-actuated tools trained in their use and carry a valid operator's card?
- Do the powder-actuated tools being used have written approval of the Division of Occupational Safety and Health?
- Is each powder-actuated tool stored in its own locked container when not being used?
- Is a sign at least 7″ by 10″ with bold type reading "POWDER-ACTUATED TOOL IN USE" conspicuously posted when the tool is being used?
- Are powder-actuated tools left unloaded until they are actually ready to be used?
- Are powder-actuated tools inspected for obstructions or defects each day before use?
- Do powder-actuated tools operators have and use appropriate personal protective equipment such as hard hats, safety goggles, safety shoes, and ear protectors?

Machine Guarding

- Is there a training program to instruct employees on safe methods of machine operation?
- Is there adequate supervision to ensure that employees are following safe machine operating procedures?
- Is there a regular program of safety inspection of machinery and equipment?
- Is all machinery and equipment kept clean and properly maintained?
- Is sufficient clearance provided around and between machines to allow for safe operations, setup and servicing, material handling, and waste removal?
- Is equipment and machinery securely placed and anchored, when necessary to prevent tipping or other movement that could result in personal injury?
- Is there a power shut-off switch within reach of the operator's position at each machine?
- Can electric power to each machine be locked out for maintenance, repair, or security?
- Are the noncurrent-carrying metal parts of electrically operated machines bonded and grounded?

▫ Are foot-operated switches guarded or arranged to prevent accidental actuation by personnel or falling.

▫ Are manually operated valves and switches controlling the operation of equipment and machines clearly identified and readily accessible?

▫ Are all emergency stop buttons colored red?

▫ Are all pulleys and belts that are within 7 feet of the floor or working level properly guarded?

▫ Are all moving chains and gears properly guarded?

▫ Are splashguards mounted on machines that use coolant, to prevent the coolant from reaching employees?

▫ Are methods provided to protect the operator and other employees in the machine area from hazards created at the point of operation, ingoing nip points, rotating parts, flying chips, and sparks?

▫ Are machinery guards secure and so arranged that they do not offer a hazard in their use?

▫ If special hand tools are used for placing and removing material, do they protect the operator's hands?

▫ Are revolving drums, barrels, and containers required to be guarded by an enclosure that is interlocked with the drive mechanism, so that revolution cannot occur unless the guard enclosure is in place, so guarded?

▫ Do arbors and mandrels have firm and secure bearings and are they free from play?

▫ Are provisions made to prevent machines from automatically starting when power is restored after a power failure or shutdown?

▫ Are machines constructed so as to be free from excessive vibration when the largest size tool is mounted and run at full speed?

▫ If machinery is cleaned with compressed air, is air pressure controlled and personal protective equipment or other safeguards used to protect operators and other workers from eye and body injury?

▫ Are fan blades protected with a guard having openings no larger than 1.5 inch, when operating within 7 feet of the floor?

▫ Are saws used for ripping, equipped with antikickback devices and spreaders?

▫ Are radial arm saws so arranged such that the cutting head will gently return to the back of the table when released?

Lockout Blockout Procedures

▫ Is all machinery or equipment capable of movement, required to be deenergized or disengaged and blocked or locked out during cleaning, servicing, adjusting, or setting up operations, whenever required?

▫ Is the locking-out of control circuits in lieu of locking-out main power disconnects prohibited?

▫ Are all equipment control valve handles provided with a means for locking-out?

▫ Does the lockout procedure require that stored energy (i.e. mechanical, hydraulic, and air) be released or blocked before equipment is locked-out for repairs?

▫ Are appropriate employees provided with individually keyed personal safety locks?

- Are employees required to keep personal control of their key(s) while they have safety locks in use?
- Is it required that employees check the safety of the lock out by attempting a start up after making sure no one is exposed?
- Where the power disconnecting means for equipment does not also disconnect the electrical control circuit:
  a. Are the appropriate electrical enclosures identified?
  b. Is means provided to assure the control circuit can also be disconnected and locked out?

Welding, Cutting, and Brazing

- Are only authorized and trained personnel permitted to use welding, cutting, or brazing equipment?
- Do all operators have a copy of the appropriate operating instructions and are they directed to follow them?
- Are compressed gas cylinders regularly examined for obvious signs of defects, deep rusting, or leakage?
- Is care used in handling and storage of cylinders, safety valves, relief valves, and the like, to prevent damage?
- Are precautions taken to prevent the mixture of air or oxygen with flammable gases, except at a burner or in a standard torch?
- Are only approved apparatus (torches, regulators, pressure-reducing valves, acetylene generators, manifolds) used?
- Are cylinders kept away from sources of heat?
- Is it prohibited to use cylinders as rollers or supports?
- Are empty cylinders appropriately marked their valves closed and valve-protection caps on?
- Are signs reading: DANGER NO-SMOKING, MATCHES, OR OPEN LIGHTS, or the equivalent posted?
- Are cylinders, cylinder valves, couplings, regulators, hoses, and apparatus kept free of oily or greasy substances?
- Is care taken not to drop or strike cylinders?
- Unless secured on special trucks, are regulators removed and valve-protection caps put in place before moving cylinders?
- Do cylinders without fixed hand wheels have keys, handles, or nonadjustable wrenches on stem valves when in service?
- Are liquefied gases stored and shipped valve-end up with valve covers in place?
- Are employees instructed to never crack a fuel-gas cylinder valve near sources of ignition?
- Before a regulator is removed, is the valve closed and gas released from the regulator?
- Is red used to identify the acetylene (and other fuel-gas) hose, green for oxygen hose, and black for inert gas and air hose?
- Are pressure-reducing regulators used only for the gas and pressures for which they are intended?

- Is open circuit (no load) voltage of arc welding and cutting machines as low as possible and not in excess of the recommended limits?
- Under wet conditions, are automatic controls for reducing no-load voltage used?
- Is grounding of the machine frame and safety ground connections of portable machines checked periodically?
- Are electrodes removed from the holders when not in use?
- Is it required that electric power to the welder be shut off when no one is in attendance?
- Is suitable fire extinguishing equipment available for immediate use?
- Are welders forbidden to coil or loop welding electrode cables around their body?
- Are wet machines thoroughly dried and tested before being used?
- Are work and electrode lead cables frequently inspected for wear and damage, and replaced when needed?
- Do means for connecting cables' lengths have adequate insulation?
- When the object to be welded cannot be moved and fire hazards cannot be removed, are shields used to confine heat, sparks, and slag?
- Are firewatchers assigned when welding or cutting is performed, in locations where a serious fire might develop?
- Are combustible floors kept wet, covered by damp sand, or protected by fire-resistant shields?
- When floors are wet down, are personnel protected from possible electrical shock?
- When welding is done on metal walls, are precautions taken to protect combustibles on the other side?
- Before hot work is begun, are used drums, barrels, tanks, and other containers so thoroughly cleaned that no substances remain that could explode, ignite, or produce toxic vapors?
- Is it required that eye protection helmets, hand shields, and goggles meet appropriate standards?
- Are employees exposed to the hazards created by welding, cutting, or bracing operations protected with personal protective equipment and clothing?
- Is a check made for adequate ventilation in and where welding or cutting is preformed?
- When working in confined places, are environmental monitoring tests taken and means provided for quick removal of welders in case of an emergency?

Compressors and Compressed Air

- Are compressors equipped with pressure relief valves, and pressure gauges?
- Are compressor air intakes installed and equipped to ensure that only clean uncontaminated air enters the compressor?
- Are air filters installed on the compressor intake?
- Are compressors operated and lubricated in accordance with the manufacturer's recommendations?
- Are safety devices on compressed air systems checked frequently?
- Before any repair work is done on the pressure system of a compressor, is the pressure bled off and the system locked out?

□ Are signs posted to warn of the automatic starting feature of the compressors?

□ Is the belt drive system totally enclosed to provide protection for the front, back, top, and sides?

□ Is it strictly prohibited to direct compressed air toward a person?

□ Are employees prohibited from using highly compressed air for cleaning purposes?

□ If compressed air is used for cleaning off clothing, is the pressure reduced to less than 10 psi?

□ When using compressed air for cleaning, do employees use personal protective equipment?

□ Are safety chains or other suitable locking devices used at couplings of high-pressure hose lines where a connection failure would create a hazard?

□ Before compressed air is used to empty containers of liquid, is the safe working pressure of the container checked?

□ When compressed air is used with abrasive blast cleaning equipment, is the operating valve a type that must be held open manually?

□ When compressed air is used to inflate auto tires, is a clip-on chuck and an inline regulator preset to 40 psi required?

□ Is it prohibited to use compressed air to clean up or move combustible dust if such action could cause the dust to be suspended in the air and cause a fire or explosion hazard?

Compressed Air Receivers

□ Is every receiver equipped with a pressure gauge and with one or more automatic, spring-loaded safety valves?

□ Is the total relieving capacity of the safety valve capable of preventing pressure in the receiver from exceeding the maximum allowable working pressure of the receiver by more than 10%?

□ Is every air receiver provided with a drainpipe and valve at the lowest point for the removal of accumulated oil and water?

□ Are compressed air receivers periodically drained of moisture and oil?

□ Are all safety valves tested frequently and at regular intervals to determine whether they are in good operating condition?

□ Is there a current operating permit issued by the Division of Occupational Safety and Health?

□ Is the inlet of air receivers and piping systems kept free of accumulated oil and carbonaceous materials?

Compressed Gas and Cylinders

□ Are cylinders with a water weight capacity over 30 pounds equipped with means for connecting a valve protector device or with a collar or recess to protect the valve?

□ Are cylinders legibly marked to clearly identify the gas contained?

□ Are compressed gas cylinders stored in areas that are protected from external heat sources such as flame impingement, intense radiant heat, electric arcs, or high-temperature lines?

□ Are cylinders located or stored in areas where they will not be damaged by passing or falling objects or subject to tampering by unauthorized persons?

□ Are cylinders stored or transported in a manner to prevent them creating a hazard by tipping, falling, or rolling?

□ Are cylinders containing liquefied fuel gas stored or transported in a position so that the safety relief device is always in direct contact with the vapor space in the cylinder?

□ Are valve protectors always placed on cylinders when the cylinders are not in use or connected for use?

□ Are all valves closed off before a cylinder is moved, when the cylinder is empty, and at the completion of each job?

□ Are low-pressure fuel-gas cylinders checked periodically for corrosion, general distortion, cracks, or any other defect that might indicate a weakness or render it unfit for service?

□ Does the periodic check of low-pressure fuel-gas cylinders include a close inspection of the cylinders' bottom?

Hoist and Auxiliary Equipment

□ Is each overhead electric hoist equipped with a limit device to stop the hook travel at its highest and lowest point of safe travel?

□ Will each hoist automatically stop and hold any load up to 125% of its rated load, if its actuating force is removed?

□ Is the rated load of each hoist legibly marked and visible to the operator?

□ Are stops provided at the safe limits of travel for trolley hoist?

□ Are the controls of hoists plainly marked to indicate the direction of travel or motion?

□ Is each cage-controlled hoist equipped with an effective warning device?

□ Are close-fitting guards or other suitable devices installed on hoist to assure hoist ropes will be maintained in the sheave grooves?

□ Are all hoist chains or ropes of sufficient length to handle the full range of movement for the application while still maintaining two full wraps on the drum at all times?

□ Are nip points or contact points between hoist ropes and sheaves that are permanently located within 7 feet of the floor, ground, or working platform guarded?

□ Is it prohibited to use chains or rope slings that are kinked or twisted?

□ Is it prohibited to use the hoist rope or chain wrapped around the load as a substitute for a sling?

□ Is the operator instructed to avoid carrying loads over people?

□ Are only employees who have been trained in the proper use of hoists allowed to operate them?

Industrial Trucks – Forklifts

□ Are only trained personnel allowed to operate industrial trucks?

□ Is substantial overhead protective equipment provided on high lift rider equipment?

□ Are the required lift truck operating rules posted and enforced?

□ Is directional lighting provided on each industrial truck that operates in an area with less than 2 feet candles per square foot of general lighting?

▫ Does each industrial truck have a warning horn, whistle, gong, or other device that can be clearly heard above the normal noise in the areas where operated?

▫ Are the brakes on each industrial truck capable of bringing the vehicle to a complete and safe stop when fully loaded?

▫ Will the industrial truck's parking brake effectively prevent the vehicle from moving when unattended?

▫ Are industrial trucks operating in areas where flammable gases or vapors, or combustible dust or ignitable fibers may be present in the atmosphere, approved for such locations?

▫ Are motorized hand and hand/rider trucks so designed that the brakes are applied and power to the drive motor shuts off when the operator releases his/her grip on the device that controls the travel?

▫ Are industrial trucks with internal combustion engine operated in buildings or enclosed areas, carefully checked to ensure such operations do not cause harmful concentration of dangerous gases or fumes?

Spraying Operations

▫ Is adequate ventilation assured before spray operations are started?

▫ Is mechanical ventilation provided when spraying operation is done in enclosed areas?

▫ When mechanical ventilation is provided during spraying operations, is it so arranged that it will not circulate the contaminated air?

▫ Is the spray area free of hot surfaces?

▫ Is the spray area at least 20 feet from flames, sparks, operating electrical motors, and other ignition sources?

▫ Are portable lamps used to illuminate spray areas suitable for use in a hazardous location?

▫ Is approved respiratory equipment provided and used when appropriate during spraying operations?

▫ Do solvents used for cleaning have a flash point of 100E F or more?

▫ Are fire control sprinkler heads kept clean?

▫ Are "NO SMOKING" signs posted in spray areas, paint rooms, paint booths, and paint storage areas?

▫ Is the spray area kept clean of combustible residue?

▫ Are spray booths constructed of metal, masonry, or other substantial noncombustible material?

▫ Are spray booth floors and baffles noncombustible and easily cleaned?

▫ Is infrared drying apparatus kept out of the spray area during spraying operations?

▫ Is the spray booth completely ventilated before using the drying apparatus?

▫ Is the electric drying apparatus properly grounded?

▫ Are lighting fixtures for spray booths located outside of the booth and the interior lighted through sealed clear panels?

▫ Are the electric motors for exhaust fans placed outside booths or ducts?

▫ Are belts and pulleys inside the booth fully enclosed?

▫ Do ducts have access doors to allow cleaning?

▫ Do all drying spaces have adequate ventilation?

Entering Confined Spaces

- Are confined spaces thoroughly emptied of any corrosive or hazardous substances, such as acids or caustics, before entry?
- Before entry, are all lines to a confined space containing inert, toxic, flammable, or corrosive materials valved off and blanked or disconnected and separated?
- Is it required that all impellers, agitators, or other moving equipment inside confined spaces be locked out if they present a hazard?
- Is either natural or mechanical ventilation provided prior to confined space entry?
- Before entry, are appropriate atmospheric tests performed to check for oxygen deficiency, toxic substance, and explosive concentrations in the confined space before entry?
- Is adequate illumination provided for the work to be performed in the confined space?
- Is the atmosphere inside the confined space frequently tested or continuously monitored during conduct of work?
- Is there an assigned safety standby employee outside of the confined space, whose sole responsibility is to watch the work in progress, sound an alarm if necessary, and render assistance?
- Is the standby employee or other employees prohibited from entering the confined space without lifelines and respiratory equipment if there are any questions as to the cause of an emergency?
- In addition to the standby employee, is there at least one other trained rescuer in the vicinity?
- Are all rescuers appropriately trained and using approved, recently inspected equipment?
- Does all rescue equipment allow for lifting employees vertically from a top opening?
- Are there trained personnel in First Aid and CPR immediately available?
- Is there an effective communication system in place whenever respiratory equipment is used and the employee in the confined space is out of sight of the standby person?
- Is approved respiratory equipment required if the atmosphere inside the confined space cannot be made acceptable?
- Is all portable electrical equipment used inside confined spaces either grounded and insulated, or equipped with ground fault protection?
- Before gas welding or burning is started in a confined space, are hoses checked for leaks, compressed gas bottles forbidden inside of the confined space, torches lighted only outside of the confined area, and the confined area tested for an explosive atmosphere each time before a lighted torch is to be taken into the confined space?
- If employees will be using oxygen-consuming equipment such as salamanders, torches, and furnaces, in a confined space, is sufficient air provided to assure combustion without reducing the oxygen concentration of the atmosphere below 19.5% by volume?
- Whenever combustion-type equipment is used in confined space, are provisions made to ensure the exhaust gases are vented outside of the enclosure?
- Is each confined space checked for decaying vegetation or animal matter, which may produce methane?
- Is the confined space checked for possible industrial waste, which could contain toxic properties?

▫ If the confined space is below the ground and near areas where motor vehicles will be operating, is it possible for vehicle exhaust or carbon monoxide to enter the space?

Environmental Controls

▫ Are all work areas properly illuminated?

▫ Are employees instructed in proper first aid and other emergency procedures?

▫ Are hazardous substances identified that may cause harm by inhalation, ingestion, skin absorption, or contact?

▫ Are employees aware of the hazards involved with the various chemicals they may be exposed to in their work environment, such as ammonia, chlorine, epoxies, and caustics?

▫ Is employees' exposure to chemicals in the workplace kept within acceptable levels?

▫ Can a less harmful method or product be used?

▫ Is the work area's ventilation system appropriate for the work being performed?

▫ Are spray painting operations done in spray rooms or booths equipped with an appropriate exhaust system?

▫ Is employee exposure to welding fumes controlled by ventilation, use of respirators, exposure time, or other means?

▫ Are welders and other workers nearby provided with flash shields during welding operations?

▫ If forklifts and other vehicles are used in buildings or other enclosed areas, are the carbon monoxide levels kept below maximum acceptable concentration?

▫ Has there been a determination that noise levels in the facilities are within acceptable levels?

▫ Are steps being taken to use engineering controls to reduce excessive noise levels?

▫ Are proper precautions being taken when handling asbestos and other fibrous materials?

▫ Are caution labels and signs used to warn of asbestos?

▫ Are wet methods used, when practicable, to prevent the emission of airborne asbestos fibers, silica dust, and similar hazardous materials?

▫ Is vacuuming with appropriate equipment used whenever possible rather than blowing or sweeping dust?

▫ Are grinders, saws, and other machines that produce respirable dusts vented to an industrial collector or central exhaust system?

▫ Are all local exhaust ventilation systems designed and operating properly such as airflow and volume necessary for the application? Are the ducts free of obstructions or the belts slipping?

▫ Is personal protective equipment provided, used, and maintained wherever required?

▫ Are there written standard operating procedures for the selection and use of respirators where needed?

▫ Are restrooms and washrooms kept clean and sanitary?

▫ Is all water provided for drinking, washing, and cooking potable?

▫ Are all outlets for water not suitable for drinking clearly identified?

▫ Are employees' physical capacities assessed before being assigned to jobs requiring heavy work?

▫ Are employees instructed in the proper manner of lifting heavy objects?

▫ Where heat is a problem, have all fixed work areas been provided with spot cooling or air conditioning?

▫ Are employees screened before assignment to areas of high heat to determine if their health condition might make them more susceptible to having an adverse reaction?

▫ Are employees working on streets and roadways where they are exposed to the hazards of traffic, required to wear bright colored (traffic orange) warning vest?

▫ Are exhaust stacks and air intakes located so that contaminated air will not be recirculated within a building or other enclosed area?

▫ Is equipment producing ultraviolet radiation properly shielded?

Flammable and Combustible Materials

▫ Are combustible scrap, debris, and waste materials (i.e. oily rags) stored in covered metal receptacles and removed from the work site promptly?

▫ Is proper storage practiced to minimize the risk of fire including spontaneous combustion?

▫ Are approved containers and tanks used for the storage and handling of flammable and combustible liquids?

▫ Are all connections on drums and combustible liquid piping, vapor, and liquid tight?

▫ Are all flammable liquids kept in closed containers when not in use (e.g. parts cleaning tanks, pans, etc.)?

▫ Are bulk drums of flammable liquids grounded and bonded to containers during dispensing?

▫ Do storage rooms for flammable and combustible liquids have explosion-proof lights?

▫ Do storage rooms for flammable and combustible liquids have mechanical or gravity ventilation?

▫ Is liquefied petroleum gas stored, handled, and used in accordance with safe practices and standards?

▫ Are liquefied petroleum storage tanks guarded to prevent damage from vehicles?

▫ Are all solvent wastes and flammable liquids kept in fire-resistant covered containers until they are removed from the work site?

▫ Is vacuuming used whenever possible rather than blowing or sweeping combustible dust?

▫ Are fire separators placed between containers of combustibles or flammables, when stacked one upon another, to assure their support and stability?

▫ Are fuel gas cylinders and oxygen cylinders separated by distance, fire-resistant barriers, or other means while in storage?

▫ Are fire extinguishers selected and provided for the types of materials in areas where they are to be used?

   a. *Class A* Ordinary combustible material fires.

   b. *Class B* Flammable liquid, gas, or grease fires.

   c. *Class C* Energized electrical equipment fires.

▫ If a Halon 1301 fire extinguisher is used, can employees evacuate within the specified time for that extinguisher?

- Are appropriate fire extinguishers mounted within 75 feet of outside areas containing flammable liquids, and within 10 feet of any inside storage area for such materials?
- Is the transfer/withdrawal of flammable or combustible liquids performed by trained personnel?
- Are fire extinguishers mounted so that employees do not have to travel more than 75 feet for a class "A" fire or 50 feet for a class "B" fire?
- Are employees trained in the use of fire extinguishers?
- Are extinguishers free from obstructions or blockage?
- Are all extinguishers serviced, maintained, and tagged at intervals not to exceed 1 year?
- Are all extinguishers fully charged and in their designated places?
- Is a record maintained of required monthly checks of extinguishers?
- Where sprinkler systems are permanently installed, are the nozzle heads directed or arranged so that water will not be sprayed into operating electrical switchboards and equipment?
- Are "NO SMOKING" signs posted where appropriate in areas where flammable or combustible materials are used or stored?
- Are "NO SMOKING" signs posted on liquefied petroleum gas tanks?
- Are "NO SMOKING" rules enforced in areas involving storage and use of flammable materials?
- Are safety cans used for dispensing flammable or combustible liquids at a point of use?
- Are all spills of flammable or combustible liquids cleaned up promptly?
- Are storage tanks adequately vented to prevent the development of excessive vacuum or pressure as a result of filling, emptying, or atmosphere temperature changes?
- Are storage tanks equipped with emergency venting that will relieve excessive internal pressure caused by fire exposure?
- Are spare portable or butane tanks, which are used by industrial trucks, stored in accordance with regulations?

Fire Protection

- Do you have a fire prevention plan?
- Does the plan describe the type of fire protection equipment and/or systems?
- Have you established practices and procedures to control potential fire hazards and ignition sources?
- Are employees aware of the fire hazards for the material and processes to which they are exposed?
- Is your local fire department well acquainted with your facilities, location, and specific hazards?
- If you have a fire alarm system, is it tested at least annually?
- If you have a fire alarm system, is it certified as required?
- If you have interior standpipes and valves, are they inspected regularly?
- If you have outside private fire hydrants, are they flushed at least once a year and on a routine preventive maintenance schedule?
- Are fire doors and shutters in good operating condition?

- Are fire doors and shutters unobstructed and protected against obstructions, including their counterweights?
- Are fire door and shutter fusible links in place?
- Are automatic sprinkler system water control valves' air and water pressures checked weekly/periodically as required?
- Is maintenance of automatic sprinkler system assigned to responsible persons or to a sprinkler contractor?
- Are sprinkler heads protected by metal guards, when exposed to physical damage?
- Is proper clearance maintained below sprinkler heads?
- Are portable fire extinguishers provided in adequate number and type?
- Are fire extinguishers mounted in readily accessible locations?
- Are fire extinguishers recharged regularly and noted on the inspection tag?
- Are employees periodically instructed in the use of extinguishers and fire protection procedures?

Hazardous Chemical Exposures

- Are employees trained in the safe handling practices of hazardous chemicals such as acids, caustics, and the like?
- Are employees aware of the potential hazards involving various chemicals stored or used in the workplace, such as acids, bases, caustics, epoxies, and phenols?
- Is employee exposure to chemicals kept within acceptable levels?
- Are eye wash fountains and safety showers provided in areas where corrosive chemicals are handled?
- Are all containers, such as vats and storage tanks, labeled as to their contents – for example, "CAUSTICS"?
- Are all employees required to use personal protective clothing and equipment when handling chemicals (i.e. gloves, eye protection, and respirators)?
- Are flammable or toxic chemicals kept in closed containers when not in use?
- Are chemical piping systems clearly marked as to their content?
- Where corrosive liquids are frequently handled in open containers or drawn from storage vessels or pipelines, is adequate means readily available for neutralizing or disposing of spills or overflows properly and safely?
- Have standard operating procedures been established and are they being followed when cleaning up chemical spills?
- Where needed for emergency use, are respirators stored in a convenient, clean, and sanitary location?
- Are respirators intended for emergency use adequate for the various uses for which they may be needed?
- Are employees prohibited from eating in areas where hazardous chemicals are present?
- Is personal protective equipment provided, used, and maintained whenever necessary?
- Are there written standard operating procedures for the selection and use of respirators where needed?

▫ If you have a respirator protection program, are your employees instructed on the correct usage and limitations of the respirators?

▫ Are the respirators NIOSH approved for this particular application?

▫ Are they regularly inspected and cleaned, sanitized, and maintained?

▫ If hazardous substances are used in your processes, do you have a medical or biological monitoring system in operation?

▫ Are you familiar with the threshold limit values or permissible exposure limits of airborne contaminants and physical agents used in your workplace?

▫ Have control procedures been instituted for hazardous materials, where appropriate, such as respirators, ventilation systems, handling practices, and the like?

▫ Whenever possible, are hazardous substances handled in properly designed and exhausted booths or similar locations?

▫ Do you use general dilution or local exhaust ventilation systems to control dusts, vapors, gases, fumes, smoke, solvents, or mists that may be generated in your workplace?

▫ Is ventilation equipment provided for removal of contaminants from such operations as production grinding, buffing, spray painting, and/or vapor decreasing, and is it operating properly?

▫ Do employees complain about dizziness, headaches, nausea, irritation, or other factors of discomfort when they use solvents or other chemicals?

▫ Is there a dermatitis problem – do employees complain about skin dryness, irritation, or sensitization?

▫ Have you considered the use of an industrial hygienist or environmental health specialist to evaluate your operation?

▫ If internal combustion engines are used, is carbon monoxide kept within acceptable levels?

▫ Is vacuuming used, rather than blowing or sweeping dusts, whenever possible for cleanup?

▫ Are materials, which give off toxic asphyxiant, suffocating or anesthetic fumes, stored in remote or isolated locations when not in use?

Hazardous Substances Communication

▫ Is there a list of hazardous substances used in your workplace?

▫ Is there a written hazard communication program dealing with material safety data sheets (MSDS) labeling, and employee training?

▫ Who is responsible for MSDS, container labeling, employee training?

▫ Is each container for a hazardous substance (i.e. vats, bottles, storage tanks, etc.) labeled with product identity and a hazard warning (communication of the specific health hazards and physical hazards)?

▫ Is there a material safety data sheet readily available for each hazardous substance used?

▫ How will you inform other employers whose employees share the same work area where the hazardous substances are used?

▫ Is there an employee training program for hazardous substances?

▫ Does this program include the following explanations:

a. What an MSDS is and how to use and obtain one?

b. MSDS contents for each hazardous substance or class of substances?

c. Employee "Right to Know"?

d. Identification of where employees can see the employer's written hazard communication program and where hazardous substances are present in their work area?

e. The physical and health hazards of substances in the work area, how to detect their presence, and specific protective measures to be used?

f. Details of the hazard communication program, including how to use the labeling system and MSDS?

g. How employees will be informed of hazards of nonroutine tasks, and hazards of unlabeled pipes?

Electrical

□ Are your workplace electricians familiar with the Cal/OSHA Electrical Safety Orders?

□ Do you specify compliance with Cal/OSHA for all contract electrical work?

□ Are all employees required to report as soon as practicable any obvious hazard to life or property observed in connection with electrical equipment or lines?

□ Are employees instructed to make preliminary inspections and/or appropriate tests to determine what conditions exist before starting work on electrical equipment or lines?

□ When electrical equipment or lines are to be serviced, maintained or adjusted, are necessary switches opened, locked out, and tagged whenever possible?

□ Are portable electrical tools and equipment grounded or of the double insulated type?

□ Are electrical appliances such as vacuum cleaners, polishers, vending machines grounded?

□ Do extension cords being used have a grounding conductor?

□ Are multiple plug adapters prohibited?

□ Are ground-fault circuit interrupters installed on each temporary 15 or 20 ampere, 120 volt AC circuit at locations where construction, demolition, modifications, alterations, or excavations are being performed?

□ Are all temporary circuits protected by suitable disconnecting switches or plug connectors at the junction with permanent wiring?

□ Is exposed wiring and cords with frayed or deteriorated insulation repaired or replaced promptly?

□ Are flexible cords and cables free of splices or taps?

□ Are clamps or other securing means provided on flexible cords or cables at plugs, receptacles, tools, and equipment and is the cord jacket securely held in place?

□ Are all cord, cable, and raceway connections intact and secure?

□ In wet or damp locations, are electrical tools and equipment appropriate for the use or location or otherwise protected?

□ Is the location of electrical power lines and cables (overhead, underground, underfloor, and other side of walls) determined before digging, drilling, or similar work is begun?

□ Are metal measuring tapes, ropes, hand lines, or similar devices with metallic thread woven into the fabric prohibited where they could come in contact with energized parts of equipment or circuit conductors?

▫ Is the use of metal ladders prohibited in area where the ladder or the person using the ladder could come in contact with energized parts of equipment, fixtures, or circuit conductors?

▫ Are all disconnecting switches and circuit breakers labeled to indicate their use or equipment served?

▫ Are disconnecting means always opened before fuses are replaced?

▫ Do all interior wiring systems include provisions for grounding metal parts of electrical raceways, equipment, and enclosures?

▫ Are all electrical raceways and enclosures securely fastened in place?

▫ Are all energized parts of electrical circuits and equipment guarded against accidental contact by approved cabinets or enclosures?

▫ Is sufficient access and working space provided and maintained about all electrical equipment to permit ready and safe operations and maintenance?

▫ Are all unused openings (including conduit knockouts) in electrical enclosures and fittings closed with appropriate covers, plugs, or plates?

▫ Are electrical enclosures such as switches, receptacles, junction boxes, and so on, provided with tight-fitting covers or plates?

▫ Are disconnecting switches for electrical motors in excess of two horsepower, capable of opening the circuit when the motor is in a stalled condition, without exploding? (Switches must be horsepower rated equal to or in excess of the motor hp rating.)

▫ Is low-voltage protection provided in the control device of motors driving machines or equipment, which could cause probably injury from inadvertent starting?

▫ Is each motor disconnecting switch or circuit breaker located within sight of the motor control device?

▫ Is each motor located within sight of its controller or the controller disconnecting means capable of being locked in the open position or is a separate disconnecting means installed in the circuit within sight of the motor?

▫ Is the controller for each motor in excess of two horsepower, rated in horsepower equal to or in excess of the rating of the motor is serves?

▫ Are employees who regularly work on or around energized electrical equipment or lines instructed in the cardiopulmonary resuscitation (CPR) methods?

▫ Are employees prohibited from working alone on energized lines or equipment over 600 volts?

Noise

▫ Are there areas in the workplace where continuous noise levels exceed 85 dBA? (To determine maximum allowable levels for intermittent or impact noise, see Title 8, Section 5097.)

▫ Are noise levels being measured using a sound level meter or an octave band analyzer and records being kept?

▫ Have you tried isolating noisy machinery from the rest of your operation?

▫ Have engineering controls been used to reduce excessive noise levels?

▫ Where engineering controls are determined not feasible, are administrative controls (i.e. worker rotation) being used to minimize individual employee exposure to noise?

- Is there an ongoing preventive health program to educate employees in safe levels of noise and exposure, effects of noise on their health, and the use of personal protection?
- Is the training repeated annually for employees exposed to continuous noise above 85 dBA?
- Have work areas where noise levels make voice communication between employees difficult been identified and posted?
- Is approved hearing protective equipment (noise attenuating devices) available to every employee working in areas where continuous noise levels exceed 85 dBA?
- If you use ear protectors, are employees properly fitted and instructed in their use and care?
- Are employees exposed to continuous noise above 85 dBA given periodic audiometric testing to ensure that you have an effective hearing protection system?

Fueling

- Is it prohibited to fuel an internal combustion engine with a flammable liquid while the engine is running?
- Are fueling operations done in such a manner that likelihood of spillage will be minimal?
- When spillage occurs during fueling operations, is the spilled fuel cleaned up completely, evaporated, or other measures taken to control vapors before restarting the engine?
- Are fuel tank caps replaced and secured before starting the engine?
- In fueling operations is there always metal contact between the container and fuel tank?
- Are fueling hoses of a type designed to handle the specific type of fuel?
- Is it prohibited to handle or transfer gasoline in open containers?
- Are open lights, open flames, or sparking or arcing equipment prohibited near fueling or transfer of fuel operations?
- Is smoking prohibited in the vicinity of fueling operations?
- Are fueling operations prohibited in building or other enclosed areas that are not specifically ventilated for this purpose?
- Where fueling or transfer of fuel is done through a gravity flow system, are the nozzles of the self-closing type?

Identification of Piping Systems

- When nonpotable water is piped through a facility, are outlets or taps posted to alert employees that it is unsafe and not to be used for drinking, washing, or other personal use?
- When hazardous substances are transported through above ground piping, is each pipeline identified at points where confusion could introduce hazards to employees?
- When pipelines are identified by color painting, are all visible parts of the line so identified?
- When pipelines are identified by color painted bands or tapes, are the bands or tapes located at reasonable intervals and at each outlet, valve, or connection?
- When pipelines are identified by color, is the color code posted at all locations where confusion could introduce hazards to employees?
- When the contents of pipelines are identified by name or name abbreviation, is the information readily visible on the pipe near each valve or outlet?

▫ When pipelines carrying hazardous substances are identified by tags, are the tags constructed of durable materials, the message carried clearly and permanently distinguishable, and are tags installed at each valve or outlet?

▫ When pipelines are heated by electricity, steam, or other external source, are suitable warning signs or tags placed at unions, valves, or other serviceable parts of the system?

Material Handling

▫ Is there a safe clearance for equipment through aisles and doorways?

▫ Are aisle ways designated, permanently marked, and kept clear to allow unhindered passage?

▫ Are motorized vehicles and mechanized equipment inspected daily or prior to use?

▫ Are vehicles shut off and brakes set prior to loading or unloading?

▫ Are containers or combustibles or flammables, when stacked while being moved, always separated by dunnage sufficient to provide stability?

▫ Are dock boards (bridge plates) used when loading or unloading operations are taking place between vehicles and docks?

▫ Are trucks and trailers secured from movement during loading and unloading operations?

▫ Are dock plates and loading ramps constructed and maintained with sufficient strength to support imposed loading?

▫ Are hand trucks maintained in safe operating condition?

▫ Are chutes equipped with sideboards of sufficient height to prevent the materials being handled from falling off?

▫ Are chutes and gravity roller sections firmly placed or secured to prevent displacement?

▫ At the delivery end of rollers or chutes, are provisions made to brake the movement of the handled materials.

▫ Are pallets usually inspected before being loaded or moved?

▫ Are hooks with safety latches or other arrangements used when hoisting materials so that slings or load attachments would not accidentally slip off the hoist hooks?

▫ Are securing chains, ropes, chockers, or slings adequate for the job to be performed?

▫ When hoisting material or equipment, are provisions made to assure no one will be passing under the suspended loads?

▫ Are material safety data sheets available to employees handling hazardous substances?

Transporting Employees and Materials

▫ Do employees who operate vehicles on public thoroughfares have valid operator's licenses?

▫ When seven or more employees are regularly transported in a van, bus, or truck, is the operator's license appropriate for the class of vehicle being driven?

▫ Is each van, bus, or truck used regularly to transport employees, equipped with an adequate number of seats?

▫ When employees are transported by truck, are provisions provided to prevent their falling from the vehicle?

▫ Are vehicles used to transport employees equipped with lamps, brakes, horns, mirrors, windshields, and turn signals in good repair?

□ Are transport vehicles provided with handrails, steps, stirrups, or similar devices, so placed and arranged that employees can safely mount or dismount?

□ Are employee transport vehicles equipped at all times with at least two reflective type flares?

□ Is a full charged fire extinguisher, in good condition, with at least 4 B:C rating maintained in each employee transport vehicle?

□ When cutting tools with sharp edges are carried in passenger compartments of employee transport vehicles, are they placed in closed boxes or containers that are secured in place?

□ Are employees prohibited from riding on top of any load, which can shift, topple, or otherwise become unstable?

Control of Harmful Substances by Ventilation

□ Is the volume and velocity of air in each exhaust system sufficient to gather the dusts, fumes, mists, vapors, or gases to be controlled, and to convey them to a suitable point of disposal?

□ Are exhaust inlets, ducts, and plenums designed, constructed, and supported to prevent collapse or failure of any part of the system?

□ Are cleanout ports or doors provided at intervals not to exceed 12 feet in all horizontal runs of exhaust ducts?

□ Where two or more different types of operations are being controlled through the same exhaust system, will the combination of substances being controlled constitute a fire, explosion, or chemical reaction hazard in the duct?

□ Is adequate makeup air provided to areas where exhaust systems are operating?

□ Is the intake for makeup air located so that only clean, fresh air, which is free of contaminates, will enter the work environment?

□ Where two or more ventilation systems are serving a work area, is their operation such that one will not offset the functions of the other?

Sanitizing Equipment and Clothing

□ Is personal protective clothing or equipment, which employees are required to wear or use, of a type capable of being easily cleaned and disinfected?

□ Are employees prohibited from interchanging personal protective clothing or equipment, unless it has been properly cleaned?

□ Are machines and equipment, which process, handle, or apply materials that could be injurious to employees, cleaned and/or decontaminated before being overhauled or placed in storage?

□ Are employees prohibited from smoking or eating in any area where contaminates are present that could be injurious if ingested?

□ When employees are required to change from street clothing into protective clothing, is a clean change room with separate storage facility for street and protective clothing provided?

□ Are employees required to shower and wash their hair as soon as possible after a known contact has occurred with a carcinogen?

□ When equipment, materials, or other items are taken into or removed from a carcinogen regulated area, is it done in a manner that will not contaminate nonregulated areas or the external environment?

Tire Inflation

▫ Where tires are mounted and/or inflated on drop center wheels, is a safe practice procedure posted and enforced?

▫ Where tires are mounted and/or inflated on wheels with split rims and/or retainer rings, is a safe practice procedure posted and enforced?

▫ Does each tire inflation hose have a clip-on chuck with at least 24 inches of hose between the chuck and an inline hand valve and gauge?

▫ Does the tire inflation control valve automatically shut off the airflow when the valve is released?

▫ Is a tire restraining device such as a cage, rack, or other effective means used while inflating tires mounted on split rims, or rims using retainer rings?

▫ Are employees strictly forbidden from taking a position directly over or in front of a tire while it is being inflated?

Emergency Action Plan

▫ Are you required to have an emergency action plan?

▫ Does the emergency action plan comply with requirements of T8CCR 3220(a)?

▫ Have emergency escape procedures and routes been developed and communicated to all employers?

▫ Do employees, who remain to operate critical plant operations before they evacuate, know the proper procedures?

▫ Is the employee alarm system that provides a warning for emergency action recognizable and perceptible above ambient conditions?

▫ Are alarm systems properly maintained and tested regularly?

▫ Is the emergency action plan reviewed and revised periodically?

▫ Do employees know their responsibilities:

    a. For reporting emergencies?

    b. During an emergency?

    c. For conducting rescue and medical duties?

Infection Control

▫ Are employees potentially exposed to infectious agents in body fluids?

▫ Have occasions of potential occupational exposure been identified and documented?

▫ Has a training and information program been provided for employees exposed to or potentially exposed to blood and/or body fluids?

▫ Have infection control procedures been instituted where appropriate, such as ventilation, universal precautions, workplace practices, and personal protective equipment?

▫ Are employees aware of specific workplace practices to follow when appropriate (e.g., hand washing, handling sharp instruments, handling of laundry, disposal of contaminated materials, and reusable equipment)?

▫ Is personal protective equipment provided to employees, and in all appropriate locations?

▫ Is the necessary equipment (i.e. mouthpieces, resuscitation bags, and other ventilation devices) provided for administering mouth-to-mouth resuscitation on potentially infected patients?

▫ Are facilities/equipment to comply with workplace practices available, such as hand-washing sinks, biohazard tags and labels, needle containers, and detergents/disinfectants to clean up spills?

▫ Are all equipment and environmental and working surfaces cleaned and disinfected after contact with blood or potentially infectious materials?

▫ Is infectious waste placed in closable, leak proof containers, bags or puncture-resistant holders with proper labels?

▫ Has medical surveillance, including HBV evaluation, antibody testing and vaccination been made available to potentially exposed employees?

▫ Is there training on universal precautions?

▫ Is there training on personal protective equipment?

▫ Is there training on workplace practices, which should include blood drawing, room cleaning, laundry handling, cleanup of blood spills?

▫ Is there training on needle stick exposure/management?

▫ Is there training on hepatitis B vaccinations?

Ergonomics

▫ Can the work be performed without eyestrain or glare to the employees?

▫ Does the task require prolonged raising of the arms?

▫ Do the neck and shoulders have to be stooped to view the task?

▫ Are there pressure points on any parts of the body (wrists, forearms, and back of thighs)?

▫ Can the work be done using the larger muscles of the body?

▫ Can the work be done without twisting or overly bending the lower back?

▫ Are there sufficient rest breaks, in addition to the regular rest breaks, to relieve stress from repetitive-motion tasks?

▫ Are tools, instruments, and machinery shaped, positioned, and handled so that tasks can be performed comfortably?

▫ Are all pieces of furniture adjusted, positioned, and arranged to minimize strain on all parts of the body?

Ventilation for Indoor Air Quality

▫ Does your HVAC system provide at least the quantity of outdoor air required by the State Building Standards Code, Title 24, Part 2 at the time the building was constructed?

▫ Is the HVAC system inspected at least annually and problems corrected?

▫ Are inspection records retained for at least 5 years?

Crane Checklist

▫ Are the cranes visually inspected for defective components prior to the beginning of any work shift?

▫ Are all electrically operated cranes effectively grounded?

▫ Is a crane preventive maintenance program established?

▫ Is the load chart clearly visible to the operator?

▫ Are operating controls clearly identified?

▫ Is a fire extinguisher provided at the operator's station?

▫ Is the rated capacity visibly marked on each crane?

▫ Is an audible warning device mounted on each crane?

▫ Is sufficient illumination provided for the operator to perform the work safely?

▫ Are cranes of such design that the boom could fall over backward equipped with boom stops?

▫ Does each crane have a certificate indicating that required testing and examinations have been performed?

▫ Are crane inspection and maintenance records maintained and available for inspection?

C.10 ELECTRICAL HAZARDS CHECKLIST

This checklist is provided to help identify hazards in office, shop, or laboratory settings.

▫ Are outlets overloaded with too many plugs?

▫ Are the outlets in good condition?

▫ Are there any switches or outlets that do not work?

▫ Have any adapters been used to plug in a 3-prong plug into a 2-prong outlet?

▫ Are all 3-prong plugs intact?

▫ Are any electrical appliances, devices, or machines used in an area where they might get wet?

▫ If so, are all the appliances, devices, or machines in the area properly grounded (including housekeeping equipment, vending machines, office equipment, etc.)?

▫ Do the electrical outlets near sinks have a ground fault circuit interrupter?

▫ Are all metallic fixed electrical items properly grounded?

▫ Is the equipment connected by cord and plug properly plugged in?

▫ Are all portable electrical hand tools, unaltered, undamaged, and properly connected?

▫ Is the wiring to all electrical appliances, devices, and machines properly insulated?

▫ Are any cords on electrical appliances, devices, or machines frayed? Do they have exposed wires or are they showing wear?

▫ Are there any cords strung through doorways, holes in walls or ceilings, under carpets or rugs, or windows that may pose a trip hazard?

▫ Are any cords, fans, or portable heaters located where they could create a trip hazard?

▫ Is only one extension cord being used in any single run?

▫ Are any extension cords being used as permanent wiring instead of fixed wiring?

▫ Is the use of extension cords being phased out of this facility?

▫ Are the switches in the electrical breaker box properly marked to identify their function and areas served?

▫ Are all junction boxes, outlets, switches, and fittings properly covered?

▫ Are flexible cords or cables fastened so that there is no direct pull on joints or terminal screws?

▫ Are there any flexible cords and cables attached to building surfaces?

▫ Are all fans and portable heaters adequately guarded?

C.11 FOOD INDUSTRY HAZARDS CHECKLIST

This checklist is provided to help identify hazards involved with food and food industry operations.

Food Delivery and Storage (Frozen, Refrigerated, and Dry)

▫ Contamination of foods with bacteria.

▫ Cross contamination of bacteria.

▫ Growth of food poisoning bacteria or their toxins.

▫ Physical contamination from outer packaging, for example, dirty vegetable boxes/wood splinters.

Food Preparation

▫ Growth of food poisoning bacteria.

▫ Contamination of foods.

Food Thawing

▫ Bacterial contamination of area.

▫ Incomplete thawing.

Food Cooking

▫ Survival of food poisoning bacteria.

Food Cooling

▫ Growth of surviving bacteria.

▫ Recontamination with bacteria.

Food Hot Holding

▫ Growth of food poisoning bacteria/toxins.

▫ Recontamination.

Food Chilled Storage

▫ Growth of food poisoning bacteria.

▫ Recontamination with bacteria.

Food Reheating

▫ Survival of food poisoning bacteria.

▫ Growth of bacteria/toxins.

Food Delivery

▫ Growth of bacteria.

◌ Contamination.

Food Rooms

▫ Harborage of dirt, dust, and debris.

▫ Dirty condition of surfaces/equipment.

▫ Access for pests.

▫ Inadequate storage of chemicals.

▫ Use of wood, glass, metal, and jewelry.

Appendix *D*

References

The following are some of the most relevant and useful resources on system safety and software safety, which have been listed by date of origination:

1. MIL-STD-882, Standard Practice for System Safety, original version 15 July 1969, version B Notice 1: 30 March 1984 contains the first software safety material Task 301, version C: 19 January 1993 contains update, version D: 10 February 2000 dropped detailed information, version E (2012) adds software safety detail back, plus additional safety tasks.
2. C. A. Ericson, *Hazard Analysis Techniques for System Safety*, 2nd ed., John Wiley & Sons, Inc., 2015.
3. C. A. Ericson, II, *Concise Encyclopedia of System Safety: Definition of Terms and Concepts*, John Wiley & Sons, Inc., 2011.
4. C. A. Ericson, II, *System Safety Primer*, CreateSpace, 2011.
5. C. A. Ericson, II, *Fault Tree Analysis Primer*, CreateSpace, 2011.
6. C. A. Ericson, II, *Hazard Analysis Primer*, CreateSpace, 2012.
7. C. A. Ericson, II, *Software Safety Primer*, CreateSpace, 2013.
8. C. A. Ericson II, Software and System Safety, *Proceedings of the 5th International System Safety Conference*, 1981.
9. T. Anderson ed., *Safe and Secure Computing Systems*, Blackwell Scientific Publications, 1989.
10. EIA SEB-6A, System Safety Engineering in Software Development, April 1990.
11. I. C. Pyle, *Developing Safety Systems: A Guide Using Ada*, Prentice Hall, 1991.
12. RTCA/DO-178, Software Considerations in Airborne Systems and Equipment Certification, 1992; version C, 2011.
13. Army TR92-2, Software System Safety Guide, 1992.
14. A. M. Neufelder, *Ensuring Software Reliability*, Marcel Dekker Inc., 1993.

Hazard Analysis Techniques for System Safety, Second Edition. Clifton A Ericson, II.
© 2016 John Wiley & Sons, Inc. Published 2016 by John Wiley & Sons, Inc.

15. W. A. Halang *et al.*, *A Safety Licensable Computing Architecture*, World Scientific Publishing, 1993.

16. IEEE Std-1228, IEEE Standard for Software Safety Plans, 17 March 1994.

17. M. A. Friedman and J. M. Voas, *Software Assessment: Reliability, Safety, Testability*, John Wiley & Sons, Inc., 1995.

18. N. G. Leveson, *Safeware: System Safety and Computers*, Addison-Wesley Publishing Company, 1995.

19. L. Hatton, *Safer C: Developing Software for High-Integrity and Safety-Critical Systems*, McGraw-Hill, 1995.

20. N. Storey, *Safety-Critical Computer Systems*, Addison-Wesley Longman, 1996.

21. SAE ARP-4754, Certification Considerations for Highly-Integrated or Complex Aircraft Systems, original 1996, version A, 2012.

22. SAE ARP-4761, Guidelines and Methods for Conducting the Safety Assessment Process on Civil Airborne Systems and Equipment, 1996.

23. NASA-GB-8719.13, NASA Software Safety Guidebook, March 31, 2004. Replaces NASA-GB-1740.13-96, NASA Guidebook for Safety-Critical Software: Analysis and Development, 1996.

24. NASA-STD-8719.13B, Software Safety Standard, NASA Technical Standard, July 8, 2004 (Ver. A, Sept. 1997).

25. J. Barnes, *High Integrity Ada: The SPARK Approach*, Addison-Wesley, 1997.

26. STANAG 4404, *Safety Design Requirements and Guidelines for Munition Related Safety-Critical Computing Systems*, NATO, 1997.

27. STANAG 4452, Safety Assessment of Munition-Related Computing Systems, contains information on software hazard analysis, draft circa 1998.

28. UL 1998, Software in Programmable Components, 29 May 1998.

29. D. S. Herrmann, *Software Safety and Reliability: Techniques, Approaches and Standards of Key Industrial Sectors*, IEEE Computer Society, 1999.

30. DoD Joint Software Systems Safety Handbook (JSSSH), original December 1999, version 1.0 August 2010.

31. S. H. Pfleeger, L. Hatton, and C. C. Howell, *Solid Software*, Prentice Hall PTR, 2002.

32. W. R. Dunn, *Practical Design of Safety-Critical Computer Systems*, Reliability Press, 2002.

33. Handbook for Software in Safety-Critical Applications, English edition, Swedish Armed Forces, Defence Material Administration, 2005.

34. AMCOM Regulation 385-17, AMCOM Software System Safety Policy, 15 March 2008.

35. AOP 52, Guidance on Software Safety Design and Assessment of Munition-Related Computing Systems, NATO, 9 Dec. 2008.

36. AMCOM Regulation 385-17, AMCOM Software System Safety Policy, 15 March 2008 (Army).

37. ANSI/GEIA-STD-0010, Standard Best Practices for System Safety Program Development and Execution, 12 February 2009 (includes software safety section).

38. Leanna Rierson, *Developing Safety-Critical Software: A Practical Guide for Aviation Software and DO-178C Compliance*, CRC Press, 2013.

39. V. Hilderman and T. Baghai, *Avionics Certification: A Complete Guide to DO-178 and DO-254*, Avionics Communications Inc., 2007.

40. F. R. Spellman and N. Whiting, *The Handbook of Safety Engineering: Principles and Applications*, Government Institutes, Division of Scarecrow Press, UK, 2009.

41. N. G. Leveson, *Engineering a Safer World: Systems Thinking Applied to Safety*, The MIT Press, 2011.

42. H. E. Roland and B. Moriarty, *System Safety Engineering and Management*, 1985, 2nd ed., John Wiley & Sons, Inc., 1990.

43. N. J. Bahr, *System Safety Engineering and Risk Assessment*, Taylor & Francis Publishing, 1997.

44. F. Redmill, M. Chudleigh, and J. Redmill, *System Safety: HAZOP and Software HAZOP*, John Wiley & Sons, Inc., 1999.

45. R. A. Stephans and W. W. Talso, *System Safety Analysis Handbook*, 2nd ed., System Safety Society, 1997.

46. Jeffrey Vincoli, *Basic Guide to System Safety*, 3rd ed., John Wiley & Sons, Inc., 2014.

47. Richard Stephans, *System Safety for the 21st Century: The Updated and Revised Edition of System Safety 2000*, John Wiley & Sons, 2004.

48. Dev Raheja and Mike Allocco, *Assurance Technologies Principles and Practices: A Product, Process, and System Safety Perspective*, John Wiley & Sons, 2006.

49. Michael V. Frank, *Choosing Safety: A Guide to Probabilistic Risk Assessment and Decision Analysis in Complex High Consequence Systems*, Resources for the Future, 2008.

50. WASH-1400, Reactor Safety Study, 1975 (mechanical and human failure rates).

51. Alphonse Chapanis, *Human Factors in Systems Engineering*, New York, John Wiley & Sons, 1996.

52. George A. Peters, and Barbara J. Peters, Human Safety, vols. I and II, 2013.

53. Terry Hardy, *Software and System Safety: Accidents, Incidents and Lesson Learned*, AuthorHouse, 2012.

54. Terry L. Hardy, *The System Safety Skeptic: Lessons Learned in Safety Management and Engineering*, 2nd ed., AuthorHouse, 2014.

55. Thomas A. Hunter, *Engineering Design for Safety*, McGraw Hill, 1992.

56. D. B. Brown, *System Analysis and Design for Safety*, Prentice-Hall, 1976.

57. E. Lloyd and W. Tye, *Systematic Safety*, Taylor Young Ltd., 1982.

58. W. Hammer, *Handbook of System and Product Safety*, Prentice-Hall, 1972.

59. W. Hammer, *Product Safety Management and Engineering*, Prentice-Hall, 1980.

60. W. Hammer, *Occupational Safety Management and Engineering*, Prentice-Hall, 1981.

61. G. R. McIntyre, *Patterns in Safety Thinking*, Ashgate Publishing, 2000.

62. W. P. Rodgers, *Introduction to System Safety Engineering*, John Wiley & Sons, Inc., 1971.

63. S. W. Malasky, *System Safety: Technology and Application*, Garland STPM Press, New York, 1982.

64. D. Layton, *System Safety: Including DOD Standards*, Weber Systems, Inc., 1989.

65. W. C. Christensen and F. A. Manuele, *Safety through Design*, National Safety Council Press, 1999.

66. D. S. Gloss and M. G. Wardel, *Introduction to Safety Engineering*, John Wiley & Sons, Inc., 1984.

67. M. Larson and S. Hann, *Safety and Reliability in System Design*, Ginn Press, Needham Height, MA, 1989.

68. H. Kumamoto and E.J. Henley, *Probabilistic Risk Assessment and Management for Engineers and Scientists*, 2nd ed., IEEE Press, 1996.

69. W. E. Vesely, F. F. Goldberg, N. H. Roberts, and D. F. Haasl, *Fault Tree Handbook*, NUREG-0492, 1981.

70. T. S. Ferry, *Modern Accident Investigation and Analysis*, John Wiley & Sons, Inc., 1988.

Index

Hazard Analysis Techniques for System Safety, Second Edition. Clifton A Ericson, II.
© 2016 John Wiley & Sons, Inc. Published 2016 by John Wiley & Sons, Inc.

Printed and bound by CPI Group (UK) Ltd, Croydon, CR0 4YY

23/04/2025

14660910-0005